Werner Linde
Probability Theory
De Gruyter Graduate

Also of interest

Probability Theory and Statistical Applications. A Profound Treatise for Self-Study
Peter Zörnig, 2016
ISBN 978-3-11-036319-7, e-ISBN (PDF) 978-3-11-040271-1,
e-ISBN (EPUB) 978-3-11-040283-4

Stochastic Finance: An Introduction in Discrete Time
Hans Föllmer, Alexander Schied, 4th Edition, 2016
ISBN 978-3-11-046344-6, e-ISBN (PDF) 978-3-11-046345-3,
e-ISBN (EPUB) 978-3-11-046346-0

Asymptotic Statistics: With a View to Stochastic Processes
Reinhard Höpfner, 2014
ISBN 978-3-11-025024-4, e-ISBN (PDF) 978-3-11-025028-2,
e-ISBN (EPUB) 978-3-11-036778-2

Brownian Motion: An Introduction to Stochastic Processes
René L.Schilling, Lothar Partzsch, 2nd Edition, 2014
ISBN 978-3-11-030729-0, e-ISBN (PDF) 978-3-11-030730-6,
e-ISBN (EPUB) 978-3-11-037398-1

Stochastics: Introduction to Probability and Statistics
Hans-Otto Georgii, 2nd Edition, 2012
ISBN 978-3-11-029254-1, e-ISBN (PDF) 978-3-11-029360-9

Werner Linde
Probability Theory

———

A First Course in Probability Theory and Statistics

DE GRUYTER

Mathematics Subject Classification 2010
Primary: 60-01, 62-01; Secondary: 60A05

Author
Prof. Dr. Werner Linde
Friedrich-Schiller-Universität Jena
Fakultät für Mathematik & Informatik
Institut für Stochastik
Prof. für Stochastische Analysis
D-07737 Jena
Werner.Linde@mathematik.uni-jena.de
and
University of Delaware
Department of Mathematical Sciences
501 Ewing Hall
Newark DE, 19716
lindew@udel.edu

ISBN 978-3-11-046617-1
e-ISBN (PDF) 978-3-11-046619-5
e-ISBN (EPUB) 978-3-11-046625-6

Library of Congress Cataloging-in-Publication Data
A CIP catalog record for this book has been applied for at the Library of Congress.

Bibliographic information published by the Deutsche Nationalbibliothek
The Deutsche Nationalbibliothek lists this publication in the Deutsche Nationalbibliografie; detailed
bibliographic data are available on the Internet at http://dnb.dnb.de.

© 2016 Walter de Gruyter GmbH, Berlin/Boston
Typesetting: Integra Software Services Pvt. Ltd.
Printing and binding: CPI books GmbH, Leck
Cover image: Werner Linde
♾ Printed on acid-free paper
Printed in Germany

www.degruyter.com

To my wife Karin

Preface

This book is intended as an introductory course for students in mathematics, physical sciences, engineering, or in other related fields. It is based on the experience of probability lectures taught during the past 25 years, where the spectrum reached from two-hour introductory courses, over Measure Theory and advanced probability classes, to such topics as Stochastic Processes and Mathematical Statistics. Until 2012 these lectures were delivered to students at the University of Jena (Germany), and since 2013 to those at the University of Delaware in Newark (USA).

The book is the completely revised version of the German edition "Stochastik für das Lehramt," which appeared in 2014 at De Gruyter. At most universities in Germany, there exist special classes in Probability Theory for students who want to become teachers of mathematics in high schools. Besides basic facts about Probability Theory, these courses are also supposed to give an introduction into Mathematical Statistics. Thus, the original main intention for the German version was to write a book that helps those students understand Probability Theory better. But soon the book turned out to also be useful as introduction for students in other fields, e.g. in mathematics, physics, and so on. Thus we decided, in order to make the book applicable for a broader audience, to provide a translation in the English language.

During numerous years of teaching I learned the following:
- Probabilistic questions are usually easy to formulate, generally have a tight relation to everyday problems, and therefore attract the interest of the audience. Every student knows the phenomena that occur when one rolls a die, plays cards, tosses a coin, or plays a lottery. Thus, an initial interest in Probability Theory exists.
- In contrast, after a short time many students have very serious difficulties with understanding the presented topics. Consequently, a common opinion among students is that Probability Theory is a very complicated topic, causing a lot of problems and troubles.

Surely there exist several reasons for the bad image of Probability Theory among students. But, as we believe, the most important one is as follows. In Probability Theory, the type of problems and questions considered, as well as the way of thinking, differs considerably from the problems, questions, and thinking in other fields of mathematics, i.e., from fields with which the students became acquainted before attending a probability course. For example, in Calculus a function has a well-described domain of definition; mostly it is defined by a concrete formula, has certain properties as continuity, differentiability, and so on. A function is something very concrete which can be made vivid by drawing its graph. In contrast, in Probability Theory functions are mostly investigated as random variables. They are defined on a completely unimportant, nonspecified sample space, and they generally do not possess a concrete formula for their definition. It may even happen that only the existence of a function (random variable) is known. The only property of a random variable which really matters is the distribution of its values. This and many other similar techniques make the whole theory something mysterious and not completely comprehensible.

Considering this observation, we organized the book in a way that tries to make probabilistic problems more understandable and that puts the focus more onto explanations of the definitions, notations, and results. The tools we use to do this are examples; we present at least one before a new definition, in order to motivate it, followed by more examples after the definition to make it comprehensible. Here we act upon the maxim expressed by Einstein's quote[1]:

Example isn't another way to teach, it is the only way to teach.

Presenting the basic results and methods in Probability Theory *without* using results, facts, and notations from Measure Theory is, in our opinion, as difficult as to square the circle. Either one restricts oneself to discrete probability measures and random variables or one has to be unprecise. There is no other choice! In some places, it is possible to avoid the use of measure theoretic facts, such as the Lebesgue integral, or the existence of infinite product measures, and so on, but the price is high.[2] Of course, I also struggled with the problem of missing facts from Measure Theory while writing this book. Therefore, I tried to include some ideas and some results about σ-fields, measures, and integrals, hoping that a few readers become interested and want to learn more about Measure Theory. For those, we refer to the books [Coh13], [Dud02], or [Bil12] as good sources.

In this context, let us make some remark about the verification of the presented results. Whenever it was possible, we tried to prove the stated results. Times have changed; when I was a student, every theorem presented in a mathematical lecture was proved – really every one. Facts and results without proof were doubtful and soon forgotten. And a tricky and elegant proof is sometimes more impressive than the proven result (at least to us). Hopefully, some readers will like some of the proofs in this book as much as we did.

One of most used applications of Probability Theory is Mathematical Statistics. When I met former students of mine, I often asked them which kind of mathematics they are mainly using now in their daily work. The overwhelming majority of them answered that one of their main fields of mathematical work is statistical problems. Therefore, we decided to include an introductory chapter about Mathematical Statistics. Nowadays, due to the existence of good and fast statistical programs, it is very easy to analyze data, to evaluate confidence regions, or to test a given hypotheses. But do those who use these programs also always know what they are doing? Since

---

1 See http://www.alberteinsteinsite.com/quotes/einsteinquotes.html

2 For example, several years ago, to avoid the use of the Lebesgue integral, I introduced the expected value of a random variable as a Riemann integral via its distribution function. This is mathematically correct, but at the end almost no students understood what the expected value really is. Try to prove that the expected value is linear using this approach!

we doubt that this is so, we stressed the focus in this chapter to the question of why the main statistical methods work and on what mathematical background they rest. We also investigate how precise statistical decisions are and what kinds of errors may occur.

The organization of this book differs a little bit from those in many other first-course books about Probability Theory. Having Measure Theory in the back of our minds causes us to think that probability measures are the most important ingredient of Probability Theory; random variables come in second. On the contrary, many other authors go exactly the other way. They start with random variables, and probability measures then occur as their distribution on their range spaces (mostly $\mathbb{R}$). In this case, a standard normal probability measure does not exist, only a standard normal distributed random variable. Both approaches have their advantages and disadvantages, but as we said, for us the probability measures are interesting in their own right, and therefore we start with them in Chapter 1, followed by random variables in Section 3.

The book also contains some facts and results that are more advanced and usually not part of an introductory course in Probability Theory. Such topics are, for example, the investigation of product measures, order statistics, and so on. We have assigned those more involved sections with a star. They may be skipped at a first reading without loss in the following chapters.

At the end of each chapter, one finds a collection of some problems related to the contents of the section. Here we restricted ourselves to a few problems in the actual task; the solutions of these problems are helpful to the understanding of the presented topics. The problems are mainly taken from our collection of homeworks and exams during the past years. For those who want to work with more problems we refer to many books, as e.g. [GS01a], [Gha05], [Pao06], or [Ros14], which contain a huge collection of probabilistic problems, ranging from easy to difficult, from natural to artificial, from interesting to boring.

Finally I want to express my thanks to those who supported my work at the translation and revision of the present book. Many students at the University of Delaware helped me to improve my English and to correct wrong phrases and wrong expressions. To mention all of them is impossible. But among them were a few students who read whole chapters and, without them, the book would have never been finished (or readable). In particular I want to mention Emily Wagner and Spencer Walker. They both did really a great job. Many thanks! Let me also express my gratitude to Colleen McInerney, Rachel Austin, Daniel Atadan, and Quentin Dubroff, all students in Delaware and attending my classes for some time. They also read whole sections of the book and corrected my broken English. Finally, my thanks go to Professor Anne Leucht from the Technical University in Braunschweig (Germany); her field of work is Mathematical Statistics, and her hints and remarks about Chapter 8 in this book were important to me.

And last but not least I want to thank the Department of Mathematical Sciences at the University of Delaware for the excellent working conditions after my retirement in Germany.

Newark, Delaware, June 6, 2016   Werner Linde

Contents

1 Probabilities

1.1 Probability Spaces

The basic concern of Probability Theory is to model experiments involving randomness, that is, experiments with nondetermined outcome, shortly called **random experiments**. The Russian mathematician A.N. Kolmogorov established the modern Probability Theory in 1933 by publishing his book (cf. [Kol33]) *Grundbegriffe der Wahrscheinlichkeitsrechnung*. In it, he postulated the following:

Random experiments are described by probability spaces $(\Omega, \mathcal{A}, \mathbb{P})$ **!**

The triple $(\Omega, \mathcal{A}, \mathbb{P})$ comprises a **sample space** Ω, a **σ-field** $\mathcal{A}$ **of events**, and a mapping $\mathbb{P}$ from $\mathcal{A}$ to $[0, 1]$, called **probability measure** or **probability distribution**.

Let us now explain the three different components of a probability space in detail. We start with the sample space.

1.1.1 Sample Spaces

> **Definition 1.1.1.** The sample space Ω is a nonempty set that contains (at least) all possible outcomes of the random experiment.

Remark 1.1.2. Due to mathematical reasons sometimes it can be useful to choose Ω larger than necessary. It is only important that the sample space contains **all** possible results.

Example 1.1.3. When rolling a die one time the natural choice for the sample space is $\Omega = \{1, \ldots, 6\}$. However, it would also be possible to take $\Omega = \{1, 2, \ldots\}$ or even $\Omega = \mathbb{R}$. In contrast, $\Omega = \{1, \ldots, 5\}$ is not suitable for the description of the experiment.

Example 1.1.4. Roll a die until the number "6" shows up for the first time. Record the number of necessary rolls until the first appearance of "6." The suitable sample space in this case is $\Omega = \{1, 2, \ldots\}$. Any finite set $\{1, 2, \ldots, N\}$ is not appropriate because, even if we choose N very large, we can never be 100% sure that the first "6" really appears during the first N rolls.

Example 1.1.5. A light bulb is switched on at time zero and burns for a certain period of time. At some random time $t > 0$ it burns out. To describe this experiment we have to take into account all possible times $t > 0$. Therefore, a natural choice for the sample space in this case is $\Omega = (0, \infty)$, or, if we do not exclude that the bulb is defective from the very beginning, then $\Omega = [0, \infty)$.

Subsets of the sample space Ω are called **events**. In other words, the powerset $\mathcal{P}(\Omega)$ is the collection of all possible events. For example, when we roll a die once there are $2^6 = 64$ possible events, as, for example,

$$\left\{ \emptyset, \{1\}, \ldots, \{6\}, \{1, 2\}, \ldots, \{1, 6\}, \{2, 3\}, \ldots, \{2, 6\}, \ldots, \{1, 2, 3, 4, 5\}, \Omega \right\}.$$

Among all events there are some of special interest, the so-called **elementary events**. These are events containing exactly one element. In Example 1.1.3 the elementary events are

$$\{1\}, \{2\}, \{3\}, \{4\}, \{5\} \quad \text{and} \quad \{6\}.$$

Remark 1.1.6. Never confuse the elementary events with the points that they contain. Look at Example 1.1.3. There we have $6 \in \Omega$ and for the generated elementary event holds $\{6\} \in \mathcal{P}(\Omega)$.

Let $A \subseteq \Omega$ be an event. After executing the random experiment one observes a result $\omega \in \Omega$. Then two cases are possible.
1. The outcome ω belongs to A. In this case we say that the event A **occurred**.
2. If ω is not in A, that is, if $\omega \in A^c$, then the event A did A **not occur**.

Example 1.1.7. Roll a die once and let $A = \{2, 4\}$. Say the outcome was number "6." Then A did not occur. But, if we obtained number "2," then A occurred.

Example 1.1.8. In Example 1.1.5 the occurrence of an event $A = [T, \infty)$ tells us that the light bulb burned out after time T or, in other words, at time T it still shone.

Let us formulate some easy **rules** for the occurrence of events.
1. By the choice of the sample space the event Ω always occurs. Therefore, Ω is also called the **certain** event.
2. The empty set never occurs. Thus it is called the **impossible** event.
3. An event A occurs if and only if the complementary event A^c does not, and vice versa, A does not occur if and only if A^c does.
4. If A and B are two events, then $A \cup B$ occurs if at least one of the two sets occurs. Hereby we do not exclude that A and B may both occur.
5. The event $A \cap B$ occurs if and only if A and B both occur.

1.1.2 σ-Fields of Events

The basic aim of Probability Theory is to assign to each event A a number $\mathbb{P}(A)$ in $[0, 1]$, which describes the likelihood of its occurrence. If the occurrence of an event A is very likely, then $\mathbb{P}(A)$ should be close to 1 while $\mathbb{P}(A)$ close to zero suggests that the appearance of A is very unlikely. The mapping $A \mapsto \mathbb{P}(A)$ must possess certain natural properties. Unfortunately, by mathematical reason it is not always possible to assign to each event A a number $\mathbb{P}(A)$ such that $A \mapsto \mathbb{P}(A)$ has the desired properties. The solution is ingenious and and one of the key observations in Kolmogorov's approach: one chooses a subset $\mathcal{A} \subseteq \mathcal{P}(\Omega)$ such that $\mathbb{P}(A)$ is only defined for $A \in \mathcal{A}$. If $A \notin \mathcal{A}$, then $\mathbb{P}(A)$ does not exist. Of course, $\mathcal{A}$ should be chosen as large as possible and, moreover, at least "ordinary" sets should belong to $\mathcal{A}$.

The collection $\mathcal{A}$ of events has to satisfy some algebraic conditions. More precisely, the following properties are supposed.

Definition 1.1.9. A collection $\mathcal{A}$ of subsets of Ω is called σ-**field** if
(1) $\emptyset \in \mathcal{A}$,
(2) if $A \in \mathcal{A}$ then $A^c \in \mathcal{A}$, and
(3) for countably many $A_1, A_2, \ldots$ in $\mathcal{A}$ follows $\bigcup_{j=1}^{\infty} A_j \in \mathcal{A}$.

Let us verify some easy properties of σ-fields.

Proposition 1.1.10. *Let $\mathcal{A}$ be a σ-field of subsets of Ω. Then the following are valid:*
(i) *$\Omega \in \mathcal{A}$.*
(ii) *If $A_1, \ldots, A_n$ are finitely many sets in $\mathcal{A}$, then $\bigcup_{j=1}^{n} A_j \in \mathcal{A}$.*
(iii) *If $A_1, A_2, \ldots$ belong to $\mathcal{A}$, then so does $\bigcap_{j=1}^{\infty} A_j$.*
(iv) *Whenever $A_1, \ldots, A_n \in \mathcal{A}$, then $\bigcap_{j=1}^{n} A_j \in \mathcal{A}$.*

Proof: Assertion (i) is a direct consequence of $\emptyset \in \mathcal{A}$ combined with property (2) of σ-fields.

To verify (ii) let $A_1, \ldots, A_n$ be in $\mathcal{A}$. Set $A_{n+1} = A_{n+2} = \cdots = \emptyset$. Then for all $j = 1, 2, \ldots$ we have $A_j \in \mathcal{A}$ and by property (3) of σ-fields also $\bigcup_{j=1}^{\infty} A_j \in \mathcal{A}$. But note that $\bigcup_{j=1}^{\infty} A_j = \bigcup_{j=1}^{n} A_j$, hence (ii) is valid.

To prove (iii) we first observe that $A_j \in \mathcal{A}$ yields $A_j^c \in \mathcal{A}$, hence $\bigcup_{j=1}^{\infty} A_j^c \in \mathcal{A}$. Another application of (2) implies $\left(\bigcup_{j=1}^{\infty} A_j^c \right)^c \in \mathcal{A}$. De Morgan's rule asserts

$$\left(\bigcup_{j=1}^{\infty} A_j^c \right)^c = \bigcap_{j=1}^{\infty} A_j,$$

which completes the proof of (iii).

Assertion (iv) may be derived from an application of (ii) to the complementary sets as we did in the proof of (iii). Or use the method in the proof of (ii), but this time we choose $A_{n+1} = A_{n+2} = \cdots = \Omega$. ∎

Corollary 1.1.11. *If sets A and B belong to a σ-field $\mathcal{A}$, then so do $A \cup B$, $A \cap B$, $A \backslash B$, and $A \Delta B$.*

The easiest examples of σ-fields are either $\mathcal{A} = \{\emptyset, \Omega\}$ or $\mathcal{A} = \mathcal{P}(\Omega)$. However, the former σ-field is much too small for applications while the latter one is generally too big, at least if the sample space is uncountably infinite. We will shortly indicate how one constructs suitable σ-fields in the case of "large" sample spaces as, for example, $\mathbb{R}$ or $\mathbb{R}^n$.

Proposition 1.1.12. *Let $\mathcal{C}$ be an arbitrary nonempty collection of subsets of Ω. Then there is a σ-field $\mathcal{A}$ possessing the following properties:*
1. *It holds $\mathcal{C} \subseteq \mathcal{A}$ or, verbally, each set $C \in \mathcal{C}$ belongs to the σ-field $\mathcal{A}$.*
2. *The σ-field $\mathcal{A}$ is the smallest one possessing this property. That is, whenever $\mathcal{A}'$ is another σ-field with $\mathcal{C} \subseteq \mathcal{A}'$, then $\mathcal{A} \subseteq \mathcal{A}'$.*

Proof: Let Φ be the collection of all σ-fields $\mathcal{A}'$ on Ω for which $\mathcal{C} \subseteq \mathcal{A}'$, that is,

$$\Phi := \{\mathcal{A}' \subseteq \mathcal{P}(\Omega) : \mathcal{C} \subseteq \mathcal{A}', \ \mathcal{A}' \text{ is a } \sigma\text{-field}\}.$$

The collection Φ is nonempty because it contains at least one element, namely the powerset of Ω. Of course, $\mathcal{P}(\Omega)$ is a σ-field and $\mathcal{C} \subseteq \mathcal{P}(\Omega)$ by trivial reason, hence $\mathcal{P}(\Omega) \in \Phi$.

Next define $\mathcal{A}$ by

$$\mathcal{A} := \bigcap_{\mathcal{A}' \in \Phi} \mathcal{A}' = \{A \subseteq \Omega : A \in \mathcal{A}', \ \forall \mathcal{A}' \in \Phi\}.$$

It is not difficult to prove that $\mathcal{A}$ is a σ-field with $\mathcal{C} \subseteq \mathcal{A}$. Indeed, if $C \in \mathcal{C}$, then $C \in \mathcal{A}'$ for all $\mathcal{A}' \in \Phi$, hence by construction of $\mathcal{A}$ we get $C \in \mathcal{A}$.

Furthermore, $\mathcal{A}$ is also the smallest σ-field containing $\mathcal{C}$. To see this, take an arbitrary σ-field $\tilde{\mathcal{A}}$ containing $\mathcal{C}$. Then $\tilde{\mathcal{A}} \in \Phi$, which implies $\mathcal{A} \subseteq \tilde{\mathcal{A}}$ because $\mathcal{A}$ is the intersection over all σ-fields in Φ. This completes the proof. ∎

Definition 1.1.13. Let $\mathcal{C}$ be an arbitrary nonempty collection of subsets of Ω. The smallest σ-field containing $\mathcal{C}$ is called the **σ-field generated by** $\mathcal{C}$. It is denoted by $\sigma(\mathcal{C})$.

Remark 1.1.14. $\sigma(\mathcal{C})$ is characterized by the three following properties:
1. $\sigma(\mathcal{C})$ is a σ-field.
2. $\mathcal{C} \subseteq \sigma(\mathcal{C})$.
3. If $\mathcal{C} \subseteq \mathcal{A}'$ for some σ-field $\mathcal{A}'$, then $\sigma(\mathcal{C}) \subseteq \mathcal{A}'$.

Definition 1.1.15. Let $C \subseteq \mathcal{P}(\mathbb{R})$ be the collection of all finite closed intervals in $\mathbb{R}$, that is,

$$C = \{[a, b] : a < b, a, b \in \mathbb{R}\}.$$

The σ-field generated by C is denoted by $\mathcal{B}(\mathbb{R})$ and is called **Borel σ-field**. If $B \in \mathcal{B}(\mathbb{R})$, then it is said to be a **Borel set**.

Remark 1.1.16. By construction every closed interval in $\mathbb{R}$ is a Borel set. Furthermore, the properties of σ-fields also imply that complements of such intervals, their countable unions, and intersections are Borel sets. One might believe that all subsets of $\mathbb{R}$ are Borel sets. This is not the case; for the construction of a non-Borel set we refer to [Gha05], Example 1.21, or [Dud02], pages 105–108.

Remark 1.1.17. There exist many other systems of subsets in $\mathbb{R}$ generating $\mathcal{B}(\mathbb{R})$. Let us only mention two of them:

$$C_1 = \{(-\infty, b] : b \in \mathbb{R}\} \quad \text{or} \quad C_2 = \{(a, \infty) : a \in \mathbb{R}\}.$$

1.1.3 Probability Measures

The occurrence of an event in a random experiment is not completely haphazardly. Although we are not able to predict the outcome of the next trial, the occurrence or nonoccurrence of an event follows certain rules. Some events are more likely to occur, others less. The degree of likelihood of an event A is described by a number $\mathbb{P}(A)$, called the probability of the occurrence of A (in short, probability of A). The most common scale for probabilities is $0 \leq \mathbb{P}(A) \leq 1$, where the larger $\mathbb{P}(A)$ is the more likely is its occurrence. One could also think of other scales as $0 \leq \mathbb{P}(A) \leq 100$. In fact, this is even quite often used; in this sense a chance of 50% equals a probability of 1/2.

What does it mean that an event A has probability $\mathbb{P}(A)$? For example, what does it tell us that an event occurs with probability 1/2? Does this mean a half-occurrence of A? Surely not.

To answer this question we have to assume that we execute an experiment not only once[1] but several, say n, times. Thereby we have to ensure that the conditions

---

[1] It does not make sense to speak of the probability of an event that can be executed only once. For example, it is (mathematically) absurd to ask for the probability that the Eiffel Tower will be in Paris for yet another 100 years.

of the experiment do not change and that the single results do not depend on each other. Let

$$a_n(A) := \text{Number of trials where } A \text{ occurs} .$$

The quantity $a_n(A)$ is called **absolute frequency** of the occurrence of A in n trials. Observe that $a_n(A)$ is a random number with $0 \le a_n(A) \le n$. Next we set

$$r_n(A) := \frac{a_n(A)}{n} \tag{1.1}$$

and name it **relative frequency** of the occurrence of A in n trials. This number is random as well, but now $0 \le r_n(A) \le 1$.

It is somehow intuitively clear[2] that these relative frequencies converge to a (nonrandom) number as $n \to \infty$. And this limit is exactly the desired probability of the occurrence of the event A. Let us express this in a different way: say we execute an experiment n times for some large n. Then, on average, we will observe $n \cdot \mathbb{P}(A)$ the occurrence of A. For example, when rolling a fair die many times, an even number will be given approximately half the cases.

Which natural properties of $A \mapsto \mathbb{P}(A)$ may be deduced from $\lim_{n\to\infty} r_n(A) = \mathbb{P}(A)$?

1. Since $0 \le r_n(A) \le 1$, we conclude $0 \le \mathbb{P}(A) \le 1$.
2. Because of $r_n(\Omega) = 1$ for each $n \ge 1$ we get $\mathbb{P}(\Omega) = 1$.
3. The property $r_n(\emptyset) = 0$ yields $\mathbb{P}(\emptyset) = 0$.
4. Let A and B be two disjoint events. Then $r_n(A \cup B) = r_n(A) + r_n(B)$, hence the limits should satisfy a similar relation, that is,

$$\mathbb{P}(A \cup B) = \mathbb{P}(A) + \mathbb{P}(B) . \tag{1.2}$$

Definition 1.1.18. A mapping $\mathbb{P}$ fulfilling eq. (1.2) for disjoint A and B is called **finitely additive.**

Remark 1.1.19. Applying eq. (1.2) successively leads to the following. If $A_1, \dots, A_n$ are disjoint, then

$$\mathbb{P}\left(\bigcup_{j=1}^{n} A_j \right) = \sum_{j=1}^{n} \mathbb{P}(A_j) .$$

Finite additivity is a very useful property of probabilities, and in the case of finite sample spaces, it completely suffices to build a fruitful theory. But as soon as the sample space is infinite it is too weak. To see this let us come back to Example 1.1.4.

---

2 We will discuss this question more precisely in Section 7.1.

Assume we want to evaluate the probability of the event $A = \{2, 4, 6, \ldots\}$, that is, the first "6" appears at an even number of trials. Then we have to split A into (infinitely) many disjoint events $\{2\}, \{4\}, \ldots$. The finite additivity of $\mathbb{P}$ does not suffice to get $\mathbb{P}(A) = \mathbb{P}(\{2\}) + \mathbb{P}(\{4\}) + \cdots$. In order to evaluate $\mathbb{P}(A)$ in this way we need the following stronger property of $\mathbb{P}$.

Definition 1.1.20. A mapping $\mathbb{P}$ is said to be σ-**additive** provided that for countably many disjoint $A_1, A_2, \ldots$ in Ω we get

$$\mathbb{P}\left(\bigcup_{j=1}^{\infty} A_j\right) = \sum_{j=1}^{\infty} \mathbb{P}(A_j)$$

Let us summarize what we have until now: a mapping $\mathbb{P}$ assigning each event its probability should possess the following natural properties:
1. For all A holds $0 \le \mathbb{P}(A) \le 1$.
2. We have $\mathbb{P}(\emptyset) = 0$ and $\mathbb{P}(\Omega) = 1$.
3. The mapping $\mathbb{P}$ has to be σ-additive.

Thus, given a sample space Ω, we look for a function $\mathbb{P}$ defined on $\mathcal{P}(\Omega)$ satisfying the previous properties. But, as already mentioned, if Ω is uncountable, for example, $\Omega = R$, then only very special[3] $\mathbb{P}$ with these properties exist.

To overcome these difficulties, in such cases we have to restrict $\mathbb{P}$ to a σ-field $\mathcal{A} \subseteq \mathcal{P}(\Omega)$.

Definition 1.1.21. Let Ω be a sample space and let $\mathcal{A}$ be a σ-field of subsets of Ω. A function $\mathbb{P} : \mathcal{A} \to [0, 1]$ is called **probability measure** or **probability distribution** on $(\Omega, \mathcal{A})$ if
1. $\mathbb{P}(\emptyset) = 0$ and $\mathbb{P}(\Omega) = 1$.
2. $\mathbb{P}$ is σ-additive, that is, for each sequence of disjoint sets $A_j \in \mathcal{A}, j = 1, 2, \ldots$, follows

$$\mathbb{P}\left(\bigcup_{j=1}^{\infty} A_j\right) = \sum_{j=1}^{\infty} \mathbb{P}(A_j). \tag{1.3}$$

Remark 1.1.22. Note that the left-hand side of eq. (1.3) is well-defined. Indeed, since $\mathcal{A}$ is a σ-field, $A_j \in \mathcal{A}$ implies $\bigcup_{j=1}^{\infty} A_j \in \mathcal{A}$ as well.

Now we are in a position to define probability spaces in the exact way.

---

3 Discrete ones as we will investigate in Section 1.3.

> **Definition 1.1.23.** A **probability space** is a triple $(\Omega, \mathcal{A}, \mathbb{P})$, where Ω is a sample space, $\mathcal{A}$ denotes a σ-field consisting of subsets of Ω and $\mathbb{P} : \mathcal{A} \to [0, 1]$ is a probability measure.

Remark 1.1.24. Given $A \in \mathcal{A}$, the number $\mathbb{P}(A)$ describes its probability or, more precisely, its probability of occurrence. Subsets A of Ω with $A \notin \mathcal{A}$ do **not** possess a probability.

Let us demonstrate a simple example on how to construct a probability space for a given random experiment. Several other examples will follow soon.

Example 1.1.25. We ask for a probability space that describes rolling a fair die one time. Of course, $\Omega = \{1, \dots, 6\}$ and $\mathcal{A} = \mathcal{P}(\Omega)$. The mapping $\mathbb{P} : \mathcal{P}(\Omega) \to [0, 1]$ is given by

$$\mathbb{P}(A) = \frac{\#(A)}{6}, \quad A \subseteq \{1, \dots, 6\}.$$

Recall that $\#(A)$ denotes the cardinality of the set A.

Remark 1.1.26. Suppose we want to find a model for some concrete random experiment. How do we do this? In most cases the sample space is immediately determined by the results we will expect. If the question about Ω is settled, the choice of the σ-field depends on the size of the sample space. Is Ω finite or countably finite, then we may choose $\mathcal{A} = \mathcal{P}(\Omega)$. If $\Omega = \mathbb{R}$ or even $\mathbb{R}^n$, we take the corresponding Borel σ-fields. The challenging task is the determination of the probability measure $\mathbb{P}$. Here the following approaches are possible.

1. **Theoretically considerations** lead quite often to the determination of $\mathbb{P}$. For example, since the faces of a fair die are all equally likely, this already describes $\mathbb{P}$ completely. Similar arguments can be used for certain games or also for lotteries.

2. If theoretical considerations are neither possible nor available then **statistically investigations** may help. This approach is based on the fact that the relative frequencies $r_n(A)$ converge to $\mathbb{P}(A)$. Thus, one executes n trials of the experiment and records the relative frequency of the occurrence of A. For example, one may question n randomly chosen persons or one does n independent measurements of the same item. Then $r_n(A)$ may be used to approximate the value of $\mathbb{P}(A)$.

3. Sometimes also **subjective or experience-based** approaches can be used to find approximative probabilities. These may be erroneous, but maybe they give some hint for the correct distribution. For example, if a new product is on the market, the distribution of its lifetime is not yet known. At the beginning one uses data of an already existing similar product. After some time data about the new product become available, the probabilities can be determined more accurately.

1.2 Basic Properties of Probability Measures

Probability measures obey many useful properties. Let us summarize the most important ones in the next proposition.

Proposition 1.2.1. *Let $(\Omega, \mathcal{A}, \mathbb{P})$ be a probability space. Then the following are valid.*
(1) $\mathbb{P}$ *is also finitely additive.*
(2) *If $A, B \in \mathcal{A}$ satisfy $A \subseteq B$, then $\mathbb{P}(B \backslash A) = \mathbb{P}(B) - \mathbb{P}(A)$.*
(3) *We have $\mathbb{P}(A^c) = 1 - \mathbb{P}(A)$ for $A \in \mathcal{A}$.*
(4) *Probability measures are **monotone**, that is, if $A \subseteq B$ for some $A, B \in \mathcal{A}$, then $\mathbb{P}(A) \leq \mathbb{P}(B)$.*
(5) *Probability measures are **subadditive**, that is, for all (not necessarily disjoint) events $A_j \in \mathcal{A}$ follows[4]*

$$\mathbb{P}\left(\bigcup_{j=1}^{\infty} A_j\right) \leq \sum_{j=1}^{\infty} \mathbb{P}(A_j). \qquad (1.4)$$

(6) *Probability measures are **continuous from below**, that is, whenever $A_j \in \mathcal{A}$ satisfy $A_1 \subseteq A_2 \subseteq \cdots$, then*

$$\mathbb{P}\left(\bigcup_{j=1}^{\infty} A_j\right) = \lim_{j \to \infty} \mathbb{P}(A_j).$$

(7) *In a similar way each probability measure is **continuous from above**: if $A_j \in \mathcal{A}$ satisfy $A_1 \supseteq A_2 \supseteq \cdots$, then*

$$\mathbb{P}\left(\bigcap_{j=1}^{\infty} A_j\right) = \lim_{j \to \infty} \mathbb{P}(A_j).$$

Proof: To prove (1) choose disjoint $A_1, \ldots, A_n$ in $\mathcal{A}$ and set $A_{n+1} = A_{n+2} = \cdots = \emptyset$. Then $A_1, A_2, \ldots$ are infinitely many disjoint events in $\mathcal{A}$, hence the σ-additivity of $\mathbb{P}$ implies

$$\mathbb{P}\left(\bigcup_{j=1}^{\infty} A_j\right) = \sum_{j=1}^{\infty} \mathbb{P}(A_j).$$

Observe that $\bigcup_{j=1}^{\infty} A_j = \bigcup_{j=1}^{n} A_j$ and $\mathbb{P}(A_j) = 0$ if $j > n$, so the previous equation reduces to

$$\mathbb{P}\left(\bigcup_{j=1}^{n} A_j\right) = \sum_{j=1}^{n} \mathbb{P}(A_j),$$

and $\mathbb{P}$ is finitely additive.

---

4 Estimate (1.4) is also known as Boole's inequality.

To prove (2) write $B = A \cup (B \backslash A)$ and observe that this is a disjoint decomposition of B. Hence, by the finite additivity of $\mathbb{P}$ we obtain

$$\mathbb{P}(B) = \mathbb{P}(A) + \mathbb{P}(B \backslash A).$$

Relocating $\mathbb{P}(A)$ to the left-hand side proves (2).

An application of (2) to Ω and A leads to

$$\mathbb{P}(A^c) = \mathbb{P}(\Omega \backslash A) = \mathbb{P}(\Omega) - \mathbb{P}(A) = 1 - \mathbb{P}(A),$$

which proves (3).

The monotonicity is an easy consequence of (2). Indeed,

$$\mathbb{P}(B) - \mathbb{P}(A) = \mathbb{P}(B \backslash A) \geq 0$$

implies $\mathbb{P}(B) \geq \mathbb{P}(A)$.

To prove inequality (1.4) choose arbitrary $A_1, A_2, \ldots$ in $\mathcal{A}$. Set $B_1 := A_1$ and, if $j \geq 2$, then

$$B_j := A_j \backslash (A_1 \cup \cdots \cup A_{j-1}).$$

Then $B_1, B_2, \ldots$ are disjoint subsets in $\mathcal{A}$ with $\bigcup_{j=1}^{\infty} B_j = \bigcup_{j=1}^{\infty} A_j$. Furthermore, by the construction holds $B_j \subseteq A_j$, hence $\mathbb{P}(B_j) \leq P(A_j)$. An application of all these properties yields

$$\mathbb{P}\left(\bigcup_{j=1}^{\infty} A_j\right) = \mathbb{P}\left(\bigcup_{j=1}^{\infty} B_j\right) = \sum_{j=1}^{\infty} \mathbb{P}(B_j) \leq \sum_{j=1}^{\infty} \mathbb{P}(A_j).$$

Thus (5) is proved.

Let us turn now to the continuity from below. Choose $A_1, A_2, \ldots$ in $\mathcal{A}$ satisfying $A_1 \subseteq A_2 \subseteq \cdots$. With $A_0 := \emptyset$ set

$$B_k := A_k \backslash A_{k-1}, \quad k = 1, 2, \ldots$$

The B_ks are disjoint and, moreover, $\bigcup_{k=1}^{\infty} B_k = \bigcup_{j=1}^{\infty} A_j$. Furthermore, because of $A_{k-1} \subseteq A_k$ from (2) we get $\mathbb{P}(B_k) = \mathbb{P}(A_k) - \mathbb{P}(A_{k-1})$. When putting this all together, it follows

$$\mathbb{P}\left(\bigcup_{j=1}^{\infty} A_j\right) = \mathbb{P}\left(\bigcup_{k=1}^{\infty} B_k\right) = \sum_{k=1}^{\infty} \mathbb{P}(B_k)$$

$$= \lim_{j \to \infty} \sum_{k=1}^{j} \mathbb{P}(B_k) = \lim_{j \to \infty} \sum_{k=1}^{j} [\mathbb{P}(A_k) - \mathbb{P}(A_{k-1})]$$

$$= \lim_{j \to \infty} [\mathbb{P}(A_j) - \mathbb{P}(A_0)] = \lim_{j \to \infty} \mathbb{P}(A_j)$$

where we used $\mathbb{P}(A_0) = \mathbb{P}(\emptyset) = 0$. This proves the continuity from below.

Thus it remains to prove (7). To this end choose $A_j \in \mathcal{A}$ with $A_1 \supseteq A_2 \supseteq \cdots$. Then the complementary sets satisfy $A_1^c \subseteq A_2^c \subseteq \cdots$. The continuity from below lets us conclude that

$$\mathbb{P}\left(\bigcup_{j=1}^{\infty} A_j^c\right) = \lim_{j\to\infty} \mathbb{P}(A_j^c) = \lim_{j\to\infty} [1 - \mathbb{P}(A_j)] = 1 - \lim_{j\to\infty} \mathbb{P}(A_j). \tag{1.5}$$

But

$$\mathbb{P}\left(\bigcup_{j=1}^{\infty} A_j^c\right) = 1 - \mathbb{P}\left(\left(\bigcup_{j=1}^{\infty} A_j^c\right)^c\right) = 1 - \mathbb{P}\left(\bigcap_{j=1}^{\infty} A_j\right),$$

and plugging this into eq. (1.5) gives

$$\mathbb{P}\left(\bigcap_{j=1}^{\infty} A_j\right) = \lim_{j\to\infty} \mathbb{P}(A_j)$$

as asserted. ∎

Remark 1.2.2. Property (2) becomes false without the assumption $A \subseteq B$. But since $B\backslash A = B\backslash(A \cap B)$ and $A \cap B \subseteq B$, we always have

$$\mathbb{P}(B\backslash A) = \mathbb{P}(B) - \mathbb{P}(A \cap B). \tag{1.6}$$

Another useful property of probability measures is as follows.

Proposition 1.2.3. *Let $(\Omega, \mathcal{A}, \mathbb{P})$ be a probability space. Then for all $A_1, A_2 \in \mathcal{A}$ it follows*

$$\mathbb{P}(A_1 \cup A_2) = \mathbb{P}(A_1) + \mathbb{P}(A_2) - \mathbb{P}(A_1 \cap A_2). \tag{1.7}$$

Proof: Write the union of the two sets as

$$A_1 \cup A_2 = A_1 \cup [A_2\backslash(A_1 \cap A_2)]$$

and note that the two sets on the right-hand side are disjoint. Because of $A_1 \cap A_2 \subseteq A_2$ property (2) of Proposition 1.2.1 applies and leads to

$$\mathbb{P}(A_1 \cup A_2) = \mathbb{P}(A_1) + \mathbb{P}(A_2\backslash(A_1 \cap A_2)) = \mathbb{P}(A_1) + [\mathbb{P}(A_2) - \mathbb{P}(A_1 \cap A_2)].$$

This completes the proof. ∎

Given $A_1, A_2, A_3 \in \mathcal{A}$ an application of the previous proposition to A_1 and $A_2 \cup A_3$ implies

$$\mathbb{P}(A_1 \cup A_2 \cup A_3) = \mathbb{P}(A_1) + \mathbb{P}(A_2 \cup A_3) - \mathbb{P}((A_1 \cap A_2) \cup (A_1 \cap A_3)) .$$

Another application of eq. (1.7) to the second and to the third term in the right-hand sum proves the following result.

Proposition 1.2.4. *Let $(\Omega, \mathcal{A}, \mathbb{P})$ be a probability space and let $A_1, A_2,$ and A_3 be in $\mathcal{A}$. Then*

$$\mathbb{P}(A_1 \cup A_2 \cup A_3) = \mathbb{P}(A_1) + \mathbb{P}(A_2) + \mathbb{P}(A_3)$$
$$- [\mathbb{P}(A_1 \cap A_2) + \mathbb{P}(A_1 \cap A_3) + \mathbb{P}(A_2 \cap A_3)] + \mathbb{P}(A_1 \cap A_2 \cap A_3) .$$

Remark 1.2.5. A generalization of Propositions 1.2.3 and 1.2.4 from 2 or 3 to an arbitrary number of sets can be found in Problem 1.5. It is the so-called inclusion–exclusion formula.

First, let us explain an easy example of how the properties of probability measures apply.

Example 1.2.6. Let $(\Omega, \mathcal{A}, \mathbb{P})$ be a probability space. Suppose two events A and B in $\mathcal{A}$ satisfy

$$\mathbb{P}(A) = 0.5, \quad \mathbb{P}(B) = 0.4 \quad \text{and} \quad \mathbb{P}(A \cap B) = 0.2 .$$

Which probabilities do $A \cup B$, $A \backslash B$, $A^c \cup B^c$, and $A^c \cap B$ possess?
  Answer: An application of Proposition 1.2.4 gives

$$\mathbb{P}(A \cup B) = \mathbb{P}(A) + \mathbb{P}(B) - \mathbb{P}(A \cap B) = 0.4 + 0.5 - 0.2 = 0.7 .$$

Furthermore, by eq. (1.6) follows

$$\mathbb{P}(A \backslash B) = \mathbb{P}(A) - \mathbb{P}(A \cap B) = 0.5 - 0.2 = 0.3 .$$

Finally, by De Morgan's rules and another application of eq. (1.6) we get

$$\mathbb{P}(A^c \cup B^c) = 1 - \mathbb{P}(A \cap B) = 0.8 \quad \text{and} \quad \mathbb{P}(A^c \cap B) = \mathbb{P}(B \backslash A) = \mathbb{P}(B) - \mathbb{P}(A \cap B) = 0.2 .$$

In summary, say one has to take two exams A and B. The probability of passing exam A is 0.5, the probability of passing B equals 0.4, and to pass both is 0.2. Then with probability 0.7 one passes at least one of the exams, with 0.3 exam A, but not B, with 0.8 one fails at least once, and, finally, the probability to pass B but not A is 0.2.

1.3 Discrete Probability Measures

We start with the investigation of **finite** sample spaces. They describe random experiments where only finitely many different results may occur, as, for example, rolling a die n times, tossing a coin finitely often, and so on. Suppose the sample space contains N different elements. Then we may enumerate these elements as follows:

$$\Omega = \{\omega_1, \ldots, \omega_N\}.$$

As σ-field we choose $\mathcal{A} = \mathcal{P}(\Omega)$.

Given an arbitrary probability measure $\mathbb{P} : \mathcal{P}(\Omega) \to \mathbb{R}$ set

$$p_j := \mathbb{P}(\{\omega_j\}), \quad j = 1, \ldots, N. \tag{1.8}$$

In this way we assign to each probability measure $\mathbb{P}$ numbers $p_1, \ldots, p_N$. Which properties do they possess? The answer to this question gives the following proposition.

Proposition 1.3.1. *If $\mathbb{P}$ is a probability measure on $\mathcal{P}(\Omega)$, then the numbers p_j defined by eq. (1.8) satisfy*

$$0 \le p_j \le 1 \quad and \quad \sum_{j=1}^{N} p_j = 1. \tag{1.9}$$

Proof: The first property is an immediate consequence of $\mathbb{P}(A) \ge 0$ for all $A \subseteq \Omega$.
The second property of the p_js follows by

$$1 = \mathbb{P}(\Omega) = \mathbb{P}\left(\bigcup_{j=1}^{N} \{\omega_j\}\right) = \sum_{j=1}^{N} \mathbb{P}(\{\omega_j\}) = \sum_{j=1}^{N} p_j. \qquad \blacksquare$$

Conclusion: Each probability measure $\mathbb{P}$ generates a sequence $(p_j)_{j=1}^{N}$ of real numbers satisfying the properties (1.9). Moreover, if $A \subseteq \Omega$, then we have

$$\mathbb{P}(A) = \sum_{\{j : \omega_j \in A\}} p_j. \tag{1.10}$$

In particular, the assignment $\mathbb{P} \to (p_j)_{j=1}^{N}$ is one-to-one.

Property (1.10) is an easy consequence of $A = \bigcup_{\{j : \omega_j \in A\}} \{\omega_j\}$. Furthermore, it tells us that $\mathbb{P}$ is uniquely determined by the p_js. Note that two probability measures $\mathbb{P}_1$ and $\mathbb{P}_2$ on $(\Omega, \mathcal{A})$ coincide if $\mathbb{P}_1(A) = \mathbb{P}_2(A)$ for all $A \in \mathcal{A}$.

Now let us look at the reverse question. Suppose we are given an **arbitrary** sequence $(p_j)_{j=1}^{N}$ of real numbers satisfying the conditions (1.9).

Proposition 1.3.2. *Define* $\mathbb{P}$ *on* $\mathcal{P}(\Omega)$ *by*

$$\mathbb{P}(A) = \sum_{\{j:\omega_j \in A\}} p_j \, . \qquad (1.11)$$

Then $\mathbb{P}$ *is a probability measure satisfying* $\mathbb{P}(\{\omega_j\}) = p_j$ *for all* $j \leq n$.

Proof: $\mathbb{P}$ has values in $[0, 1]$ and $\mathbb{P}(\Omega) = 1$ by $\sum_{j=1}^{N} p_j = 1$. Since the summation over the empty set equals zero, $\mathbb{P}(\emptyset) = 0$.

Thus it remains to be shown that $\mathbb{P}$ is σ-additive. Take disjoint subsets $A_1, A_2, \ldots$ of Ω. Since Ω is finite, there are at most finitely many of the A_js nonempty. Say, for simplicity, these are the first n sets $A_1, \ldots, A_n$. Then we get

$$\mathbb{P}\left(\bigcup_{k=1}^{\infty} A_k \right) = \mathbb{P}\left(\bigcup_{k=1}^{n} A_k \right) = \sum_{\{j:\omega_j \in \bigcup_{k=1}^{n} A_k\}} p_j$$

$$= \sum_{k=1}^{n} \sum_{\{j:\omega_j \in A_k\}} p_j = \sum_{k=1}^{\infty} \sum_{\{j:\omega_j \in A_k\}} p_j = \sum_{k=1}^{\infty} \mathbb{P}(A_k),$$

hence $\mathbb{P}$ is σ-additive.

By the construction $\mathbb{P}(\{\omega_j\}) = p_j$, which completes the proof. ∎

Summary: If $\Omega = \{\omega_1, \ldots, \omega_N\}$, then probability measures $\mathbb{P}$ on $\mathcal{P}(\Omega)$ can be identified with sequences $(p_j)_{j=1}^{N}$ satisfying conditions (1.9).

!

$$\left\{ \text{Probability measures } \mathbb{P} \text{ on } \mathcal{P}(\Omega) \right\} \iff \left\{ \text{Sequences } (p_j)_{j=1}^{N} \text{ with (1.9)} \right\}$$

Hereby the assignment from the left- to the right-hand side goes via $p_j = \mathbb{P}(\{\omega_j\})$ while in the other direction $\mathbb{P}$ is given by eq. (1.11).

Example 1.3.3. Assume $\Omega = \{1, 2, 3\}$. Then each probability measure $\mathbb{P}$ on $\mathcal{P}(\Omega)$ is uniquely determined by the three numbers $p_1 = \mathbb{P}(\{1\})$, $p_2 = \mathbb{P}(\{2\})$, and $p_3 = \mathbb{P}(\{3\})$. These numbers satisfy $p_1, p_2, p_3 \geq 0$ and $p_1 + p_2 + p_3 = 1$. Conversely, any three numbers p_1, p_2, and p_3 with these properties generate a probability measure on $\mathcal{P}(\Omega)$ via (1.11). For example, if $A = \{1, 3\}$, then $\mathbb{P}(A) = p_1 + p_3$.

Next we treat **countably infinite** sample spaces, that is, $\Omega = \{\omega_1, \omega_2, \ldots\}$. Also here we may take $\mathcal{P}(\Omega)$ as σ-field and as in the case of finite sample spaces, given a probability measure $\mathbb{P}$ on $\mathcal{P}(\Omega)$, we set

$$p_j := \mathbb{P}(\{\omega_j\}), \quad j = 1, 2, \ldots$$

Then $(p_j)_{j=1}^{\infty}$ obeys the following properties:

$$p_j \geq 0 \quad \text{and} \quad \sum_{j=1}^{\infty} p_j = 1. \tag{1.12}$$

The proof is the same as in the finite case. The only difference is that here we have to use the σ-additivity of $\mathbb{P}$ because this time $\Omega = \bigcup_{j=1}^{\infty}\{\omega_j\}$. By the same argument follows for $A \subseteq \Omega$ that

$$\mathbb{P}(A) = \sum_{\{j \geq 1\,:\,\omega_j \in A\}} p_j.$$

Hence, again the p_js determine $\mathbb{P}$ completely.

Conversely, let $(p_j)_{j=1}^{\infty}$ be an **arbitrary** sequence of real numbers with properties (1.12).

Proposition 1.3.4. *The mapping* $\mathbb{P}$ *defined by*

$$\mathbb{P}(A) = \sum_{\{j \geq 1\,:\,\omega_j \in A\}} p_j. \tag{1.13}$$

is a probability measure on $\mathcal{P}(\Omega)$ *with* $\mathbb{P}(\{\omega_j\}) = p_j, 1 \leq j < \infty$.

Proof: The proof is analogous to Proposition 1.3.2 with one important exception. In the case $\#(\Omega) < \infty$ we used that there are at most finitely many disjoint nonempty subsets. This is no longer valid. Thus a different argument is needed.

Given disjoint subsets $A_1, A_2, \ldots$ in Ω set

$$I_k = \{j \geq 1 : \omega_j \in A_k\}.$$

Then $I_k \cap I_l = \emptyset$ if $k \neq l$,

$$\mathbb{P}(A_k) = \sum_{j \in I_k} p_j \quad \text{and} \quad \mathbb{P}\left(\bigcup_{k=1}^{\infty} A_k\right) = \sum_{j \in I} p_j$$

where $I = \bigcup_{k=1}^{\infty} I_k$.

Since $p_j \geq 0$, Remark A.5.6 applies and leads to

$$\mathbb{P}\left(\bigcup_{k=1}^{\infty} A_k\right) = \sum_{j \in I} p_j = \sum_{k=1}^{\infty} \sum_{j \in I_k} p_j = \sum_{k=1}^{\infty} \mathbb{P}(A_k).$$

Thus $\mathbb{P}$ is σ-additive.

The equality $\mathbb{P}(\{\omega_j\}) = p_j, 1 \leq j < \infty$, is again a direct consequence of the definition of $\mathbb{P}$. ∎

Summary: If $\Omega = \{\omega_1, \omega_2, \ldots\}$, then probability measures $\mathbb{P}$ on $\mathcal{P}(\Omega)$ can be identified with (infinite) sequences $(p_j)_{j=1}^{\infty}$ possessing the properties (1.12).

!

$$\left\{ \text{Probability measures } \mathbb{P} \text{ on } \mathcal{P}(\Omega) \right\} \quad \Longleftrightarrow \quad \left\{ \text{Sequences } (p_j)_{j=1}^{\infty} \text{ with (1.12)} \right\}$$

Again, the assignment from the left-hand to the right-hand side goes via $p_j = \mathbb{P}(\{\omega_j\})$ while the other direction rests upon eq. (1.13).

Example 1.3.5. For $\Omega = \mathbb{N}$ and $j \geq 1$ let $p_j = 2^{-j}$. These p_js satisfy conditions (1.12) (check this!). The generated probability measure $\mathbb{P}$ on $\mathcal{P}(\mathbb{N})$ is then given by

$$\mathbb{P}(A) := \sum_{j \in A} \frac{1}{2^j} \, .$$

For example, if $A = \{2, 4, 6, \ldots\}$ then we get

$$\mathbb{P}(A) = \sum_{j \in A} \frac{1}{2^j} = \sum_{k=1}^{\infty} \frac{1}{2^{2k}} = \frac{1}{1 - 1/4} - 1 = \frac{1}{3} \, .$$

Example 1.3.6. Let $\Omega = \mathbb{Z} \setminus \{0\}$, that is, $\Omega = \{1, -1, 2, -2, \ldots\}$. With $c > 0$ specified later on assume

$$p_k = \frac{c}{k^2}, \quad k \in \Omega.$$

The number $c > 0$ has to be chosen so that the conditions (1.12) are satisfied, hence it has to satisfy

$$1 = c \sum_{k \in \mathbb{Z} \setminus \{0\}} \frac{1}{k^2} = 2c \sum_{k=1}^{\infty} \frac{1}{k^2} \, .$$

But as is well known[5]

$$\sum_{k=1}^{\infty} \frac{1}{k^2} = \frac{\pi^2}{6} \, ,$$

which implies $c = \frac{3}{\pi^2}$. Thus $\mathbb{P}$ on $\mathcal{P}(\Omega)$ is uniquely described by

$$\mathbb{P}(\{k\}) = \frac{3}{\pi^2} \frac{1}{k^2}, \quad k \in \mathbb{Z} \setminus \{0\} \, .$$

---

[5] We refer to [Mor16], where one can find an easy proof of this fact. The problem to compute the value of the sum is known as "Basel problem." The first solution was found in 1734 by Leonhard Euler. Note that $\sum_{k \geq 1} 1/k^2 = \zeta(2)$ with Riemann's ζ-function.

For example, if $A = \mathbb{N}$, then

$$\mathbb{P}(A) = \frac{3}{\pi^2} \sum_{k=1}^{\infty} \frac{1}{k^2} = \frac{3}{\pi^2} \frac{\pi^2}{6} = \frac{1}{2} .$$

Or if $A = \{2, 4, 6, \ldots\}$, it follows

$$\mathbb{P}(A) = \frac{3}{\pi^2} \sum_{k=1}^{\infty} \frac{1}{(2k)^2} = \frac{1}{4} \mathbb{P}(\mathbb{N}) = \frac{1}{8} .$$

For later purposes we want to combine the two cases of finite and countably infinite sample spaces and thereby introduce a slight generalization.

Let Ω be an arbitrary sample space. A probability measure $\mathbb{P}$ is said to be **discrete** if there is an at most countably infinite set $D \subseteq \Omega$ (i.e., either D is finite or countably infinite) such that $\mathbb{P}(D) = 1$. Then for $A \subseteq \Omega$ follows

$$\mathbb{P}(A) = \mathbb{P}(A \cap D) = \sum_{\omega \in D} \mathbb{P}(\{\omega\}) .$$

Since $\mathbb{P}(D^c) = 0$, this says that $\mathbb{P}$ is **concentrated** on D. Of course, all previous results for finite or countably infinite sample space carry over to this more general setting.

Discrete probability measures $\mathbb{P}$ are concentrated on an at most countably infinite set D. They are uniquely determined by the values $\mathbb{P}(\{\omega\})$, where $\omega \in D$. !

Of course, if the sample space is either finite or countably infinite, then **all** probability measures on this space are discrete. Nondiscrete probability measures will be introduced and investigated in Section 1.5

Example 1.3.7. We once more model the one-time rolling of a die, but now we take as sample space $\Omega = \mathbb{R}$. Define $\mathbb{P}(\{\omega\}) = \frac{1}{6}$ if $\omega = 1, \ldots, 6$ and $\mathbb{P}(\{\omega\}) = 0$ otherwise. If $D = \{1, \ldots, 6\}$, then $\mathbb{P}(D) = 1$, hence $\mathbb{P}$ is discrete. Given $A \subseteq \mathbb{R}$, it follows

$$\mathbb{P}(A) = \frac{\#(A \cap D)}{6} .$$

For example, we have $\mathbb{P}([-2, 2]) = \frac{1}{3}$ or $\mathbb{P}([3, \infty)) = \frac{2}{3}$.

1.4 Special Discrete Probability Measures

1.4.1 Dirac Measure

The simplest discrete probability measure is the one concentrated at a single point. That is, there exists an $\omega_0 \in \Omega$ such that $\mathbb{P}(\{\omega_0\}) = 1$. This probability measure is denoted by δ_{ω_0}. Consequently, for each $A \in \mathcal{P}(\Omega)$ one has

$$\delta_{\omega_0}(A) = \begin{cases} 1 : \omega_0 \in A \\ 0 : \omega_0 \notin A \end{cases} \tag{1.14}$$

Definition 1.4.1. The probability measure δ_{ω_0} defined by eq. (1.14) is called **Dirac measure** or **point measure** at ω_0 .

Which random experiment does $(\Omega, \mathcal{P}(\Omega), \delta_{\omega_0})$ model? It describes the experiment where with probability one the value ω_0 occurs. Thus, in fact it is a deterministic experiment, not random.

Dirac measures are useful tools to represent general discrete probability measures. Assume $\mathbb{P}$ is concentrated on $D = \{\omega_1, \omega_2, \ldots\}$ and let $p_j = \mathbb{P}(\{\omega_j\})$. Then we may write

$$\mathbb{P} = \sum_{j=1}^{\infty} p_j \, \delta_{\omega_j} . \tag{1.15}$$

Conversely, if a measure $\mathbb{P}$ is represented as in eq. (1.15) with certain $\omega_j \in \Omega$ and numbers $p_j \geq 0$, $\sum_{j=1}^{\infty} p_j = 1$, then $\mathbb{P}$ is discrete with $\mathbb{P}(D) = 1$, where $D = \{\omega_1, \omega_2, \ldots\}$.

1.4.2 Uniform Distribution on a Finite Set

The sample space is finite, say $\Omega = \{\omega_1, \ldots, \omega_N\}$, and we assume that all elementary events are equally likely, that is,

$$\mathbb{P}(\{\omega_1\}) = \cdots = \mathbb{P}(\{\omega_N\}) .$$

A typical example is a fair die, where $\Omega = \{1, \ldots, 6\}$.

Since $1 = \mathbb{P}(\Omega) = \sum_{j=1}^{N} \mathbb{P}(\{\omega_j\})$ we immediately get $\mathbb{P}(\{\omega_j\}) = 1/N$ for all $j \leq N$. If $A \subseteq \Omega$, an application of eq. (1.11) leads to

$$\mathbb{P}(A) = \frac{\#(A)}{N} = \frac{\#(A)}{\#(\Omega)} . \tag{1.16}$$

> **Definition 1.4.2.** The probability measure $\mathbb{P}$ defined by eq. (1.16) is called **uniform distribution** or **Laplace distribution** on the finite set Ω.

The following formula may be helpful for remembrance:

$$\mathbb{P}(A) = \frac{\text{Number of cases favorable for } A}{\text{Number of possible cases}}$$

!

Example 1.4.3. In a lottery, 6 numbers are chosen out of 49 and each number appears only once. What is the probability that the chosen numbers are exactly the six ones on my lottery coupon?

Answer: Let us give two different approaches to answer this question.

Approach 1: We record the chosen numbers in the order they show up. As a sample space we may take

$$\Omega := \{(\omega_1, \dots, \omega_6) : \omega_i \in \{1, \dots, 49\}, \ \omega_i \neq \omega_j \text{ if } i \neq j\}.$$

Then the number of possible cases is

$$\#(\Omega) = 49 \cdot 48 \cdot 47 \cdot 46 \cdot 45 \cdot 44 = \frac{49!}{43!}.$$

Let A be the event that the numbers on my lottery coupon appear. Which cardinality does A possess?

Say, for simplicity, in our coupon are the numbers $1, 2, \dots, 6$. Then it is favorable for A if these numbers appear in this order. But it is also favorable if $(2, 1, 3, \dots, 6)$ shows up, that is, any permutation of $1, \dots, 6$ is favorable. Hence $\#(A) = 6!$ which leads to[6]

$$\mathbb{P}(A) = \frac{6!}{49 \cdots 44} = \frac{1}{\binom{49}{6}} = 7.15112 \times 10^{-8}.$$

Approach 2: We assume that the chosen numbers are already ordered by their size (as they are published in a newspaper). In this case our sample space is

$$\Omega := \{(\omega_1, \dots, \omega_6) : 1 \leq \omega_1 < \cdots < \omega_6 \leq 49\}$$

---

6 To get an impression about the size of this number assume we buy lottery coupons with all possible choices of the six numbers. If each coupon is 0.5 mm thick, then all coupons together have a size of 6.992 km, which is about 4.3 miles. And in this row of 4.3 miles there exists exactly one coupon with the six numbers chosen in the lottery.

and now

$$\#(\Omega) = \binom{49}{6}.$$

Why? Any set of six different numbers may be written exactly in one way in increasing order and thus, to choose six ordered numbers is exactly the same as to choose a (nonordered) set of six numbers. And there are $\binom{49}{6}$ possibilities to choose six numbers. In this setting we have $\#(A) = 1$, thus also here we get

$$\mathbb{P}(A) = \frac{1}{\binom{49}{6}}.$$

Example 1.4.4. A fair coin is labeled with "0" and "1." Toss it n times and record the sequence of 0s and 1s in the order of their appearance. Thus,

$$\Omega := \{0,1\}^n = \{(\omega_1, \ldots, \omega_n) : \omega_i \in \{0,1\}\},$$

and $\#(\Omega) = 2^n$. The coin is assumed to be fair, hence each sequence of 0s and 1s is equally likely. Therefore, whenever $A \subseteq \Omega$, then

$$\mathbb{P}(A) = \frac{\#(A)}{2^n}.$$

Take, for example, the event A where for some fixed $i \leq n$ the ith toss equals "0," that is,

$$A = \{(\omega_1, \ldots, \omega_n) : \omega_i = 0\},$$

Then $\#(A) = 2^{n-1}$ leads to the (not surprising) result

$$\mathbb{P}(A) = \frac{2^{n-1}}{2^n} = \frac{1}{2}.$$

Or let A occur if we observe for some given $k \leq n$ exactly k times the number "1." Then $\#(A) = \binom{n}{k}$ and we get

$$\mathbb{P}(A) = \binom{n}{k} \cdot \frac{1}{2^n}.$$

Example 1.4.5. We have k particles that we distribute randomly into n boxes. All possible distributions of the particles are assumed to be equally likely. How do we get $\mathbb{P}(A)$ for a given event A?

Answer: In this formulation the question is not asked correctly because we did not fix when two distributions of particles coincide.

Let us illustrate this problem in the case of two particles and two boxes. If the particles are **not distinguishable (anonymous)** then there are three different ways to distribute the particles into the two boxes. Thus, assuming that all distributions are equally likely, each elementary event has probability 1/3.

On the other hand, if the particles are **distinguishable**, that is, they carry names, then there exist four different ways of distributing them (check this!), hence each elementary event has probability 1/4.

Let us answer the above question in the two cases (distinguishable and anonymous) separately.

Distinguishable particles: Here we may enumerate the particles from 1 to k and each distribution of particles is uniquely described by a sequence $(a_1, \dots, a_k)$, where $a_j \in \{1, \dots, n\}$. For example, $a_1 = 3$ means that particle one is in box 3. Hence, a suitable sample space is

$$\Omega = \{(a_1, \dots, a_k) : 1 \le a_i \le n\}.$$

Since $\#(\Omega) = n^k$ for events $A \subseteq \Omega$ follows

$$\mathbb{P}(A) = \frac{\#(A)}{n^k}.$$

Anonymous particles: We record how many of the k particles are in box 1, how many are in box 2 up to box n. Thus as sample space we may choose

$$\Omega = \{(k_1, \dots, k_n) : k_j = 0, \dots, k, \ k_1 + \cdots + k_n = k\}.$$

The sequence $(k_1, \dots, k_n)$ occurs if box 1 contains k_1 particles, box 2 contains k_2, and so on. From the results in case 3 of Section A.3.2 we derive

$$\#(\Omega) = \binom{n+k-1}{k}.$$

Hence, if $A \subseteq \Omega$, then

$$\mathbb{P}(A) = \frac{\#(A)}{\#(\Omega)} = \#(A) \frac{k!\,(n-1)!}{(n+k-1)!}$$

Summary: If we distribute k particles and assume that **all partitions are equally likely**[7], then in the case of distinguishable or of anonymous particles

$$\mathbb{P}(A) = \frac{\#(A)}{n^k} \quad \text{or} \quad \mathbb{P}(A) = \#(A) \frac{k!\,(n-1)!}{(n+k-1)!},$$

respectively.

---

7 Compare Example 1.4.16 and the following remark.

Let us evaluate $\mathbb{P}(A)$ for some concrete event A in both cases. Suppose $k \le n$ and select k of the n boxes. Set

$$A := \{\text{In each of the chosen } k \text{ boxes is exactly one particle}\}. \tag{1.17}$$

To simplify the notation assume that the first k boxes have been chosen. The general case is treated in a similar way. Then in the "distinguishable case" the event A occurs if and only if for some permutation $\pi \in S_k$ the sequence $(\pi(1), \dots, \pi(k), 0 \dots, 0)$ appears. Thus $\#(A) = k!$ and

$$\mathbb{P}(A) = \frac{k!}{n^k} . \tag{1.18}$$

In the "anonymous case" it follows $\#(A) = 1$ (why?). Hence here we obtain

$$\mathbb{P}(A) = \frac{k! \, (n-1)!}{(n+k-1)!} . \tag{1.19}$$

Additional question: For $k \le n$ define B by

$$B := \{\text{Each of the } n \text{ boxes contains at most 1 particle}\}$$

Find $\mathbb{P}(B)$ in both cases.

Answer: The event B is the (disjoint) union of the following events: the k particles are distributed in a given collection of k boxes. The probability of this event was calculated in eqs. (1.18) and (1.19), respectively. Since there are $\binom{n}{k}$ possibilities to choose k boxes of the n we get $\mathbb{P}(B) = \binom{n}{k}\mathbb{P}(A)$ with A as defined by (1.17), that is,

$$\mathbb{P}(B) = \binom{n}{k} \cdot \frac{k!}{n^k} = \frac{n!}{(n-k)! \, n^k} \quad \text{and}$$

$$\mathbb{P}(B) = \binom{n}{k} \cdot \frac{k! \, (n-1)!}{(n+k-1)!} = \frac{n! \, (n-1)!}{(n-k)! \, (n+k-1)!},$$

respectively.

1.4.3 Binomial Distribution

The sample space is $\Omega = \{0, 1, \dots, n\}$ for some $n \ge 1$ and p is a real number with $0 \le p \le 1$.

Proposition 1.4.6. *There exists a unique probability measure $B_{n,p}$ on $\mathcal{P}(\Omega)$ satisfying*

$$B_{n,p}(\{k\}) = \binom{n}{k} p^k (1-p)^{n-k}, \quad k = 0, \dots, n. \tag{1.20}$$

Proof: In order to use Proposition 1.3.2 we have to verify $B_{n,p}(\{k\}) \geq 0$ and $\sum_{k=0}^{n} B_{n,p}(\{k\}) = 1$. The first property is obvious because of $0 \leq p \leq 1$ and $0 \leq 1 - p \leq 1$. To prove the second one we apply the binomial theorem (Proposition A.3.7) with $a = p$ and with $b = 1 - p$. This leads to

$$\sum_{k=0}^{n} B_{n,p}(\{k\}) = \sum_{k=0}^{n} \binom{n}{k} p^k (1-p)^{n-k} = (p + (1-p))^n = 1.$$

Hence the assertion follows by Proposition 1.3.2 with $p_k = B_{n,p}(\{k\})$, $k = 0, \ldots, n$. ■

Definition 1.4.7. The probability measure $B_{n,p}$ defined by eq. (1.20) is called **binomial distribution** with parameters n and p.

Remark 1.4.8. Observe that $B_{n,p}$ acts as follows. If $A \subseteq \{0, \ldots, n\}$, then

$$B_{n,p}(A) = \sum_{k \in A} \binom{n}{k} p^k (1-p)^{n-k}.$$

Furthermore, for $p = 1/2$ we get

$$B_{n,1/2}(\{k\}) = \binom{n}{k} \frac{1}{2^n}.$$

As we saw in Example 1.4.4 this probability describes the k-fold occurrence of "1" when tossing a fair coin n times.

Which random experiment describes the binomial distribution? To answer this question let us first look at the case $n = 1$. Here we have $\Omega = \{0, 1\}$ with

$$B_{n,p}(\{0\}) = 1 - p \quad \text{and} \quad B_{n,p}(\{1\}) = p.$$

If we identify "0" with **failure** and "1" with **success**, then the binomial distribution describes an experiment where either success or failure may occur, and the success probability is p. Now we execute the same experiment n times and every time we may observe either failure or success. If we have k times success, then there are $\binom{n}{k}$ ways to obtain these k successes during the n trials. The probability for success is p and for failure $1 - p$. By the independence of the single trials, the probability for the sequence is $p^k (1-p)^{n-k}$. By multiplying this probability with the number of different positions of successes we finally arrive at $\binom{n}{k} p^k (1-p)^{n-k}$, the value of $B_{n,p}(\{k\})$.

Summary: The binomial distribution describes the following experiment. We execute n times independently the same experiment where each time either success or failure may appear. The **success probability** is p. Then $B_{n,p}(\{k\})$ is the probability **to observe exactly k times success** or, equivalently, $n - k$ times failure.

Example 1.4.9. An exam consists of 100 problems where each of the question may be answered either with "yes" or "no." To pass the exam at least 60 questions have to be answered correctly. Let p be the probability to answer a single question correctly. How big has p to be in order to pass the exam with a probability greater than 75% ?

Answer: The number p has to be chosen such that the following estimate is satisfied:

$$\sum_{k=60}^{100} \binom{100}{k} p^k (1-p)^{100-k} \geq 0.75 .$$

Numerical calculations show that this is valid if and only if $p \geq 0.62739$.

Example 1.4.10. In an auditorium there are N students. Find the probability that at least two of them have their birthday on April 1.

Answer: We do not take leap years into account and assume that there are no twins among the students. Finally, we make the (probably unrealistic) assumption that all days of a year are equally likely as birthdays. Say success occurs if a student has birthday on April 1. Under the above assumptions the success probability is 1/365. Hence the number of students having birthday on April 1 is binomially distributed with parameters N and $p = 1/365$. We ask for the probability of $A = \{2, 3, \ldots, N\}$. This may be evaluated by

$$\sum_{k=2}^{N} \binom{N}{k} \left(\frac{1}{365}\right)^k \left(\frac{364}{365}\right)^{N-k} = 1 - B_{N,1/365}(\{0\}) - B_{N,1/365}(\{1\})$$

$$= 1 - \left(\frac{364}{365}\right)^N - \frac{N}{365} \left(\frac{364}{365}\right)^{N-1} .$$

For example, $N = 500$ this probability is approximately 0.397895.

1.4.4 Multinomial Distribution

Given natural numbers n and m, the sample space for the multinomial distribution is[8]

$$\Omega := \{(k_1, \ldots, k_m) \in \mathbb{N}_0^m : k_1 + \cdots + k_m = n\} .$$

With certain non-negative real numbers $p_1, \ldots, p_m$ satisfying $p_1 + \cdots + p_m = 1$ set

$$\mathbb{P}(\{(k_1, \ldots, k_m)\}) := \binom{n}{k_1, \ldots, k_m} p_1^{k_1} \cdots p_m^{k_m}, \quad (k_1, \ldots, k_m) \in \Omega . \tag{1.21}$$

---

8 By case 3 in Section A.3.2 the cardinality of Ω is $\binom{n+m-1}{n}$.

Recall that the multinomial coefficients appearing in eq. (1.21) were defined in eq. (A.15) as

$$\binom{n}{k_1, \ldots, k_m} = \frac{n!}{k_1! \cdots k_m!}.$$

The next result shows that eq. (1.21) defines a probability measure.

Proposition 1.4.11. *There is a unique probability measure $\mathbb{P}$ on $\mathcal{P}(\Omega)$ such that (1.21) holds for all $(k_1, \ldots, k_m) \in \Omega$.*

Proof: An application of the multinomial theorem (Proposition A.3.18) implies

$$\sum_{(k_1, \ldots, k_m) \in \Omega} \mathbb{P}(\{(k_1, \ldots, k_m)\}) = \sum_{\substack{k_1 + \cdots + k_m = n \\ k_j \geq 0}} \binom{n}{k_1, \ldots, k_m} p_1^{k_1} \cdots p_m^{k_m}$$

$$= (p_1 + \cdots + p_m)^n = 1^n = 1.$$

Since $\mathbb{P}(\{(k_1, \ldots, k_m)\}) \geq 0$ the assertion follows by Proposition 1.3.2. ∎

In view of the preceding proposition, the following definition is justified.

Definition 1.4.12. The probability measure $\mathbb{P}$ defined by eq. (1.21) is called **multinomial distribution** with parameters n, m, and $p_1, \ldots, p_m$.

Remark 1.4.13. Sometimes it is useful to regard the multinomial distribution on the larger sample space $\Omega = \mathbb{N}_0^m$. In this case we have to modify eq. (1.21) slightly as follows:

$$\mathbb{P}(\{(k_1, \ldots, k_m)\}) = \begin{cases} \binom{n}{k_1, \ldots, k_m} p_1^{k_1} \cdots p_m^{k_m} & : k_1 + \cdots + k_m = n \\ 0 & : k_1 + \cdots + k_m \neq n \end{cases}$$

Which random experiment does the multinomial distribution describe? To answer this question let us recall the model for the binomial distribution. In an urn are balls of two different colors, say white and red. The proportion of the white balls is p, hence $1 - p$ of the red ones. If we choose n balls with replacement, then $B_{n,p}(\{k\})$ is the probability to observe exactly k white balls.

What happens if in the urn are balls of more than two different colors, say of m ones, and the proportions of the colored balls are $p_1, \ldots, p_m$ with $p_1 + \cdots + p_m = 1$?

As in the model for the binomial distribution we choose n balls with replacement. Given integers $k_j \geq 0$ one asks now for the probability of the following event: balls of color 1 showed up k_1 times, those of color 2 k_2 times, and so on. Of course, this probability is zero whenever $k_1 + \cdots + k_m \neq n$. But if the sum is n, then $p_1^{k_1} \cdots p_m^{k_m}$ is the probability for k_j balls of color j in some fixed order. There are $\binom{n}{k_1, \ldots, k_m}$ ways to order the balls without changing the frequency of the colors. Thus the desired probability equals $\binom{n}{k_1, \ldots, k_m} p_1^{k_1} \cdots p_m^{k_m}$.

Summary: Suppose in an experiment are m different results possible (e.g., m colors) and assume that each time the jth result occurs with probability p_j. If we execute the experiment n times, then the multinomial distribution describes the probability of the following event: the first result occurs k_1 times, the second k_2 times, and so on.

Remark 1.4.14. If $m = 2$, then $p_2 = 1 - p_1$ as well as $\binom{n}{k_1, k_2} = \binom{n}{k_1, n-k_1} = \binom{n}{k_1}$. Consequently, in this case the multinomial distribution coincides with the binomial distribution B_{n, p_1}.

Remark 1.4.15. Suppose that all m possible different outcomes of the experiment are equally likely, that is, we have

$$p_1 = \cdots = p_m = \frac{1}{m}.$$

Under this assumption it follows

$$\mathbb{P}(\{(k_1, \ldots, k_m)\}) = \binom{n}{k_1, \ldots, k_m} \frac{1}{m^n}, \quad k_1 + \cdots + k_m = n. \tag{1.22}$$

Example 1.4.16. Suppose we have m boxes $B_1, \ldots, B_m$ and n particles that we place successively into these boxes. Thereby p_j is the probability to place a single particle into box B_j. What is the probability that after distributing all n particles there are k_1 particles in the first box, k_2 in the second up to k_m in the last one?

Answer: This probability is given by formula (1.21), that is,

$$\mathbb{P}\{k_1 \text{ particles are in } B_1, \ldots, k_m \text{ particles are in } B_m\} = \binom{n}{k_1, \ldots, k_m} p_1^{k_1} \cdots p_m^{k_m}.$$

Suppose now $n \leq m$ and that all boxes are chosen with probability $1/m$. Find the probability that each of the first n boxes $B_1, \ldots, B_n$ contains exactly one particle.

Answer: By eq. (1.22) follows

$$\mathbb{P}(\{(\underbrace{1, \ldots, 1}_{n}, 0 \ldots, 0)\}) = \binom{n}{\underbrace{1, \ldots, 1, 0, \ldots, 0}_{n}} \frac{1}{m^n} = \frac{n!}{m^n}. \tag{1.23}$$

Remark 1.4.17. From a different point of view we investigated the last problem already in Example 1.4.5. But why do we get in eq. (1.23) the same answer as in the case of distinguishable particles although the n distributed ones are anonymous?

Answer: The crucial point is that we assumed in the anonymous case that all partitions of the particles are equally likely. And this is not valid when distributing the particles successively. To see this, assume $n = m = 2$. Then there exist three different ways to distribute the particles, but they have different probabilities.

$$\mathbb{P}(\{(0, 2)\}) = \mathbb{P}(\{(2, 0)\}) = \frac{1}{4} \quad \text{while} \quad \mathbb{P}(\{(1, 1)\}) = \frac{1}{2}.$$

Thus, although the distributed particles are not distinguishable, they get names due to the successive distribution (first particle, second particle, etc.).

Example 1.4.18. Six people randomly enter a train with three coaches. Each person chooses his wagon independently of the others and all coaches are equally likely to be chosen. Find the probability that there are two people in each coach.
Answer: We have $m = 3$, $n = 6$, and $p_1 = p_2 = p_3 = \frac{1}{3}$. Hence the probability we are looking for is

$$\mathbb{P}(\{(2, 2, 2)\}) = \binom{6}{2, 2, 2} \frac{1}{3^6} = \frac{6!}{2! \, 2! \, 2!} \frac{1}{3^6} = \frac{10}{81} = 0.12345679.$$

Example 1.4.19. In a country are 40% of the cars gray, 20% are black, and 10% are red. The remaining cars have different colors. Now we observe by random 10 cars. What is the probability to see two gray cars, four black, and one red?
Answer: By assumption $m = 4$ (gray, black, red, and others), $p_1 = 2/5$, $p_2 = 1/5$, $p_3 = 1/10$, and $p_4 = 3/10$. Thus the probability of the vector $(2, 4, 1, 3)$ is given by

$$\binom{10}{2, 4, 1, 3} \left(\frac{2}{5}\right)^2 \left(\frac{1}{5}\right)^4 \left(\frac{1}{10}\right)^1 \left(\frac{3}{10}\right)^3 = \frac{10!}{2! \, 4! \, 1! \, 3!} \cdot \frac{2^2}{5^2} \cdot \frac{1}{5^4} \cdot \frac{1}{10} \cdot \frac{3^3}{10^3}$$

$$= 0.00870912.$$

1.4.5 Poisson Distribution

The sample space for this distribution is $\mathbb{N}_0 = \{0, 1, 2, \ldots\}$. Furthermore, $\lambda > 0$ is a given parameter.

Proposition 1.4.20. *There exists a unique probability measure* $Pois_\lambda$ *on* $\mathcal{P}(\mathbb{N}_0)$ *such that*

$$Pois_\lambda(\{k\}) = \frac{\lambda^k}{k!} \, e^{-\lambda}, \quad k \in \mathbb{N}_0. \tag{1.24}$$

Proof: Because of $e^{-\lambda} > 0$ follows $Pois_\lambda(\{k\}) > 0$. Thus it suffices to verify

$$\sum_{k=0}^{\infty} \frac{\lambda^k}{k!} \, e^{-\lambda} = 1.$$

But this is a direct consequence of

$$\sum_{k=0}^{\infty} \frac{\lambda^k}{k!} e^{-\lambda} = e^{-\lambda} \sum_{k=0}^{\infty} \frac{\lambda^k}{k!} = e^{-\lambda} e^{\lambda} = 1 .$$

■

Definition 1.4.21. The probability measure Pois_λ on $\mathcal{P}(\mathbb{N}_0)$ satisfying eq. (1.24) is called **Poisson distribution** with parameter $\lambda > 0$.

The Poisson distribution describes experiments where the number of trials is big, but the single success probability is small. More precisely, the following limit theorem holds.

Proposition 1.4.22 (Poisson's limit theorem). *Let $(p_n)_{n=1}^{\infty}$ be a sequence of numbers with $0 < p_n \le 1$ and*

$$\lim_{n \to \infty} n p_n = \lambda$$

for some $\lambda > 0$. Then for all $k \in \mathbb{N}_0$ follows

$$\lim_{n \to \infty} B_{n,p_n}(\{k\}) = \text{Pois}_\lambda(\{k\}) .$$

Proof: Write

$$B_{n,p_n}(\{k\}) = \binom{n}{k} p_n^k (1 - p_n)^{n-k}$$

$$= \frac{1}{k!} \frac{n (n-1) \cdots (n-k+1)}{n^k} (n p_n)^k (1 - p_n)^n (1 - p_n)^{-k}$$

$$= \frac{1}{k!} \left[\frac{n}{n} \cdot \frac{n-1}{n} \cdots \frac{n-k+1}{n} \right] (n p_n)^k (1 - p_n)^n (1 - p_n)^{-k} ,$$

and investigate the behavior of the different parts of the last equation separately. Each of the fractions in the bracket tends to 1, hence the whole bracket tends to 1. By assumption we have $n p_n \to \lambda$, thus, $\lim_{n \to \infty} (n p_n)^k = \lambda^k$. Moreover, since $n p_n \to \lambda$ with $\lambda > 0$ we get $p_n \to 0$, which implies $\lim_{n \to \infty} (1 - p_n)^{-k} = 1$.

Thus, it remains to determine the behavior of $(1 - p_n)^n$ as $n \to \infty$. Proposition A.5.1 asserts that if a sequence of real numbers $(x_n)_{n \ge 1}$ converges to $x \in \mathbb{R}$, then

$$\lim_{n \to \infty} \left(1 + \frac{x_n}{n} \right)^n = e^x .$$

Setting $x_n := -n\,p_n$ by assumption $x_n \to -\lambda$, hence

$$\lim_{n\to\infty} (1 - p_n)^n = \lim_{n\to\infty} \left(1 + \frac{x_n}{n}\right)^n = e^{-\lambda}.$$

If we combine all the different parts, then this completes the proof by

$$\lim_{n\to\infty} B_{n,p_n}(\{k\}) = \frac{1}{k!}\,\lambda^k\,e^{-\lambda} = \mathrm{Pois}_\lambda(\{k\}). \qquad \blacksquare$$

The previous theorem allows two **conclusions**.
(1) Whenever n is large and p is small, without hesitation one may replace $B_{n,p}$ by Pois_λ, where $\lambda = n\,p$. In this way one avoids the (sometimes) difficult evaluation of the binomial coefficients.

Example 1.4.23. In Example 1.4.10 we found the probability that among N students there are at least two having their birthday on April 1. We then used the binomial distribution with parameters N and $p = 1/365$. Hence the approximating Poisson distribution has parameter $\lambda = N/365$ and the corresponding probability is given by

$$\mathrm{Pois}_\lambda(\{2, 3, \ldots\}) = 1 - (1 + \lambda)e^{-\lambda} = 1 - \left(1 + \frac{N}{365}\right) e^{-N/365}.$$

If again $N = 500$, hence $\lambda = 500/365$, the approximative probability equals 0.397719. Compare this value with the "precise" probability 0.397895 obtained in Example 1.4.10.

(2) Poisson's limit theorem explains why the Poisson distribution describes experiments with many trials and small success probability. For example, if we look for a model for the number of car accidents per year, then the Poisson distribution is a good choice. There are many cars, but the probability[9] that a single driver is involved in an accident is quite small.

Later on we will investigate other examples where the Poisson distribution appears in a natural way.

1.4.6 Hypergeometric Distribution

Among N delivered machines are M defective. One chooses n of the N machines randomly and checks them. What is the probability to observe m defective machines in the sample of size n?

---

[9] To call it "success" probability in this case is perhaps not quite appropriate.

First note that there are $\binom{N}{n}$ ways to choose n machines for checking. In order to observe m defective ones these have to be taken from the M defective. The remaining $n - m$ machines are nondefective, hence they must be chosen from the $N - M$ nondefective ones. There are $\binom{M}{m}$ ways to take the defective machines and $\binom{N-M}{n-m}$ possibilities for the nondefective ones.

Thus the following approach describes this experiment:

$$H_{N,M,n}(\{m\}) := \frac{\binom{M}{m}\binom{N-M}{n-m}}{\binom{N}{n}}, \quad 0 \le m \le n. \tag{1.25}$$

Recall that in Section A.3.1 we agreed that $\binom{n}{k} = 0$ whenever $k > n$. This turns out be useful in the definition of $H_{N,M,n}$. For example, if $m > M$, then the probability to observe m defective machines is of course zero.

We want to prove now that eq. (1.25) defines a probability measure.

Proposition 1.4.24. *There exists a unique probability measure $H_{N,M,n}$ on the powerset of $\{0, \ldots , n\}$ satisfying eq. (1.25).*

Proof: Vandermonde's identity (cf. Proposition A.3.8) asserts that for all k, m, and n in $\mathbb{N}_0$

$$\sum_{j=0}^{k} \binom{n}{j}\binom{m}{k-j} = \binom{n+m}{k}. \tag{1.26}$$

Now replace n by M, next m by $N - M$, then k by n, and, finally, j by m. Doing so eq. (1.26) leads to

$$\sum_{m=0}^{n} \binom{M}{m}\binom{N-M}{n-m} = \binom{N}{n}.$$

But this implies

$$\sum_{m=0}^{n} H_{N,M,n}(\{m\}) = \frac{1}{\binom{N}{n}} \cdot \sum_{m=0}^{n} \binom{M}{m}\binom{N-M}{n-m} = \frac{1}{\binom{N}{n}} \cdot \binom{N}{n} = 1.$$

Clearly, $H_{N,M,n}(\{m\} \ge 0$, which completes the proof by virtue of Proposition 1.41. ∎

Definition 1.4.25. The probability measure $H_{N,M,n}$ defined by eq. (1.25) is called **hypergeometric distribution** with parameters N, M, and n.

Example 1.4.26. A retailer gets a delivery of 100 machines; 10 of them are defective. He chooses by random eight machines and tests them. Find the probability that two or more of the tested machines are defective.

Answer: The desired probability is

$$\sum_{m=2}^{8} \frac{\binom{10}{m}\binom{90}{8-m}}{\binom{100}{8}} = 0.18195 \, .$$

Remark 1.4.27. In the daily practice the reversed question is more important. The size N of the delivery is known and, of course, also the size of the tested sample. The number M of defective machines is unknown. Now suppose we observed m defective machines among the n tested. Does this (random) number m lead to some information about the number M of defective machines in the delivery? We will investigate this problem in Proposition 8.5.15.

Example 1.4.28. In a pond are 200 fish. One day the owner of the pond catches 20 fish, marks them, and puts them back into the pond. After a while the owner catches once more 20 fish. Find the probability that among these fish there is exactly one marked.

Answer: We have $N = 200$, $M = 20$, and $n = 20$. Hence the desired probability is

$$H_{200,20,20}(\{1\}) = \frac{\binom{20}{1}\binom{180}{19}}{\binom{200}{20}} = 0.26967 \, .$$

Remark 1.4.29. The previous example is not very realistic because in general the number N of fish is unknown. Known are M and n, the (random) number m was observed. Also here one may ask whether the knowledge of m leads to some information about N. This question will be investigated later in Proposition 8.5.17.

Example 1.4.30. In a lottery 6 numbers are chosen randomly out of 49. Suppose we bought a lottery coupon with six numbers. What is the probability that exactly k, $k = 0, \ldots, 6$, of our numbers appear in the drawing?

Answer: There are $n = 6$ numbers randomly chosen out of $N = 49$. Among the 49 numbers are $M = 6$ "defective." These are the six numbers on our coupon, and we ask for the probability that k of the "defective" are among the chosen six. The question is answered by the hypergeometric distribution $H_{49,6,6}$, that is, the probability of k correct numbers on our coupon is given by

$$H_{49,6,6}(\{k\}) = \frac{\binom{6}{k}\binom{43}{6-k}}{\binom{49}{6}} \, , \quad k = 0, \ldots, 6 \, .$$

The numerical values of these probabilities for $k = 0, \ldots, 6$ are

k	Probability
0	0.435965
1	0.413019
2	0.132378
3	0.0176504
4	0.00096862
5	0.0000184499
6	$7.15112 \cdot 10^{-8}$

Remark 1.4.31. Another model for the hypergeometric distribution is as follows: in an urn are N balls, M of them are white, the remaining $N - M$ are red. Choose n balls out of the urn **without replacing** the chosen ones. Then $H_{N,M,n}(\{m\})$ is the probability to observe m white balls among the n chosen.

If we do the same experiment, but now **replacing** the chosen balls, then this is described by the binomial distribution. The success probability for a white ball is $p = M/N$, hence now the probability for m white balls is given by

$$B_{n,M/N}(\{m\}) = \binom{n}{m} \left(\frac{M}{N}\right)^m \left(1 - \frac{M}{N}\right)^{n-m}.$$

It is intuitively clear that for large N and M (and comparable small n) the difference between both models (replacing and nonreplacing) is insignificant. Imagine there are 10^6 white and also 10^6 red balls in an urn. Choosing two balls it does not matter a lot whether the first ball was replaced or not.

The next proposition makes the previous observation more precise.

Proposition 1.4.32. *If $0 \le m \le n$ and $0 \le p \le 1$, then follows*

$$\lim_{\substack{N,M \to \infty \\ M/N \to p}} H_{N,M,n}(\{m\}) = B_{n,p}(\{m\}).$$

Proof: Suppose first $0 < p < 1$. Then the definition of the hypergeometric distribution yields

$$\lim_{\substack{N,M \to \infty \\ M/N \to p}} H_{N,M,n}(\{m\}) = \lim_{\substack{N,M \to \infty \\ M/N \to p}} \frac{\frac{M \cdots (M-m+1)}{m!} \frac{(N-M) \cdots (N-M-(n-m)+1)}{(n-m)!}}{\frac{N(N-1) \cdots (N-n+1)}{n!}}$$

$$= \lim_{\substack{N,M \to \infty \\ M/N \to p}} \binom{n}{m} \frac{\left[\frac{M}{N} \cdots \left(\frac{M-m+1}{N}\right)\right]\left[\left(1 - \frac{M}{N}\right) \cdots \left(1 - \frac{M-(n-m)+1}{N}\right)\right]}{\left(1 - \frac{1}{N}\right) \cdots \left(1 - \frac{n+1}{N}\right)} \qquad (1.27)$$

$$= \binom{n}{m} p^m (1-p)^{n-m} = B_{n,p}(\{m\}).$$

Note that if either $m = 0$ or $m = n$, then the first or the second bracket in eq. (1.27) become 1, thus they do not appear.

The cases $p = 0$ and $p = 1$ have to be treated separately. For example, if $p = 0$, the fraction in eq. (1.27) converges to zero provided that $m \geq 1$. If $m = 0$, then

$$\lim_{\substack{N,M\to\infty \\ M/N\to 0}} \frac{(1 - \frac{M}{N}) \cdots (1 - \frac{M-n+1}{N})}{(1 - \frac{1}{N}) \cdots (1 - \frac{n+1}{N})} = 1 = B_{n,0}(\{0\}).$$

The case $p = 1$ is treated similarly. Hence, the proposition is also valid in the border cases. ∎

Example 1.4.33. Suppose there are $N = 200$ balls in an urn, $M = 80$ of them are white. Choosing $n = 10$ balls with or without replacement we get the following numerical values. Note that $p = M/N = 2/5$.

m	$H_{N,M,n}(\{m\})$	$B_{n,p}(\{m\})$
1	0.0372601	0.0403108
2	0.118268	0.120932
3	0.217696	0.214991
4	0.257321	0.250823
5	0.204067	0.200658
6	0.10995	0.111477
7	0.0397376	0.0424673
8	0.00921879	0.0106168
9	0.0012395	0.00157286

1.4.7 Geometric Distribution

At a first glance the model for the geometric distribution looks as that for the binomial distribution. In each single trial we may observe "0" or "1," that is, failure or success. Again the success probability is a fixed number p. While in the case of the binomial distribution, we executed a fixed number of trials, now this number is random. More precisely, we execute the experiment until we observe success for the first time. Recorded is the number of necessary trials until this first success shows up. Or, in other words, a number $k \geq 1$ occurs if and only if the first $k - 1$ trials were all failures and the kth one success, that is, we observe the sequence $(\underbrace{0, \ldots, 0}_{k-1}, 1)$. Since failure appears with probability $1-p$ and success shows up with probability p, the following approach is plausible:

$$G_p(\{k\}) := p\,(1 - p)^{k-1}, \quad k \in \mathbb{N}. \tag{1.28}$$

Proposition 1.4.34. *If $0 < p < 1$, then (1.28) defines a probability measure on $\mathcal{P}(\mathbb{N})$.*

Proof: Because of $p(1-p)^{k-1} > 0$ it suffices to verify $\sum_{k=1}^{\infty} G_p(\{k\}) = 1$. Using the formula for the sum of a geometric series this follows directly by

$$\sum_{k=1}^{\infty} p(1-p)^{k-1} = p\sum_{k=0}^{\infty}(1-p)^k = p\,\frac{1}{1-(1-p)} = 1.$$

Observe that by assumption $1-p < 1$, thus the formula for geometric series applies. ∎

> **Definition 1.4.35.** The probability measure G_p on $\mathcal{P}(\mathbb{N})$ defined by eq. (1.28) is called **geometric distribution** with parameter p.

If $p = 0$, then success will never show up, thus, G_p is not a probability measure. On the other hand, for $p = 1$, success appears with probability one in the first trial, that is, $G_p = \delta_1$. Therefore, this case is of no interest.

Example 1.4.36. Given a number $n \in \mathbb{N}$, let $A_n = \{k \in \mathbb{N} : k > n\}$. Find $G_p(A_n)$.

Answer: We answer this question by two different approaches.

At first we remark that A_n occurs if and only if the first occurrence of success shows up strictly after n trials or, equivalently, if and only if the first n trials were all failures. But this event has probability $B_{n,p}(\{0\}) = (1-p)^n$, hence $G_p(A_n) = (1-p)^n$.

In the second approach we use eq. (1.28) directly and obtain

$$G_p(A_n) = \sum_{k=n+1}^{\infty} G_p(\{k\}) = p\sum_{k=n+1}^{\infty}(1-p)^{k-1} = p(1-p)^n\sum_{k=0}^{\infty}(1-p)^k$$

$$= p(1-p)^n\,\frac{1}{1-(1-p)} = (1-p)^n\,.$$

Example 1.4.37. Roll a die until number "6" occurs for the first time. What is the probability that this happens in roll k?

Answer: The success probability is $1/6$, hence the probability of first occurrence of "6" in the kth trial is $(1/6)(5/6)^{k-1}$.

k	Probability
1	0.166667
2	0.138889
3	0.115741
.	.
.	.
.	.
12	0.022431
13	0.018693

Example 1.4.38. Roll a die until the first "6" shows up. What is the probability that this happens at an even number of trials?

Answer: The first "6" has to appear in the second or fourth or sixth, and so on, trial. Hence, the probability of this event is

$$\sum_{k=1}^{\infty} G_{1/6}(\{2k\}) = \frac{1}{6} \sum_{k=1}^{\infty} (5/6)^{2k-1} = \frac{5}{36} \sum_{k=1}^{\infty} (5/6)^{2k-2} = \frac{5}{36} \frac{1}{1-(5/6)^2} = \frac{5}{11}.$$

Example 1.4.39. Play a series of games where p is the chance of winning. Whenever you put M dollars into the pool you get back $2M$ dollars if you win. If you lose, then the M dollars are lost.

Apply the following strategy. After losing double the amount in the pool in the next game. Say you start with $1 and lose, then next time put $2 into the pool, then $4, and so on until you win for the first time. As easily seen, in the kth game the stakes is 2^{k-1} dollars.

Suppose for some $k \geq 1$ you lost $k - 1$ games and won the kth one. How much money did you lose? If $k = 1$, then you lost nothing, while for $k \geq 2$ you spent

$$1 + 2 + 4 + \cdots + 2^{k-2} = 2^{k-1} - 1$$

dollars. Note that $2^{k-1} - 1 = 0$ if $k = 1$, hence for all $k \geq 1$ the total lost is $2^{k-1} - 1$ dollars.

On the other hand, if you win the kth game, you gain 2^{k-1} dollars. Consequently, no matter what the results are, you will always win $2^{k-1} - (2^{k-1} - 1) = 1$ dollars[10].

Let $X(k)$ be the amount of money needed in the case that one wins for the first time in the kth game. One needs 1 dollar to play the first game, $1 + 2 = 3$ dollars to play the second, until

$$1 + 2 + 4 + \cdots + 2^{k-1} = 2^k - 1$$

to play the kth game. Thus, $X(k) = 2^k - 1$ and

$$\mathbb{P}\{X = 2^k - 1\} = \mathbb{P}\{\text{First win in game } k\} = p(1-p)^{k-1}, \quad k = 1, 2 \ldots.$$

In particular, if $p = 1/2$ then this probability equals 2^{-k}. For example, if one starts the game with $127 = 2^7 - 1$ dollars in the pocket, then one goes bankrupt if the first success appears after game 7. The probability for this equals $\sum_{k=8}^{\infty} 2^{-k} = 2^{-7} = 0.0078125$.

1.4.8 Negative Binomial Distribution

The geometric distribution describes the probability for having the first success in trial k. Given a fixed $n \geq 1$ we ask now for the probability that in trial k success appears

---

10 Starting the first game with x dollars one will always win x dollars no matter what happens.

not for the first but for the nth time. Of course, this question makes only sense if $k \geq n$. But how to determine this probability for those k?

Thus, take $k \geq n$ and suppose we had success in trial k. When is this the nth one? This is the case if and only if we had $n - 1$ times success during the first $k - 1$ trials or, equivalently, $k - n$ failures. There exist $\binom{k-1}{k-n}$ possibilities to distribute the $k - n$ failures among the first $k - 1$ trials. Furthermore, the probability for n times success is p^n and for $k - n$ failures it is $(1 - p)^{k-n}$, hence the probability for observing the nth success in trial k is given by

$$\bar{B}_{n,p}(\{k\}) := \binom{k-1}{k-n} p^n (1 - p)^{k-n}, \quad k = n, n + 1, \ldots . \tag{1.29}$$

We still have to verify that there is a probability measure satisfying eq. (1.29).

Proposition 1.4.40. *By*

$$\bar{B}_{n,p}(\{k\}) = \binom{k-1}{k-n} p^n (1 - p)^{k-n}, \quad k = n, n + 1, \ldots ,$$

a probability measure $\bar{B}_{n,p}$ on $\mathcal{P}(\{n, n + 1, \ldots\})$ is defined.

Proof: Of course, $\bar{B}_{n,p}(\{k\}) \geq 0$. Hence it remains to show

$$\sum_{k=n}^{\infty} \bar{B}_{n,p}(\{k\}) = 1 \quad \text{or, equivalently,} \quad \sum_{k=0}^{\infty} \bar{B}_{n,p}(\{k + n\}) = 1. \tag{1.30}$$

Because of Lemma A.3.9 we get

$$\bar{B}_{n,p}(\{k + n\}) = \binom{n + k - 1}{k} p^n (1 - p)^k$$

$$= \binom{-n}{k} p^n (-1)^k (1 - p)^k = \binom{-n}{k} p^n (p - 1)^k, \tag{1.31}$$

where the generalized binomial coefficient is defined in eq. (A.13) as

$$\binom{-n}{k} = \frac{-n(-n - 1) \cdots (-n - k + 1)}{k!} .$$

In Proposition A.5.2 we proved for $|x| < 1$

$$\sum_{k=0}^{\infty} \binom{-n}{k} x^k = \frac{1}{(1 + x)^n} . \tag{1.32}$$

Note that $0 < p < 1$, hence eq. (1.32) applies with $x = p - 1$ and leads to

$$\sum_{k=0}^{\infty} \binom{-n}{k} (p-1)^k = \frac{1}{p^n}. \tag{1.33}$$

Combining eqs. (1.31) and (1.33) implies

$$\sum_{k=0}^{\infty} B_{n,p}^-(\{k+n\}) = p^n \sum_{k=0}^{\infty} \binom{-n}{k} (p-1)^k = p^n \frac{1}{p^n} = 1,$$

thus the equations in (1.30) are valid and this completes the proof. ∎

Definition 1.4.41. The probability measure $B_{n,p}^-$ with

$$B_{n,p}^-(\{k\}) := \binom{k-1}{k-n} p^n (1-p)^{k-n} = \binom{k-1}{n-1} p^n (1-p)^{k-n}, \quad k = n, n+1, \dots$$

is called **negative binomial distribution** with parameters $n \geq 1$ and $p \in (0, 1)$. Of course, $B_{1,p}^- = G_p$.

Remark 1.4.42. We saw in eq. (1.31) that

$$B_{n,p}^-(\{k+n\}) = \binom{n+k-1}{k} p^n (1-p)^k = \binom{-n}{k} p^n (p-1)^k. \tag{1.34}$$

Alternatively one may define the negative binomial distribution also via eq. (1.34). Then it describes the event that the nth success appears in trial $n + k$. The advantage of this approach is that now $k \in \mathbb{N}_0$, that is, the restriction $k \geq n$ is no longer needed. Its disadvantage is that we are interested in what happens in trial k, not in trial $k + n$.

Example 1.4.43. Roll a die successively. Determine the probability that in the 20th trial number "6" appears for the fourth time.

Answer: We have $p = 1/6$, $n = 4$, and $k = 20$. Therefore, the probability for this event is given by

$$B_{4,1/6}^-(\{20\}) = \binom{19}{16} \left(\frac{1}{6}\right)^4 \left(\frac{5}{6}\right)^{16} = 0.0404407.$$

Let us ask now for the probability that the fourth success appears (strictly) before trial 21. This probability is given by

$$\sum_{k=4}^{20} \binom{k-1}{3} \left(\frac{1}{6}\right)^4 \left(\frac{5}{6}\right)^{k-4} = 0.433454.$$

Example 1.4.44. There are two urns, say U_0 and U_1, each containing N balls. Choose one of the two urns by random and take out a ball. Hereby U_0 is chosen with probability $1 - p$, hence U_1 with probability p. Repeat the procedure until we choose the last (the Nth) ball out of one of the urns. What is the probability that there are left m balls in the other urn, where $m = 1, \ldots, N$?[11]

Answer: For $m = 1, \ldots, N$ let A_m be the event that there are still m balls in one of the urns when choosing the last ball out of the other. Then A_m splits into the disjoint events $A_m = A_m^0 \cup A_m^1$, where

A_m^0 occurs if we take the last ball out of U_0 and U_1 contains m balls and

A_m^1 occurs if choosing the Nth ball out of U_1 and there are m balls in U_0.

Let us start with evaluating the probability of A_m^1. Say success occurs if we choose urn U_1. Thus, if we take out the last ball of urn U_1, then success occurred for the Nth time. On the other hand, if there are still m balls in U_0, then failure had occurred $N - m$ times. Consequently, there are still m balls left in urn U_0 if and only if the Nth success shows up in trial $N + (N - m) = 2N - m$. Therefore, we get

$$\mathbb{P}(A_m^1) = B_{N,p}^-(\{2N - m\}) = \binom{2N - m - 1}{N - m} p^N (1 - p)^{N-m} . \tag{1.35}$$

The probability of A_m^0 may be derived from that of A_m^1 by interchanging p and $1 - p$ (success occurs now with probability $1 - p$). This yields

$$\mathbb{P}(A_m^0) = B_{N,1-p}^-(\{2N - m\}) = \binom{2N - m - 1}{N - m} p^{N-m} (1 - p)^N . \tag{1.36}$$

Adding eqs. (1.35) and (1.36) leads to

$$\mathbb{P}(A_m) = \binom{2N - m - 1}{N - m} \left[p^N (1 - p)^{N-m} + p^{N-m} (1 - p)^N \right] ,$$

$m = 1, \ldots, N$.

If $p = 1/2$, that is, both urns are equally likely, the previous formula simplifies to

$$\mathbb{P}(A_m) = \binom{2N - m - 1}{N - m} 2^{-2N+m+1} . \tag{1.37}$$

Remark 1.4.45. The case $p = 1/2$ in the previous problem is known as *Banach's matchbox problem*. In each of two matchboxes are N matches. One chooses randomly a matchbox (both boxes are equally likely) and takes out a match. What is the probability that there are still m matches left in the other box when taking the last match out of one of the boxes? The answer is given by eq. (1.37).

---

11 There exists a slightly different version of this problem. What is the probability that there are m balls left in one of the urns when choosing for the first time an empty one. Note that in this setting also $m = 0$ may occur.

Example 1.4.46. We continue Example 1.4.44 with $0 < p < 1$ and ask the following question: What is the probability that U_1 becomes empty before U_0?

Answer: This happens if and only if U_0 is nonempty when choosing U_1 for the Nth time, that is, when in U_0 there are m balls left for some $m = 1, \ldots, N$. Because of eq. (1.35), this probability is given by

$$\sum_{m=1}^{N} \mathbb{P}(A_m^1) = p^N \sum_{m=1}^{N} \binom{2N - m - 1}{N - m} (1 - p)^{N-m}$$

$$= p^N \sum_{k=0}^{N-1} \binom{N + k - 1}{k} (1 - p)^k. \tag{1.38}$$

Remark 1.4.47. Formula (1.38) leads to an interesting (known) property of the binomial coefficients. Since $\sum_{m=1}^{N} \mathbb{P}(A_m) = 1$, by eqs. (1.38) and (1.36) we obtain

$$\sum_{k=0}^{N-1} \binom{N + k - 1}{k} \left[p^N (1 - p)^k + (1 - p)^N p^k \right] = 1$$

or, setting $n = N - 1$, to

$$\sum_{k=0}^{n} \binom{n + k}{k} \left[p^{n+1} (1 - p)^k + (1 - p)^{n+1} p^k \right] = 1.$$

In particular, if $p = 1/2$, this yields

$$\sum_{k=0}^{n} \binom{n + k}{k} \frac{1}{2^k} = 2^n$$

for $n = 0, 1, \ldots$.

1.5 Continuous Probability Measures

Discrete probability measures are inappropriate for the description of random experiments where uncountably many different results may appear. Typical examples of such experiments are the lifetime of an item, the duration of a phone call, the measuring result of workpiece, and so on.

Discrete probability measures are concentrated on a finite or countably infinite set of points. An extension to larger sets is impossible. For example, there is no[12] probability measure $\mathbb{P}$ on $[0, 1]$ with $\mathbb{P}(\{t\}) > 0$ for $t \in [0, 1]$.

---

12 Compare Problem 1.31.

Consequently, in order to describe random experiments with "many" possible different outcomes another approach is needed. To explain this "new" approach let us shortly recall how we evaluated $\mathbb{P}(A)$ in the discrete case. If Ω is either finite or countably infinite and if $p(\omega) = \mathbb{P}(\{\omega\})$, $\omega \in \Omega$, then with this $p : \Omega \to \mathbb{R}$ we have

$$\mathbb{P}(A) = \sum_{\omega \in A} p(\omega). \tag{1.39}$$

If the sample space is $\mathbb{R}$ or $\mathbb{R}^n$, then such a representation is no longer possible. Indeed, if $\mathbb{P}$ is not discrete, then, we will have $p(\omega) = 0$ for all possible observations ω. Therefore, the sum in eq. (1.39) has to be replaced by an integral over a more general function. We start with introducing functions p, which may be used for representing $\mathbb{P}(A)$ via an integral.

> **Definition 1.5.1.** A Riemann integrable function $p : \mathbb{R} \to \mathbb{R}$ is called **probability density function** or simply **density function** if
>
> $$p(x) \geq 0, \ x \in \mathbb{R}, \quad \text{and} \quad \int_{-\infty}^{\infty} p(x)\,dx = 1. \tag{1.40}$$

Remark 1.5.2. Let us more precise formulate the second condition about p in the previous definition. For all finite intervals $[a, b]$ in $\mathbb{R}$ the function p is Riemann integrable on $[a, b]$ and, moreover,

$$\lim_{\substack{a \to -\infty \\ b \to +\infty}} \int_a^b p(x)\,dx = 1.$$

The density functions we will use later on are either continuous or piecewise continuous, that is they are the composition of finitely many continuous functions. These functions are Riemann integrable, hence in this case it remains to verify the two conditions (1.40).

Example 1.5.3. Define p on $\mathbb{R}$ by $p(x) = 0$ if $x < 0$ and by $p(x) = e^{-x}$ if $x \geq 0$. Then p is piecewise continuous, $p(x) \geq 0$ if $x \in \mathbb{R}$ and satisfies

$$\int_{-\infty}^{\infty} p(x)dx = \lim_{b \to \infty} \int_0^b e^{-x}\,dx = \lim_{b \to \infty} \left[-e^{-x} \right]_0^b = 1 - \lim_{b \to \infty} e^{-b} = 1.$$

Hence, p is a density function.

Definition 1.5.4. Let p be a probability density function. Given a finite interval $[a, b]$, its probability (of occurrence) is defined by

$$\mathbb{P}([a, b]) := \int_a^b p(x)dx.$$

A graphical presentation of the previous definition is as follows. The probability $\mathbb{P}([a, b])$ is the area under the graph of the density p, taken from a to b.

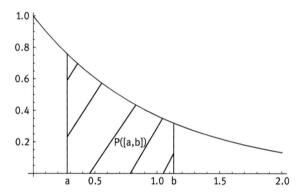

Figure 1.1: Probability of the interval $[a, b]$.

Let us illustrate Definition 1.5.4 by the density function regarded in Example 1.5.3. Then

$$\mathbb{P}([a, b]) = \int_a^b e^{-x}dx = \left[-e^{-x} \right]_a^b = e^{-a} - e^{-b}$$

whenever $0 \le a < b < \infty$. On the other hand, if $a < b < 0$, then $\mathbb{P}([a, b]) = 0$ while for $a < 0 \le b$ the probability of $[a, b]$ is calculated by

$$\mathbb{P}([a, b]) = \mathbb{P}([0, b]) = 1 - e^{-b}.$$

Remark 1.5.5. Definition 1.5.4 of the probability measure $\mathbb{P}$ does not fit into the scheme presented in Section 1.1.3. Why? Probability measures are defined on σ-fields. But the collection of finite intervals in $\mathbb{R}$ is not a σ-field. It is neither closed under taking complements nor is the union of intervals in general again an interval. Furthermore, it is far from being clear in which sense $\mathbb{P}$ should be σ-additive.

The next result justifies the approach in Definition 1.5.4. Its proof rests upon an extension theorem in Measure Theory (cf. [Coh13] or [Dud02]).

Proposition 1.5.6. *Let $\mathcal{B}(\mathbb{R})$ be the σ-field of Borel sets introduced in Definition 1.1.15. Then for each density function p there exists a unique probability measure $\mathbb{P} : \mathcal{B}(\mathbb{R}) \to [0, 1]$ such that*

$$\mathbb{P}([a, b]) = \int_a^b p(x)\, dx \qquad \text{for all} \quad a < b\,. \tag{1.41}$$

Definition 1.5.7. A probability measure $\mathbb{P}$ on $\mathcal{B}(\mathbb{R})$ is said to be **continuous**[13] provided that there exists a density function p such that for $a < b$

$$\mathbb{P}([a, b]) = \int_a^b p(x)\, dx\,. \tag{1.42}$$

The function p is called **density function** or simply **density** of $\mathbb{P}$.

Remark 1.5.8. Note that changing the density function at finitely many points does not change the generated probability measure. For instance, if we define $p(x) = 0$ if $x \le 0$ and $p(x) = e^{-x}$ if $x > 0$, then this density function is different from that in Example 1.5.3 but, of course, generates the same probability measure

Moreover, observe that eq. (1.42) is valid for all $a < b$ if and only if for each $t \in \mathbb{R}$

$$\mathbb{P}((-\infty, t]) = \int_{-\infty}^t p(x)\, dx\,. \tag{1.43}$$

Consequently, $\mathbb{P}$ is continuous if and only if there is a density p with eq. (1.43) for $t \in \mathbb{R}$.

Proposition 1.5.9. *Let $\mathbb{P} : \mathcal{B}(\mathbb{R}) \to [0, 1]$ be a continuous probability measure with density p. Then the following are valid:*
1. *$\mathbb{P}(\mathbb{R}) = 1$.*
2. *For each $t \in \mathbb{R}$ follows $\mathbb{P}(\{t\}) = 0$. More generally, if $A \subset \mathbb{R}$ is either finite or countably infinite, then $\mathbb{P}(A) = 0$.*
3. *For all $a < b$ we have*

$$\mathbb{P}((a, b)) = \mathbb{P}((a, b]) = \mathbb{P}([a, b)) = \mathbb{P}([a, b]) = \int_a^b p(x)\, dx\,.$$

Proof: Let us start with proving $\mathbb{P}(\mathbb{R}) = 1$. For $n \ge 1$ set $A_n := [-n, n]$ and note that $A_1 \subseteq A_2 \subseteq \cdots$ as well as $\bigcup_{n=1}^\infty A_n = \mathbb{R}$. Thus we may use that $\mathbb{P}$ is continuous from below and by the properties of the density p we obtain

---

[13] The mathematical correct notation would be "absolutely continuous." But since we do not treat so-called "singularly continuous" probability measures, there is no need to distinguish between them, and we may shorten the notation to "continuous."

$$\mathbb{P}(\mathbb{R}) = \lim_{n\to\infty} \mathbb{P}(A_n) = \lim_{n\to\infty} \int_{-n}^{n} p(x)\, dx = \int_{-\infty}^{\infty} p(x)\, dx = 1\,.$$

To verify the second property fix $t \in \mathbb{R}$ and define for $n \geq 1$ intervals B_n by $B_n :=$ $[t, t + \frac{1}{n}]$. Now we have $B_1 \supseteq B_2 \supseteq \cdots$ and $\bigcap_{n=1}^{\infty} B_n = \{t\}$. Use this time the continuity from above. Then we get

$$\mathbb{P}(\{t\}) = \lim_{n\to\infty} \mathbb{P}(B_n) = \lim_{n\to\infty} \int_{t}^{t+\frac{1}{n}} p(x)\, dx = 0\,.$$

If $A = \{t_1, t_2, \dots\}$, then the σ-additivity of $\mathbb{P}$ together with $\mathbb{P}(\{t_j\}) = 0$ give

$$\mathbb{P}(A) = \sum_{j=1}^{\infty} \mathbb{P}(\{t_j\}) = 0$$

as asserted.

The third property is an immediate consequence of the second one. Observe

$$[a, b] = (a, b) \cup \{a\} \cup \{b\}\,,$$

hence $\mathbb{P}([a, b]) = \mathbb{P}((a, b)) + \mathbb{P}(\{a\}) + \mathbb{P}(\{b\})$ proving (1.42) by $\mathbb{P}(\{a\}) = \mathbb{P}(\{b\}) = 0$. ■

Remark 1.5.10. Say a set $C \subseteq \mathbb{R}$ can be represented as $C = \bigcup_{j=1}^{\infty} I_j$ with disjoint (open or half-open or closed) intervals I_j, then

$$\mathbb{P}(C) = \sum_{j=1}^{\infty} \int_{I_j} p(x)\, dx := \int_{C} p(x)\, dx\,.$$

More generally, if a set B may be written as $B = \bigcap_{n=1}^{\infty} C_n$ where the C_ns are a union of disjoint intervals, and satisfy $C_1 \supseteq C_2 \supseteq \cdots$, then

$$\mathbb{P}(B) = \lim_{n\to\infty} \mathbb{P}(C_n)\,.$$

In this way, one may evaluate $\mathbb{P}(B)$ for a large class of subsets $B \subseteq \mathbb{R}$.

1.6 Special Continuous Distributions

1.6.1 Uniform Distribution on an Interval

Let $I = [\alpha, \beta]$ be a finite interval of real numbers. Define a function $p : \mathbb{R} \to \mathbb{R}$ by

$$p(x) := \begin{cases} \frac{1}{\beta - \alpha} & : x \in [\alpha, \beta] \\ 0 & : x \notin [\alpha, \beta] \end{cases} \tag{1.44}$$

Proposition 1.6.1. *The mapping p defined by eq. (1.44) is a probability density function.*

Proof: Note that p is piecewise continuous, hence Riemann integrable. Moreover, $p(x) \geq 0$ for $x \in \mathbb{R}$ and

$$\int_{-\infty}^{\infty} p(x)\,dx = \int_{\alpha}^{\beta} \frac{1}{\beta - \alpha}\,dx = \frac{1}{\beta - \alpha}(\beta - \alpha) = 1.$$

Consequently, p is a probability density. ∎

Definition 1.6.2. The probability measure $\mathbb{P}$ generated by the density in eq. (1.44) is called **uniform distribution** on the interval $I = [\alpha, \beta]$.

How is $\mathbb{P}([a, b])$ evaluated for some interval $[a, b]$? Let us first treat the case $[a, b] \subseteq I$. Then

$$\mathbb{P}([a, b]) = \int_{a}^{b} \frac{1}{\beta - \alpha}\,dx = \frac{b - a}{\beta - \alpha} = \frac{\text{Length of } [a, b]}{\text{Length of } [\alpha, \beta]}. \tag{1.45}$$

This explains why $\mathbb{P}$ is called "uniform distribution." The probability of an interval $[a, b] \subseteq I$ depends only on its length, not on its position. Shifting $[a, b]$ inside I does not change its probability of occurrence.

If $[a, b]$ is arbitrary, not necessarily contained in I, then $\mathbb{P}([a, b])$ can be easily calculated by

$$\mathbb{P}([a, b]) = \mathbb{P}([a, b] \cap I).$$

Example 1.6.3. Let $\mathbb{P}$ be the uniform distribution on $[0, 1]$. Which probabilities have $[-1, 0.5]$, $[0, 0.25] \cup [0.75, 1]$, $(-\infty, t]$ if $t \in \mathbb{R}$, and $A \subseteq \mathbb{R}$, where $A = \bigcup_{n=1}^{\infty} \left[\frac{1}{2^{n+1/2}}, \frac{1}{2^n}\right]$?
Answer: The first two intervals have probability $\frac{1}{2}$. If $t \in \mathbb{R}$, then

$$\mathbb{P}((-\infty, t]) = \begin{cases} 0 : & t < 0 \\ t : & 0 \leq t \leq 1 \\ 1 : & t > 1 \end{cases}$$

Finally, observe that the intervals $\left[\frac{1}{2^{n+1/2}}, \frac{1}{2^n}\right]$ are disjoint subsets of $[0, 1]$. Hence we get

$$\mathbb{P}(A) = \sum_{n=1}^{\infty} \left[\frac{1}{2^n} - \frac{1}{2^{n+1/2}}\right] = \left(1 - 2^{-1/2}\right) \sum_{n=1}^{\infty} \frac{1}{2^n} = 1 - 2^{-1/2}$$

Example 1.6.4. A stick of length L is randomly broken into two pieces. Find the probability that the size of one piece is at least twice that of the other one.

Answer: This event happens if and only if the point at which the stick is broken is either in $[0, L/3]$ or in $[2L/3, L]$. Assuming that the point at which the stick is broken is uniformly distributed on $[0, L]$, the desired probability is $\frac{2}{3}$. Another way to get this result is as follows. The size of each piece is less than twice as that of the other one if the point at which the stick is broken is in $[L/3, 2L/3]$. Hence, the probability of the complementary event is $1/3$ leading again to $2/3$ for the desired probability.

Example 1.6.5. Let $C_0 := [0, 1]$. Extract from C_0 the interval $(\frac{1}{3}, \frac{2}{3})$, thus it remains $C_1 = [0, \frac{1}{3}] \cup [\frac{2}{3}, 1]$. To construct C_2 extract from C_1 the two middle intervals $(\frac{1}{9}, \frac{2}{9})$ and $(\frac{7}{9}, \frac{8}{9})$, hence $C_2 = [0, \frac{1}{9}] \cup [\frac{2}{9}, \frac{1}{3}] \cup [\frac{2}{3}, \frac{7}{9}] \cup [\frac{8}{9}, 1]$.

Suppose that through this method we already got sets C_n which are the union of 2^n disjoint closed intervals of length 3^{-n}. In order to construct C_{n+1}, split each of the 2^n intervals into three intervals of length 3^{-n-1} and erase the middle one of these three. In this way we get C_{n+1}, which consists of 2^{n+1} disjoint intervals of length 3^{-n-1}. Finally, one defines

$$C = \bigcap_{n=1}^{\infty} C_n.$$

The set C is known as the **Cantor set**. Let $\mathbb{P}$ be the uniform distribution on $[0, 1]$. Which value does $\mathbb{P}(C)$ have?

Answer: First observe that $C_0 \supset C_1 \supset C_2 \supset \cdots$, hence, using that $\mathbb{P}$ is continuous from above, it follows

$$\mathbb{P}(C) = \lim_{n \to \infty} \mathbb{P}(C_n). \tag{1.46}$$

The sets C_n are the disjoint union of 2^n intervals of length 3^{-n}. Consequently, it follows $\mathbb{P}(C_n) = \frac{2^n}{3^n}$, which by eq. (1.46) implies $\mathbb{P}(C) = 0$.

One might conjecture that $C = \emptyset$. On the contrary, C is even uncountably infinite. To see this we have to make the construction of the Cantor set a little bit more precise. Given $n \geq 1$ let

$$A_n = \{\alpha = (\alpha_1, \ldots, \alpha_n) : \alpha_1, \ldots, \alpha_{n-1} \in \{0, 2\}, \alpha_n = 1\}.$$

If $\alpha = (\alpha_1, \ldots, \alpha_n) \in A_n$, set $x_\alpha = \sum_{k=1}^{n} \frac{\alpha_k}{3^k}$ and $I_\alpha = (x_\alpha, x_\alpha + \frac{1}{3^n})$. In this notation

$$I_{(1)} = \left(\frac{1}{3}, \frac{2}{3}\right), \quad I_{(0,1)} = \left(\frac{1}{9}, \frac{2}{9}\right), \quad I_{(2,1)} = \left(\frac{7}{9}, \frac{8}{9}\right) \quad \text{and} \quad I_{(0,0,1)} = \left(\frac{1}{27}, \frac{2}{27}\right).$$

Then, if $C_0 = [0, 1]$, for $n \geq 1$ we have

$$C_n = C_{n-1} \setminus \bigcup_{\alpha \in A_n} I_\alpha, \quad \text{hence} \quad C = [0, 1] \setminus \bigcup_{n=1}^{\infty} \bigcup_{\alpha \in A_n} I_\alpha .$$

Take now any sequence $x_1, x_2, \ldots$ with $x_k \in \{0, 2\}$ and set $x = \sum_{k=1}^{\infty} \frac{x_k}{3^k}$. Then x cannot belong to any I_α because otherwise at least one of the x_ks should satisfy $x_k = 1$. Thus $x \in C$, and the number of x that may be represented by x_ks with $x_k \in \{0, 2\}$ is uncountably infinite.

1.6.2 Normal Distribution

This section is devoted to the most important probability measure, the normal distribution. Before we can introduce it, we need the following result.

Proposition 1.6.6. *We have*

$$\int_{-\infty}^{\infty} e^{-x^2/2} \, dx = \sqrt{2\pi} .$$

Proof: Set

$$a := \int_{-\infty}^{\infty} e^{-x^2/2} \, dx$$

and note that $a > 0$. Then we get

$$a^2 = \left(\int_{-\infty}^{\infty} e^{-x^2/2} \, dx \right) \left(\int_{-\infty}^{\infty} e^{-y^2/2} \, dy \right) = \int_{-\infty}^{\infty} \int_{-\infty}^{\infty} e^{-(x^2+y^2)/2} \, dx \, dy .$$

Change the variables in the right-hand double integral as follows: $x := r \cos \theta$ and $y := r \sin \theta$, where $0 < r < \infty$ and $0 \leq \theta < 2\pi$. Observe that

$$dx \, dy = |D(r, \theta)| \, dr \, d\theta$$

with

$$D(r, \theta) = \det \begin{pmatrix} \frac{\partial x}{\partial r} & \frac{\partial x}{\partial \theta} \\ \frac{\partial y}{\partial r} & \frac{\partial y}{\partial \theta} \end{pmatrix} = \det \begin{pmatrix} \cos \theta & -r \sin \theta \\ \sin \theta & r \cos \theta \end{pmatrix} = r \cos^2 \theta + r \sin^2 \theta = r .$$

Using $x^2 + y^2 = r^2 \cos^2 \theta + r^2 \sin^2 \theta = r^2$, this change of variables leads to

$$a^2 = \int_0^{2\pi} \int_0^{\infty} r e^{-r^2/2} \, dr \, d\theta = \int_0^{2\pi} \left[-e^{-r^2/2} \right]_0^{\infty} d\theta = 2\pi ,$$

which by $a > 0$ implies $a = \sqrt{2\pi}$. This completes the proof. ∎

Given $\mu \in \mathbb{R}$ and $\sigma > 0$ let

$$p_{\mu,\sigma}(x) := \frac{1}{\sqrt{2\pi}\sigma}\, e^{-(x-\mu)^2/2\sigma^2}, \quad x \in \mathbb{R}. \tag{1.47}$$

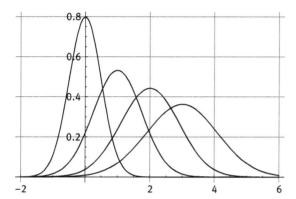

Figure 1.2: The function $p_{\mu,\sigma}$ with parameters $\mu = 0, 1, 2, 3$ and $\sigma = 0.5, 0.75, 0.9, 1.1$.

Proposition 1.6.7. *If $\mu \in \mathbb{R}$ and $\sigma > 0$, then $p_{\mu,\sigma}$ is a probability density function.*

Proof: We have to verify

$$\int_{-\infty}^{\infty} p_{\mu,\sigma}(x)\,dx = 1 \quad \text{or} \quad \int_{-\infty}^{\infty} e^{(x-\mu)^2/2\sigma^2}\,dx = \sqrt{2\pi}\,\sigma.$$

Setting $u := (x - \mu)/\sigma$ it follows $dx = \sigma du$, hence Proposition 1.6.6 leads to

$$\int_{-\infty}^{\infty} e^{(x-\mu)^2/2\sigma^2}\,dx = \sigma \int_{-\infty}^{\infty} e^{-u^2/2}\,du = \sigma\sqrt{2\pi}.$$

This completes the proof. ∎

Definition 1.6.8. The probability measure generated by $p_{\mu,\sigma}$ is called **normal distribution** with expected value μ and variance σ^2. It is denoted by $\mathcal{N}(\mu, \sigma^2)$, that is, for all $a < b$

$$\mathcal{N}(\mu, \sigma^2)([a, b]) = \frac{1}{\sqrt{2\pi}\sigma} \int_a^b e^{-(x-\mu)^2/2\sigma^2}\,dx.$$

Remark 1.6.9. In the moment the numbers $\mu \in \mathbb{R}$ and $\sigma > 0$ are nothing else than parameters. Why they are called "expected value" and "variance" will become clear in Section 5.

Definition 1.6.10. The probability measure $\mathcal{N}(0, 1)$ is called **standard normal distribution**. It is given by

$$\mathcal{N}(0, 1)([a, b]) = \frac{1}{\sqrt{2\pi}} \int_a^b e^{-x^2/2} \, dx \, .$$

Example 1.6.11. For example, we have

$$\mathcal{N}(0, 1)([-1, 1]) = \frac{1}{\sqrt{2\pi}} \int_{-1}^1 e^{-x^2/2} \, dx = 0.682689 \quad \text{or}$$

$$\mathcal{N}(0, 1)([2, 4]) = \frac{1}{\sqrt{2\pi}} \int_2^4 e^{-x^2/2} \, dx = 0.0227185 \, .$$

1.6.3 Gamma Distribution

Euler's **gamma function** is a mapping from $(0, \infty)$ to $\mathbb{R}$ defined by

$$\Gamma(x) := \int_0^\infty s^{x-1} e^{-s} ds \, , \quad x > 0 \, .$$

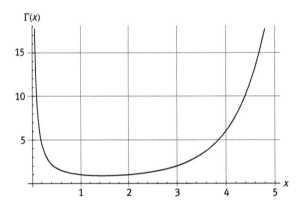

Figure 1.3: Graph of the gamma function.

Let us summarize the main properties of the gamma function.

Proposition 1.6.12.
1. Γ *maps* $(0, \infty)$ *continuously to* $(0, \infty)$ *and possesses continuous derivatives of any order.*
2. *If* $x > 0$, *then*

$$\Gamma(x + 1) = x \, \Gamma(x) \, . \tag{1.48}$$

3. *For* $n \in \mathbb{N}$ *follows* $\Gamma(n) = (n - 1)!$. *In particular,* $\Gamma(1) = \Gamma(2) = 1$ *and* $\Gamma(3) = 2$.
4. $\Gamma(1/2) = \sqrt{\pi}$.

Proof: For the proof of the continuity and differentiability we refer to [Art64]
The proof of eq. (1.48) is carried out by integration by parts as follows:

$$\Gamma(x + 1) = \int_0^\infty s^x e^{-s} ds = \left[-s^x e^{-s} \right]_0^\infty + \int_0^\infty x s^{x-1} e^{-s} ds \,. = x\,\Gamma(x)\,.$$

Note that $s^x e^{-s} = 0$ if $s = 0$ or $s \to \infty$.
From

$$\Gamma(1) = \int_0^\infty e^{-s} ds = 1$$

and eq. (1.48) follows, as claimed,

$$\Gamma(n) = (n - 1)\Gamma(n - 1) = (n - 1)(n - 2)\Gamma(n - 2) = \cdots = (n - 1) \cdots 1 \cdot \Gamma(1) = (n - 1)! \,.$$

To prove the fourth assertion we use Proposition 1.6.6. Because of

$$\sqrt{2\pi} = \int_{-\infty}^\infty e^{-t^2/2} dt = 2 \int_0^\infty e^{-t^2/2} dt$$

it follows that

$$\int_0^\infty e^{-t^2/2} dt = \sqrt{\frac{\pi}{2}} \,. \tag{1.49}$$

Substituting $s = t^2/2$, thus $ds = t\,dt$, by eq. (1.49) the integral for $\Gamma(1/2)$ transforms to

$$\Gamma(1/2) = \int_0^\infty s^{-1/2} e^{-s} ds = \int_0^\infty \frac{\sqrt{2}}{t} e^{-t^2/2} t\, dt = \sqrt{2} \int_0^\infty e^{-t^2/2} dt = \sqrt{\pi} \,.$$

This completes the proof. ■

If $x \to \infty$, then $\Gamma(x)$ increases very rapidly. More precisely, the following is valid (cf. [Art64]):

Proposition 1.6.13 (Stirling's formula for the Γ-function). *For* $x > 0$ *there exists a number* $\theta \in (0, 1)$ *such that*

$$\Gamma(x) = \sqrt{\frac{2\pi}{x}} \left(\frac{x}{e}\right)^x e^{\theta/12x} \,. \tag{1.50}$$

Corollary 1.6.14 (Stirling's formula for *n*-factorial). *In view of*

$$n! = \Gamma(n+1) = n\Gamma(n)$$

formula (1.50) leads to[14]

$$n! = \sqrt{2\pi n} \left(\frac{n}{e}\right)^n e^{\theta/12n} \tag{1.51}$$

for some $\theta \in (0,1)$ depending on n. In particular,

$$\lim_{n\to\infty} \frac{e^n}{n^{n+1/2}} \, n! = \sqrt{2\pi}.$$

Our next aim is to introduce a continuous probability measure with density tightly related to the Γ-function. Given $\alpha, \beta > 0$ define $p_{\alpha,\beta}$ from $\mathbb{R}$ to $\mathbb{R}$ by

$$p_{\alpha,\beta}(x) := \begin{cases} 0 & : x \le 0 \\ \frac{1}{\alpha^\beta \, \Gamma(\beta)} x^{\beta-1} e^{-x/\alpha} & : x > 0 \end{cases} \tag{1.52}$$

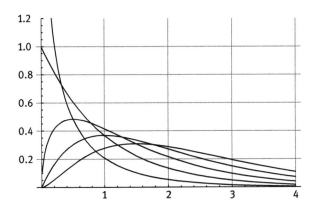

Figure 1.4: The functions $p_{1,\beta}$ with $\beta = 0.5, 1, 1.5, 2$ and $\beta = 2.5$.

Proposition 1.6.15. *For all $\alpha, \beta > 0$ the function $p_{\alpha,\beta}$ in eq. (1.52) is a probability density.*

Proof: Of course, $p_{\alpha,\beta}(x) \ge 0$. Thus it remains to verify

$$\int_{-\infty}^{\infty} p_{\alpha,\beta}(x) \, dx = 1. \tag{1.53}$$

---

14 cf. also [Spi08], Chapter 27, Problem 19.

By the definition of $p_{\alpha,\beta}$ we have

$$\int_{-\infty}^{\infty} p_{\alpha,\beta}(x)\, dx = \frac{1}{\alpha^{\beta}\, \Gamma(\beta)} \int_{0}^{\infty} x^{\beta-1} e^{-x/\alpha}\, dx\,.$$

Substituting in the right-hand integral $u := x/\alpha$, thus $dx = \alpha\, du$, the right-hand side becomes

$$\frac{1}{\Gamma(\beta)} \int_{0}^{\infty} u^{\beta-1} e^{-u}\, du = \frac{1}{\Gamma(\beta)}\, \Gamma(\beta) = 1\,.$$

Hence eq. (1.53) is valid, and $p_{\alpha,\beta}$ is a probability density function. ∎

Definition 1.6.16. The probability measure $\Gamma_{\alpha,\beta}$ with density function $p_{\alpha,\beta}$ is called **gamma distribution** with parameters α and β. For all $0 \le a < b < \infty$

$$\Gamma_{\alpha,\beta}([a, b]) = \frac{1}{\alpha^{\beta}\, \Gamma(\beta)} \int_{a}^{b} x^{\beta-1} e^{-x/\alpha}\, dx\,. \tag{1.54}$$

Remark 1.6.17. Since $p_{\alpha,\beta}(x) = 0$ for $x \le 0$ it follows that $\Gamma_{\alpha,\beta}((-\infty, 0]) = 0$. Hence, if $a < b$ are arbitrary, then

$$\Gamma_{\alpha,\beta}([a, b]) = \Gamma_{\alpha,\beta}([0, \infty) \cap [a, b])\,.$$

Remark 1.6.18. If $\beta \notin \mathbb{N}$, then the integral in eq. (1.54) cannot be expressed by elementary functions. Only numerical evaluations are possible.

1.6.4 Exponential Distribution

An important special gamma distribution is the exponential distribution. This probability measure is defined as follows.

Definition 1.6.19. For $\lambda > 0$ let $E_{\lambda} := \Gamma_{\lambda^{-1}, 1}$ be the **exponential distribution** with parameter $\lambda > 0$.

Remark 1.6.20. The probability density function p_{λ} of E_{λ} is given by

$$p_{\lambda}(x) = \begin{cases} 0 & : x \le 0 \\ \lambda\, e^{-\lambda x} & : x > 0 \end{cases}$$

Consequently, if $0 \le a < b < \infty$, then the probability of $[a, b]$ can be evaluated by

$$E_\lambda([a, b]) = e^{-\lambda a} - e^{-\lambda b}.$$

Moreover,

$$E_\lambda([t, \infty)) = e^{-\lambda t}, \quad t \ge 0.$$

Remark 1.6.21. The exponential distribution plays an important role for the description of lifetimes. For instance, it is used to determine the probability that the lifetime of a component part or the duration of a phone call exceeds a certain time $T > 0$. Furthermore, it is applied to describe the time between the arrivals of customers at a counter or in a shop.

Example 1.6.22. Suppose that the duration of phone calls is exponentially distributed with parameter $\lambda = 0.1$. What is the probability that a call lasts less than two time units? Or what is the probability that it lasts between one and two units? Or more than five units?

Answer: These probabilities are evaluated by

$$E_{0.1}([0, 2]) = 1 - e^{-0.2} = 0.181269, \; E_{0.1}([1, 2]) = e^{-0.1} - e^{-0.2} = 0.08611 \text{ and}$$

$$E_{0.1}([5, \infty)) = e^{-0.5} = 0.60653.$$

1.6.5 Erlang Distribution

Another important class of gamma distributions is that of Erlang distributions defined as follows.

Definition 1.6.23. For $\lambda > 0$ and $n \in \mathbb{N}$ let $E_{\lambda,n} := \Gamma_{\lambda^{-1},n}$. This probability measure is called **Erlang distribution** with parameters λ and n.

Remark 1.6.24. The density $p_{\lambda,n}$ of the Erlang distribution is

$$p_{\lambda,n}(x) = \begin{cases} 0 & : x \le 0 \\ \frac{\lambda^n}{(n-1)!} x^{n-1} e^{-\lambda x} & : x > 0 \end{cases}$$

Of course, $E_{\lambda,1} = E_\lambda$. Thus the Erlang distribution may be viewed as generalized exponential distribution.

An important property of the Erlang distribution is as follows.

Proposition 1.6.25. *If $t > 0$, then*

$$E_{\lambda,n}([t, \infty)) = \sum_{j=0}^{n-1} \frac{(\lambda t)^j}{j!} e^{-\lambda t}.$$

Proof: We have to show that for $t > 0$

$$\int_t^\infty p_{\lambda,n}(x)\,dx = \frac{\lambda^n}{(n-1)!} \int_t^\infty x^{n-1} e^{-\lambda x}\,dx = \sum_{j=0}^{n-1} \frac{(\lambda t)^j}{j!} e^{-\lambda t}. \tag{1.55}$$

This is done by induction over n.

If $n = 1$ then eq. (1.55) is valid by

$$\int_t^\infty p_{\lambda,1}(x)\,dx = \int_t^\infty \lambda e^{-\lambda x}\,dx = e^{-\lambda t}.$$

Suppose now eq. (1.55) is proven for some $n \geq 1$. Next, we have to show that it is also valid for $n + 1$. Thus, we know

$$\frac{\lambda^n}{(n-1)!} \int_t^\infty x^{n-1} e^{-\lambda x}\,dx = \sum_{j=0}^{n-1} \frac{(\lambda t)^j}{j!} e^{-\lambda t} \tag{1.56}$$

and want

$$\frac{\lambda^{n+1}}{n!} \int_t^\infty x^n e^{-\lambda x}\,dx = \sum_{j=0}^{n} \frac{(\lambda t)^j}{j!} e^{-\lambda t}. \tag{1.57}$$

Let us integrate the integral in eq. (1.57) by parts as follows. Set $u := x^n$, hence $u' = n x^{n-1}$, and $v' = e^{-\lambda x}$, thus $v = -\lambda^{-1} e^{-\lambda x}$. Doing so and using eq. (1.56) the left-hand side of eq. (1.57) becomes

$$\frac{\lambda^{n+1}}{n!} \int_t^\infty x^n e^{-\lambda x}\,dx = \left[-\frac{\lambda^n}{n!} x^n e^{-\lambda x} \right]_t^\infty + \frac{\lambda^n}{(n-1)!} \int_t^\infty x^{n-1} e^{-\lambda x}\,dx$$

$$= \frac{(\lambda t)^n}{n!} e^{-\lambda t} + \sum_{j=0}^{n-1} \frac{(\lambda t)^j}{j!} e^{-\lambda t} = \sum_{j=0}^{n} \frac{(\lambda t)^j}{j!} e^{-\lambda t}.$$

This proves eq. (1.57) and, consequently, eq. (1.55) is valid for all $n \geq 1$. ■

1.6.6 Chi-Squared Distribution

Another important class of gamma distributions is that of χ^2-distributions. These probability measures play a crucial role in Mathematical Statistics (cf. Chapter 8).

Definition 1.6.26. For $n \geq 1$ let

$$\chi_n^2 := \Gamma_{2,n/2} \, .$$

This probability measure is called χ^2-**distribution** with n degrees of freedom.

Remark 1.6.27. In the moment the integer $n \geq 1$ in Definition 1.6.26 is only a parameter. The notation "degree of freedom" will become clear when we apply the χ^2-distribution for statistical problems.

Remark 1.6.28. The density p of a χ_n^2-distribution is given by

$$p(x) = \begin{cases} 0 & : x \leq 0 \\ \frac{x^{n/2-1}e^{-x/2}}{2^{n/2}\Gamma(n/2)} & : x > 0 \, , \end{cases}$$

i.e., if $0 \leq a < b$, then

$$\chi_n^2([a, b]) = \frac{1}{2^{n/2}\Gamma(n/2)} \int_a^b x^{n/2-1} e^{-x/2} \, dx \, .$$

1.6.7 Beta Distribution

Tightly connected with the gamma function is Euler's **beta function** B. It maps $(0, \infty) \times (0, \infty)$ to $\mathbb{R}$ and is defined by

$$B(x, y) := \int_0^1 s^{x-1}(1 - s)^{y-1} \, ds \, , \quad x, y > 0 \, . \tag{1.58}$$

The link between gamma and beta function is the following important identity:

$$B(x, y) = \frac{\Gamma(x) \cdot \Gamma(y)}{\Gamma(x + y)} \, , \quad x, y > 0 \, . \tag{1.59}$$

For a proof of eq. (1.59) we refer to Problem 1.27.

Further properties of the beta function are either easy to prove or follow via eq. (1.59) by those of the gamma function.

1. The beta function is continuous on $(0, \infty) \times (0, \infty)$ with values in $(0, \infty)$.
2. For $x, y > 0$ one has $B(x, y) = B(y, x)$.
3. If $x, y > 0$, then

$$B(x + 1, y) = \frac{x}{x + y} B(x, y) \, . \tag{1.60}$$

4. For $x > 0$ follows $B(x, 1) = 1/x$.

5. if $n, m \geq 1$ are integers, then

$$B(n, m) = \frac{(n-1)!\,(m-1)!}{(n+m-1)!}.$$

6. $B\left(\frac{1}{2}, \frac{1}{2}\right) = \pi.$

Definition 1.6.29. Let $\alpha, \beta > 0$. The probability measure $\mathcal{B}_{\alpha,\beta}$ defined by

$$\mathcal{B}_{\alpha,\beta}([a, b]) := \frac{1}{B(\alpha, \beta)} \int_a^b x^{\alpha-1}(1-x)^{\beta-1}\,dx, \quad 0 \leq a < b \leq 1,$$

is called **beta distribution** with parameters α and β.

The density function $q_{\alpha,\beta}$ of $\mathcal{B}_{\alpha,\beta}$ is given by

$$q_{\alpha,\beta}(x) = \begin{cases} \frac{1}{B(\alpha,\beta)}\,x^{\alpha-1}(1-x)^{\beta-1} & : \ 0 < x < 1 \\ 0 & : \text{otherwise} \end{cases}$$

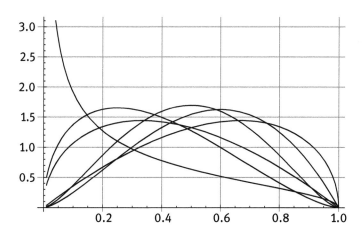

Figure 1.5: Density functions of the beta distribution with parameters $(0.5, 1.5)$, $(1.5, 2.5)$, $(2.5, 2)$, $(1.5, 2)$, $(2, 1.5)$, and $(2.5, 2.5)$.

Remark 1.6.30. It is easy to see that $q_{\alpha,\beta}$ is a density function.

$$\int_{-\infty}^{\infty} q_{\alpha,\beta}(x)\,dx = \frac{1}{B(\alpha, \beta)} \int_0^1 x^{\alpha-1}(1-x)^{\beta-1}\,dx = \frac{B(\alpha, \beta)}{B(\alpha, \beta)} = 1.$$

Furthermore, since $q_{\alpha,\beta}(x) = 0$ if $x \notin [0, 1]$, the probability measure $\mathcal{B}_{\alpha,\beta}$ is concentrated on $[0, 1]$, that is, $\mathcal{B}_{\alpha,\beta}([0, 1]) = 1$ or, equivalently, $\mathcal{B}_{\alpha,\beta}(\mathbb{R}\backslash[0, 1]) = 0$.

Example 1.6.31. Choose independently n numbers $x_1, \ldots, x_n$ in $[0, 1]$ according to the uniform distribution. Ordering these numbers by their size we get $x_1^* \leq \cdots \leq x_n^*$. In Example 3.7.8 we will show that the kth largest number x_k^* is $\mathcal{B}_{k,n-k+1}$-distributed. In other words, if $0 \leq a < b \leq 1$, then

$$\mathbb{P}\{a \leq x_k^* \leq b\} = \mathcal{B}_{k,n-k+1}([a, b]) = \frac{n!}{(k-1)!\,(n-k)!} \int_a^b x^{k-1}(1-x)^{n-k}\,dx\,.$$

1.6.8 Cauchy Distribution

We start with the following statement.

Proposition 1.6.32. *The function p defined by*

$$p(x) = \frac{1}{\pi} \cdot \frac{1}{1+x^2}\,, \qquad x \in \mathbb{R}\,, \tag{1.61}$$

is a probability density.

Proof: Of course, $p(x) > 0$ for $x \in \mathbb{R}$. Let us now investigate $\int_{-\infty}^{\infty} p(x)\,dx$. Because of $\lim_{b \to \infty} \arctan(b) = \pi/2$ and $\lim_{a \to -\infty} \arctan(a) = -\pi/2$ follows

$$\int_{-\infty}^{\infty} p(x)\,dx = \frac{1}{\pi} \lim_{a \to -\infty} \lim_{b \to \infty} \int_a^b \frac{1}{1+x^2}\,dx = \frac{1}{\pi} \lim_{a \to -\infty} \lim_{b \to \infty} \Big[\arctan x\Big]_a^b = 1\,.$$

Thus, as asserted, p is a probability density. ∎

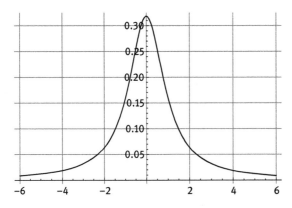

Figure 1.6: The density function of the Cauchy distribution.

Definition 1.6.33. The probability measure $\mathbb{P}$ with density p from eq. (1.61) is called **Cauchy distribution**. In other words, the Cauchy distribution $\mathbb{P}$ is characterized by

$$\mathbb{P}([a, b]) = \frac{1}{\pi} \int_a^b \frac{1}{1 + x^2} \, dx = \frac{1}{\pi} \left[\arctan(b) - \arctan(a) \right].$$

1.7 Distribution Function

In this section we always assume that the sample space is $\mathbb{R}$, even if the random experiment has only finitely or countably infinite many different outcomes. For example, rolling a die once is modeled by $(\mathbb{R}, \mathcal{P}(\mathbb{R}), \mathbb{P})$, where $\mathbb{P}(\{k\}) = 1/6$, $k = 1, \ldots, 6$, and $\mathbb{P}(\{x\}) = 0$ whenever $x \notin \{1, \ldots, 6\}$.

Thus, let $\mathbb{P}$ be a probability measure either defined on $\mathcal{B}(\mathbb{R})$ (continuous case) or on $\mathcal{P}(\mathbb{R})$ (discrete case) .

Definition 1.7.1. The function $F : \mathbb{R} \to [0, 1]$ defined by

$$F(t) := \mathbb{P}((-\infty, t]), \quad t \in \mathbb{R},$$

is called[15] **(cumulative) distribution function** of $\mathbb{P}$.

Remark 1.7.2. If $\mathbb{P}$ is discrete, that is, $\mathbb{P}(D) = 1$ for some $D = \{x_1, x_2, \ldots\}$, then its distribution function can be evaluated by

$$F(t) = \sum_{x_j \leq t} \mathbb{P}(\{x_j\}) = \sum_{x_j \leq t} p_j,$$

where $p_j = \mathbb{P}(\{x_j\})$, while for continuous $\mathbb{P}$ with probability density p

$$F(t) = \int_{-\infty}^t p(x) \, dx.$$

Example 1.7.3. Let $\mathbb{P}$ be the uniform distribution on $\{1, \ldots, 6\}$. Then

$$F(t) = \begin{cases} 0 : t < 1 \\ \frac{k}{6} : k \leq t < k + 1, \ k \in \{1, \ldots, 5\} \\ 1 : t \geq 6. \end{cases}$$

---

[15] To shorten the notation, mostly we will call it "distribution function" instead of, as often used in the literature, "cumulative distribution function" or, abbreviated, CDF.

Example 1.7.4. The distribution function of the binomial distribution $B_{n,p}$ is given by

$$F(t) = \sum_{0 \le k \le t} \binom{n}{k} p^k (1-p)^{n-k}, \ 0 \le t < \infty,$$

and $F(t) = 0$ if $t < 0$.

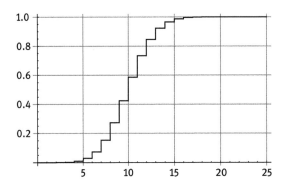

Figure 1.7: Distribution function of the binomial distribution $B_{25,0.4}$.

Example 1.7.5. The distribution function of the exponential distribution E_λ equals

$$F(t) = \begin{cases} 0 & : t < 0 \\ 1 - e^{-\lambda t} & : t \ge 0 \end{cases}$$

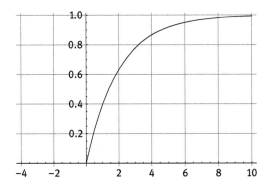

Figure 1.8: Distribution function of $E_{0.5}$.

Example 1.7.6. The distribution function of the standard normal distribution is de-noted[16] by Φ, therefore also called **Gaussian Φ-function**.

$$\Phi(t) = \frac{1}{\sqrt{2\pi}} \int_{-\infty}^{t} e^{-x^2/2} \, dx, \quad t \in \mathbb{R}. \tag{1.62}$$

---

[16] Sometimes also denoted as "norm$(\cdot)$."

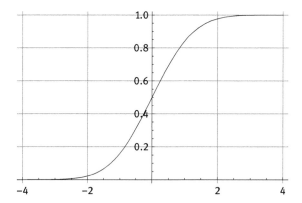

Figure 1.9: Distribution function of the standard normal distribution (Φ-function).

Remark 1.7.7. The Gaussian Φ-function is tightly related to the **Gaussian error function** defined by

$$\mathrm{erf}(t) = \frac{2}{\sqrt{\pi}} \int_0^t e^{-x^2} dx, \quad t \in \mathbb{R}.$$

Observe that $\mathrm{erf}(-t) = -\mathrm{erf}(t)$. The link between the Φ and the error function is

$$\Phi(t) = \frac{1}{2}\left[1 + \mathrm{erf}\left(\frac{t}{\sqrt{2}}\right)\right] \quad \text{and} \quad \mathrm{erf}(t) = 2\Phi(\sqrt{2}\,t) - 1, \quad t \in \mathbb{R}. \tag{1.63}$$

Example 1.7.8. Let $\mathbb{P}$ be the uniform distribution on the interval $[\alpha, \beta]$. Then its distribution function is

$$F(t) = \begin{cases} 0 & : \quad t < \alpha \\ \frac{t-\alpha}{\beta-\alpha} & : \quad \alpha \le t \le \beta \\ 1 & : \quad t > \beta \end{cases}$$

In particular, for the uniform distribution on $[0, 1]$ one obtains

$$F(t) = \begin{cases} 0 : & t < 0 \\ t : & 0 \le t \le 1 \\ 1 : & t > 1 \end{cases}$$

The next proposition lists the main properties of distribution functions.

Proposition 1.7.9. *Let F be the distribution function of a probability measure $\mathbb{P}$ on $\mathbb{R}$, discrete or continuous. Then F possesses the following properties.*
(1) *F is nondecreasing.*
(2) *$F(-\infty) = \lim_{t \to -\infty} F(t) = 0$ and $F(\infty) = \lim_{t \to \infty} F(t) = 1$.*
(3) *F is continuous from the right.*

Proof: Suppose $s < t$. This implies $(-\infty, s] \subset (-\infty, t]$, hence, since $\mathbb{P}$ is monotone, we obtain

$$F(s) = \mathbb{P}((-\infty, s]) \leq \mathbb{P}((-\infty, t]) = F(t).$$

Thus F is nondecreasing.

Take any sequence $(t_n)_{n \geq 1}$ that decreases monotonely to $-\infty$. Set $A_n := (-\infty, t_n]$. Then $A_1 \supseteq A_2 \supseteq \cdots$ as well as $\bigcap_{n=1}^{\infty} A_n = \emptyset$. Since $\mathbb{P}$ is continuous from above it follows

$$\lim_{n \to \infty} F(t_n) = \lim_{n \to \infty} \mathbb{P}(A_n) = \mathbb{P}(\emptyset) = 0.$$

This being true for any sequence $(t_n)_{n \geq 1}$ tending to $-\infty$ implies $F(-\infty) = 0$.

The proof for $F(\infty) = 1$ is very similar. Now $(t_n)_{n \geq 1}$ increases monotonely to ∞. If as before $A_n := (-\infty, t_n]$, this time $A_1 \subseteq A_2 \subseteq \cdots$ and $\bigcup_{n=1}^{\infty} A_n = \mathbb{R}$. By the continuity of $\mathbb{P}$ from below now we obtain

$$\lim_{n \to \infty} F(t_n) = \lim_{n \to \infty} \mathbb{P}(B_n) = \mathbb{P}(\mathbb{R}) = 1.$$

Again, since the t_ns were arbitrary, $F(\infty) = 1$.

Thus it remains to prove that F is continuous from the right. To do this, we take $t \in \mathbb{R}$ and a decreasing sequence $(t_n)_{n \geq 1}$ tending to t. We have to show that if $n \to \infty$, then $F(t_n) \to F(t)$.

As before set $A_n := (-\infty, t_n]$. Again $A_1 \supseteq A_2 \supseteq \cdots$, but now $\bigcap_{n=1}^{\infty} A_n = (-\infty, t]$. Another application of the continuity from above implies

$$F(t) = \mathbb{P}((-\infty, t]) = \lim_{n \to \infty} \mathbb{P}(A_n) = \lim_{n \to \infty} F(t_n).$$

This is valid for each $t \in \mathbb{R}$, hence F is continuous from the right. ∎

Properties (1), (2), and (3) in Proposition 1.7.9 characterize distribution functions. More precisely, the following result is true. Its proof is based on an extension theorem in Measure Theory. Therefore, we can show here only its main ideas.

Proposition 1.7.10. *Let $F : \mathbb{R} \to \mathbb{R}$ be an arbitrary function possessing the properties stated in Proposition 1.7.9. Then there exists a unique probability measure $\mathbb{P}$ on $\mathcal{B}(\mathbb{R})$ such that*

$$F(t) = \mathbb{P}((-\infty, t]), \quad t \in \mathbb{R}.$$

Idea of the proof: If $a < b$ set

$$\mathbb{P}_0((a, b]) := F(b) - F(a).$$

In this way we get a mapping $\mathbb{P}_0$ defined on the collection of all half-open intervals $\{(a, b] : a < b\}$. The key point is to verify that $\mathbb{P}_0$ can be uniquely extended to a probability measure $\mathbb{P}$ on $\mathcal{B}(\mathbb{R})$. One way to do this is to introduce a so-called outer measure $\mathbb{P}^*$ defined on $\mathcal{P}(\mathbb{R})$ by

$$\mathbb{P}^*(B) := \inf \left\{ \sum_{i=1}^{\infty} \mathbb{P}_0((a_i, b_i]) : B \subseteq \bigcup_{i=1}^{\infty} (a_i, b_i] \right\}.$$

Generally, this outer measure is not σ-additive. Therefore, one restricts $\mathbb{P}^*$ to $\mathcal{B}(\mathbb{R})$. If $\mathbb{P}$ denotes this restriction, the most difficult part of the proof is to verify that $\mathbb{P}$ is σ-additive. After this has been done, by the construction, $\mathbb{P}$ is the probability measure possessing distribution function F.

The uniqueness of $\mathbb{P}$ follows by a general uniqueness theorem for probability measures asserting the following.

Let $\mathbb{P}_1$ and $\mathbb{P}_2$ be two probability measures on $(\Omega, \mathcal{A})$ and let $\mathcal{E} \subseteq \mathcal{A}$ be a collection of events closed under taking intersections and generating $\mathcal{A}$. If $\mathbb{P}_1(E) = \mathbb{P}_2(E)$ for all $E \in \mathcal{E}$, then $\mathbb{P}_1 = \mathbb{P}_2$. In our case $\mathcal{E} = \{(-\infty, t] : t \in \mathbb{R}\}$.

Conclusion: If the outcomes of a random experiment are real numbers, then this experiment can also be described by a function $F : \mathbb{R} \to \mathbb{R}$ possessing the properties in Proposition 1.7.9. Then $F(t)$ is the probability to observe a result that is less than or equal to t.

Let us state further properties of distribution functions.

Proposition 1.7.11. *If F is the distribution function of a probability measure $\mathbb{P}$, then for all $a < b$*

$$F(b) - F(a) = \mathbb{P}((a, b]).$$

Proof: Observing that $(-\infty, a] \subseteq (-\infty, b]$ this is an immediate consequence of

$$F(b) - F(a) = \mathbb{P}((-\infty, b]) - \mathbb{P}((-\infty, a]) = \mathbb{P}((-\infty, b] \setminus (-\infty, a]) = \mathbb{P}((a, b]). \qquad \blacksquare$$

Since F is nondecreasing and bounded, for each $t \in \mathbb{R}$ the left-hand limit

$$F(t - 0) := \lim_{\substack{s \to t \\ s < t}} F(s)$$

exists and, moreover, $F(t - 0) \leq F(t)$. Furthermore, by the right continuity of F one has $F(t - 0) = F(t)$ if and only if F is continuous at the point t.

If this is not so, then $h = F(t) - F(t - 0) > 0$, that is, F possesses at t a **jump** of height $h > 0$. This height is directly connected with the value of $\mathbb{P}(\{t\})$.

Proposition 1.7.12. *The distribution function F of a probability measure $\mathbb{P}$ has a jump of height $h \geq 0$ at $t \in \mathbb{R}$ if and only if $\mathbb{P}(\{t\}) = h$.*

Proof: Let $(t_n)_{n\geq1}$ be a sequence of real numbers increasing monotonely to t. Then, using that $\mathbb{P}$ is continuous from above, it follows

$$h = F(t) - F(t - 0) = \lim_{n \to \infty} [F(t) - F(t_n)] = \lim_{n \to \infty} \mathbb{P}((t_n, t]) = \mathbb{P}(\{t\}).$$

Observe that $\bigcap_{n=1}^{\infty} (t_n, t] = \{t\}$. This proves the assertion. ∎

Corollary 1.7.13. *The function F is continuous at $t \in \mathbb{R}$ if and only if $\mathbb{P}(\{t\}) = 0$.*

Example 1.7.14. Suppose the function F is defined by

$$F(t) = \begin{cases} 0 & : \quad t < -1 \\ 1/3 & : \quad -1 \leq t < 0 \\ 1/2 & : \quad 0 \leq t < 1 \\ 2/3 & : \quad 1 \leq t < 2 \\ 1 & : \quad t \geq 2. \end{cases}$$

Then F fulfils the assumptions of Proposition 1.7.10. Hence there is a probability measure $\mathbb{P}$ with $F(t) = \mathbb{P}\{(-\infty, t]\}$. What does $\mathbb{P}$ look like?

Answer: The function F has jumps at $-1, 0, 1$, and 2 with heights $1/3, 1/6, 1/6$, and $1/3$. Therefore,

$$\mathbb{P}(\{-1\}) = 1/3, \quad \mathbb{P}(\{0\}) = 1/6, \quad \mathbb{P}(\{1\}) = 1/6 \text{ and } \mathbb{P}(\{2\}) = 1/3,$$

hence $\mathbb{P}$ is the discrete probability measure concentrated on $D = \{-1, 0, 1, 2\}$ with $\mathbb{P}(\{t\})$, $t \in D$, given above.

Suppose now that $\mathbb{P}$ is continuous with density function p. Recall that then

$$F(t) = \mathbb{P}((-\infty, t]) = \int_{-\infty}^{t} p(x)\, dx, \quad t \in \mathbb{R}. \tag{1.64}$$

In particular, since F is the function of the upper bound in an integral, it is continuous.

Next we investigate the question whether we may evaluate the density p knowing F.

Proposition 1.7.15. *Suppose p is continuous at some $t \in \mathbb{R}$. Then F is differentiable at t with*

$$F'(t) = \frac{d}{dt} F(t) = p(t).$$

Proof: This follows immediately by an application of the fundamental theorem of Calculus to representation (1.64) of F.　　　　　　　　　　　　　■

Remark 1.7.16. Let F be the distribution function of a probability measure $\mathbb{P}$. If F is continuous, then $\mathbb{P}(\{t\}) = 0$ for all $t \in \mathbb{R}$. But does this also imply that $\mathbb{P}$ is continuous, that is, that $\mathbb{P}$ has a density? The answer is negative. There exist probability measures $\mathbb{P}$ on $(\mathbb{R}, \mathcal{B}(\mathbb{R}))$ with a continuous distribution function but without possessing a density. Such probability measures are called **singularly continuous**.

To get an impression of how such probability measures look, let us shortly sketch the construction of an example. Let C be the Cantor set introduced in Example 1.6.5. The basic idea is to transfer the uniform distribution on $[0, 1]$ to a probability measure $\mathbb{P}$ with $\mathbb{P}(C) = 1$. The transformation is done by the function f defined as follows. If $x \in [0, 1]$ is represented as $x = \sum_{k=1}^{\infty} \frac{x_k}{2^k}$ with $x_k \in \{0, 1\}$, then $f(x) = \sum_{k=1}^{\infty} \frac{2x_k}{3^k}$. Note that f maps $[0, 1]$ into C. If $\tilde{\mathbb{P}}$ denotes the uniform distribution on $[0, 1]$, define the probability measure $\mathbb{P}$ by

$$\mathbb{P}(B) = \tilde{\mathbb{P}}\{x \in [0, 1] : f(x) \in B\}.$$

Then for all $t \in \mathbb{R}$ we have $\mathbb{P}(\{t\}) = 0$, but since $\mathbb{P}(C) = 1$, $\mathbb{P}$ cannot have a density. Indeed, such a density should vanish outside C. But, as we saw, the probability of C with respect to the uniform distribution is zero. Hence the only possible density would be $p(t) = 0$, $t \in \mathbb{R}$. This contradiction shows that $\mathbb{P}$ is not continuous in our sense.

Assuming a little bit more than the continuity of F, the corresponding probability measure possesses a density (cf. [Coh13]).

Proposition 1.7.17. *Let F be the distribution function of a probability measure $\mathbb{P}$. If F is continuous and continuously differentiable with the exception of at most finitely many points, then $\mathbb{P}$ is continuous. That is, there is a density function p such that*

$$F(t) = \mathbb{P}((-\infty, t]) = \int_{-\infty}^{t} p(x)\, dx, \quad t \in \mathbb{R}.$$

Remark 1.7.18. Proposition 1.7.15 implies $p(t) = F'(t)$ for those t where $F'(t)$ exists. If F is not differentiable at t, define $p(t)$ somehow, for example, $p(t) = 0$.

Example 1.7.19. For some $\alpha, \beta > 0$ define F by

$$F(t) = \begin{cases} 0 & : t \le 0 \\ 1 - e^{-\alpha t^{\beta}} & : t > 0 \end{cases}$$

It is easy to see that this function satisfies the properties of Proposition 1.7.9. Moreover, it is continuous and continuously differentiable on $\mathbb{R}\backslash\{0\}$. By Proposition 1.7.17 the corresponding probability measure $\mathbb{P}$ is continuous and since

$$F'(t) = \begin{cases} 0 & : t < 0 \\ \alpha\beta t^{\beta-1} e^{-\alpha t^{\beta}} & : t > 0. \end{cases}$$

a suitable density function is $p(t) = F'(t)$, $t \neq 0$, and $p(0) = 0$.

1.8 Multivariate Continuous Distributions

1.8.1 Multivariate Density Functions

In this section we suppose that $\Omega = \mathbb{R}^n$. A subset $Q \subset \mathbb{R}^n$ is called a (closed, n-dimensional) **box**[17] provided that for some real numbers $a_i < b_i$, $1 \leq i \leq n$,

$$Q = \{(x_1, \ldots, x_n) \in \mathbb{R}^n : a_i \leq x_i \leq b_i, \ 1 \leq i \leq n\}. \tag{1.65}$$

Definition 1.8.1. A Riemann integrable function $p : \mathbb{R}^n \to \mathbb{R}$ is said to be an n-dimensional **probability density function** or shorter n-dimensional **density function** if $p(x) \geq 0$ for $x \in \mathbb{R}^n$ and, furthermore,

$$\int_{\mathbb{R}^n} p(x)\,dx := \int_{-\infty}^{\infty} \cdots \int_{-\infty}^{\infty} p(x_1, \ldots, x_n)\,dx_n \cdots dx_1 = 1.$$

Suppose a box Q is represented with certain $a_i < b_i$ as in eq. (1.65). Then we set

$$\mathbb{P}(Q) = \int_Q p(x)\,dx = \int_{a_1}^{b_1} \cdots \int_{a_n}^{b_n} p(x_1, \ldots, x_n)\,dx_n \cdots dx_1. \tag{1.66}$$

In analogy to Definition 1.1.15 we introduce now the Borel σ-field $\mathcal{B}(\mathbb{R}^n)$.

Definition 1.8.2. Let $\mathcal{C}$ be the collection of all boxes in $\mathbb{R}^n$. Then $\sigma(\mathcal{C}) := \mathcal{B}(\mathbb{R}^n)$ denotes the **Borel σ-field**[18]. In other words, $\mathcal{B}(\mathbb{R}^n)$ is the smallest σ-field containing all (closed) boxes in $\mathbb{R}^n$. Sets in $\mathcal{B}(\mathbb{R}^n)$ are called (n-dimensional) **Borel sets.**

---

17 Also called "hyper-rectangle."
18 Recall that the existence of $\sigma(\mathcal{C})$ was proven in Proposition 1.1.12.

Remark 1.8.3. As in the univariate case there exist several other collections of subsets in $\mathbb{R}^n$ generating $\mathcal{B}(\mathbb{R}^n)$. For example, one may choose the collection of open boxes or the sets, which may be written as

$$(-\infty, t_1] \times \cdots \times (-\infty, t_n], \quad t_1, \ldots, t_n \in \mathbb{R}.$$

With the previous notations, the following multivariate extension theorem is valid. Compare Proposition 1.5.6 for the univariate case.

Proposition 1.8.4. *Let $\mathbb{P}$ be defined on boxes by eq. (1.66). Then $\mathbb{P}$ admits a unique extension to a probability measure $\mathbb{P}$ on $\mathcal{B}(\mathbb{R}^n)$.*

Definition 1.8.5. A probability measure $\mathbb{P}$ on $\mathcal{B}(\mathbb{R}^n)$ is called **continuous** provided that there exists a probability density $p : \mathbb{R}^n \to \mathbb{R}$ such that $\mathbb{P}(Q) = \int_Q p(x)\,dx$ for all boxes $Q \subseteq \mathbb{R}^n$. The function p is said to be the **density function** or simply **density** of $\mathbb{P}$.

Remark 1.8.6. It is easy to see that the validity of eq. (1.66) for all boxes is equivalent to the following. If $t_j \in \mathbb{R}$ and $B_{t_1,\ldots,t_n} := (-\infty, t_1] \times \cdots \times (-\infty, t_n]$, then

$$\mathbb{P}(B_{t_1,\ldots,t_n}) = \int_{B_{t_1,\ldots,t_n}} p(x)\,dx = \int_{-\infty}^{t_1} \cdots \int_{-\infty}^{t_n} p(x_1, \ldots, x_n)\,dx_n \cdots dx_1. \qquad (1.67)$$

Thus $\mathbb{P}$ is continuous if and only if eq. (1.67) is satisfied for all $t_j \in \mathbb{R}$.

Let us first give an example of a multivariate probability density function.

Example 1.8.7. Regard $p : \mathbb{R}^3 \to \mathbb{R}$ defined by

$$p(x_1, x_2, x_3) = \begin{cases} 48\, x_1 x_2 x_3 & : 0 \le x_1 \le x_2 \le x_3 \le 1 \\ 0 & : \quad \text{otherwise} \end{cases}$$

Of course, $p(x) \ge 0$ for $x \in \mathbb{R}^3$. Moreover,

$$\int_{\mathbb{R}^3} p(x)\,dx = 48 \int_0^1 \int_0^{x_3} \int_0^{x_2} x_1 x_2 x_3 \, dx_1 dx_2 dx_3$$

$$= 48 \int_0^1 \int_0^{x_3} \frac{x_3 x_2^3}{2}\,dx_2\,dx_3 = 48 \int_0^1 \frac{x_3^5}{8}\,dx_3 = 1\,,$$

hence it is a density function on $\mathbb{R}^3$. For example, if $\mathbb{P}$ is the generated probability measure, then

$$\mathbb{P}([0, 1/2]^3) = 48 \int_0^{1/2} \int_0^{x_3} \int_0^{x_2} x_1 x_2 x_3 \, dx_1 dx_2 dx_3 = \frac{1}{2^6} = \frac{1}{64}.$$

1.8.2 Multivariate Uniform Distribution

Our next aim is the introduction and the investigation of a special multivariate distribution, the *uniform distribution* on a set K in $\mathbb{R}^n$. To do so we remember as we defined the uniform distribution on an interval I in $\mathbb{R}$. Its density p is given by

$$p(s) = \begin{cases} \frac{1}{|I|} & : s \in I \\ 0 & : s \notin I \end{cases}$$

Here $|I|$ denotes the length of the interval I. Let now $K \subset \mathbb{R}^n$ be bounded. In order to introduce a similar density for the uniform distribution on K, the length of the underlying set has to be replaced by the n-dimensional volume, which we will denote by $\mathrm{vol}_n(K)$. But how is this volume defined?

To answer this question let us first investigate a box Q represented as in eq. (1.65). It is immediately clear that its n-dimensional volume is evaluated by

$$\mathrm{vol}_n(Q) = \prod_{i=1}^n (b_i - a_i).$$

If $n = 1$, then Q is an interval and its one-dimensional volume is nothing else as its length. For $n = 2$ the box Q is the rectangle $[a_1, b_1] \times [a_2, b_2]$ and

$$\mathrm{vol}_2(Q) = (b_1 - a_1)(b_2 - a_2)$$

coincides with the area of Q. If $n = 3$, then $\mathrm{vol}_3(Q)$ is the ordinary volume of bodies in $\mathbb{R}^3$.

For arbitrary $K \subset \mathbb{R}^n$ the definition of its volume $\mathrm{vol}_n(K)$ is more involved. Let us shortly sketch one way how this can be done. Setting

$$\mathrm{vol}_n(K) := \inf \left\{ \sum_{j=1}^\infty \mathrm{vol}_n(Q_j) : K \subseteq \bigcup_{j=1}^\infty Q_j, \, Q_j \text{ box} \right\}, \tag{1.68}$$

at least for Borel sets $K \subseteq \mathbb{R}^n$ a suitable volume is defined. In the case of "ordinary" sets as balls, ellipsoids, or similar bodies this approach leads to the known values. Background is the basic formula

$$\mathrm{vol}_n(K) = \int_K \cdots \int 1 \, dx_n \cdots dx_1 \tag{1.69}$$

valid for Borel sets $K \subseteq \mathbb{R}^n$. For example, if K is the cube in $\mathbb{R}^2$ with corner points $(1, 0)$, $(0, 1)$, $(-1, 0)$, and $(0, -1)$, then

$$\mathrm{vol}_2(K) = \iint_K 1\,dx_2 dx_1 = \int_{-1}^{0} \int_{-x_1-1}^{1+x_1} dx_2 dx_1 + \int_{0}^{1} \int_{x_1-1}^{1-x_1} dx_2 dx_1$$

$$= 2 \int_{-1}^{0} (x_1 + 1)dx_1 + 2 \int_{0}^{1}(1 - x_1)dx_1 = 2\left[\frac{x_1^2}{2} + x_1\right]_{-1}^{0} + 2\left[x_1 - \frac{x_1^2}{2}\right]_{0}^{1} = 2.$$

Example 1.8.8. Let $K_n(r)$ be the n-dimensional ball of radius $r > 0$, that is,

$$K_n(r) = \{x \in \mathbb{R}^n : |x| \leq r\} = \{(x_1, \ldots, x_n) \in \mathbb{R}^n : x_1^2 + \cdots + x_n^2 \leq r^2\}.$$

If

$$V_n(r) := \mathrm{vol}_n(K_n(r)), \quad r > 0,$$

denotes the n-dimensional volume of this ball, an easy change of variables implies $V_n(r) = V_n \cdot r^n$, where $V_n = V_n(1)$. But for $K_n = K_n(1)$ eq. (1.69) gives

$$V_n = \int_{K_n} \cdots \int 1\, dx_n \cdots dx_1 = \int_{-1}^{1}\left[\int \cdots \int_{\{x_2^2 + \cdots + x_n^2 \leq 1 - x_1^2\}} 1\, dx_n \cdots dx_2\right] dx_1$$

$$= \int_{-1}^{1} V_{n-1}\left(\sqrt{1 - x_1^2}\right) dx_1 = \int_{-1}^{1} V_{n-1}\left(\sqrt{1 - s^2}\right) ds.$$

Hence, by $V_{n-1}(r) = r^{n-1} V_{n-1}(1) = r^{n-1} V_{n-1}$ we obtain

$$V_n = V_{n-1} \cdot \int_{-1}^{1} (1 - s^2)^{(n-1)/2}\, ds = 2 V_{n-1} \cdot \int_{0}^{1} (1 - s^2)^{(n-1)/2}\, ds.$$

The change of the variables $s = y^{1/2}$, thus $ds = \frac{1}{2} y^{-1/2}\, dy$, yields

$$V_n = V_{n-1} \cdot \int_{0}^{1} y^{-1/2}(1 - y)^{(n-1)/2}\, dy = V_{n-1} B\left(\frac{1}{2}, \frac{n+1}{2}\right)$$

$$= \sqrt{\pi}\, V_{n-1} \frac{\Gamma\left(\frac{n+1}{2}\right)}{\Gamma\left(\frac{n}{2} + 1\right)}.$$

Hereby we used eq. (1.59) as well as $\Gamma(1/2) = \sqrt{\pi}$. Starting with $V_1 = 2$, a recursive application of the last formula finally leads to

$$\mathrm{vol}_n(K_n(r)) = V_n(r) = \frac{\pi^{n/2}}{\Gamma\left(\frac{n}{2} + 1\right)} r^n = \frac{2\pi^{n/2}}{n\Gamma\left(\frac{n}{2}\right)} r^n, \quad r > 0.$$

If we distinguish between even and odd dimensions, properties of the Γ-function imply

$$V_{2k}(r) = \frac{\pi^k}{k!} r^{2k} \quad \text{and} \quad V_{2k+1}(r) = \frac{2^{k+1}\pi^k}{(2k+1)!!} r^{2k+1}$$

where $(2k+1)!! = 1 \cdot 3 \cdot 5 \cdots (2k-1)(2k+1)$.

After the question about the volume is settled we are now in the position to introduce the uniform distribution on bounded Borel sets in $\mathbb{R}^n$. Thus let $K \subseteq \mathbb{R}^n$ be a bounded Borel set in $\mathbb{R}^n$ with volume $\text{vol}_n(K)$. Define $p : \mathbb{R}^n \to \mathbb{R}$ by

$$p(x) := \begin{cases} \frac{1}{\text{vol}_n(K)} & : x \in K \\ 0 & : x \notin K. \end{cases} \tag{1.70}$$

Proposition 1.8.9. *The function p defined by eq. (1.70) is an (n-dimensional) probability density function.*

Proof: By virtue of eq. (1.69) follows

$$\int_{\mathbb{R}^n} p(x)\,dx = \int_K \frac{1}{\text{vol}_n(K)}\,dx = \frac{1}{\text{vol}_n(K)} \underbrace{\int \cdots \int}_{K} 1\,dx_n \cdots dx_1$$

$$= \frac{\text{vol}_n(K)}{\text{vol}_n(K)} = 1.$$

Since $p(x) \geq 0$ if $x \in \mathbb{R}^n$, as asserted, p is a probability density function. $\blacksquare$

Definition 1.8.10. The probability measure $\mathbb{P}$ on $(\mathbb{R}^n, \mathcal{B}(\mathbb{R}^n))$ with density p given by eq. (1.70) is said to be the (multivariate) **uniform distribution on** K.

Let $\mathbb{P}$ be the uniform distribution on K. How do we get $\mathbb{P}(B)$ for a Borel set B? Let us first assume $B \subseteq K$. Then

$$\mathbb{P}(B) = \int_B p(x)\,dx = \frac{1}{\text{vol}_n(K)} \underbrace{\int \cdots \int}_{B} 1\,dx_n \cdots dx_1 = \frac{\text{vol}_n(B)}{\text{vol}_n(K)}.$$

If $B \subseteq \mathbb{R}^n$ is arbitrary, that is, B is not necessarily a subset of K, by $\mathbb{P}(B) = \mathbb{P}(B \cap K)$ it follows that[19]

---

[19] This is an alternative way to introduce the uniform distribution on K.

!

$$P(B) = \frac{\mathrm{vol}_n(B \cap K)}{\mathrm{vol}_n(K)} \; .$$

If $n = 1$ and K is an interval the last formula coincides with eq. (1.45).

Example 1.8.11. Two friends agree to meet each other in a restaurant between 1 and 2 pm. Both friends go to the restaurant randomly during this hour. After they arrive they wait 20 minutes each. What is the probability that they meet each other?

Answer: Let t_1 be the moment where the first of the two friends enters the restaurant, while t_2 is the arrival time of the second one. They arrive independently of each other, thus we may assume that the point $t := (t_1, t_2)$ is uniformly distributed in the square $Q := [1, 2]^2$. Observing that 20 minutes are a third of an hour, they meet each other if and only if $|t_1 - t_2| \le 1/3$.

Setting $B := \{(t_1, t_2) \in \mathbb{R}^2 : |t_1 - t_2| \le 1/3\}$, it is easy to see that $\mathrm{vol}_2(B \cap Q) = 5/9$. Hence, if $\mathbb{P}$ is the uniform distribution on Q, because of $\mathrm{vol}_2(Q) = 1$ it follows $\mathbb{P}(B) = 5/9$. Therefore, the probability that the friends meet each other equals 5/9.

Example 1.8.12. Suppose n particles are uniformly distributed in a ball K_R of radius $R > 0$. Let K_r be a smaller ball of radius $r > 0$ contained in K_R. Find the probability that exactly k of the n particles are inside K_r for some $k = 0, \dots, n$.

Answer: In a first step we determine the probability that a single particle is in K_r. Since we assumed that the particles are uniformly distributed in K_R, this probability equals

$$p := \frac{\mathrm{vol}_3(K_r)}{\mathrm{vol}_3(K_R)} = \frac{(4/3)\pi r^3}{(4/3)\pi R^3} = \left(\frac{r}{R}\right)^3 .$$

For each of the n particles this p is the "success" probability to be inside K_r, hence the number of particles in K_r is $B_{n,p}$-distributed with $p = (r/R)^3$. Thus,

$$\mathbb{P}\{k \text{ particles in } K_r\} = B_{n,p}(\{k\}) = \binom{n}{k} \left(\frac{r}{R}\right)^{3k} \left(\frac{R-r}{R}\right)^{3(n-k)} , \quad k = 0, \dots, n .$$

If the number n of particles is big and r is much smaller than R, then the number of particles in K_r is approximately Pois_λ distributed, where $\lambda = np = \frac{nr^3}{R^3}$. In other words,

$$\mathbb{P}\{k \text{ particles in } K_r\} \approx \frac{1}{k!} \left(\frac{nr^3}{R^3}\right)^k e^{-nr^3/R^3} .$$

Example 1.8.13 (Buffon's needle test). Take a needle of length $a < 1$ and throw it randomly on a lined sheet of paper. Say the distance between two lines on the paper is 1. Find the probability that the needle cuts a line.

Answer: What is random in this experiment? Choose the two lines such that between them the midpoint of the needle lies. Let $x \in [0, 1]$ be the distance of the midpoint of the needle to the lower line. Furthermore, denote by $\theta \in [-\pi/2, \pi/2]$ the angle of the needle to a line perpendicularly to the lines on the paper. For example, if $\theta = 0$, then the needle is perpendicular to the lines on the paper while for $\theta = \pm\pi/2$ it lies parallel.

Hence, to throw a needle randomly is equivalent to choosing a point (θ, x) uniformly distributed in $K = [-\pi/2, \pi/2] \times [0, 1]$.

The needle cuts the lower line if and only if $\frac{a}{2} \cos \theta \geq x$ and it cuts the upper line provided that $\frac{a}{2} \cos \theta \geq 1 - x$.

If

$$A = \{(\theta, x) \in [-\pi/2, \pi/2] \times [0, 1] : x \leq \frac{a}{2} \cos \theta \quad \text{or} \quad 1 - x \leq \frac{a}{2} \cos \theta\},$$

then we get

$$\mathbb{P}\{\text{The needle cuts a line}\} = \mathbb{P}(A) = \frac{\text{vol}_2(A)}{\text{vol}_2(K)} = \frac{\text{vol}_2(A)}{\pi}.$$

But it follows

$$\text{vol}_2(A) = 2 \int_{-\pi/2}^{\pi/2} \frac{a}{2} \cos \theta \, d\theta = 2a,$$

hence

$$\mathbb{P}(A) = \frac{2a}{\pi}.$$

Remark 1.8.14. Suppose we throw the same needle n times. Let r_n be the relative frequency of the occurrence of A, that is,

$$r_n = \frac{\text{Number of throws where the needle cuts a line}}{n}.$$

As mentioned in Section 1.1.3, if $n \to \infty$, then r_n approaches $\mathbb{P}(A) = \frac{2a}{\pi}$. Thus for large n we have $r_n \approx \frac{2a}{\pi}$ or, equivalently, $\pi \approx \frac{2a}{r_n}$. Consequently, throwing the needle sufficiently often, $\frac{2a}{r_n}$ should be close to π.

1.9 *Products of Probability Spaces

1.9.1 Product σ-Fields and Measures

Suppose we execute n (maybe different) random experiments so that the outcomes do not depend on each other. In order to describe these n experiments two different approaches are possible. Firstly, we record each single result separately, that is, we have n (maybe different) probability spaces $(\Omega_1, \mathcal{A}_1, \mathbb{P}_1)$ to $(\Omega_n, \mathcal{A}_n, \mathbb{P}_n)$ modeling the outcomes of the first up to the nth experiment.

A second possible approach is that we combine the n experiments into a single one. Thus, instead of n different outcomes ω_1 to ω_n, we observe now a vector $\omega = (\omega_1, \ldots, \omega_n)$. The sample space in this approach is given by $\Omega = \Omega_1 \times \cdots \times \Omega_n$.

Example 1.9.1. When rolling a die n times the outcome is a series of n numbers ω_1 to ω_n, each in $\{1, \ldots, 6\}$. Now, imagine we have a die with 6^n equally likely faces. On these faces, all possible sequences of length n with entries from $\{1, \ldots, 6\}$ are written. Roll this die **once**. The first experiment may be described by n probability spaces, one for each roll. The second experiment involves only one probability space. Nevertheless, both experiments lead to the same result, a random sequence of numbers from 1 to 6.

It is intuitively clear that both approaches to this experiment (rolling a die n times) are equivalent; they differ only by the point of view. But how to come from one model to the other? One direction is immediately clear. If the random result is a vector $\omega = (\omega_1, \ldots, \omega_n)$, then its coordinates may be taken as the results of the single experiments[20]. But how about the other direction? That is, we are given n probability spaces $(\Omega_1, \mathcal{A}_1, \mathbb{P}_1), \ldots, (\Omega_n, \mathcal{A}_n, \mathbb{P}_n)$ and have to construct a model for the joint execution of these experiments.

Of course, the "new" sample space is

$$\Omega = \Omega_1 \times \cdots \times \Omega_n, \tag{1.71}$$

but what are $\mathcal{A}$ and $\mathbb{P}$? We start with the construction of the product σ-field.

Definition 1.9.2. Let $\mathcal{A}_j$ be σ-fields on Ω_j, $1 \le j \le n$. Set $\Omega = \Omega_1 \times \cdots \times \Omega_n$. Then

$$\mathcal{A} = \sigma\{A_1 \times \cdots \times A_n : A_j \in \mathcal{A}_j\}$$

is called the **product σ-field** of $\mathcal{A}_1$ to $\mathcal{A}_n$. It is denoted by $\mathcal{A} = \mathcal{A}_1 \otimes \cdots \otimes \mathcal{A}_n$.

---

[20] Of course, one still has to verify that the distribution of the coordinates is the same as in the single experiments. But before we can do this we need a probability measure describing the distribution of the vectors (cf. Proposition 1.9.8).

Remark 1.9.3. In other words, $\mathcal{A}$ is the smallest σ-field containing measurable rectangle sets, that is, sets of the form $A_1 \times \cdots \times A_n$ with $A_j \in \mathcal{A}_j$, $1 \le j \le n$.

It is easy to see that $\mathcal{P}(\Omega_1) \otimes \cdots \otimes \mathcal{P}(\Omega_n) = \mathcal{P}(\Omega)$. A more complicated example is as follows.

Proposition 1.9.4. *Suppose $\Omega_1 = \cdots = \Omega_n = \mathbb{R}$, hence $\Omega = \mathbb{R}^n$. Then the σ-field $\mathcal{B}(\mathbb{R}^n)$ of Borel sets in $\mathbb{R}^n$ is the n-fold product of the σ-fields $\mathcal{B}(\mathbb{R})$ of Borel sets in $\mathbb{R}$, that is,*

$$\mathcal{B}(\mathbb{R}^n) = \underbrace{\mathcal{B}(\mathbb{R}) \otimes \cdots \otimes \mathcal{B}(\mathbb{R})}_{n \ times} .$$

Proof: We only give a sketch of the proof. Let Q be a box as in eq. (1.65). Then $Q = A_1 \times \cdots \times A_n$, where the A_js are intervals, hence in $\mathcal{B}(\mathbb{R})$. By the construction of the product σ-field it follows that $Q \in \mathcal{B}(\mathbb{R}) \otimes \cdots \otimes \mathcal{B}(\mathbb{R})$. But $\mathcal{B}(\mathbb{R}^n)$ is the smallest σ-field containing all boxes, which lets us conclude

$$\mathcal{B}(\mathbb{R}^n) \subseteq \mathcal{B}(\mathbb{R}) \otimes \cdots \otimes \mathcal{B}(\mathbb{R}) .$$

The inclusion in the other direction may be proved as follows: fix $a_2 < b_2$ to $a_n < b_n$ and let

$$C_1 = \{C \in \mathcal{B}(\mathbb{R}) : C \times [a_2, b_2] \times \cdots [a_n, b_n] \in \mathcal{B}(\mathbb{R}^n)\} .$$

It is not difficult to prove that C_1 is a σ-field. If $C = [a_1, b_1]$, then $C \times [a_2, b_2] \times \cdots [a_n, b_n]$ is a box, thus in $\mathcal{B}(\mathbb{R}^n)$. Consequently, C_1 contains closed intervals, hence, since $\mathcal{B}(\mathbb{R})$ is the smallest σ-field with this property, it follows $C_1 = \mathcal{B}(\mathbb{R})$. This tells us that for all $B_1 \in \mathcal{B}(\mathbb{R})$ and all $a_j < b_j$

$$B_1 \times [a_2, b_2] \times \cdots \times [a_n, b_n] \in \mathcal{B}(\mathbb{R}^n) .$$

In a next step fix $B_1 \in \mathcal{B}(\mathbb{R})$ and $a_3 < b_3$ up to $a_n < b_n$ and set

$$C_2 = \{C \in \mathcal{B}(\mathbb{R}) : B_1 \times C \times [a_3, b_3] \times \cdots [a_n, b_n] \in \mathcal{B}(\mathbb{R}^n)\} .$$

By the same arguments as before, but now using the first step, we get $C_2 = \mathcal{B}(\mathbb{R})$, that is,

$$B_1 \times B_2 \times [a_3, b_3] \times \cdots [a_n, b_n] \in \mathcal{B}(\mathbb{R}^n)$$

for all $B_1, B_2 \in \mathcal{B}(\mathbb{R})$ and $a_j < b_j$.

Iterating further we finally obtain that for all $B_j \in \mathcal{B}(\mathbb{R})$ it follows that

$$B_1 \times \cdots \times B_n \in \mathcal{B}(\mathbb{R}^n) .$$

Since $\mathcal{B}(\mathbb{R}) \otimes \cdots \otimes \mathcal{B}(\mathbb{R}) = \sigma\{B_1 \times \cdots \times B_n : B_j \in \mathcal{B}(\mathbb{R})\}$ is the smallest σ-field containing sets $B_1 \times \cdots \times B_n$, this implies

$$\mathcal{B}(\mathbb{R}) \otimes \cdots \otimes \mathcal{B}(\mathbb{R}) \subseteq \mathcal{B}(\mathbb{R}^n)$$

and completes the proof. ∎

Let us now turn to the probability measure $\mathbb{P}$ on $(\Omega, \mathcal{A})$ that describes the combined experiment.

Definition 1.9.5. Let $(\Omega_1, \mathcal{A}_1, \mathbb{P}_1)$ to $(\Omega_n, \mathcal{A}_n, \mathbb{P}_n)$ be n probability spaces. Define Ω by eq. (1.71) and endow it with the product σ-field $\mathcal{A} = \mathcal{A}_1 \otimes \cdots \otimes \mathcal{A}_n$. A probability measure $\mathbb{P}$ on $(\Omega, \mathcal{A})$ is called the **product measure** of $\mathbb{P}_1, \ldots, \mathbb{P}_n$ if

$$\mathbb{P}(A_1 \times \cdots \times A_n) = \mathbb{P}_1(A_1) \cdots \mathbb{P}_n(A_n) \quad \text{for all} \quad A_j \in \mathcal{A}_j. \tag{1.72}$$

We write $\mathbb{P} = \mathbb{P}_1 \otimes \cdots \otimes \mathbb{P}_n$ and if $\mathbb{P}_1 = \cdots \mathbb{P}_n = \mathbb{P}_0$ set

$$\mathbb{P}_0^{\otimes n} := \underbrace{\mathbb{P}_0 \otimes \cdots \otimes \mathbb{P}_0}_{n \text{ times}}.$$

It is not clear at all whether product measures exist, and if this is so, whether condition (1.72) determines them uniquely. The next result shows that the answer to both questions is affirmative. Unfortunately, the proof is too complicated to be presented here. The idea is quite similar to that used in the introduction of volumes in eq. (1.68). The boxes appearing there have to be replaced by rectangle sets $A_1 \times \cdots \times A_n$ with $A_j \in \mathcal{A}_j$ and the volume of the boxes by $\mathbb{P}_1(A_1) \cdots \mathbb{P}_n(A_n)$. We refer to [Dur10], Section 1.7, or [Coh13], for a detailed proof for the existence (and uniqueness) of product measures.

Proposition 1.9.6. *Let* $(\Omega_1, \mathcal{A}_1, \mathbb{P}_1), \ldots, (\Omega_n, \mathcal{A}_n, \mathbb{P}_n)$ *be probability spaces. Define* Ω *by eq. (1.71) and let* $\mathcal{A}$ *be the product* σ-field *of the* $\mathcal{A}_j$. *Then there is a unique probability measure* $\mathbb{P}$ *on* $(\Omega, \mathcal{A})$ *satisfying eq. (1.72). Hence, the product measure* $\mathbb{P} = \mathbb{P}_1 \otimes \cdots \otimes \mathbb{P}_n$ *always exists and is uniquely determined by eq. (1.72).*

Corollary 1.9.7. *Let* $\mathbb{P}_1, \ldots, \mathbb{P}_n$ *be probability measures on* $(\mathbb{R}, \mathcal{B}(\mathbb{R}))$. *Then there is a unique probability measure* $\mathbb{P}$ *on* $(\mathbb{R}^n, \mathcal{B}(\mathbb{R}^n))$ *such that*

$$\mathbb{P}(B_1 \times \cdots \times B_n) = \mathbb{P}_1(B_1) \cdots \mathbb{P}_n(B_n) \quad \text{for all} \quad B_j \in \mathcal{B}(\mathbb{R}).$$

Proof: The proof is a direct consequence of Propositions 1.9.6 and 1.9.4. Indeed, take $\mathbb{P} = \mathbb{P}_1 \otimes \cdots \otimes \mathbb{P}_n$ and observe that $\mathcal{B}(\mathbb{R}^n) = \mathcal{B}(\mathbb{R}) \otimes \cdots \otimes \mathcal{B}(\mathbb{R})$. ∎

Let us shortly come back to the question asked at the beginning of this section. Suppose we observe a vector $\omega = (\omega_1, \ldots, \omega_n)$. How are the coordinates distributed?

Proposition 1.9.8. *Let $(\Omega, \mathcal{A}, \mathbb{P})$ be the product probability space of $(\Omega_1, \mathcal{A}_1, \mathbb{P}_1)$ to $(\Omega_n, \mathcal{A}_n, \mathbb{P}_n)$. If $j \leq n$ and $A \in \mathcal{A}_j$, then*

$$\mathbb{P}\{(\omega_1, \ldots, \omega_n) \in \Omega : \omega_j \in A\} = \mathbb{P}_j(A).$$

Proof: Observe that

$$\{(\omega_1, \ldots, \omega_n) \in \Omega : \omega_j \in A\} = \Omega_1 \times \cdots \Omega_{j-1} \times A \times \Omega_{j+1} \times \cdots \Omega_n,$$

thus eq. (1.72) implies

$$\mathbb{P}\{(\omega_1, \ldots, \omega_n) \in \Omega : \omega_j \in A\} = \mathbb{P}_1(\Omega_1) \cdots \mathbb{P}_{j-1}(\Omega_{j-1}) \cdot \mathbb{P}_j(A) \cdots \mathbb{P}_n(\Omega_n) = \mathbb{P}_j(A)$$

as asserted. ∎

How do we get product measures in concrete cases? We answer this question for discrete and continuous probability measures separately.

1.9.2 Product Measures: Discrete Case

Let Ω_1 to Ω_n be either finite or countably infinite sets. Given probability measures $\mathbb{P}_j$ defined on $\mathcal{P}(\Omega_j)$, $1 \leq j \leq n$, the following result characterizes the product measure of the $\mathbb{P}_j$s.

Proposition 1.9.9. $\mathbb{P}$ *is the product measure of $\mathbb{P}_1, \ldots, \mathbb{P}_n$ if and only if*

$$\mathbb{P}(\{\omega\}) = \mathbb{P}_1(\{\omega_1\}) \cdots \mathbb{P}_n(\{\omega_n\}) \quad \text{for all} \quad \omega = (\omega_1, \ldots, \omega_n) \in \Omega. \tag{1.73}$$

Proof: One direction is easy. Indeed, if $\mathbb{P} = \mathbb{P}_1 \otimes \cdots \otimes \mathbb{P}_n$, given $\omega = (\omega_1, \ldots, \omega_n) \in \Omega$ set $A = \{\omega\}$ and $A_j = \{\omega_j\}$. Then $A = A_1 \times \cdots \times A_n$, hence

$$\mathbb{P}(\{\omega\}) = \mathbb{P}(A) = \mathbb{P}_1(A_1) \cdots \mathbb{P}_n(A_n) = \mathbb{P}_1(\{\omega_1\}) \cdots \mathbb{P}_n(\{\omega_n\})$$

proving eq. (1.73).

To verify the other implication let $\mathbb{P}$ be a probability measure on $(\Omega, \mathcal{P}(\Omega))$ satisfying (1.73). We have to show that $\mathbb{P}$ fulfills eq. (1.72). Thus choose arbitrary $A_j \subseteq \Omega_j$ and set $A = A_1 \times \cdots \times A_n$. By applying eq. (1.73) it follows

$$\mathbb{P}(A) = \sum_{\omega \in A} \mathbb{P}(\{\omega\}) = \sum_{(\omega_1, \ldots, \omega_n) \in A} \mathbb{P}(\{(\omega_1, \ldots, \omega_n)\})$$

$$= \sum_{\omega_1 \in A_1, \ldots \omega_n \in A_n} P_1(\{\omega_1\}) \cdots P_n(\{\omega_n\})$$

$$= \sum_{\omega_1 \in A_1} P_1(\{\omega_1\}) \cdots \sum_{\omega_n \in A_n} P_n(\{\omega_n\}) = P_1(A_1) \ldots P_n(A_n).$$

This being true for all $A_j \subseteq \Omega_j$ shows that $\mathbb{P} = P_1 \otimes \cdots \otimes P_n$, and the proof is complete. ∎

Summary: In the discrete case the product measure is characterized as follows. Given $A \subseteq \Omega$, then

---

!

$$(\mathbb{P}_1 \otimes \cdots \otimes \mathbb{P}_n)(A) = \sum_{(\omega_1, \ldots, \omega_n) \in A} P_1(\{\omega_1\}) \cdots P_n(\{\omega_n\}).$$

Example 1.9.10. Suppose two players, say U and V, each toss simultaneously a biased coin. At both coins appears "0" (failure) with probability $1 - p$ and "1" (success) with probability p. The pair $(k, l) \in \mathbb{N}^2$ occurs if player U has his first success in trial k and player V in trial l. Each single experiment is described by the geometric distribution G_p, hence the model for the combined experiment is $(\mathbb{N}^2, \mathcal{P}(\mathbb{N}^2), G_p^{\otimes 2})$. Here

$$G_p^{\otimes 2}(A) = \sum_{(k,l) \in A} G_p(\{k\}) G_p(\{l\}) = \sum_{(k,l) \in A} p^2 (1 - p)^{k+l-2}, \quad A \subseteq \mathbb{N}^2.$$

For example, if $A = \{(k, k) : k \geq 1\}$, then

$$G_p^{\otimes 2}(A) = p^2 \sum_{k=1}^{\infty} (1 - p)^{2k-2} = \frac{p^2}{1 - (1 - p)2} = \frac{p}{2 - p}.$$

Thus in the case of a fair coin, the probability that both players have their first success at the same time equals 1/3.

Example 1.9.11. Toss a biased coin n times. Say the coin is labeled with "0" and "1" and $p \in [0, 1]$ is the probability of the occurrence of "1." Recording each single result separately the describing probability spaces are $(\{0, 1\}, \mathcal{P}(\{0, 1\}), \mathbb{P}_j)$, $1 \leq j \leq n$, with $\mathbb{P}_j(\{1\}) = p$. Which probability space does the combined result describe?

Answer: Of course, the sample space is $\{0, 1\}^n$ with σ-field $\mathcal{P}(\Omega)$. Let $\omega = (\omega_1, \ldots, \omega_n)$ be an arbitrary vector in Ω. Then by Proposition 1.9.9 the product measure $\mathbb{P}$ of the $\mathbb{P}_j$s is characterized by

$$\mathbb{P}(\{\omega\}) = \mathbb{P}_1(\{\omega_1\}) \cdots \mathbb{P}_n(\{\omega\}) = p^k(1 - p)^{n-k}$$

where $k = \#\{j \leq n : \omega_j = 1\} = \sum_{j=1}^{n} \omega_j$.

For example, tossing the coin five times, the sequence $(0, 0, 1, 1, 0)$ occurs with probability $p^2 (1 - p)^3$.

1.9.3 Product Measures: Continuous Case

Here we assume $\Omega_1 = \cdots = \Omega_n = \mathbb{R}$, hence the product sample space is $\Omega = \mathbb{R}^n$. Furthermore, each $\Omega_j = \mathbb{R}$ is endowed with the Borel σ-field. Because of Proposition 1.9.4 the product σ-field on $\Omega = \mathbb{R}^n$ is given by $\mathcal{B}(\mathbb{R}^n)$.

The next proposition characterizes the product measure of continuous probability measures.

Proposition 1.9.12. *Let $\mathbb{P}_1, \ldots, \mathbb{P}_n$ be probability measures on $(\mathbb{R}, \mathcal{B}(\mathbb{R}))$ with respective density functions $p_1, \ldots, p_n$, that is,*

$$\mathbb{P}_j([a, b]) = \int_a^b p_j(x)dx, \quad 1 \leq j \leq n.$$

Define $p : \mathbb{R}^n \to [0, \infty)$ by

$$p(x) = p_1(x_1) \cdots p_n(x_n), \quad x = (x_1, \ldots, x_n) \in \mathbb{R}^n. \tag{1.74}$$

The product measure $\mathbb{P}_1 \otimes \cdots \otimes \mathbb{P}_n$ is continuous with (n-dimensional) density p defined by (1.74). In other words, for each Borel set $A \subseteq \mathbb{R}^n$ holds

$$(\mathbb{P}_1 \otimes \cdots \otimes \mathbb{P}_n)(A) = \underbrace{\int \cdots \int}_{A} p_1(x_1) \cdots p_n(x_n) \, dx_n \cdots dx_1 = \int_A p(x) \, dx.$$

Proof: First note that p is a density of the product measure $\mathbb{P}_1 \otimes \cdots \otimes \mathbb{P}_n$ if

$$(\mathbb{P}_1 \otimes \cdots \otimes \mathbb{P}_n)(Q) = \int_Q p(x) \, dx$$

for all boxes $Q = [a_1, b_1] \times \cdots \times [a_n, b_n]$. But this is an immediate consequence of

$$\int_Q p(x)\, dx = \int_{a_1}^{b_1} \cdots \int_{a_n}^{b_n} p_1(x_1) \cdots p_n(x_n)\, dx_n \cdots dx_1$$

$$= \left(\int_{a_1}^{b_1} p_1(x_1)\, dx_1 \right) \cdots \left(\int_{a_n}^{b_n} p_n(x_n)\, dx_n \right) = \mathbb{P}_1([a_1, b_1]) \cdots \mathbb{P}_n([a_n, b_n])$$

$$= (\mathbb{P}_1 \otimes \cdots \otimes \mathbb{P}_n)([a_1, b_1] \times \cdots \times [a_n, b_n]) = (\mathbb{P}_1 \otimes \cdots \otimes \mathbb{P}_n)(Q) .$$

This completes the proof. ∎

Because of its importance let us explain through several examples how Proposition 1.9.12 applies. Further applications, for example, the characterization of independent random variables, will follow in Sections 3 and 8.

Example 1.9.13. Let the probability measures $\mathbb{P}_j$, $1 \le j \le n$, be uniform distributions on $[\alpha_j, \beta_j]$. Thus

$$p_j(x) = \begin{cases} \frac{1}{\beta_j - \alpha_j} & : \alpha_j \le x \le \beta_j \\ 0 & : \text{otherwise} \end{cases}$$

henceforth, if $x = (x_1, \ldots, x_n)$, then

$$p(x) = p_1(x_1) \cdots p_n(x_n) = \begin{cases} \frac{1}{\prod_{j \le n}(\beta_j - \alpha_j)} & : \quad x \in K \\ 0 & : \text{otherwise} \end{cases}$$

Here $K \subseteq \mathbb{R}^n$ is the box $[\alpha_1, \beta_1] \times \cdots \times [\alpha_n, \beta_n]$. Since $\prod_{j \le n}(\beta_j - \alpha_j) = \text{vol}_n(K)$, it follows that the product measure $\mathbb{P}_1 \otimes \cdots \otimes \mathbb{P}_n$ is nothing else as the (n-dimensional) uniform distribution on K as introduced in[21] Definition 1.8.10.

Summary: The product measure of n uniform distributions on intervals $[\alpha_j, \beta_j]$ is the uniform distribution on the box $[\alpha_1, \beta_1] \times \cdots \times [\alpha_n, \beta_n]$.

Example 1.9.14. Assume now $\mathbb{P}_1 = \cdots = \mathbb{P}_n = E_\lambda$, that is, we want to describe the product of n exponential distributions with parameter $\lambda > 0$. Since $p_j(s) = \lambda e^{-\lambda s}$ if $s \ge 0$ and $p_j(s) = 0$ if $s < 0$, their product $E_\lambda^{\otimes n}$ possesses the density

$$p(s_1, \ldots, s_n) = \begin{cases} \lambda^n\, e^{-\lambda(s_1 + \cdots + s_n)} & : s_1, \ldots, s_n \ge 0 \\ 0 & : \quad \text{otherwise} \end{cases}$$

Which random experiment does $E_\lambda^{\otimes n}$ describe? Suppose we have n light bulbs of the same type with lifetime distributed according to E_λ. Switch on all n bulbs at once and

---

21 This result was already used in Example 1.8.11. Indeed, the arrival times t_1 and t_2 were described by the uniform distributions on $[1, 2]$, thus the pair $t = (t_1, t_s)$ is distributed according to the product measure, which is the uniform distribution on $[1, 2] \times [1, 2]$.

record the times $t_1, \ldots, t_n$ where the first bulb, the second, and so on burns out. If $t = (t_1, \ldots, t_n) \in \mathbb{R}^n$ denotes the generated vector of these times, then for Borel sets $A \subseteq [0, \infty)^n$,

$$\mathbb{P}\{t \in A\} = E_\lambda^{\otimes n}(A) = \lambda^n \int_A e^{-\lambda(s_1 + \cdots + s_n)} ds_1 \ldots ds_n \, .$$

For example, if we want to compute the probability for

$$A := \{(t_1, \ldots, t_n) : 0 \le t_1 \le \cdots \le t_n\} \, ,$$

that is, the second bulb burns longer than the first one, the third longer than the second, and so on, then

$$E_\lambda^{\otimes n}(A) = \lambda^n \int_0^\infty e^{-\lambda s_n} \int_0^{s_n} e^{-\lambda s_{n-1}} \int_0^{s_{n-1}} \cdots \int_0^{s_3} e^{-\lambda s_2} \int_0^{s_2} e^{-\lambda s_1} ds_1 \ldots ds_n \, .$$

Iterative integration leads to $E_\lambda^{\otimes n}(A) = 1/n!$. This is more or less obvious by the following observation. Each order of the times of failure is equally likely. And since there are $n!$ different ways to order these times, each order has probability $1/n!$. In particular, this is true for the order $t_1 \le \cdots \le t_n$.

Next is given another example of a product measure that will play a crucial role in Sections 6 and 8.

Example 1.9.15. Let $\mathbb{P}_1, \ldots, \mathbb{P}_n$ be standard normal distributions. The corresponding densities are

$$p_j(x_j) = \frac{1}{\sqrt{2\pi}} e^{-x_j^2/2}, \quad 1 \le j \le n \, .$$

Thus, by eq. (1.74) the density p of their product $\mathcal{N}(0, 1)^{\otimes n}$ coincides with

$$p(x) = \frac{1}{(2\pi)^{n/2}} e^{-\sum_{j=1}^n x_j^2/2} = \frac{1}{(2\pi)^{n/2}} e^{-|x|^2/2} \, ,$$

where $|x| = \left(\sum_{j=1}^n x_j^2 \right)^{1/2}$ denotes the Euclidean distance of the vector x to 0 (compare Section A.4).

Definition 1.9.16. The probability measure $\mathcal{N}(0, 1)^{\otimes n}$ on $\mathcal{B}(\mathbb{R}^n)$ is called the *n*-dimensional or **multivariate standard normal distribution**. It is described by

$$\mathcal{N}(0, 1)^{\otimes n}(B) = \frac{1}{(2\pi)^{n/2}} \int_B e^{-|x|^2/2} dx \, .$$

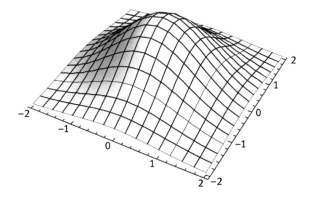

Figure 1.10: The density of the two-dimensional standard normal distribution.

Example 1.9.17. Finally we describe the n-fold product measure of the normal distribution $\mathcal{N}(\mu, \sigma^2)$ with $\mu \in \mathbb{R}$ and $\sigma^2 > 0$. The densities are

$$p_j(x_j) = \frac{1}{\sqrt{2\pi}\,\sigma}\, e^{-(x_j-\mu)^2/2\sigma^2},$$

hence, as in Example 1.9.15, setting with $\bar{\mu} = (\mu, \dots, \mu) \in \mathbb{R}^n$, the product $\mathcal{N}(\mu, \sigma^2)^{\otimes n}$ may be represented as

$$\mathcal{N}(\mu, \sigma^2)^{\otimes n}(B) = \frac{1}{(2\pi)^{n/2}\sigma^n} \int_B e^{-|x-\bar{\mu}|^2/2\sigma^2}\, dx, \quad B \in \mathcal{B}(\mathbb{R}^n). \tag{1.75}$$

1.10 Problems

Problem 1.1. Let A, B, and C be three events in a sample space Ω. Express the following events in terms of these sets:

- Only A occurs.
- At least one of the three events occurs.
- At most one of the three events occurs.
- Exactly two of the events occur.

- A and B occur, but C does not.
- At least two of the events occur.
- None of the events occurs.
- Not more than two of the events occur.

Problem 1.2. Suppose an urn contains black and white balls. Successively one draws n balls out of the urn. The event A_j occurs if the ball drawn in the jth trial is white. Hereby $1 \le j \le n$. Express the following events $B_1, \dots, B_4$ in terms of the A_js:

$B_1 = \{$All drawn balls are white$\}$

$B_2 = \{$At least one of the balls is white$\}$

$B_3 = \{$Exactly one of the drawn balls is white$\}$

$B_4 = \{$All n balls possess the same color$\}$

Determine the cardinalities $\#(B_j), j = 1, \ldots, 4.$

Problem 1.3. Let $\mathbb{P}$ be a probability measure on $(\Omega, \mathcal{A})$. Given $A, B \in \mathcal{A}$ show that

$$\mathbb{P}(A \Delta B) = \mathbb{P}(A) + \mathbb{P}(B) - 2\mathbb{P}(A \cap B).$$

Problem 1.4. The events A and B possess the probabilities $\mathbb{P}(A) = 1/3$ and $\mathbb{P}(B) = 1/4$. Moreover, we know that $\mathbb{P}(A \cap B) = 1/6$. Compute $\mathbb{P}(A^c)$, $\mathbb{P}(A^c \cup B)$, $\mathbb{P}(A \cup B^c)$, $\mathbb{P}(A \cap B^c)$, $\mathbb{P}(A \Delta B)$, and $\mathbb{P}(A^c \cup B^c)$.

Problem 1.5 (Inclusion–exclusion formula). Let $(\Omega, \mathcal{A}, \mathbb{P})$ be a probability space and let $A_1, \ldots, A_n \in \mathcal{A}$ be some (not necessarily disjoint) events. Prove that

$$\mathbb{P}\left(\bigcup_{j=1}^{n} A_j\right) = \sum_{k=1}^{n} (-1)^{k+1} \sum_{1 \leq j_1 < \cdots < j_k \leq n} \mathbb{P}(A_{j_1} \cap \cdots \cap A_{j_k}).$$

Hint: One way to prove this is by induction over n, thereby using Proposition 1.2.3.

Problem 1.6. Use Problem 1.5 to investigate the following question: The numbers from 1 to n are ordered randomly. All orderings are equally likely. What is the probability that there exists an integer $m \leq n$ so that m is at position m of the ordering? Determine the limit of this probability as $n \to \infty$.

Still another version of this problem. Suppose n persons attend a Christmas party. Each of the n participants brings a present with him. These presents are collected, mixed, and then randomly distributed among the guests. Compute the probability that at least one of the participants gets his own present.

Problem 1.7. Suppose in an urn are N balls; k are white, l are red, and m are black. Thus, $k + l + m = N$. Choose n balls out of the urn. Find a formula for the probability that among the n chosen balls are those of all three colors. Investigate this problem if
1. the chosen ball is always replaced and
2. if $n \leq N$ and the balls are not replaced.

Hint: If A is the event that all three colors appear then compute $\mathbb{P}(A^c)$. To this end write $A^c = A_1 \cup A_2 \cup A_3$ with suitable A_js and apply Proposition 1.2.4.

Problem 1.8. Suppose events A and B occur both with probability 1/2. Prove that then

$$\mathbb{P}(A \cup B) = \mathbb{P}(A^c \cup B^c). \tag{1.76}$$

Does (1.76) remain valid assuming $\mathbb{P}(A) + \mathbb{P}(B) = 1$ instead of $\mathbb{P}(A) = \mathbb{P}(B) = \frac{1}{2}$?

Problem 1.9. Three men and three women sit down randomly on six chairs in a row. Find the probability that the three men and the three women sit side by side. What is the probability that next to each woman sits a man (to the right or to the left)?

Problem 1.10. Let $(\Omega, \mathcal{A}, \mathbb{P})$ be a probability space. Prove the following: Whenever events $A_1, A_2, \ldots$ in $\mathcal{A}$ satisfy $\mathbb{P}(A_1) = \mathbb{P}(A_2) = \cdots = 1$, then this implies

$$\mathbb{P}\left(\bigcap_{j=1}^{\infty} A_j\right) = 1.$$

Problem 1.11. (Paradox of Chevalier de Méré). Chevalier de Méré mentioned that when rolling three fair nondistinguishable dice there are 6 different possibilities for obtaining either 11 or 12 as the sum. Thus he concluded that both events (sum equals 11 or sum equals 12) should be equally likely. But experiments showed that this is not the case. Why he was wrong and what are the correct probabilities for both events?

Problem 1.12. A man has forgotten an important phone number. He only remembers that the seven-digit number contained three times "1" and "4" and "6" twice each. He dials the seven numbers in random order. Find the probability that he dialed the correct one.

Problem 1.13. In an urn are n black and m red balls. One draws successively **all** $n + m$ balls (without replacement). What is the probability that the ball chosen last is red?

Problem 1.14. A man has in his pocket n keys to open a door. Only one of the keys fits. He tries the keys one after the other until he has chosen the correct one. Given an integer k compute the probability that the correct key is the one chosen in the kth trial.
   Evaluate this probability in each of the two following cases:

- The man always discards wrong keys.
- The man does not discard them, that is, he puts back wrong keys.

Problem 1.15 (Monty Hall problem). At the end of a quiz the winner has the choice between three doors, say A, B, and C. Behind two of the doors there is a goat, behind the third one a car. His prize is what is behind the chosen door.
   Say the winner has chosen door A. Then the quizmaster (who knows what is behind each of the three doors) opens one of the two remaining doors (in our case either

door B or door C) and shows that there is a goat behind it. After that the quizmaster asks the candidate whether or not he wants to revise his decision, that is, for example, if B was opened, to switch from A to C, or if he furthermore chooses door A.

Find the probabilities to win the car in both cases (switching or nonswitching).

Problem 1.16. In a lecture room are N students. Evaluate the probability that at least two of the students were born at the same day of a year (day and month of their births are the same, but not necessarily the year). Hereby disregard leap years and assume that all days in a year are equally likely. How big must N be in order that this probability is greater than 1/2?

Problem 1.17. In an urn are balls labeled from 0 to 6 so that all numbers are equally likely. Choose successively and with replacement three balls. Find the probability that the three observed numbers sum up to 6.

Problem 1.18. When sending messages from A to B on average 3% are transmitted falsely. Suppose 300 messages are sent. What is the probability that at least three messages are transmitted falsely? Evaluate the exact probability by using the binomial distribution as well as the approximate probability by using the Poisson distribution. Compute the probability (exact and approximative one) that all messages arrive correctly.

Problem 1.19. The number of accidents in a city per week is assumed to be Poisson distributed with parameter 5. Find the probability that next week there will be either two or three accidents. How likely is that there will be no accidents?

Problem 1.20. In a room are 12 men and 8 women. One randomly chooses 5 of the 20 persons. Given $k \in \{0, \ldots, 5\}$, what is the probability that among the five chosen are exactly k women? How likely is it that among the five persons are more women than men?

Problem 1.21. Two players A and B take turns rolling a die. The first to roll a "6" wins. Player A starts. Find the probability that A wins. Suppose now there is a third player C and the order of rolling the die is given by $ABCABCA \cdots$. Find each players probability of winning.

Problem 1.22. Two players, say A and B, toss a biased coin where "head" appears with probability $0 < p < 1$. Winner is who gets the first "head". A starts, then B tosses twice, then again A once, B twice, and so on. Determine the number p for which the game is fair, that is, the probability that A (or B) wins is 1/2.

Problem 1.23. In an urn are 50 white and 200 red balls.

(1) Take out 10 balls **with** replacement. What is the probability to observe four white balls? Give the exact value via the binomial distribution as well as the approximated one using the related Poisson distribution.
(2) Next choose 10 balls **without** replacement. What is the probability to get four white balls in this case?
(3) The number of balls in the urn is as above. But now we choose the balls with replacement until for the first time a white ball shows up. Find the probability of the following events:
  (a) The first white ball shows up in the fourth trial.
  (b) The first white ball appears strictly after the third trial.
  (c) The first white ball is observed in an even number of trials, that is, in the second or in the fourth or in the sixth, and so on trial.

Problem 1.24. Place successively and independently four particles into five boxes. Thereby each box is equally likely. Find the probabilities of the following events:
$A := \{$Each box contains at most one particle$\}$ and $B := \{$All 4 particles are in the same box$\}$.

Problem 1.25. Investigate the following generalization of Example 1.4.44: in urn U_0 are M balls and in urn U_1 are N balls for some $N, M \geq 1$. Choose U_0 with probability $1-p$ and U_1 with probability p, and take out a ball from the chosen urn. Given $1 \leq m \leq M$, find the probability that there are m balls left in U_0 when choosing the last ball out of U_1. How do these probabilities change when $1 \leq m \leq N$, and we assume that there are m balls in U_1 when choosing the last ball from U_0 ?

Problem 1.26. Use properties of the Γ-function to compute for $n \in \mathbb{N}$

$$\int_0^\infty x^{2n} e^{-x^2/2} dx \quad \text{and} \quad \int_0^\infty x^{2n+1} e^{-x^2/2} dx.$$

Problem 1.27. Prove formula (1.59) that relates the beta and the Γ-function.
  Hint: Start with

$$\Gamma(x)\Gamma(y) = \int_0^\infty \int_0^\infty u^{x-1} v^{y-1} e^{-u-v} du dv$$

and change the variables as follows: $u = f(z,t) = zt$ and $v = g(z,t) = z(1-t)$, where $0 \leq z < \infty$ and $0 \leq t \leq 1$.

Problem 1.28. Prove that for $0 \leq k \leq n$

$$\binom{n}{k} = \frac{1}{(n+1)B(n-k+1, k+1)}$$

where $B(\cdot, \cdot)$ denotes Euler's beta function (cf. formula (1.58)).

Problem 1.29. Write $x \in [0, 1)$ as decimal fraction $x = 0.x_1x_2 \cdots$ with $x_j \in \{0, \dots, 9\}$. Let

$$A_j = \{x \in [0, 1) : x_j = 1\}.$$

If $\mathbb{P}$ denotes the uniform distribution on $[0, 1]$, compute $\mathbb{P}(A_j)$ as well as $\mathbb{P}\left(\bigcap_{j=1}^{\infty} A_j\right)$.

Compute the same probabilities if $A_j = \{x \in [0, 1) : x_j = m\}$ for some fixed $m \in \{0, \dots 9\}$.

Problem 1.30. Compute the distribution function of the Cauchy distribution (cf. Definition 1.6.33).

Problem 1.31. Let $F : \mathbb{R} \to [0, 1]$ be the distribution function of a probability measure. Show that F possesses at most countably many points of discontinuity. Conclude from this and Proposition 1.7.12 the following: If $\mathbb{P}$ is a probability measure on $\mathcal{B}(\mathbb{R})$, then there are at most countably infinite many $t \in \mathbb{R}$ such that $\mathbb{P}(\{t\}) > 0$.

Problem 1.32. Let Φ be the distribution function of the standard normal distribution introduced in eq. (1.62). Show the following properties of Φ.
1. For $t \in \mathbb{R}$ holds $\Phi(-t) = 1 - \Phi(t)$.
2. If $a > 0$, then

$$\mathcal{N}(0, 1)([-a, a]) = 2\Phi(a) - 1.$$

3. Prove formulas (1.63), that is,

$$\Phi(t) = \frac{1}{2}\left[1 + \mathrm{erf}\left(\frac{t}{\sqrt{2}}\right)\right] \quad \text{and} \quad \mathrm{erf}(t) = 2\Phi(\sqrt{2}t) - 1, \quad t \in \mathbb{R}.$$

4. Compute

$$\lim_{t \to \infty} \frac{1 - \Phi(t)}{t^{-1}e^{-t^2/2}}.$$

Hint: Use l'Hôpitale's rule.

Problem 1.33 (Bertrand paradox). Consider an equilateral triangle inscribed in a circle of radius $r > 0$. Suppose a chord of the circle is chosen at random. What is the probability that the chord is longer than a side of the triangle?

In this form the problem allows different answers. Why? Because we did not define in which way the random chord is chosen.
1. The "random endpoints" method: Choose independently two uniformly distributed random points on the circumference of the circle and draw the chord joining them.

2. The "random radius" method: Choose a radius of the circle, that is, choose a random angle in $[0, 2\pi]$, choose independently a point on the radius according to the uniform distribution on $[0, r]$, and construct the chord through this point and perpendicular to the radius.
3. The "random midpoint" method: Choose a point within the circle according to the uniform distribution on the circle and construct a chord with the chosen point as its midpoint.

Answer the above question about the length of the chord in each of the three cases.

Problem 1.34. A stick of length $L > 0$ is randomly broken into three pieces. Hereby we assume that both points of break are uniformly distributed on $[0, L]$ and independent of each other. What is the probability that these three parts piece together to a triangle?

2 Conditional Probabilities and Independence

2.1 Conditional Probabilities

In order to motivate the definition of conditional probabilities, let us start with the following easy example.

Example 2.1.1. Roll a fair die twice. The probability of the event "sum of both rolls equals 5" is 1/9. Suppose now we were told that the first roll was an even number. Does this additional information make the event "sum equals 5" more likely? Or does it even diminish the probability of its occurrence? To answer this question, we apply the so-called technique of "restricting the sample space." Since we know that the event B = {First roll is even} had occurred, we may rule out elements in B^c and restrict our sample space. Choose B as new sample space. Its cardinality is 18. Moreover, under this condition, an event A occurs if and only if $A \cap B$ does so. Hence, the "new" probability of A under condition B, written $\mathbb{P}(A|B)$, is given by

$$\mathbb{P}(A|B) = \frac{\#(A \cap B)}{\#(B)} = \frac{\#(A \cap B)}{18} . \tag{2.1}$$

In the question above, we asked for $\mathbb{P}(A|B)$, where

$$A = \{\text{Sum of both rolls equals } 5\} = \{(1, 4), (2, 3), (3, 2), (4, 1)\} .$$

Since $A \cap B = \{(2, 3), (4, 1)\}$, we obtain $\mathbb{P}(A|B) = 2/18 = 1/9$. Consequently, in this case, condition B does not change the probability of the occurrence of A.

Define now A as a set of pairs adding to 6. Then $\mathbb{P}(A) = 5/36$, while the conditional probability remains 1/9. Note that now $A \cap B = \{(2, 4), (4, 2)\}$. Thus, in this case, condition B makes the occurrence of A less likely.

Before we state the definition of conditional probabilities in the general case, let us rewrite eq. (2.1) as follows:

$$\mathbb{P}(A|B) = \frac{\#(A \cap B)}{\#(B)} = \frac{\#(A \cap B)/36}{\#(B)/36} = \frac{\mathbb{P}(A \cap B)}{\mathbb{P}(B)} . \tag{2.2}$$

Equation (2.2) gives us a hint to introduce conditional probabilities in the general setting.

Definition 2.1.2. Let $(\Omega, \mathcal{A}, \mathbb{P})$ be a probability space. Given events $A, B \in \mathcal{A}$ with $\mathbb{P}(B) > 0$, the **probability of A under condition B** is defined by

$$\mathbb{P}(A|B) = \frac{\mathbb{P}(A \cap B)}{\mathbb{P}(B)} . \tag{2.3}$$

Remark 2.1.3. If we know the values of $\mathbb{P}(A \cap B)$ and $\mathbb{P}(B)$, then formula (2.3) allows us to evaluate $\mathbb{P}(A|B)$. Sometimes, it happens that we know the values of $\mathbb{P}(B)$ and $\mathbb{P}(A|B)$ and want to calculate $\mathbb{P}(A \cap B)$. In order to do this, we rewrite eq. (2.3) as

$$\mathbb{P}(A \cap B) = \mathbb{P}(B)\,\mathbb{P}(A|B). \qquad (2.4)$$

In this way, we get the desired value of $\mathbb{P}(A \cap B)$. Formula (2.4) is called the **law of multiplication**.

The next two examples show how this law applies.

Example 2.1.4. In an urn are two white and two black balls. Choose two balls without replacing the first one. We want to evaluate the probability of occurrence of a black ball in the first draw **and** of a white in the second one. Let us first find a suitable mathematical model that describes this experiment. The sample space is given by $\Omega = \{(b, b), (b, w), (w, b), (w, w)\}$, and we regard the events

$$A := \{\text{Second ball is white}\} = \{(b, w), (w, w)\} \quad \text{as well as}$$
$$B := \{\text{First ball is black}\} = \{(b, b), (b, w)\}.$$

The event of interest is then $A \cap B = \{(b, w)\}$.

Which probabilities can be directly determined? Of course, the probability of occurrence of B equals 1/2 because the number of white and black balls is the same. Furthermore, if B occurred, then in the urn remained two white balls and one black ball. Under this condition, event A occurs with probability 2/3, that is, $\mathbb{P}(A|B) = 2/3$. Using eq. (2.4), we obtain

$$\mathbb{P}(\{(b, w)\}) = \mathbb{P}(A \cap B) = \mathbb{P}(B) \cdot \mathbb{P}(A|B) = \frac{1}{2} \cdot \frac{2}{3} = \frac{1}{3}.$$

Example 2.1.5. Among three non-distinguishable coins are two fair and one is biased. Tossing the biased coin "head" appears with probability 1/3, hence "tail" appears with probability 2/3. We choose by random one of the three coins and toss it. Find the probability to observe "tail" at the biased coin.

To solve this problem, let us first mention that the sample space $\Omega = \{H, T\}$ is not adequate to describe that experiment. Why? Because the event $\{H\}$ may have different probabilities depending on occurrence at a biased or at a fair coin. We have to distinguish between the appearance of "head" or "tail" at the different types of coins. Hence, an adequate choice of the sample space is

$$\Omega := \{(H, B), (T, B), (H, F), (T, F)\}.$$

Here, B stands for biased and F assigns that the coin was fair. The event of interest is $\{(T, B)\}$. Set

$$T := \{(T, B), (T, F)\} \quad \text{as well as} \quad B := \{(H, B), (T, B)\} .$$

Then T occurs if "tail" appears regardless of the type of the coin while B occurs if we have chosen the biased coin. Of course, it follows that $\{(T, B)\} = T \cap B$. Since only one of the three coins is biased, we have $\mathbb{P}(B) = 1/3$. By assumption $\mathbb{P}(T|B) = 2/3$, hence an application of eq. (2.4) leads to

$$\mathbb{P}(\{(T, B)\}) = \mathbb{P}(B) \, \mathbb{P}(T|B) = \frac{1}{3} \cdot \frac{2}{3} = \frac{2}{9} .$$

Next, we present two examples where formula (2.3) applies directly.

Example 2.1.6. Roll a die twice. One already knows that the first number is not "6." What is the probability that the sum of both rolls is greater than or equal to "10?"

Answer: The model for this experiment is $\Omega = \{1, \ldots, 6\}^2$ endowed with the uniform distribution $\mathbb{P}$ on $\mathcal{P}(\Omega)$. The event $B := \{\text{First result is not "6"}\}$ contains the 30 elements

$$\{(1, 1), \ldots, (5, 1), \ldots, (1, 6), \ldots, (5, 6)\},$$

and if A consists of pairs with sum equal to or larger than 10, then

$$A = \{(4, 6), (5, 6), (6, 6), (5, 5), (6, 5), (6, 4)\} , \quad \text{hence } A \cap B = \{(4, 6), (5, 6), (5, 5)\} .$$

Therefore, it follows

$$\mathbb{P}(A|B) = \frac{\mathbb{P}(A \cap B)}{\mathbb{P}(B)} = \frac{3/36}{30/36} = \frac{1}{10} .$$

In the case that all elementary events are equally likely, there exists a more direct way to evaluate $\mathbb{P}(A|B)$. We reduce the sample space as we already did in Example 2.1.1.

Proposition 2.1.7 (Reduction of the sample space). *Suppose the sample space Ω is finite and let $\mathbb{P}$ be the uniform distribution on $\mathcal{P}(\Omega)$. Then for all events A and non-empty B in Ω, we have*

$$\mathbb{P}(A|B) = \frac{\#(A \cap B)}{\#(B)} . \tag{2.5}$$

Proof: This easily follows from

$$\mathbb{P}(A|B) = \frac{\mathbb{P}(A \cap B)}{\mathbb{P}(B)} = \frac{\#(A \cap B)/\#(\Omega)}{\#(B)/\#(\Omega)} = \frac{\#(A \cap B)}{\#(B)} . \qquad \blacksquare$$

Example 2.1.8. We want to investigate Example 2.1.6 once more, this time using formula (2.5) directly. Since $\#(A \cap B) = 3$ and $\#(B) = 30$, we get as before

$$\mathbb{P}(A|B) = \frac{\#(A \cap B)}{\#(B)} = \frac{3}{30} = \frac{1}{10}.$$

Remark 2.1.9. It is important to state that Proposition 2.1.7 becomes false for general probabilities $\mathbb{P}$ on $\mathcal{P}(\Omega)$. Formula (2.5) is **only valid in the case that $\mathbb{P}$ is the uniform distribution** on $\mathcal{P}(\Omega)$.

Example 2.1.10. The duration of a telephone call is exponentially distributed with parameter $\lambda > 0$. Find the probability that a call does not last more than 5 minutes provided it already lasted 2 minutes.
Solution: Let A be the event that the call does not last more than 5 minutes, that is, $A = [0, 5]$. We know it already lasted 2 minutes, hence event $B = [2, \infty)$ has occurred. Thus, under condition B, it follows

$$E_\lambda(A|B) = \frac{E_\lambda(A \cap B)}{E_\lambda(B)} = \frac{E_\lambda([2, 5])}{E_\lambda([2, \infty))} = \frac{e^{-2\lambda} - e^{-5\lambda}}{e^{-2\lambda}} = 1 - e^{-3\lambda}.$$

Note the interesting fact that this conditional probability equals $E_\lambda([0, 3])$. What does this tell us? It says that the probability that a call lasts no more than another 3 minutes is independent of the fact that it already lasted 2 minutes. This means that the duration of a call did not "become older." Independent of the fact that it already lasted 2 minutes, the probability for talking no more than another 3 minutes remains the same.

Let us come back to the general case. Fix an event $B \in \mathcal{A}$ with $\mathbb{P}(B) > 0$. Then

$$A \mapsto \mathbb{P}(A|B), \quad A \in \mathcal{A},$$

is a well-defined mapping from $\mathcal{A}$ to $[0, 1]$. Its main properties are summarized in the next proposition.

Proposition 2.1.11. *Let $(\Omega, \mathcal{A}, \mathbb{P})$ be an arbitrary probability space. Then for each $B \in \mathcal{A}$ with $\mathbb{P}(B) > 0$, the mapping $A \mapsto \mathbb{P}(A|B)$ is a probability measure on $\mathcal{A}$. It is concentrated on B, that is,*

$$\mathbb{P}(B|B) = 1 \quad \text{or, equivalently,} \quad \mathbb{P}(B^c|B) = 0.$$

Proof: Of course, one has

$$\mathbb{P}(\emptyset|B) = \mathbb{P}(\emptyset \cap B)/\mathbb{P}(B) = 0 \text{ and } \mathbb{P}(\Omega|B) = \mathbb{P}(\Omega \cap B)/\mathbb{P}(B) = \mathbb{P}(B)/\mathbb{P}(B) = 1.$$

Thus, it remains to prove that $\mathbb{P}(\cdot|B)$ is σ-additive. To this end, choose disjoint $A_1, A_2, \ldots$ in $\mathcal{A}$. Then also $A_1 \cap B, A_2 \cap B, \ldots$ are disjoint and using the σ-additivity of $\mathbb{P}$ leads to

$$\mathbb{P}\left(\bigcup_{j=1}^{\infty} A_j \middle| B\right) = \frac{\mathbb{P}\left(\left[\bigcup_{j=1}^{\infty} A_j\right] \cap B\right)}{\mathbb{P}(B)} = \frac{\mathbb{P}\left(\bigcup_{j=1}^{\infty}(A_j \cap B)\right)}{\mathbb{P}(B)}$$

$$= \frac{\sum_{j=1}^{\infty} \mathbb{P}(A_j \cap B)}{\mathbb{P}(B)} = \sum_{j=1}^{\infty} \frac{\mathbb{P}(A_j \cap B)}{\mathbb{P}(B)} = \sum_{j=1}^{\infty} \mathbb{P}(A_j | B).$$

Consequently, as asserted, $\mathbb{P}(\cdot \,|B)$ is a probability. Since the identity $\mathbb{P}(B|B) = 1$ is obvious, this ends the proof. ∎

Definition 2.1.12. The mapping $\mathbb{P}(\cdot \,|B)$ is called **conditional probability** or also **conditional distribution** (under condition B).

Remark 2.1.13. The main advantage of Proposition 2.1.11 is that it implies that conditional probabilities share all the properties of "ordinary" probability measures. For example, it holds

$$\mathbb{P}(A_2 \backslash A_1 | B) = \mathbb{P}(A_2 | B) - \mathbb{P}(A_1 | B) \quad \text{provided that} \quad A_1 \subseteq A_2$$

or

$$\mathbb{P}(A_1 \cup A_2 | B) = \mathbb{P}(A_1 | B) + \mathbb{P}(A_2 | B) - \mathbb{P}(A_1 \cap A_2 | B).$$

We come now to the so-called law of total probability. It allows us to evaluate the probability of an event A knowing only its conditional probabilities $\mathbb{P}(A|B_j)$ for certain $B_j \in \mathcal{A}$. More precisely, the following is valid.

Proposition 2.1.14 (Law of total probability). *Let $(\Omega, \mathcal{A}, \mathbb{P})$ be a probability space and let $B_1, \ldots, B_n$ in $\mathcal{A}$ be disjoint with $\mathbb{P}(B_j) > 0$ and $\bigcup_{j=1}^{n} B_j = \Omega$. Then for each $A \in \mathcal{A}$ holds*

$$\mathbb{P}(A) = \sum_{j=1}^{n} \mathbb{P}(B_j)\, \mathbb{P}(A|B_j). \tag{2.6}$$

Proof: Let us start with the investigation of the right-hand side of eq. (2.6). By the definition of the conditional probability, this expression may be rewritten as

$$\sum_{j=1}^{n} \mathbb{P}(B_j)\, \mathbb{P}(A|B_j) = \sum_{j=1}^{n} \mathbb{P}(B_j)\, \frac{\mathbb{P}(A \cap B_j)}{\mathbb{P}(B_j)} = \sum_{j=1}^{n} \mathbb{P}(A \cap B_j). \tag{2.7}$$

The sets $B_1, \ldots, B_n$ are disjoint, hence so are $A \cap B_1, \ldots, A \cap B_n$. Thus, the finite additivity of $\mathbb{P}$ implies

$$\sum_{j=1}^{n} \mathbb{P}(A \cap B_j) = \mathbb{P}\left(\bigcup_{j=1}^{n}(A \cap B_j) \right) = \mathbb{P}\left(\left(\bigcup_{j=1}^{n} B_j \right) \cap A \right) = \mathbb{P}(\Omega \cap A) = \mathbb{P}(A).$$

Together with eq. (2.7), this proves eq. (2.6). ∎

Example 2.1.15. A fair coin is tossed four times. Suppose we observe exactly k times "heads" for some $k = 0, \ldots, 4$. According to the observed k, we take k dice and roll them. Find the probability that number "6" does not appear. Note that $k = 0$ means that we do not roll a die, hence in this case "6" cannot appear.

Solution: As sample space, we choose $\Omega = \{(k, Y), (k, N) : k = 0, \ldots, 4\}$, where (k, Y) means that we rolled k dice and at least at one of them we got a "6." In the same way (k, N) stands for k dice and no "6." Let $N = \{(0, N), \ldots, (4, N)\}$ and $B_k = \{(k, Y), (k, N)\}$, $k = 0, \ldots, 4$. Then B_k occurs if we observed k "heads." The conditional probabilities equal

$$\mathbb{P}(N|B_0) = 1, \;\; \mathbb{P}(N|B_1) = 5/6, \ldots, \mathbb{P}(N|B_4) = (5/6)^4,$$

while

$$\mathbb{P}(B_k) = \binom{4}{k} \frac{1}{2^4}, \quad k = 0, \ldots, 4.$$

The events $B_0, \ldots, B_4$ satisfy the assumptions of Proposition 2.1.14, thus eq. (2.6) applies and leads to

$$\mathbb{P}(A) = \frac{1}{2^4} \sum_{k=0}^{4} \binom{4}{k} (5/6)^k = \frac{1}{2^4} \left(\frac{5}{6} + 1 \right)^4 = \left(\frac{11}{12} \right)^4 = 0.706066743.$$

Example 2.1.16. Three different machines, M_1, M_2 and M_3, produce light bulbs. In a single day, M_1 produces 500 bulbs, M_2 200 and M_3 100. The quality of the produced bulbs depends on the machines: Among the light bulbs produced by M_1 are 5% defective, M_2 10% and M_3 only 2%. At the end of a day, a controller chooses by random one of the 800 produced light bulbs and tests it. Determine the probability that the checked bulb is defective.

Solution: The probabilities that the checked bulb was produced by M_1, M_2 or M_3 are 5/8, 1/4 and 1/8, respectively. The conditional probabilities for choosing a defective bulb produced by M_1, M_2 or M_3 were given as 1/20, 1/10 and 1/50, respectively. If D is the event that the tested bulb was defective, then the law of total probability yields

$$\mathbb{P}(D) = \frac{5}{8} \cdot \frac{1}{20} + \frac{1}{4} \cdot \frac{1}{10} + \frac{1}{8} \cdot \frac{1}{50} = \frac{47}{800} = 0.05875.$$

Let us look at Example 2.1.16 from a different point of view. When choosing a light bulb out of the 800 produced, there were certain fixed probabilities whether it was produced by M_1, M_2 or M_3, namely with probabilities 5/8, 1/4 and 1/8. These are the probabilities **before** checking a bulb. Therefore, they are called *a priori* probabilities. After checking a bulb, we obtained the additional information that it was defective. Does this additional information change the probabilities which of the M_1, M_2 or M_3 produced it? More precisely, if as above D occurs if the tested bulb is defective, then we now ask for the conditional probabilities $\mathbb{P}(M_1|D)$, $\mathbb{P}(M_2|D)$ and $\mathbb{P}(M_3|D)$. To understand that these probabilities may differ considerably from the *a priori* probabilities, imagine that, for example, M_1 produces almost no defective bulbs. Then it will be very unlikely that the tested bulb has been produced by M_1, although $\mathbb{P}(M_1)$ may be big.

Because $\mathbb{P}(M_1|D)$, $\mathbb{P}(M_2|D)$ and $\mathbb{P}(M_3|D)$ are the probabilities **after** executing the random experiment (choosing and testing the bulb), they are called *a posteriori* probabilities.

Let us now introduce the exact and general definition of *a priori* and *a posteriori* probabilities.

Definition 2.1.17. Suppose there is a probability space $(\Omega, \mathcal{A}, \mathbb{P})$ and there are disjoint events $B_1, \ldots, B_n \in \mathcal{A}$ satisfying $\Omega = \bigcup_{j=1}^{n} B_j$. Then we call $\mathbb{P}(B_1), \ldots, \mathbb{P}(B_n)$ the **a priori** probabilities of $B_1, \ldots, B_n$. Let $A \in \mathcal{A}$ with $\mathbb{P}(A) > 0$ be given. Then the conditional probabilities $\mathbb{P}(B_1|A), \ldots, \mathbb{P}(B_n|A)$ are said to be the **a posteriori** probabilities, that is, those after the occurrence of A.

To calculate the *a posteriori* probabilities, the next proposition turns out to be very useful.

Proposition 2.1.18 (Bayes' formula). *Suppose we are given disjoint events B_1 to B_n satisfying $\bigcup_{j=1}^{n} B_j = \Omega$ and $\mathbb{P}(B_j) > 0$. Let A be an event with $\mathbb{P}(A) > 0$. Then for each $j \leq n$ the following equation holds:*

$$\mathbb{P}(B_j|A) = \frac{\mathbb{P}(B_j)\,\mathbb{P}(A|B_j)}{\sum_{i=1}^{n}\mathbb{P}(B_i)\mathbb{P}(A|B_i)}\,. \tag{2.8}$$

Proof: Proposition 2.1.14 implies

$$\sum_{i=1}^{n}\mathbb{P}(B_i)\mathbb{P}(A|B_i) = \mathbb{P}(A)\,.$$

Hence, the right-hand side of eq. (2.8) may also be written as

$$\frac{\mathbb{P}(B_j)\mathbb{P}(A|B_j)}{\mathbb{P}(A)} = \frac{\mathbb{P}(B_j)\frac{\mathbb{P}(A\cap B_j)}{\mathbb{P}(B_j)}}{\mathbb{P}(A)} = \frac{\mathbb{P}(A\cap B_j)}{\mathbb{P}(A)} = \mathbb{P}(B_j|A)$$

and the proposition is proven. ∎

Remark 2.1.19. In the case $\mathbb{P}(A)$ is already known, Bayes' formula simplifies to

$$\mathbb{P}(B_j|A) = \frac{\mathbb{P}(B_j)\mathbb{P}(A|B_j)}{\mathbb{P}(A)}, \quad j = 1, \dots, n. \tag{2.9}$$

Remark 2.1.20. Let us treat the special case of two sets partitioning Ω. If $B_1 = B$, then necessarily $B_2 = B^c$, hence $\Omega = B \cup B^c$. Then formula (2.8) looks as follows:

$$\mathbb{P}(B|A) = \frac{\mathbb{P}(B)\mathbb{P}(A|B)}{\mathbb{P}(B)\mathbb{P}(A|B) + \mathbb{P}(B^c)\mathbb{P}(A|B^c)} \tag{2.10}$$

and

$$\mathbb{P}(B^c|A) = \frac{\mathbb{P}(B^c)\mathbb{P}(A|B^c)}{\mathbb{P}(B)\mathbb{P}(A|B) + \mathbb{P}(B^c)\mathbb{P}(A|B^c)}. \tag{2.11}$$

Again, if the probability of A is known, the denominators in eqs. (2.10) and (2.11) may be replaced by $\mathbb{P}(A)$.

Example 2.1.21. Let us use Bayes' formula to calculate the *a posteriori* probabilities in Example 2.1.16. Recall that D occurred if the tested bulb was defective. We already know $\mathbb{P}(D) = 47/800$, hence we may apply eq. (2.9). Doing so, we get

$$\mathbb{P}(M_1|D) = \frac{\mathbb{P}(M_1)\mathbb{P}(D|M_1)}{\mathbb{P}(D)} = \frac{5/8 \cdot 1/20}{47/800} = 25/47$$

$$\mathbb{P}(M_2|D) = \frac{\mathbb{P}(M_2)\mathbb{P}(D|M_2)}{\mathbb{P}(D)} = \frac{1/4 \cdot 1/10}{47/800} = 20/47$$

$$\mathbb{P}(M_3|D) = \frac{\mathbb{P}(M_3)\mathbb{P}(D|M_3)}{\mathbb{P}(D)} = \frac{1/8 \cdot 1/50}{47/800} = 2/47.$$

By assignment of the problem, the *a priori* probabilities were given by $\mathbb{P}(M_1) = 5/8$, $\mathbb{P}(M_2) = 1/4$ and $\mathbb{P}(M_3) = 1/8$. In the case that the tested light bulb was defective, these probabilities change to 25/47, 20/47 and 2/47. This tells us that it becomes less likely that the tested bulb was produced by M_1 or M_3; their probabilities diminish by 0.0930851 and 0.0824468, respectively. On the other hand, the probability of M_2 increases by 0.175532.

Finally, note that Proposition 2.1.11 implies that the sum of the *a posteriori* probabilities has to be 1. Because of $25/47 + 20/47 + 2/47 = 1$, this is true in that example.

Example 2.1.22. In order to figure out whether or not a person suffers from a certain disease, say disease X, a test is assumed to give a clue. If the tested person is sick, then the test is positive in 96% of cases. If the person is well, then with 94% accuracy the test will be negative. Furthermore, it is known that 0.4% of the population suffers from X.

Now a person, chosen by random, is tested. Suppose the result was positive. Find the probability that this person really suffers from X.

Solution: As sample space, we may choose $\Omega = \{(X, p), (X, n), (X^c, p), (X^c, n)\}$, where, for example, (X, n) means the person suffers from X and the test was negative. Set $A := \{(X, p), (X^c, p)\}$. Then A occurs if and only if the test turned out to be positive. Furthermore, event $B := \{(X, p), (X, n)\}$ occurs in the case that the tested person suffers from X. Known are

$$\mathbb{P}(A|B) = 0.96, \quad \mathbb{P}(A|B^c) = 0.06 \quad \text{and} \quad \mathbb{P}(B) = 0.004, \quad \text{hence} \quad \mathbb{P}(B^c) = 0.996.$$

Therefore, by eq. (2.10), the probability we asked for can be calculated as follows:

$$\mathbb{P}(B|A) = \frac{\mathbb{P}(B)\mathbb{P}(A|B)}{\mathbb{P}(B)\mathbb{P}(A|B) + \mathbb{P}(B^c)\mathbb{P}(A|B^c)}$$
$$= \frac{0.004 \cdot 0.96}{0.004 \cdot 0.96 + 0.996 \cdot 0.06} = \frac{0.00384}{0.0636} = 0.0603774.$$

That tells us that it is quite unlikely that a randomly chosen person with A positive test is really sick. The chance for this being true is only about 6%.

2.2 Independence of Events

What does it mean that two events are independent or, more precisely, that they occur independently of each other? To get an idea, let us look at the following example.

Example 2.2.1. Roll a fair die twice. Event B occurs if the first number is even while event A consists of all pairs (x_1, x_2), where $x_2 = 5$ or $x_2 = 6$. It is intuitively clear that these two events occur independently of each other. But how to express this mathematically? To answer this question, think about the probability of A under the condition B. The fact whether or not B occurred has no influence on the occurrence of A. For the occurrence or nonoccurrence of A, it is completely insignificant what happened in the first roll. Mathematically this means that $\mathbb{P}(A|B) = \mathbb{P}(A)$. Let us check whether this is true in this concrete case. Indeed, it holds $\mathbb{P}(A) = 1/3$ as well as

$$\mathbb{P}(A|B) = \frac{\mathbb{P}(A \cap B)}{\mathbb{P}(B)} = \frac{6/36}{1/2} = 1/3.$$

The previous example suggests that independence of A of B could be described by

$$\mathbb{P}(A) = \mathbb{P}(A|B) = \frac{\mathbb{P}(A \cap B)}{\mathbb{P}(B)}. \tag{2.12}$$

But formula (2.12) has a disadvantage, namely we have to assume $\mathbb{P}(B) > 0$ to ensure that $\mathbb{P}(A|B)$ exists. To overcome this problem, rewrite eq. (2.12) as

$$\mathbb{P}(A \cap B) = \mathbb{P}(A)\,\mathbb{P}(B). \tag{2.13}$$

In this form, we may take eq. (2.13) as a basis for the definition of independence.

> **Definition 2.2.2.** Let $(\Omega, \mathcal{A}, \mathbb{P})$ be a probability space. Two events A and B in $\mathcal{A}$ are said to be (stochastically) **independent** provided that
>
> $$\mathbb{P}(A \cap B) = \mathbb{P}(A) \cdot \mathbb{P}(B) . \tag{2.14}$$
>
> In the case that eq. (2.14) does not hold, the events A and B are called (stochastically) **dependent**.

Remark 2.2.3. In the sequel, we use the notations "independent" and "dependent" without adding the word "stochastically." Since we will not use other versions of independence, there should be no confusion.

Example 2.2.4. A fair die is rolled twice. Event A occurs if the first roll is either "1" or "2" while B occurs if the sum of both rolls equals 7. Are A and B independent?

Answer: It holds $\mathbb{P}(A) = 1/3$, $\mathbb{P}(B) = 1/6$ as well as $\mathbb{P}(A \cap B) = 2/36 = 1/18$. Hence, we get $\mathbb{P}(A \cap B) = \mathbb{P}(A) \cdot \mathbb{P}(B)$ and A and B are independent.

Question: Are A and B also independent if A is as before and B is defined as a set of pairs with sum 4?

Example 2.2.5. In an urn, there are n, $n \geq 2$, white balls and also n black balls. One chooses two balls without replacing the first one. Let A be the event that the second ball is black while B occurs if the first ball was white. Are A and B independent?

Answer: The probability of B equals $1/2$. To calculate $\mathbb{P}(A)$, we use Proposition 2.1.14. Then we get

$$\mathbb{P}(A) = \mathbb{P}(B)\mathbb{P}(A|B) + \mathbb{P}(B^c)\mathbb{P}(A|B^c) = \frac{1}{2} \cdot \frac{n}{2n-1} + \frac{1}{2} \cdot \frac{n-1}{2n-1} = \frac{1}{2} ,$$

hence, $\mathbb{P}(A) \cdot \mathbb{P}(B) = 1/4$.

On the other hand, we have

$$\mathbb{P}(A \cap B) = \mathbb{P}(B)\mathbb{P}(A|B) = \frac{1}{2} \cdot \frac{n}{2n-1} = \frac{n}{4n-2} \neq \frac{1}{4} .$$

Consequently, A and B are dependent.

Remark 2.2.6. Note that, if $n \to \infty$, then

$$\mathbb{P}(A \cap B) = \frac{n}{4n-2} \to \frac{1}{4} = \mathbb{P}(A)\,\mathbb{P}(B) .$$

This tells us the following: if n is big, then A and B are "almost" independent or, equivalently, the degree of dependence between A and B is very small. This question will be investigated more thoroughly in Chapter 5 when a measure for the degree of dependence is available.

Next, we prove some properties of independent events.

Proposition 2.2.7. *Let $(\Omega, \mathcal{A}, \mathbb{P})$ be a probability space.*
1. *For any $A \in \mathcal{A}$, the events A and $\emptyset$ as well as A and Ω are independent[1].*
2. *If A and B are independent, then so are A and B^c as well as A^c and B^c.*

Proof: We have

$$\mathbb{P}(A \cap \emptyset) = \mathbb{P}(\emptyset) = 0 = \mathbb{P}(A) \cdot 0 = \mathbb{P}(A) \cdot \mathbb{P}(\emptyset),$$

hence, A and $\emptyset$ are independent.

In the same way follows the independence of A and Ω by

$$\mathbb{P}(A \cap \Omega) = \mathbb{P}(A) = \mathbb{P}(A) \cdot 1 = \mathbb{P}(A) \cdot \mathbb{P}(\Omega).$$

To prove the second part, assume that A and B are independent. Our aim is to show that A and B^c are independent as well. We know that

$$\mathbb{P}(A \cap B) = \mathbb{P}(A) \mathbb{P}(B)$$

and we want to show that

$$\mathbb{P}(A \cap B^c) = \mathbb{P}(A) \mathbb{P}(B^c).$$

Let us start with the right-hand side of the last equation. Using the independence of A and B and $A \cap B \subseteq B$, it follows that

$$\mathbb{P}(A) \mathbb{P}(B^c) = \mathbb{P}(A)\big(1 - \mathbb{P}(B)\big) = \mathbb{P}(A) - \mathbb{P}(A) \cdot \mathbb{P}(B)$$
$$= \mathbb{P}(A) - \mathbb{P}(A \cap B) = \mathbb{P}\big(A \backslash (A \cap B)\big). \qquad (2.15)$$

Since $A \backslash (A \cap B) = A \backslash B = A \cap B^c$ from eq. (2.15), we derive

$$\mathbb{P}(A) \cdot \mathbb{P}(B^c) = \mathbb{P}(A \cap B^c).$$

Consequently, as asserted, A and B^c are independent.

If A and B are independent, then so are B and A, and as seen above, so are B and A^c. Another application of the first step, this time with A^c and B shows that also A^c and B^c are independent. This completes the proof. ∎

Suppose we are given n events $A_1, \ldots, A_n$ in $\mathcal{A}$. We want to figure out when they are independent. A first possible approach could be as follows.

---

[1] For a more general result, compare Problem 2.10.

Definition 2.2.8. Events $A_1, \ldots, A_n$ are said to be **pairwise independent** if, whenever $i \neq j$, then

$$\mathbb{P}(A_i \cap A_j) = \mathbb{P}(A_i) \cdot \mathbb{P}(A_j).$$

In other words, for all $1 \leq i < j \leq 1$ the events A_i and A_j are independent.

Unfortunately, for many purposes, the property of pairwise independence is too weak. For example, as we will see next, in general it does not imply the important equation

$$\mathbb{P}(A_1 \cap \cdots \cap A_n) = \mathbb{P}(A_1) \cdots \mathbb{P}(A_n). \tag{2.16}$$

Example 2.2.9. Roll a die twice and define events A_1, A_2 and A_3 as follows:

$$A_1 := \{2, 4, 6\} \times \{1, \ldots, 6\}$$

$$A_2 := \{1, \ldots, 6\} \times \{1, 3, 5\}$$

$$A_3 := \{2, 4, 6\} \times \{1, 3, 5\} \cup \{1, 3, 5\} \times \{2, 4, 6\}.$$

Verbally this says that A_1 occurs if the first roll is even, A_2 occurs if the second one is odd and A_3 occurs if either the first number is odd and the second is even or vice versa. Direct calculations give $\mathbb{P}(A_1) = \mathbb{P}(A_2) = \mathbb{P}(A_3) = 1/2$ as well as

$$\mathbb{P}(A_1 \cap A_2) = \mathbb{P}(A_1 \cap A_3) = \mathbb{P}(A_2 \cap A_3) = \frac{1}{4}.$$

Hence, A_1, A_2 and A_3 are pairwise independent.

Since

$$A_1 \cap A_2 \cap A_3 = A_1 \cap A_2$$

it follows

$$\mathbb{P}(A_1 \cap A_2 \cap A_3) = \mathbb{P}(A_1 \cap A_2) = \frac{1}{4} \neq \frac{1}{8} = \mathbb{P}(A_1) \cdot \mathbb{P}(A_2) \cdot \mathbb{P}(A_3).$$

So, we found three pairwise independent events for which eq. (2.16) is not valid.

After mentioning that pairwise independence of $A_1, \ldots, A_n$ does not imply

$$\mathbb{P}(A_1 \cap \cdots \cap A_n) = \mathbb{P}(A_1) \cdots \mathbb{P}(A_n), \tag{2.17}$$

it makes sense to ask whether or not pairwise independence can be derived from eq. (2.17). The next example shows that, in general, this is also not true.

Example 2.2.10. Let $\Omega = \{1, \ldots, 12\}$ be endowed with the uniform distribution $\mathbb{P}$, that is, for any $A \subseteq \Omega$ we have $\mathbb{P}(A) = \#(A)/12$. Define events A_1, A_2 and A_3 as $A_1 := \{1, \ldots, 9\}$, $A_2 := \{6, 7, 8, 9\}$ and $A_3 := \{9, 10, 11, 12\}$. Direct calculations give

$$\mathbb{P}(A_1) = \frac{9}{12} = \frac{3}{4}, \ \mathbb{P}(A_2) = \frac{4}{12} = \frac{1}{3} \ \text{ and } \ \mathbb{P}(A_3) = \frac{4}{12} = \frac{1}{3}.$$

Moreover, we have

$$\mathbb{P}(A_1 \cap A_2 \cap A_3) = \mathbb{P}(\{9\}) = \frac{1}{12} = \frac{3}{4} \cdot \frac{1}{3} \cdot \frac{1}{3} = \mathbb{P}(A_1) \cdot \mathbb{P}(A_2) \cdot \mathbb{P}(A_3),$$

hence eq. (2.17) is valid. But, because of

$$\mathbb{P}(A_1 \cap A_2) = \mathbb{P}(A_2) = \frac{1}{3} \neq \frac{1}{4} = \mathbb{P}(A_1) \cdot \mathbb{P}(A_2),$$

the events A_1, A_2 and A_3 are **not** pairwise independent.

Remark 2.2.11. Summing up, Examples 2.2.9 and 2.2.10 show that neither pairwise independence nor eq. (2.17) are suitable to define the independence of more than two events. Why? On the one hand, independence should yield eq. (2.17) and, on the other hand, whenever $A_1, \ldots, A_n$ are independent, then so should be any subcollection of them. In particular, independence should imply pairwise independence.

A reasonable definition of independence of n events is as follows.

Definition 2.2.12. The events $A_1, \ldots, A_n$ are said to be **independent** provided that for each subset $I \subseteq \{1, \ldots, n\}$ we have

$$\mathbb{P}\left(\bigcap_{i \in I} A_i\right) = \prod_{i \in I} \mathbb{P}(A_i). \tag{2.18}$$

Remark 2.2.13. Of course, it suffices that eq. (2.18) is valid for sets $I \subseteq \{1, \ldots, n\}$ satisfying $\#(I) \geq 2$. Indeed, if $\#(I) = 1$, then eq. (2.18) holds by trivial reason.

Remark 2.2.14. Another way to introduce independence is as follows: For all $m \geq 2$ and all $1 \leq i_1 < \cdots < i_m \leq n$, it follows

$$\mathbb{P}(A_{i_1} \cap \cdots \cap A_{i_m}) = \mathbb{P}(A_{i_1}) \cdots \mathbb{P}(A_{i_m}).$$

Identify I with $\{i_1, \ldots, i_m\}$ to see that both definitions are equivalent.

At a first glance, Definition 2.2.12 looks complicated; in fact, it is not. To see this, let us once more investigate the case $n = 3$. Here exist exactly four different subsets $I \subseteq \{1, 2, 3\}$ with $\#(I) \geq 2$. These are $I = \{1, 2\}$, $I = \{1, 3\}$, $I = \{2, 3\}$ and $I = \{1, 2, 3\}$. Consequently, three events A_1, A_2 and A_3 are independent if and only if the four following conditions hold **at once:**

$$\mathbb{P}(A_1 \cap A_2) = \mathbb{P}(A_1) \cdot \mathbb{P}(A_2)$$
$$\mathbb{P}(A_1 \cap A_3) = \mathbb{P}(A_1) \cdot \mathbb{P}(A_3)$$
$$\mathbb{P}(A_2 \cap A_3) = \mathbb{P}(A_2) \cdot \mathbb{P}(A_3) \quad \text{as well as}$$
$$\mathbb{P}(A_1 \cap A_2 \cap A_3) = \mathbb{P}(A_1) \cdot \mathbb{P}(A_2) \cdot \mathbb{P}(A_3) \,.$$

Examples 2.2.9 and 2.2.10 show that all four equations are really necessary. None of them is a consequence of the other three ones.

Independence of n events possesses the following properties:

Proposition 2.2.15.
1. *Let $A_1, \ldots, A_n$ be independent. For any $J \subseteq \{1, \ldots n\}$, the events $\{A_j : j \in J\}$ are independent as well. In particular, independence implies pairwise independence.*
2. *For each permutation π of $\{1, \ldots, n\}$, the independence of $A_1, \ldots, A_n$ implies that of*[2] *$A_{\pi(1)}, \ldots, A_{\pi(n)}$.*
3. *Suppose for each $1 \leq j \leq n$ holds either $B_j = A_j$ or $B_j = A_j^c$. Then the independence of $A_1, \ldots, A_n$ implies that of $B_1, \ldots, B_n$.*

Proof: The first two properties are an immediate consequence of the definition of independence.

To prove the third assertion, reorder $A_1, \ldots, A_n$ such that[3] $B_1 = A_1^c$. In a first step, we show that $A_1^c, A_2, \ldots, A_n$ are independent as well, that is, we have $B_1 = A_1^c$, $B_2 = A_2$ and so on. Given $I \subseteq \{1, \ldots, n\}$, it has to hold

$$\mathbb{P}\left(\bigcap_{i \in I} B_i\right) = \prod_{i \in I} \mathbb{P}(B_i) \,.$$

In the case $1 \notin I$, this follows by the independence of $A_1, \ldots, A_n$. If $1 \in I$, we apply Proposition 2.2.7 with[4] A_1 and $C = \bigcap_{i \in I \setminus \{1\}} A_i = \bigcap_{i \in I \setminus \{1\}} B_i$. Then $A_1^c = B_1$ and C are independent as well. Hence, by the independence of $A_2, \ldots, A_n$, we get

$$\mathbb{P}\left(\bigcap_{i \in I} B_i\right) = \mathbb{P}(B_1 \cap C) = \mathbb{P}(B_1) \cdot \mathbb{P}(C) = \mathbb{P}(B_1) \cdot \prod_{i \in I \setminus \{1\}} \mathbb{P}(B_i) = \prod_{i \in I} \mathbb{P}(B_i) \,.$$

---

2 For example, in the case $n = 3$ with A_1, A_2, A_3 also A_3, A_2, A_1 or A_2, A_3, A_1 are independent.
3 If all $B_j = A_j$, there is nothing to prove.
4 Why are A_1 and C independent? Give a short proof.

The general case then follows by reordering the A_js and by an iterative application of the first step. This is exactly the procedure we did in the proof of Proposition 2.2.7 when verifying the independence of A^c and B^c for independent A and B. ∎

The next two examples show how independence of more than two events appears in a natural way.

Example 2.2.16. Toss a fair coin n times. Let us assume that the coin is labeled with "0" and "1." Choose a fixed sequence $(a_j)_{j=1}^n$ of numbers in $\{0, 1\}$ and suppose that the event A_j occurs if in the jth trial a_j comes up.

We claim now that $A_1, \ldots, A_n$ are independent. To verify this, choose a subset $I \subseteq \{1, \ldots, n\}$ with $\#(I) = k$ for some $k = 2, \ldots, n$. The cardinality of $\bigcap_{i \in I} A_i$ equals 2^{n-k}. Why? At k positions the values of the tosses are fixed; at $n - k$ positions, they still may be either "0" or "1." Consequently,

$$\mathbb{P}\left(\bigcap_{i \in I} A_i\right) = \frac{2^{n-k}}{2^n} = 2^{-k}. \tag{2.19}$$

The same argument as before gives $\#(A_j) = 2^{n-1}$, hence $\mathbb{P}(A_j) = 1/2, 1 \le j \le n$. Consequently, it follows

$$\prod_{i \in I} \mathbb{P}(A_i) = \left(\frac{1}{2}\right)^{\#(I)} = 2^{-k}. \tag{2.20}$$

Combining eqs. (2.19) and (2.20) gives

$$\mathbb{P}\left(\bigcap_{i \in I} A_i\right) = \prod_{i \in I} \mathbb{P}(A_i),$$

and since I was arbitrary, the sets $A_1, \ldots, A_n$ are independent.

Remark 2.2.17. Even the simple Example 2.2.16 shows that it might be rather complicated to verify the independence of n given events. For example, if we modify the previous example by taking a biased coin, then the A_js remain independent, but the proof becomes more complicated.

Example 2.2.18. A machine consists of n components. These components break down with certain probabilities $p_1, \ldots, p_n$. Moreover, we assume that they break down independently of each other. Find the probability that a chosen machine stops working. Before answering this question, we have to determine the conditions.

Case 1: The machine stops working provided at least one component breaks down.

Let M be the event that the machine stops working. If $j \le n$, assume A_j occurs if component j breaks down. By assumption, $\mathbb{P}(A_j) = p_j$. Since

$$M = \bigcup_{j=1}^{n} A_j,$$

by the independence[5] it follows that

$$\mathbb{P}(M) = 1 - \mathbb{P}(M^c) = 1 - \mathbb{P}\left(\bigcap_{j=1}^{n} A_j^c\right) = 1 - \prod_{j=1}^{n} \mathbb{P}(A_j^c) = 1 - \prod_{j=1}^{n}(1 - p_j). \tag{2.21}$$

Case 2: The machine stops working provided all n components break down.

Using the same notation as in case 1, we now have

$$M = \bigcap_{j=1}^{n} A_j.$$

Hence, by the independence we obtain

$$\mathbb{P}(M) = \mathbb{P}\left(\bigcap_{j=1}^{n} A_j\right) = \prod_{j=1}^{n} p_j. \tag{2.22}$$

Remark 2.2.19. Formula (2.21) tells us the following: If among the n components there is one of bad quality, say the component j_0, then p_{j_0} is close to one; hence, $1 - p_{j_0}$ is close to zero, and so is $\prod_{j=1}^{n}(1 - p_j)$. Because of eq. (2.21), $\mathbb{P}(M)$ is large, and so the machine breaks down with large probability.

In the second case, the conclusion is as follows: if among the n components there is one of high quality, say component j_0, then p_{j_0} is small and so is $\prod_{j=1}^{n} p_j$. By eq. (2.22), $\mathbb{P}(M)$ is also small, hence it is very unlikely that the machine stops working.

2.3 Problems

Problem 2.1. The chance to win a certain game is 50%. One plays six games. Find the probability to win exactly four games. Evaluate the probability of this event under the condition to win at least two games. Suppose one had won exactly one of the two first games. Which probability has the event "winning 4 games" under this condition?

---

5 In fact, we also have to use Proposition 2.2.15.

Problem 2.2. Toss a fair coin six times. Define events A and B as follows:

$$A = \{\text{"Head" appears exactly 3 times}\}$$
$$B = \{\text{The first and the second toss are "head"}\}$$

Evaluate $\mathbb{P}(A)$, $\mathbb{P}(A|B)$ and $\mathbb{P}(A|B^c)$.

Problem 2.3. Let A and B be as in Problem 1.24, that is, A occurs if each box contains at most one particle while B occurs if all four particles are in the same box.
Find now $\mathbb{P}(A|C)$ and $\mathbb{P}(B|C)$ with $C = \{\text{The first box remains empty}\}$.

Problem 2.4. Justify why Propositions 2.1.14 and 2.1.18 (Law of total probability and Bayes' formula) remain valid for **infinitely many** disjoint sets $B_1, B_2, \ldots$ satisfying $\mathbb{P}(B_j) > 0$ and $\bigcup_{j=1}^{\infty} B_j = \Omega$.
Prove that Proposition 2.1.14 also holds without assuming $\bigcup_{j=1}^{n} B_j = \Omega$. But then we have to suppose $A \subseteq \bigcup_{j=1}^{n} B_j$.

Problem 2.5. To go to work, a man can either use the train, the bus or his car. He chooses the train 50%, the bus 30% and the car 20% of work days. If he takes the train, he arrives on time with probability 0.95. By bus, he is on time with probability 0.8, and by car with probability 0.7.
1. Evaluate the probability that the man is at work on time.
2. How big is this probability given the man does **not** use the car?
3. Assume the man arrived at work on time. What are then the probabilities that he came by train, bus or car?

Problem 2.6. Let U_1, U_2 and U_3 be three urns containing each five balls. Urn U_1 contains four white balls and one black ball, U_2 has three white balls and two black balls and, finally, U_3 contains two white balls and three black balls. Choose one urn by random (each urn is equally likely) and without replacing the first ball take two balls out of the chosen urn.
1. Give a suitable sample space for this random experiment.
2. Find the probability to observe two balls of different color.
3. Assume the chosen balls were of different color. What are the probabilities that the balls were taken out of U_1, U_2 or U_3?

Problem 2.7. Suppose we have three nondistinguishable dice. Two of them are fair, the other one is biased. There the number "6" appears with probability 1/5 while all other numbers have probability 4/25. We choose by random one of the dice and roll it.
1. Find a suitable sample space for the description of this experiment.
2. Give the probability of occurrence of $\{1\}$ to $\{6\}$ in that experiment.
3. Suppose we have observed the number "2" on the chosen die. Find the probability that this die was the biased one.

Problem 2.8.

1. Let $(\Omega, \mathcal{A}, \mathbb{P})$ be a probability space. Given events $A_1, \ldots, A_n$ prove the following *chain rule* for conditional probabilities:

$$\mathbb{P}(A_1 \cap \cdots \cap A_n) = \mathbb{P}(A_1)\mathbb{P}(A_2|A_1) \cdots \mathbb{P}(A_n|A_1 \cap A_2 \cap \cdots \cap A_{n-1}).$$

Hereby, we assume that all conditional probabilities are well defined.

2. Choose by random three numbers out of 1 to 10 without replacement. Find the probability that the first number is even, the second one is odd and the third one is again even.

3. Compare this probability with that of the following event: among three randomly chosen numbers in $\{1, \ldots, 10\}$ are exactly two even and one odd.

Problem 2.9. Three persons, say X, Y and Z, stand randomly in a row. All ordering are assumed to be equally likely. Event A occurs if Y stands on the right-hand side of X while B occurs in the case that Z is on the right-hand side of X. Hereby, we do not suppose that Y and X nor that Z and X stand directly next to each other. Are events A and B independent or dependent?

Problem 2.10. Prove the following generalization of part 1 in Proposition 2.2.7. Let $A \in \mathcal{A}$ be an event with either $\mathbb{P}(A) = 0$ or $\mathbb{P}(A) = 1$. Then for any $B \in \mathcal{A}$, the events A and B are independent.

Problem 2.11. Let $(\Omega, \mathcal{A}, \mathbb{P})$ be a probability space. Given independent events $A_1, \ldots, A_n$ in $\mathcal{A}$ prove that

$$\mathbb{P}\left(\bigcup_{j=1}^{n} A_j\right) = 1 - \prod_{j=1}^{n}\left(1 - \mathbb{P}(A_j)\right). \tag{2.23}$$

Use $1 - x \le e^{-x}, x \ge 0$, to derive from eq. (2.23) the following:
If independent events[6] $A_1, A_2, \ldots$ satisfy $\sum_{j=1}^{\infty} \mathbb{P}(A_j) = \infty$, then $\mathbb{P}\left(\bigcup_{j=1}^{\infty} A_j\right) = 1$.

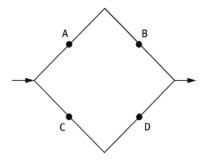

---

6 Compare Definition 7.1.17: for each $n \in \mathbb{N}$, the events $A_1, \ldots, A_n$ are independent.

Problem 2.12. An electric circuit (see the above figure) contains four switches A, B, C and D. Each of the switches is independently open or closed (then electricity flows). The switches are open with probability $1 - p$ and closed with probability p. Here, $0 \leq p \leq 1$ is given. Find the probability that electricity flows from the left-hand side to the right-hand one.

Problem 2.13. Let $(\Omega, \mathcal{A}, \mathbb{P})$ be a probability space. Suppose A and B are disjoint events with $\mathbb{P}(A) > 0$ and $P(B) > 0$. Is it possible that A and B are independent?

Problem 2.14. Let A, B and C be three independent events.
1. Show that $A \cap B$ and C are independent as well.
2. Even more, show that the independence of A, B and C implies that of $A \cup B$ and C.

Problem 2.15.
1. Suppose that A and C as well as B and C are independent. Furthermore, assume $A \cap B = \emptyset$. Show that $A \cup B$ and C are independent as well.
2. Give an example that shows that the preceding assertion becomes false without the assumption $A \cap B = \emptyset$.

Remark: To construct such an example, because of Problem 2.14, the events A, B and C cannot be chosen to be independent. Therefore, the sets defined in Example 2.2.9 are natural candidates for such an example.

Problem 2.16. Suppose $\mathbb{P}(A|B) = \mathbb{P}(A|B^c)$ for some events A and B with $0 < \mathbb{P}(B) < 1$. Does this imply that A and B are independent?

Problem 2.17. Is it possible that an event A is independent of itself? If yes, which events A have this property? Similarly, which A are independent of A^c ?

Problem 2.18. Let A, B and C be three **independent** events with

$$\mathbb{P}(A) = \mathbb{P}(B) = \mathbb{P}(C) = \frac{1}{3}.$$

Evaluate

$$\mathbb{P}\big((A \cap B) \cup (A \cap C)\big).$$

3 Random Variables and Their Distribution

3.1 Transformation of Random Values

Assume the probability space $(\Omega, \mathcal{A}, \mathbb{P})$ describes a certain random experiment, for example, rolling a die or tossing a coin. If the experiment is executed, a random result $\omega \in \Omega$ shows up. In a second step we transform this observed result via a mapping $X : \Omega \to \mathbb{R}$. In this way we obtain a (random) real number $X(\omega)$. Let us point out that X is a fixed, nonrandom function from Ω into $\mathbb{R}$; the randomness of $X(\omega)$ stems from the input $\omega \in \Omega$.

Example 3.1.1. Toss a fair coin, labeled on one side with "0" and on the other side with "1", exactly n times. The appropriate probability space is $(\Omega, \mathcal{P}(\Omega), \mathbb{P})$, where $\Omega = \{0, 1\}^n$ and $\mathbb{P}$ is the uniform distribution on Ω. The result of the experiment is a vector $\omega = (\omega_1, \ldots, \omega_n)$ with $\omega_j = 0$ or $\omega_j = 1$. Let X from $\Omega \to \mathbb{R}$ be defined by

$$X(\omega) = X(\omega_1, \ldots, \omega_n) = \omega_1 + \cdots + \omega_n.$$

Then $X(\omega)$ tells us how often "1" occurred, but we do no longer know in which order this happened. Of course, $X(\omega)$ is random because, if one tosses the coin another n times, it is very likely that X attains a value different from that in the first trial.

Here we state the most important question in this topic: how are the values of X distributed? In our case X attains the values $k \leq n$ with probabilities $\binom{n}{k} 2^{-n}$.

Example 3.1.2. Roll a fair die twice. The sample space describing this experiment consists of pairs $\omega = (\omega_1, \omega_2)$, where $\omega_1, \omega_2 \in \{1, \ldots, 6\}$. Now define the mapping $X : \Omega \to \mathbb{R}$ by $X(\omega) := \max\{\omega_1, \omega_2\}$. Thus, instead of recording the values of both rolls, we are only interested in the larger one.

Other possible transformations are, for example, $X_1(\omega) := \min\{\omega_1, \omega_2\}$ or also $X_2(\omega_1, \omega_2) := \omega_1 + \omega_2$.

Let $A \in \mathcal{A}$ be an event. Recall that this event A occurs if and only if we observe an $\omega \in A$. Suppose now $X : \Omega \to \mathbb{R}$ is a given mapping from Ω into $\mathbb{R}$ and let $B \subseteq \mathbb{R}$ be some event. When do we observe an $\omega \in \Omega$ for which we have $X(\omega) \in B$ or, equivalently, when does the event

$$\{X \in B\} := \{\omega \in \Omega : X(\omega) \in B\}$$

occur? To answer this question, let us recall the definition of the preimage of B with respect to X as given in eq. (A.1):

$$X^{-1}(B) := \{\omega \in \Omega : X(\omega) \in B\}.$$

We observe an $\omega \in \Omega$ for which $X(\omega) \in B$ if and only if $\omega \in X^{-1}(B)$. In other words, the event $\{X \in B\}$ occurs if and only if $X^{-1}(B)$ does so. Consequently, the probability to observe an $\omega \in \Omega$ with $X(\omega) \in B$ should be $\mathbb{P}(X^{-1}(B))$. But to this end, we have to know that $X^{-1}(B) \in \mathcal{A}$; otherwise $\mathbb{P}(X^{-1}(B))$ is not defined at all. Thus, a natural condition for X is $X^{-1}(B) \in \mathcal{A}$ for "sufficiently many" subsets $B \subseteq \mathbb{R}$. The precise mathematical condition reads as follows.

> **Definition 3.1.3.** Let $(\Omega, \mathcal{A}, \mathbb{P})$ be a probability space. A mapping $X : \Omega \to \mathbb{R}$ is called a (real-valued) **random variable** (sometimes also called **random real number**), provided that it satisfies the following condition:
>
> $$B \in \mathcal{B}(\mathbb{R}) \quad \Longrightarrow \quad X^{-1}(B) \in \mathcal{A}. \qquad (3.1)$$
>
> Verbally, this condition says that for each Borel set $B \subseteq \mathbb{R}$, its preimage $X^{-1}(B)$ has to be an element of the σ-field $\mathcal{A}$.

Remark 3.1.4. Condition (3.1) is purely technical and will not be important later on. But, in general, it cannot be avoided, at least if $\mathcal{A} \neq \mathcal{P}(\Omega)$. On the contrary, if $\mathcal{A} = \mathcal{P}(\Omega)$, for example, if either Ω is finite or countably infinite, then **every** mapping $X : \Omega \to \mathbb{R}$ is a random variable. Indeed, by trivial reason, in this case the condition $X^{-1}(B) \in \mathcal{A}$ is always satisfied.

Remark 3.1.5. In order to verify that a given mapping $X : \Omega \to \mathbb{R}$ is a random variable, it is not necessary to show $X^{-1}(B) \in \mathcal{A}$ for all Borel sets $B \subseteq \mathbb{R}$. It suffices to prove this only for some special Borel sets B. More precisely, the following proposition holds.

Proposition 3.1.6. *A function $X : \Omega \to \mathbb{R}$ is a random variable if and only if, for all $t \in \mathbb{R}$, we have*

$$X^{-1}\big((-\infty, t]\big) = \{\omega \in \Omega : X(\omega) \leq t\} \in \mathcal{A}. \qquad (3.2)$$

The assertion remains valid when we replace the intervals $(-\infty, t]$ with intervals of the form $(-\infty, t)$, or we may take intervals $[t, \infty)$ and also (t, ∞).

Proof: Suppose first that X is a random variable. Given $t \in \mathbb{R}$, the interval $(-\infty, t]$ is a Borel set, hence $X^{-1}\big((-\infty, t]\big) \in \mathcal{A}$. Thus, each random variable satisfies condition (3.2).

To prove the converse implication, let X be a mapping from Ω to $\mathbb{R}$ satisfying condition (3.2) for each $t \in \mathbb{R}$. Set

$$\mathcal{C} := \{C \in \mathcal{B}(\mathbb{R}) : X^{-1}(C) \in \mathcal{A}\}.$$

In a first step, one proves[1] that C is a σ-field. Moreover, property (3.2) implies $(-\infty, t] \in$ C for each $t \in \mathbb{R}$. But $\mathcal{B}(\mathbb{R})$ is the smallest σ-field containing all these intervals. Since C is another σ-field containing the intervals $(-\infty, t]$, it has to be larger[2] than the smallest one, that is, we have $C \supseteq \mathcal{B}(\mathbb{R})$. In other words, every Borel set belongs to C or, equivalently, for all $B \in \mathcal{B}(\mathbb{R})$ it follows $X^{-1}(B) \in \mathcal{A}$. Thus, as asserted, X is a random variable. The proof for intervals of the other types goes along the same line. Here one has to use that these intervals generate the σ-field of Borel sets as well. ∎

3.2 Probability Distribution of a Random Variable

Suppose we are given a random variable $X : \Omega \to \mathbb{R}$. We define now a mapping $\mathbb{P}_X$ from $\mathcal{B}(\mathbb{R})$ to $[0, 1]$ as follows:

$$\mathbb{P}_X (B) := \mathbb{P}(X^{-1}(B)) = \mathbb{P}\{\omega \in \Omega : X(\omega) \in B\}, \quad B \in \mathcal{B}(\mathbb{R}).$$

Observe that $\mathbb{P}_X$ is well-defined. Indeed, since X is a random variable, for all Borel sets $B \subseteq \mathbb{R}$ we have $X^{-1}(B) \in \mathcal{A}$, hence $\mathbb{P}(X^{-1}(B))$ makes sense.

To simplify the notation, given $B \in \mathcal{B}(\mathbb{R})$, we will often write

$$\mathbb{P}\{X \in B\} = \mathbb{P}\{\omega \in \Omega : X(\omega) \in B\}.$$

That is generally used and does not lead to any confusion. Having said this we may now define $\mathbb{P}_X$ also by

$$\mathbb{P}_X(B) = \mathbb{P}\{X \in B\}.$$

A first easy example shows how $\mathbb{P}_X$ is calculated in concrete cases. Other, more interesting examples will follow after some necessary preliminary considerations.

Example 3.2.1. Toss a fair coin, labeled on one side by "0" and on the other side by "1," three times. The sample space is $\Omega = \{0, 1\}^3$ with the uniform distribution $\mathbb{P}$ describing probability measure. Let the random variable X on Ω be defined by

$$X(\omega) := \omega_1 + \omega_2 + \omega_3 \quad \text{whenever} \quad \omega = (\omega_1, \omega_2, \omega_3) \in \Omega.$$

---

1 Use Proposition A.2.1 to verify this.

2 By the construction of C, it even coincides with $\mathcal{B}(\mathbb{R})$.

It follows

$$\mathbb{P}_X(\{0\}) = \mathbb{P}\{X = 0\} = \mathbb{P}(\{(0, 0, 0)\}) = \frac{1}{8}$$

$$\mathbb{P}_X(\{1\}) = \mathbb{P}\{X = 1\} = \mathbb{P}(\{(1, 0, 0), (0, 1, 0), (0, 0, 1)\}) = \frac{3}{8}$$

$$\mathbb{P}_X(\{2\}) = \mathbb{P}\{X = 2\} = \mathbb{P}(\{(1, 1, 0), (0, 1, 1), (1, 0, 1)\}) = \frac{3}{8}$$

$$\mathbb{P}_X(\{3\}) = \mathbb{P}\{X = 3\} = \mathbb{P}(\{(1, 1, 1)\}) = \frac{1}{8} \; .$$

Of course, these values describe the distribution of X completely. Indeed, whenever $B \subseteq \mathbb{R}$, then

$$\mathbb{P}_X(B) = \sum_{\substack{k=0 \\ k \in B}}^{3} \mathbb{P}_X(\{k\}) \; .$$

The proof of the next result heavily depends on properties of the preimage proved in Proposition A.2.1.

Proposition 3.2.2. *Let $(\Omega, \mathcal{A}, \mathbb{P})$ be a probability space. For each random variable $X : \Omega \to \mathbb{R}$ the mapping $\mathbb{P}_X : \mathcal{B}(\mathbb{R}) \to [0, 1]$ is a probability measure.*

Proof: Using property (1) in Proposition A.2.1 one easily gets

$$\mathbb{P}_X(\emptyset) = \mathbb{P}(X^{-1}(\emptyset)) = \mathbb{P}(\emptyset) = 0$$

as well as

$$\mathbb{P}_X(\mathbb{R}) = \mathbb{P}(X^{-1}(\mathbb{R})) = \mathbb{P}(\Omega) = 1 \; .$$

Thus it remains to verify the σ-additivity of $\mathbb{P}_X$. Take any sequence of disjoint Borel sets $B_1, B_2, \ldots$ in $\mathbb{R}$. Then also $X^{-1}(B_1), X^{-1}(B_2), \ldots$ are disjoint subsets of Ω. To see this apply Proposition A.2.1, which, if $i \neq j$, implies

$$X^{-1}(B_i) \cap X^{-1}(B_j) = X^{-1}(B_i \cap B_j) = X^{-1}(\emptyset) = \emptyset \; .$$

Another application of Proposition A.2.1 and of the σ-additivity of $\mathbb{P}$ finally gives

$$\mathbb{P}_X\left(\bigcup_{j=1}^{\infty} B_j\right) = \mathbb{P}\left(X^{-1}\left(\bigcup_{j=1}^{\infty} B_j\right)\right) = \mathbb{P}\left(\bigcup_{j=1}^{\infty} X^{-1}(B_j)\right)$$

$$= \sum_{j=1}^{\infty} \mathbb{P}(X^{-1}(B_j)) = \sum_{j=1}^{\infty} \mathbb{P}_X(B_j) \; .$$

Hence, $\mathbb{P}_X$ is a probability measure as asserted. ∎

Definition 3.2.3. The probability measure $\mathbb{P}_X$ on $(\mathbb{R}, \mathcal{B}(\mathbb{R}))$ defined by

$$\mathbb{P}_X(B) := \mathbb{P}\big(X^{-1}(B)\big) = \mathbb{P}\{\omega \in \Omega : X(\omega) \in B\} = \mathbb{P}\{X \in B\}, \quad B \in \mathcal{B}(\mathbb{R}),$$

is called **probability distribution** of X (with respect to $\mathbb{P}$) or, in short, only **distribution** of X.

Remark 3.2.4. The distribution $\mathbb{P}_X$ is the most important characteristic of a random variable X. In general, it is completely unimportant how a random variable is defined analytically; only its distribution matters. Thus, two random variables with identical distributions may be regarded as equivalent because they describe the same random experiment.

Remark 3.2.4 leads us to the following definition:

Definition 3.2.5. Two random variables X_1 and X_2 are said to be **identically distributed** provided that $\mathbb{P}_{X_1} = \mathbb{P}_{X_2}$. Hereby, it is not necessary that X_1 and X_2 are defined on the same sample space. Only their distributions have to coincide. In the case of identically distributed X_1 and X_2 one writes $X_1 \overset{d}{=} X_2$.

Example 3.2.6. Toss a fair coin, labeled on each side by "0" or "1," twice. Let X_1 be the value of the first toss and X_2 that of the second one. Then

$$\mathbb{P}\{X_1 = 0\} = \mathbb{P}\{X_2 = 0\} = \frac{1}{2} = \mathbb{P}\{X_1 = 1\} = \mathbb{P}\{X_2 = 1\} .$$

Hence, X_1 and X_2 are identically distributed or $X_1 \overset{d}{=} X_2$. Both random variables describe the same experiment, namely to toss a fair coin one time. Now, toss the coin a third time and let X_3 be the result of the third trial. Then we also have $X_1 \overset{d}{=} X_3$, but note that X_1 and X_3 are defined on different sample spaces.

Next, we state and prove some general rules for evaluating the probability distribution of a given random variable. Here we have to distinguish between two different types of random variables, namely between discrete and continuous ones. Let us start with the discrete case.

Definition 3.2.7. A random variable X is **discrete** provided there exists an at most countably infinite set $D \subset \mathbb{R}$ such that $X : \Omega \to D$.

In other words, a random variable is discrete if it attains at most countably infinite many different values.

Remark 3.2.8. If a random variable X is discrete with values in $D \subset \mathbb{R}$, then, of course,

$$\mathbb{P}_X(D) = \mathbb{P}\{X \in D\} = 1.$$

Consequently, in this case its probability distribution $\mathbb{P}_X$ is a discrete probability measure on $\mathbb{R}$. In general, the converse is not valid as the next example shows.

Example 3.2.9. We model the experiment of rolling a fair die by the probability space $(\mathbb{R}, \mathcal{P}(\mathbb{R}), \mathbb{P})$, where $\mathbb{P}(\{1\}) = \cdots = \mathbb{P}(\{6\}) = 1/6$ and $\mathbb{P}(\{x\}) = 0$ whenever $x \neq 1, \ldots, 6$. If $X : \mathbb{R} \to \mathbb{R}$ is defined by $X(s) = s^2$ then, of course, $\mathbb{P}_X$ is discrete. Indeed, we have $\mathbb{P}_X(D) = 1$, where $D = \{1, 4, 9, 16, 25, 36\}$. On the other hand, X does not attain values in a countably infinite set; its range is $[0, \infty)$.

Remark 3.2.10. If we look at Example 3.2.9 more thoroughly, then it becomes immediately clear that the values of X outside of $\{1, \ldots, 6\}$ are completely irrelevant. With a small change of X, it will attain values in D. More precisely, let $\tilde{X}(\omega) = 1$ if $\omega \neq 1, \ldots, 6$ and $\tilde{X}(k) = k^2$, $k = 1, \ldots, 6$; then $X \stackrel{d}{=} \tilde{X}$ and $\tilde{X}$ has values in $\{1, 4, 9, 16, 25, 36\}$.

This procedure is also possible in general: if $\mathbb{P}_X$ is discrete with $\mathbb{P}_X(D) = 1$ for some countable set D, then we may change X to $\tilde{X}$ such that $X \stackrel{d}{=} \tilde{X}$ and $\tilde{X} : \Omega \to D$. Indeed, choose some fixed $d_0 \in D$ and set $\tilde{X}(\omega) = X(\omega)$ if $\omega \in X^{-1}(D)$ and $\tilde{X}(\omega) = d_0$ otherwise. Then $\mathbb{P}_X = \mathbb{P}_{\tilde{X}}$ and $\tilde{X}$ has values in D.

Convention 3.1. *Without losing generality we may always assume the following: if a random variable X has a discrete probability distribution, that is, $\mathbb{P}\{X \in D\} = 1$ for some finite or countably infinite set D, then X attains values in D.*

The second type of random variables we investigate is that of continuous ones.[3]

> **Definition 3.2.11.** A random variable X is said to be **continuous** provided that its distribution $\mathbb{P}_X$ is a continuous probability measure. That is, $\mathbb{P}_X$ possesses a density p. This function p is called the **density function** or, in short, **density** of the random variable X.

Remark 3.2.12. One should not confuse the continuity of a random variable with the continuity of a function as taught in Calculus. The latter is an (analytic) property of a function, while the former is a property of its distribution. Moreover, whether or not

---

[3] The precise notation would be "absolutely continuous"; but for simplicity let us call them "continuous."

a random variable X is continuous depends not only on X, but also on the underlying probability space.

Remark 3.2.13. Another way to express that a random variable is continuous is as follows: there exists a function $p : \mathbb{R} \to [0, \infty)$ (the density of X) such that

$$\mathbb{P}\{\omega \in \Omega : X(\omega) \le t\} = \mathbb{P}\{X \le t\} = \int_{-\infty}^{t} p(x)\,dx, \quad t \in \mathbb{R},$$

or, equivalently, for all real numbers $a < b$,

$$\mathbb{P}\{\omega \in \Omega : a \le X(\omega) \le b\} = \mathbb{P}\{a \le X \le b\} = \int_{a}^{b} p(x)\,dx.$$

How do we determine the probability distribution of a given random variable? To answer this question, let us first consider the case of **discrete** random variables.

Thus, let X be discrete with values in $D = \{x_1, x_2, \dots\} \subset \mathbb{R}$. Then, as observed above, it follows $\mathbb{P}_X(D) = 1$, and, consequently, $\mathbb{P}_X$ is uniquely determined by the numbers

$$p_j := \mathbb{P}_X(\{x_j\}) = \mathbb{P}\{X = x_j\} = \mathbb{P}\{\omega \in \Omega : X(\omega) = x_j\}, \quad j = 1, 2, \dots . \tag{3.3}$$

Moreover, for any $B \subseteq \mathbb{R}$ it follows that

$$\mathbb{P}\{\omega \in \Omega : X(\omega) \in B\} = \mathbb{P}_X(B) = \sum_{x_j \in B} p_j .$$

Consequently, in order to determine $\mathbb{P}_X$ for discrete X it completely suffices to determine the p_js defined by eq. (3.3). If we know $(p_j)_{j\ge1}$, then the probability distribution $\mathbb{P}_X$ of X is completely described.

Remark 3.2.14. In the literature, quite often, one finds a slightly different approach for the description of $\mathbb{P}_X$. Define $p : \mathbb{R} \to [0, 1]$ by

$$p(x) = \mathbb{P}\{X = x\}, \quad x \in \mathbb{R}. \tag{3.4}$$

This function p is then called the **probability mass function** of X. Note that $p(x) = 0$ whenever $x \notin D$. This function p satisfies $p(x) \ge 0$, $\sum_{x \in \mathbb{R}} p(x) = 1$ and

$$\mathbb{P}\{X \in B\} = \sum_{x \in B} p(x).$$

In this setting, the numbers p_j in eq. (3.3) coincide with $p(x_j)$.

Example 3.2.15. Roll a fair die twice. Let X on $\{1, \ldots, 6\}^2$ be defined by

$$X(\omega) = X(\omega_1, \omega_2) := \omega_1 + \omega_2, \quad \omega = (\omega_1, \omega_2).$$

Which distribution does X possess?

Answer: The very first question one has to answer is always about the possible values of X. In our case, X attains values in $D = \{2, \ldots, 12\}$, thus it suffices to determine

$$\mathbb{P}_X(\{k\}) = \mathbb{P}\{X = k\} = \mathbb{P}\{(\omega_1, \omega_2) \in \Omega : X(\omega_1, \omega_2) = k\}, \quad k = 2, \ldots, 12.$$

One easily gets

$$\mathbb{P}_X(\{2\}) = \mathbb{P}\{(\omega_1, \omega_2) : \omega_1 + \omega_2 = 2\} = \frac{\#(\{(1, 1)\})}{36} = \frac{1}{36}$$

$$\mathbb{P}_X(\{3\}) = \mathbb{P}\{(\omega_1, \omega_2) : \omega_1 + \omega_2 = 3\} = \frac{\#(\{(1, 2), (2, 1)\})}{36} = \frac{2}{36}$$

$$\cdot$$
$$\cdot$$

$$\mathbb{P}_X(\{7\}) = \mathbb{P}\{(\omega_1, \omega_2) : \omega_1 + \omega_2 = 7\} = \frac{\#(\{(1, 6), \ldots, (6, 1)\})}{36} = \frac{6}{36}$$

$$\cdot$$
$$\cdot$$

$$\mathbb{P}_X(\{12\}) = \mathbb{P}\{(\omega_1, \omega_2) : \omega_1 + \omega_2 = 12\} = \frac{\#(\{(6, 6)\})}{36} = \frac{1}{36},$$

hence $\mathbb{P}_X$ is completely described. For example, it follows that

$$\mathbb{P}\{X \leq 4\} = \mathbb{P}_X((-\infty, 4]) = \mathbb{P}_X(\{2\}) + \mathbb{P}_X(\{3\}) + \mathbb{P}_X(\{4\}) = \frac{1}{36} + \frac{2}{36} + \frac{3}{36} = \frac{1}{6}.$$

Example 3.2.16. A coin is labeled on one side by by "0" and and on the other side by "1" and biased as follows: for some $p \in [0, 1]$, number "1" shows up with probability p, thus "0" with probability $1 - p$. We toss the coin n times. The result is a sequence $\omega = (\omega_1, \ldots, \omega_n)$, where $\omega_i \in \{0, 1\}$, hence the describing sample space is

$$\Omega = \{0, 1\}^n = \{\omega = (\omega_1, \ldots, \omega_n) : \omega_i \in \{0, 1\}\}.$$

For $i \leq n$ let $X_i : \Omega \to \mathbb{R}$ be defined by $X_i(\omega) := \omega_i$. That is, $X_i(\omega)$ is the value of the ith trial. What distribution does X_i possess?

Answer: In Example 1.9.11 we determined the probability measure $\mathbb{P}$ on $\mathcal{P}(\Omega)$, which describes the n-fold tossing of a biased coin. This probability measure was given by

$$\mathbb{P}(\{\omega\}) = p^k (1 - p)^{n-k}, \quad k = \sum_{j=1}^n \omega_j \quad \text{where} \quad \omega = (\omega_1, \ldots, \omega_n). \tag{3.5}$$

The random variable X_i only attains the values "0" and "1." Thus, in order to determine $\mathbb{P}_{X_i}$ it suffices to evaluate $\mathbb{P}_{X_i}(\{0\}) = \mathbb{P}\{\omega \in \Omega : \omega_i = 0\}$. Let $\omega \in \Omega$ be a sequence with $\omega_i = 0$. Then it may contain the value "1" at most $n - 1$ times. Given $k \le n - 1$, there are exactly $\binom{n-1}{k}$ such sequences ω with $\omega_i = 0$ and with k times "1." Therefore, we obtain

$$\mathbb{P}_{X_i}(\{0\}) = \mathbb{P}\{\omega \in \Omega : \omega_i = 0\} = \sum_{k=0}^{n-1} \mathbb{P}\{\omega \in \Omega : \omega_i = 0, \, \omega_1 + \cdots + \omega_n = k\}$$

$$= \sum_{k=0}^{n-1} \binom{n-1}{k} p^k (1-p)^{n-k} = (1-p) \sum_{k=0}^{n-1} \binom{n-1}{k} p^k (1-p)^{n-1-k}$$

$$= (1-p)[p + (1-p)]^{n-1} = 1 - p \, .$$

Of course, this also implies $\mathbb{P}_{X_i}(\{1\}) = p$.

Remark 3.2.17. Note that all $X_1, \ldots, X_n$ possess the same distribution, that is,

$$X_1 \stackrel{d}{=} \cdots \stackrel{d}{=} X_n \, .$$

Summary: Let $X : \Omega \to \mathbb{R}$ be a **discrete** random variable. In order to describe its distribution $\mathbb{P}_X$, two things have to be done:
(1) Determine the finite or countably infinite set $D \subset \mathbb{R}$ for which $X : \Omega \to D$.
(2) For each $x \in D$ evaluate

$$\mathbb{P}_X(\{x\}) = \mathbb{P}\{X = x\} = \mathbb{P}\{\omega \in \Omega : X(\omega) = x\} \, .$$

If $B \subseteq \mathbb{R}$, then it follows

$$\mathbb{P}\{X \in B\} = \sum_{x \in B \cap D} \mathbb{P}_X(\{x\}) = \sum_{x \in B \cap D} \mathbb{P}\{X = x\} \, .$$

How do we determine the probability distribution of a random variable if it is **continuous**? For each $x \in \mathbb{R}$, $\mathbb{P}\{X = x\} = 0$, hence the values of $\mathbb{P}\{X = x\}$ cannot be used to describe $\mathbb{P}_X$ as they did in the discrete case. Consequently, a different approach is needed, and this approach is based on the use of distribution functions.

Definition 3.2.18. Let X be a random variable, either discrete or continuous. Then its **(cumulative) distribution function** $F_X : \mathbb{R} \to [0, 1]$ is defined by

$$F_X(t) := \mathbb{P}_X((-\infty, t]) = \mathbb{P}\{X \le t\}, \quad t \in \mathbb{R}. \tag{3.6}$$

Remark 3.2.19. Observe that for discrete and continuous random variables the distribution function equals

$$F_X(t) = \sum_{x_j \le t} p_j \quad \text{and} \quad F_X(t) = \int_{-\infty}^{t} p(x)\, dx,$$

respectively. Here, in the discrete case, the x_js and p_js are as in eq. (3.3), while p denotes the density of X in the continuous case.

Furthermore, note that F_X is nothing else than the distribution function of the probability measure $\mathbb{P}_X$, as it was introduced in Definition 1.7.1. Consequently, it possesses all properties of a "usual" distribution function as stated in Proposition 1.7.9.

Proposition 3.2.20. *Let F_X be defined by eq. (3.6). Then it possesses the following properties.*
(1) *F_X is nondecreasing.*
(2) *It follows $F_X(-\infty) = 0$ as well as $F_X(\infty) = 1$.*
(3) *F_X is continuous from the right.*

Furthermore, if $t \in \mathbb{R}$, then

$$\mathbb{P}\{X = t\} = F_X(t) - F_X(t - 0).$$

In particular, if X is continuous, then F_X is a continuous function from $\mathbb{R}$ to $[0, 1]$.

Remark 3.2.21. Note that the converse of the last implication does not hold. Indeed, there exist random variables X for which F_X is continuous, but X does not possess a density. Such random variables are said to be **singularly continuous**. These are exactly those random variables for which the probability measure $\mathbb{P}_X$ is singularly continuous in the sense of Remark 1.7.16.

The next result shows that under slightly stronger conditions about F_X a density of X exists.

Proposition 3.2.22. *Let F_X be continuous and continuously differentiable with the exception of at most finitely many points. Then X is continuous with density $p(t) = \frac{d}{dt} F_X(t)$. Hereby the values of p may be chosen arbitrarily at points where the derivative does not exist; for example, set $p(t) = 0$ for those points.*

Proof: The proof follows from the corresponding properties of distribution functions for probability measures. Recall that F_X is the distribution function of $\mathbb{P}_X$. ∎

The previous proposition provides us with a method to determine the density of a given random variable X. Determine the distribution function F_X and differentiate it. The obtained derivative is the density function we are looking for.

The next three examples demonstrate how this method applies.

Example 3.2.23. Let $\mathbb{P}$ be the uniform distribution on a sphere K of radius 1. That is, for each Borel set $B \in \mathcal{B}(\mathbb{R}^2)$ we have

$$\mathbb{P}(B) = \frac{\text{vol}_2(B \cap K)}{\text{vol}_2(K)} = \frac{\text{vol}_2(B \cap K)}{\pi} .$$

Define the random variable $X : \mathbb{R}^2 \to \mathbb{R}$ by $X(x_1, x_2) := x_1$. Of course, we have $F_X(t) = 0$ whenever $t < -1$ and $F_X(t) = 1$ when $t > 1$. Thus, it suffices to determine $F_X(t)$ if $-1 \le t \le 1$. For those $t \in \mathbb{R}$ we obtain

$$F_X(t) = \frac{\text{vol}_2(S_t \cap K)}{\pi}$$

where S_t is the half-space $\{(x_1, x_2) \in \mathbb{R}^2 : x_1 \le t\}$.

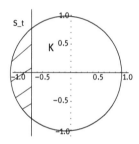

Figure 3.1: The intersecting set between K and the half-space S_t.

If $|t| \le 1$, then

$$\text{vol}_2(S_t \cap K) = 2 \int_{-1}^{t} \sqrt{1 - x^2} \, dx ,$$

hence,

$$F_X(t) = \frac{2}{\pi} \int_{-1}^{t} \sqrt{1 - x^2} \, dx ,$$

and by the fundamental theorem of Calculus, we finally get

$$p(t) = \frac{d}{dt} F_X(t) = \frac{2}{\pi} \sqrt{1 - t^2}, \quad |t| \le 1.$$

Summing up, the random variable X has the density p with

$$p(t) = \begin{cases} \frac{2}{\pi}\sqrt{1-t^2} & : |t| \leq 1 \\ 0 & : |t| > 1. \end{cases} \tag{3.7}$$

Example 3.2.24. The probability space is the same as in Example 3.2.23, but this time we define X by

$$X(x_1, x_2) := \sqrt{x_1^2 + x_2^2}, \quad (x_1, x_2) \in \mathbb{R}^2.$$

Of course, it follows $F_X(t) = 0$ if $t < 0$ while $F_X(t) = 1$ if $t > 1$. Take $t \in [0, 1]$. Then

$$F_X(t) = \frac{\text{vol}_2(K(t))}{\text{vol}_2(K(1))} = \frac{t^2\pi}{\pi} = t^2,$$

where $K(t)$ denotes a sphere of radius t. Differentiating F_X with respect to t gives the density

$$p(t) = \begin{cases} 2t & : \ 0 \leq t \leq 1 \\ 0 & : \ \text{otherwise} \end{cases}$$

Example 3.2.25. Let $\mathbb{P}$ be the uniform distribution on $[0, 1]$ and define the random variable X by $X(s) = \min\{s, 1 - s\}$, $s \in \mathbb{R}$. Find the probability distribution of X.

Answer: It is not difficult to see that

$$\mathbb{P}\{X \leq t\} = 0 \ \text{ if } \ t < 0 \quad \text{and} \quad \mathbb{P}\{X \leq t\} = 1 \ \text{ if } \ t > 1/2.$$

Thus it remains to evaluate $F_X(t)$ for $0 \leq t \leq 1/2$. Here we obtain

$$F_X(t) = \mathbb{P}\{X \leq t\} = \mathbb{P}\{s \in [0, 1] : 0 \leq s \leq t \text{ or } 1 - t \leq s \leq 1\}$$
$$= \mathbb{P}\{s \in [0, 1] : 0 \leq s \leq t\} + \mathbb{P}\{s \in [0, 1] : 1 - t \leq s \leq 1\} = 2t.$$

Differentiating gives $F_X'(t) = 2$ if $0 \leq t \leq 1/2$ and $F_X(t) = 0$ otherwise. Hence $\mathbb{P}_X$ is the uniform distribution on $[0, 1/2]$.

Summary: To determine the density of a **continuous** random variable, proceed as follows:
(1) Determine $F_X(t) = \mathbb{P}\{X \leq t\}$.
(2) Differentiate F_X. Then the derivative $p(t) = F_X'(t)$ is the desired density.

3.3 Special Distributed Random Variables

We agree upon the following notation: a random variable X is said to be *ABC-distributed* (or distributed according to *ABC*) if its probability distribution is a probability measure of type *ABC*. For example, a random variable is $\mathbf{B_{n,p}}$-**distributed** (or distributed according to $B_{n,p}$) if $\mathbb{P}_X = B_{n,p}$, that is, if

$$\mathbb{P}\{X = k\} = \binom{n}{k} p^k (1 - p)^{n-k}, \quad k = 0, \ldots, n.$$

In this way we define the following random variables of special type: X is
1. **uniformly distributed** on $\{x_1, \ldots, x_N\}$ if

$$\mathbb{P}\{X = x_1\} = \cdots = \mathbb{P}\{X = x_N\} = \frac{1}{N},$$

2. **Poisson distributed** or Pois_λ-distributed if

$$\mathbb{P}\{X = k\} = \frac{\lambda^k}{k!} e^{-\lambda}, \quad k = 0, 1, \ldots$$

3. **hypergeometric distributed** if

$$\mathbb{P}\{X = m\} = \frac{\binom{M}{m} \binom{N-M}{n-m}}{\binom{N}{n}}, \quad m = 0, \ldots, n,$$

4. $\mathbf{G_p}$-**distributed** or geometric distributed if

$$\mathbb{P}\{X = k\} = p\,(1 - p)^{k-1}, \quad k = 1, 2, \ldots,$$

5. $\mathbf{B_{n,p}^-}$-**distributed** or negatively binomial distributed if

$$\mathbb{P}\{X = k\} = \binom{k-1}{k-n} p^n (1 - p)^{k-n} = \binom{k-1}{n-1} p^n (1 - p)^{k-n}, \quad k = n, n+1, \ldots.$$

Remark 3.3.1. In view of Convention 3.1 we may suppose that all random variables of the preceding type are discrete. More precisely, we even may assume that X has values in the (at most countably infinite) set D with $\mathbb{P}_X(D) = 1$. For example, if X is $B_{n,p}$-distributed we may suppose that X has values in $\{0, \ldots, n\}$.

In quite similar way we denote special distributed continuous random variables. A real-valued random variable X is said to be

1. **uniformly distributed** on $[\alpha, \beta]$ if $\mathbb{P}_X$ is the uniform distribution on $[\alpha, \beta]$. That is, if $[a, b] \subseteq [\alpha, \beta]$, then

$$\mathbb{P}\{a \le X \le b\} = \frac{b-a}{\beta - \alpha},$$

2. **normally distributed** or $\mathcal{N}(\mu, \sigma^2)$-distributed if

$$\mathbb{P}\{a \le X \le b\} = \frac{1}{\sqrt{2\pi}\sigma} \int_a^b e^{-(x-\mu)^2/2\sigma^2} \, dx,$$

3. **standard normally distributed** if it is $\mathcal{N}(0, 1)$-distributed, that is,

$$\mathbb{P}\{a \le X \le b\} = \frac{1}{\sqrt{2\pi}} \int_a^b e^{-x^2/2} \, dx,$$

4. **gamma distributed** or $\Gamma_{\alpha,\beta}$-distributed if for $0 \le a < b < \infty$

$$\mathbb{P}\{a \le X \le b\} = \frac{1}{\alpha^\beta \, \Gamma(\beta)} \int_a^b x^{\beta-1} e^{-x/\alpha} \, dx,$$

5. **$E_{\lambda,n}$-distributed** or Erlang distributed if it is $\Gamma_{\lambda^{-1},n}$-distributed, that is, for $0 \le a < b < \infty$

$$\mathbb{P}\{a \le X \le b\} = \frac{\lambda^n}{(n-1)!} \int_a^b x^{n-1} e^{-\lambda x} \, dx,$$

6. **E_λ-distributed** or exponentially distributed if for $0 \le a < b < \infty$

$$\mathbb{P}\{a \le X \le b\} = \lambda \int_a^b e^{-\lambda x} \, dx = e^{-\lambda a} - e^{-\lambda b},$$

7. **Cauchy distributed** if

$$\mathbb{P}\{a \le X \le b\} = \frac{1}{\pi} \int_a^b \frac{dx}{1+x^2} = \frac{1}{\pi} [\arctan b - \arctan a].$$

Remark 3.3.2. If a random variable X possesses a special distribution, then all properties of $\mathbb{P}_X$ carry over to X. For example, in this language we may now formulate Poisson's limit theorem (Proposition 1.4.22) as follows.

Let X_n be B_{n,p_n}-distributed and suppose that $n p_n \to \lambda > 0$ as $n \to \infty$. Then

$$\lim_{n\to\infty} \mathbb{P}\{X_n = k\} = \mathbb{P}\{X = k\}, \quad k = 0, 1, \ldots,$$

where X is Pois_λ-distributed.

Or if X is gamma distributed, then $\mathbb{P}\{X > 0\} = \mathbb{P}_X((0, \infty)) = 1$, and so on.

Remark 3.3.3. A common question is **how does one get a random variable X possessing a certain given distribution**. For example, how do we construct a binomial or a normally distributed random variable? Suppose we want to model the rolling of a die by a random variable X, which is uniformly distributed on $\{1, \ldots, 6\}$. The easiest solution is to take $\Omega = \{1, \ldots, 6\}$ endowed with the uniform distribution $\mathbb{P}$ and define X by $X(\omega) = \omega$. But this is not the only way to get such a random variable. One may also roll the die n times and choose X as the value of the first (or of the second, etc.) roll. In a similar way random variables with other probability distribution may be constructed. Further possibilities to model random variables will be investigated in Section 4.4.

Summary: There are two ways to model a random experiment. The **classical approach** is to construct a **probability space** that describes this experiment. For example, if we toss a fair coin n times and record the number of "heads," then this may be described by the sample space $\{0, \ldots, n\}$ endowed with the probability measure $B_{n,1/2}$. **Another way** to model a certain random experiment is to choose a **random variable** X so that the probability of the occurrence of an event $B \subseteq \mathbb{R}$ equals $\mathbb{P}\{X \in B\}$. For example, the above experiment of tossing a coin may also be described by a binomial distributed random variable X (with parameters n and $1/2$). The great advantage of the second approach is that random variables allow algebraic operations. For example, they can be added, multiplied, or linearly combined. We will use this advantage extensively in the following sections.

3.4 Random Vectors

Suppose we are given n random variables $X_1, \ldots, X_n$ defined on a sample space Ω. Our objective is to combine these n variables into a single variable. More precisely, we will investigate the following type of vector-valued mappings.

Definition 3.4.1. Let $\vec{X}$ be a mapping from $\Omega \to \mathbb{R}^n$ represented as

$$\vec{X}(\omega) = \left(X_1(\omega), \ldots, X_n(\omega)\right), \quad \omega \in \Omega.$$

Then, $\vec{X}$ is said to be an (n-dimensional) **random vector** or **vector valued random variable**, provided that each of the X_js is a (real-valued) random variable. The random variables X_j, $1 \le j \le n$, are called **coordinate mappings** of $\vec{X}$.

Instead of $\vec{X}$, we may also write $(X_1, \ldots, X_n)$, that is,

$$(X_1, \ldots, X_n)(\omega) = (X_1(\omega), \ldots, X_n(\omega)), \quad \omega \in \Omega.$$

A random vector $\vec{X}$ maps Ω into $\mathbb{R}^n$, that is, we assign to each observed $\omega \in \Omega$ a vector $\vec{X}(\omega)$. The mapping $\vec{X}$ is again fixed and nonrandom. The randomness of $\vec{X}(\omega)$ is caused by the input.

Example 3.4.2. Roll a die two times. Let X_1 be the maximum value, X_2 the minimum and X_3 the sum of both rolls. The three-dimensional vector $\vec{X} = (X_1, X_2, X_3)$ maps $\Omega = \{1, \ldots, 6\}^2$ into $\mathbb{R}^3$. For example, the pair $(2, 5)$ is mapped to $(5, 2, 7)$ or $(5, 6)$ to $(6, 5, 11)$.

Example 3.4.3. Suppose there are N people in an auditorium. Enumerate them from 1 to N and choose one person according to the uniform distribution on $\{1, \ldots, N\}$. Say we have chosen person k. Let $X_1(k)$ be the height of this person and $X_2(k)$ his or her weight. As a result, we get a random two-dimensional vector $(X_1(k), X_2(k))$.

Example 3.4.4. We place n balls into m urns successively. Hereby, each urn is equally likely. If X_j denotes the number of balls in urn j, then we get an m-dimensional vector $\vec{X} = (X_1, \ldots, X_m)$. Observe that the values of $\vec{X}$ lie in the set $D = \{(k_1, \ldots, k_m) : k_1 + \cdots + k_m = n\} \subset \mathbb{N}_0^m$.

Remark 3.4.5. The preceding examples suggest that the values of the coordinate mappings depend on each other. For instance, in Example 3.4.3 larger values of X_1 make also those of X_2 more likely and vice versa. A basic aim of the following sections is to confirm this guess, that is, we want to find a mathematical formulation that describes whether or not two or more random variables are dependent or independent.

3.5 Joint and Marginal Distributions

The values of the vector $\vec{X}$ are randomly distributed in $\mathbb{R}^n$. Consequently, as in the case of random variables, events of the form $\{\vec{X} \in B\}$ occur with certain probabilities. But, in contrast to the case of random variables, the event B is now a subset of $\mathbb{R}^n$, not of $\mathbb{R}$ as before. More precisely, for events $B \subseteq \mathbb{R}^n$ we are interested in the following quantity[4]:

$$\mathbb{P}\{\omega \in \Omega : \vec{X}(\omega) \in B\} = \mathbb{P}\{\omega \in \Omega : (X_1(\omega), \ldots, X_n(\omega)) \in B\}. \tag{3.8}$$

The next proposition gives the exact formulation of the problem.

---

4 For random vectors $\vec{X}$ and $B \in \mathcal{B}(\mathbb{R}^n)$ it follows that $\vec{X}^{-1}(B) \in \mathcal{A}$. This can be proved by similar methods as we used in the proof of Proposition 3.1.6. Thus, if $B \in \mathcal{B}(\mathbb{R}^n)$, then eq. (3.8) and also eq. (3.9) are well-defined.

Definition 3.5.1. Let $\vec{X} : \Omega \rightarrow \mathbb{R}^n$ be a random vector with coordinate mappings $X_1, \ldots, X_n$. For each Borel set $B \in \mathcal{B}(\mathbb{R}^n)$ we set

$$\mathbb{P}_{\vec{X}}(B) = \mathbb{P}_{(X_1, \ldots, X_n)}(B) = \mathbb{P}\{\vec{X} \in B\}. \tag{3.9}$$

The mapping $\mathbb{P}_{\vec{X}}$ from $\mathcal{B}(\mathbb{R}^n)$ into $[0, 1]$ is said to be the **probability distribution**, or, in short, the **distribution** of $\vec{X}$. Often, $\mathbb{P}_{\vec{X}} = \mathbb{P}_{(X_1, \ldots, X_n)}$ will also be called the **joint distribution** of $X_1, \ldots, X_n$.

In eq. (3.9) we used the shorter expression

$$\mathbb{P}\{\vec{X} \in B\} = \mathbb{P}\{\omega \in \Omega : \vec{X}(\omega) \in B\}.$$

As for random variables the following is also valid in the case of random vectors.

Proposition 3.5.2. *The mapping $\mathbb{P}_{\vec{X}}$ is a probability measure defined on $\mathcal{B}(\mathbb{R}^n)$.*

Proof: The proof is completely analogous to that of Proposition 3.2.2. Therefore, we decide not to present it here. ∎

Let us evaluate $\mathbb{P}_{\vec{X}}(B)$ for special Borel sets $B \subseteq \mathbb{R}^n$. If Q is a box in $\mathbb{R}^n$ as in eq. (1.65), that is, for certain real numbers $a_i < b_i$ we have

$$Q = [a_1, b_1] \times \cdots \times [a_n, b_n],$$

then it follows that

$$\mathbb{P}_{\vec{X}}(Q) = \mathbb{P}\{\vec{X} \in Q\} = \mathbb{P}\{\omega \in \Omega : a_1 \leq X_1(\omega) \leq b_1, \ldots, a_n \leq X_n(\omega) \leq b_n\}.$$

The last expression may also be written as

$$\mathbb{P}\{a_1 \leq X_1 \leq b_1, \ldots, a_n \leq X_n \leq b_n\}.$$

Hence, for each box $Q = [a_1, b_1] \times \cdots \times [a_n, b_n]$, we obtain

$$\mathbb{P}_{\vec{X}}(Q) = \mathbb{P}\{a_1 \leq X_1 \leq b_1, \ldots, a_n \leq X_n \leq b_n\}.$$

Thus the quantity $\mathbb{P}_{\vec{X}}(Q)$ is the probability of the occurrence of the following event: X_1 attains a value in $[a_1, b_1]$, and **at the same time** X_2 attains a value in $[a_2, b_2]$, and so on up to X_n attains a value in $[a_n, b_n]$.

Example 3.5.3. Roll a fair die three times. Let X_1, X_2, and X_3 be the observed values in the first, second, and third roll. If $Q = [1, 2] \times [0, 1] \times [3, 4]$, then it follows

$$\mathbb{P}_{\vec{X}}(Q) = \mathbb{P}\{X_1 \in \{1, 2\}, X_2 = 1, X_3 \in \{3, 4\}\} = \frac{1}{54}.$$

Remark 3.5.4. The previous considerations can easily be generalized to sets $B \subseteq \mathbb{R}^n$ of the form $B = B_1 \times \cdots \times B_n$ with $B_j \in \mathcal{B}(\mathbb{R})$. Then

$$\mathbb{P}_{\vec{X}}(B) = \mathbb{P}\{X_1 \in B_1, \ldots, X_n \in B_n\} \tag{3.10}$$

Next we introduce the notion of marginal distributions of a random vector.

> **Definition 3.5.5.** Let $\vec{X} = (X_1, \ldots, X_n)$ be a random vector. The n probability measures $\mathbb{P}_{X_1}$ to $\mathbb{P}_{X_n}$ are called the **marginal distributions** of $\vec{X}$.

Observe that each marginal distribution $\mathbb{P}_{X_j}$ is a probability measure on $\mathcal{B}(\mathbb{R})$, while the joint distribution $\mathbb{P}_{(X_1, \ldots, X_n)}$ is a probability measure defined on $\mathcal{B}(\mathbb{R}^n)$.

In this context, the following important question arises: does the joint distribution determine the marginal distributions and/or can the joint distribution be derived from the marginal ones?

The next proposition gives the first answer.

Proposition 3.5.6. *Let $\vec{X} = (X_1, \ldots, X_n)$ be a random vector. If $1 \le j \le n$ and $B \in \mathcal{B}(\mathbb{R})$, then it follows*

$$\mathbb{P}_{X_j}(B) = \mathbb{P}_{(X_1, \ldots, X_n)}(\mathbb{R} \times \cdots \times \underbrace{B}_{j} \times \cdots \times \mathbb{R}).$$

In particular, the joint distribution determines the marginal ones.

Proof: The proof is a direct consequence of formula (3.10). Let us apply it to $B_i = \mathbb{R}$ if $i \ne j$ and to $B_j = B$. Then, as asserted,

$$\mathbb{P}_{(X_1, \ldots, X_n)}(\mathbb{R} \times \cdots \times \underbrace{B}_{j} \times \cdots \times \mathbb{R})$$

$$= \mathbb{P}\{X_1 \in \mathbb{R}, \ldots, X_j \in B, \ldots, X_n \in \mathbb{R}\} = \mathbb{P}\{X_j \in B\} = \mathbb{P}_{X_j}(B). \qquad \blacksquare$$

The question whether or not the marginal distributions determine the joint distribution is postponed for a moment. It will be investigated in Example 3.5.8 and, more thoroughly, in Section 3.6. Before investigating this problem, let us derive some concrete formulas to evaluate the marginal distributions. Here we consider the two cases of discrete and continuous random variables separately.

3.5.1 Marginal Distributions: Discrete Case

To make the results in this subsection easier to understand, we only consider the case of two-dimensional vectors. That is, we investigate two random variables and show how their distributions may be derived from their joint one. We indicate later on how this approach extends to more than two random variables.

In order to avoid confusing notations with many indices, given a two-dimensional random vector, we denote its coordinate mappings by X and Y and not by X_1 and X_2. This should not lead to mix-ups. Thus, we investigate the random vector (X, Y) with joint distribution $\mathbb{P}_{(X,Y)}$ and marginal distributions $\mathbb{P}_X$ and $\mathbb{P}_Y$. This vector acts as

$$(X, Y)(\omega) = (X(\omega), Y(\omega)), \quad \omega \in \Omega.$$

Suppose now that X and Y are discrete. Then, there are finite or countably infinite sets $D = \{x_1, x_2, \ldots\}$ and $E = \{y_1, y_2, \ldots\}$ such that $X : \Omega \to D$ as well as $Y : \Omega \to E$. Consequently, the vector (X, Y) maps Ω into the (at most countably infinite) set $D \times E \subset \mathbb{R}^2$. Observe that

$$D \times E = \{(x_i, y_j) : i, j = 1, 2, \ldots\},$$

hence $\mathbb{P}_{(X,Y)}$ is discrete as well and uniquely described by the numbers

$$p_{ij} := \mathbb{P}_{(X,Y)}(\{(x_i, y_j)\}) = \mathbb{P}(X = x_i, Y = y_j), \quad i, j = 1, 2, \ldots . \tag{3.11}$$

More precisely, given $B \subseteq \mathbb{R}^2$, then

$$\mathbb{P}_{(X,Y)}(B) = \mathbb{P}\{(X, Y) \in B\} = \sum_{\{(i,j):(x_i,y_j)\in B\}} p_{ij} .$$

We turn now to the description of the marginal distributions $\mathbb{P}_X$ and $\mathbb{P}_Y$. These are uniquely determined by the numbers

$$q_i := \mathbb{P}_X(\{x_i\}) = \mathbb{P}\{X = x_i\} \quad \text{and} \quad r_j := \mathbb{P}_Y(\{y_j\}) = \mathbb{P}\{Y = y_j\}. \tag{3.12}$$

In other words, if $B, C \subseteq \mathbb{R}$, then it follows

$$\mathbb{P}_X(B) = \mathbb{P}\{X \in B\} = \sum_{\{i:x_i\in B\}} q_i \quad \text{and} \quad \mathbb{P}_Y(C) = \mathbb{P}\{Y \in C\} = \sum_{\{j:y_j\in C\}} r_j .$$

The next proposition is nothing else than a reformulation of Proposition 3.5.6 in the case of discrete random variables.

Proposition 3.5.7. *Let the probabilities p_{ij}, q_i, and r_j be defined by eqs. (3.11) and (3.12), respectively. Then the q_is and r_js may be evaluated by the following equations:*

$$q_i = \sum_{j=1}^{\infty} p_{ij} \quad for \ i = 1, 2, \ldots \quad and \quad r_j = \sum_{i=1}^{\infty} p_{ij} \quad for \ j = 1, 2, \ldots .$$

Proof: As already mentioned, Proposition 3.5.7 is a direct consequence of Proposition 3.5.6. But for better understanding we prefer to give a direct proof.

By virtue of the σ-additivity of $\mathbb{P}$ it follows

$$q_i = \mathbb{P}\{X = x_i\} = \mathbb{P}\{X = x_i, Y \in E\} = \mathbb{P}\left\{X = x_i, Y \in \bigcup_{j=1}^{\infty}\{y_j\}\right\}$$

$$= \sum_{j=1}^{\infty} \mathbb{P}\left\{X = x_i, Y \in \{y_j\}\right\} = \sum_{j=1}^{\infty} \mathbb{P}\left\{X = x_i, Y = y_j\right\} = \sum_{j=1}^{\infty} p_{ij} .$$

This proves the first part. The proof for the r_js follows exactly along the same line. Here, one uses

$$r_j = \mathbb{P}\{Y = y_j\} = \mathbb{P}\{X \in D, Y = y_j\} = \sum_{i=1}^{\infty} \mathbb{P}\{X = x_i, Y = y_j\} = \sum_{i=1}^{\infty} p_{ij} .$$

This completes the proof. ∎

The equations in Proposition 3.5.7 may be represented in table form as follows:

$Y\backslash X$	x_1	x_2	x_3	$\cdot$	$\cdot$	
y_1	p_{11}	p_{21}	p_{31}	$\cdot$	$\cdot$	r_1
y_2	p_{12}	p_{22}	p_{32}	$\cdot$	$\cdot$	r_2
y_3	p_{13}	p_{23}	p_{33}	$\cdot$	$\cdot$	r_3
$\cdot$		$\cdot$	$\cdot$	$\cdot$	$\cdot$	$\cdot$
$\cdot$		$\cdot$	$\cdot$	$\cdot$	$\cdot$	$\cdot$
	q_1	q_2	q_3	$\cdot$	$\cdot$	1

The entries in the above matrix are the corresponding probabilities. For example, the entry p_{32} is put into the row marked by x_3 and into the column where one finds y_2 at the left-hand side. This tells us p_{32} is the probability that X attains the value x_3 and, **at the**

same time, Y equals y_2. At the right and at the lower margins,[5] one finds the corresponding sums of the columns and of the rows, respectively. These numbers describe the marginal distributions (that of X at the bottom and that of Y at the right margin). Finally, the number "1" at the right lower corner says that both the right column and bottom row have to add up to "1."

Example 3.5.8. There are four balls in an urn, two labeled with "0" and another two labeled with "1." Choose two balls without replacing the first one. Let X be the value of the first ball and Y that of the second. Direct calculations (use the law of multiplication) lead to

$$\mathbb{P}\{X = 0, Y = 0\} = \frac{1}{6}, \quad \mathbb{P}\{X = 0, Y = 1\} = \frac{1}{3}$$

$$\mathbb{P}\{X = 1, Y = 0\} = \frac{1}{3}, \quad \mathbb{P}\{X = 1, Y = 1\} = \frac{1}{6}$$

In tabular form this result reads as follows:

$Y\backslash X$	0	1	
0	$\frac{1}{6}$	$\frac{1}{3}$	$\frac{1}{2}$
1	$\frac{1}{3}$	$\frac{1}{6}$	$\frac{1}{2}$
	$\frac{1}{2}$	$\frac{1}{2}$	1

Now suppose that we replace the first ball. This time we denote the values of the first and second ball by X' and Y', respectively. The corresponding table may now be written as follows:

$Y'\backslash X'$	0	1	
0	$\frac{1}{4}$	$\frac{1}{4}$	$\frac{1}{2}$
1	$\frac{1}{4}$	$\frac{1}{4}$	$\frac{1}{2}$
	$\frac{1}{2}$	$\frac{1}{2}$	1

Let us look at Example 3.5.8 more thoroughly. In both cases (nonreplacing and replacing) the marginal distributions coincide, that is, $\mathbb{P}_X = \mathbb{P}_{X'}$ and $\mathbb{P}_Y = \mathbb{P}_{Y'}$. But, on the other hand, the joint distributions are different, that is, we have $\mathbb{P}_{(X,Y)} \neq \mathbb{P}_{(X',Y')}$.

Conclusion: The marginal distributions do **not**, in general, determine the joint distribution. Recall that Proposition 3.5.6 asserts the converse implication: The marginal distributions can be derived from the joint distribution.

---

5 This explains the name "marginal" for the distribution of the coordinate mappings.

Example 3.5.9. Roll a fair die twice. Let X be the minimum value of both rolls and Y the maximum. Then, if $k, l = 1, \ldots, 6$, it is easy to see that

$$\mathbb{P}\{X = k, Y = l\} = \begin{cases} 0 & : \quad k > l \\ \frac{1}{36} & : \quad k = l \\ \frac{1}{18} & : \quad k < l \end{cases}$$

Hence, the joint distribution in table form looks as follows:

$Y\backslash X$	1	2	3	4	5	6	
1	$\frac{1}{36}$	0	0	0	0	0	$\frac{1}{36}$
2	$\frac{1}{18}$	$\frac{1}{36}$	0	0	0	0	$\frac{3}{36}$
3	$\frac{1}{18}$	$\frac{1}{18}$	$\frac{1}{36}$	0	0	0	$\frac{5}{36}$
4	$\frac{1}{18}$	$\frac{1}{18}$	$\frac{1}{18}$	$\frac{1}{36}$	0	0	$\frac{7}{36}$
5	$\frac{1}{18}$	$\frac{1}{18}$	$\frac{1}{18}$	$\frac{1}{18}$	$\frac{1}{36}$	0	$\frac{9}{36}$
6	$\frac{1}{18}$	$\frac{1}{18}$	$\frac{1}{18}$	$\frac{1}{18}$	$\frac{1}{18}$	$\frac{1}{36}$	$\frac{11}{36}$
	$\frac{11}{36}$	$\frac{9}{36}$	$\frac{7}{36}$	$\frac{5}{36}$	$\frac{3}{36}$	$\frac{1}{36}$	1

If, for example, $B = \{(4, 5), (5, 4), (6, 5), (5, 6)\}$, then the values in the table imply $\mathbb{P}_{(X,Y)}(B) = 1/9$. In the same way, one gets $\mathbb{P}\{2 \leq X \leq 4\} = (9 + 7 + 5)/36 = 7/12$.

To finish, we shortly go into the case of more than two discrete random variables. Thus, let $X_1, \ldots, X_n$ be random variables with $X_j : \Omega \to D_j$, where the sets D_j are either finite or countably infinite. The set D defined by

$$D = D_1 \times \cdots \times D_n = \{(x_1, \ldots, x_n), \; x_j \in D_j\}$$

is at most countably infinite and $\vec{X} : \Omega \to D$. Consequently, $\mathbb{P}_{\vec{X}}$ is uniquely described by the probabilities

$$p_{x_1, \ldots, x_n} = \mathbb{P}\{X_1 = x_1, \ldots, X_n = x_n\}, \quad x_j \in D_j.$$

Proposition 3.5.10. *For $1 \leq j \leq n$ and $x \in D_j$,*

$$\mathbb{P}\{X_j = x\} = \sum_{x_1 \in D_1} \cdots \sum_{x_{j-1} \in D_{j-1}} \sum_{x_{j+1} \in D_{j+1}} \cdots \sum_{x_n \in D_n} p_{x_1, \ldots, x_{j-1}, x, x_{j+1} \ldots x_n}.$$

Proof: The proof is exactly the same as that of Proposition 3.5.7. Therefore, we omit it. ∎

Next, we want to state an important example that shows how Proposition 3.5.10 applies. To do so we need the following definition.

Definition 3.5.11. Let n and m be integers with $m \geq 2$ and let $p_1, \ldots, p_m$ be certain success probabilities satisfying $p_j \geq 0$ and $p_1 + \cdots + p_m = 1$. An m-dimensional random vector $\vec{X} = (X_1, \ldots, X_m)$ is called **multinomial distributed** with parameters n and $p_1, \ldots, p_m$ if, whenever $k_1 + \cdots + k_m = n$, then

$$\mathbb{P}\{X_1 = k_1, \ldots, X_m = k_m\} = \binom{n}{k_1, \ldots, k_m} p_1^{k_1} \cdots p_m^{k_m}.$$

Equivalently, a random vector $\vec{X}$ is multinomial distributed if and only if its probability distribution $\mathbb{P}_{\vec{X}}$ is a multinomial distribution as introduced in Definition 1.4.12.

Remark 3.5.12. The m-dimensional random vector $\vec{X}$ in Example 3.4.4 is multinomial distributed with parameters n and $p_j = 1/m$. That means

$$\mathbb{P}\{X_1 = k_1, \ldots, X_m = k_m\} = \binom{n}{k_1, \ldots, k_m} \left(\frac{1}{m}\right)^n, \quad k_1 + \cdots + k_m = n.$$

Example 3.5.13. Let $\vec{X} = (X_1, \ldots, X_m)$ be a multinomial random vector with parameters n and $p_1, \ldots, p_m$. What are the marginal distributions of $\vec{X}$?

Answer: To simplify the calculations, we only determine the probability distribution of X_m. The other cases follow in the same way. First note that in the notation of Proposition 3.5.10

$$p_{k_1, \ldots, k_m} = \begin{cases} \binom{n}{k_1, \ldots, k_m} p_1^{k_1} \cdots p_m^{k_m} & : k_1 + \cdots + k_m = n \\ 0 & : k_1 + \cdots + k_m \neq n \end{cases}$$

Consequently, Proposition 3.5.10 leads to

$$\mathbb{P}\{X_m = k\} = \sum_{k_1=0}^{n} \cdots \sum_{k_{m-1}=0}^{n} p_{k_1, \ldots, k_{m-1}, k}$$

$$= \sum_{k_1 + \cdots + k_{m-1} = n-k} \frac{n!}{k_1! \cdots k_{m-1}! \, k!} p_1^{k_1} \cdots p_{m-1}^{k_{m-1}} p_m^{k}$$

$$= \frac{n!}{k! \, (n-k)!} p_m^{k} \sum_{k_1 + \cdots + k_{m-1} = n-k} \frac{(n-k)!}{k_1! \cdots k_{m-1}!} p_1^{k_1} \cdots p_{m-1}^{k_{m-1}}$$

$$= \binom{n}{k} p_m^{k} \sum_{k_1 + \cdots + k_{m-1} = n-k} \binom{n-k}{k_1, \ldots, k_{m-1}} p_1^{k_1} \cdots p_{m-1}^{k_{m-1}}$$

$$= \binom{n}{k} p_m^{k} (p_1 + \cdots + p_{m-1})^{n-k} = \binom{n}{k} p_m^{k} (1 - p_m)^{n-k}.$$

Hereby, in the last step, we used the multinomial theorem (Proposition A.16) with $m-1$ summands, with power $n-k$ and entries $p_1, \ldots, p_{m-1}$.

Thus X_m is binomial distributed with parameters n and p_m. In the same way one gets that each X_j is B_{n,p_j}-distributed.

Remark 3.5.14. The previous result can also be seen more directly without using Proposition 3.5.10. Assume we place n particles into m boxes, where p_j is the probability to put a single particle into box j. Fix some $j \le m$ and and let success occur if a particle is placed into box j. Then X_j equals the number of successes, hence it is B_{n,p_j}-distributed. Note that failure occurs if the particle is not placed into box j, and the probability for this is given by $1 - p_j = \sum_{\substack{i=1 \\ i \ne j}}^{n} p_i$.

3.5.2 Marginal Distributions: Continuous Case

Let us turn now to the continuous case. Analogous to Definition 3.2.7, a random vector is said to be continuous whenever it possesses a density.[6] More precisely, we suppose that a random vector shares the following property.

Definition 3.5.15. A random vector $\vec{X} = (X_1, \dots, X_n)$ is said to be **continuous** if there is a function $p : \mathbb{R}^n \to \mathbb{R}$ such that, for all numbers, $a_j < b_j, 1 \le j \le n$,

$$\mathbb{P}\{a_1 \le X_1 \le b_1, \dots, a_n \le X_n \le b_n\} = \int_{a_1}^{b_1} \cdots \int_{a_n}^{b_n} p(x_1, \dots, x_n)\, dx_n \dots dx_1.$$

Equivalently, for all real numbers $t_1, \dots, t_n$,

$$\mathbb{P}\{X_1 \le t_1, \dots, X_n \le t_n\} = \int_{-\infty}^{t_1} \cdots \int_{-\infty}^{t_n} p(x_1, \dots, x_n)\, dx_n \dots dx_1.$$

The function p is called the **density function** of $\vec{X}$ or also the **joint density** of $X_1, \dots, X_n$.

Remark 3.5.16. Observe that a random vector $\vec{X}$ is continuous if and only if its probability distribution $\mathbb{P}_{\vec{X}}$ is so, that is, the joint distribution of $X_1, \dots X_n$, is a continuous probability measure on $\mathcal{B}(\mathbb{R}^n)$ in the sense of Definition 1.8.5. Moreover, its density function coincides with the density of $\mathbb{P}_{\vec{X}}$.

---

6 The following is true: for continuous random variables the generated vector possesses a density. The proof is far above the scope of this book. Furthermore, we do not need this assertion because we assume $\vec{X}$ to be continuous, not the X_js.

In the case of continuous random variables, the marginal distributions are evaluated by the following rule.

Proposition 3.5.17. *If a random vector* $\vec{X} = (X_1, \ldots, X_n)$ *has density* $p : \mathbb{R}^n \to \mathbb{R}$, *then for each* $j \leq n$ *the random variable* X_j *is continuous with density*

$$p_j(x_j) = \underbrace{\int_{-\infty}^{\infty} \cdots \int_{-\infty}^{\infty}}_{n-1 \, integrals} p(\ldots, x_{j-1}, x_j, x_{j+1} \ldots) dx_n \ldots dx_{j+1} \, dx_{j-1} \ldots dx_1 . \qquad (3.13)$$

If $n = 2$, *the above formula reads as*

$$p_1(x_1) = \int_{-\infty}^{\infty} p(x_1, x_2) \, dx_2 \quad and \quad p_2(x_2) = \int_{-\infty}^{\infty} p(x_1, x_2) \, dx_1 .$$

Proof: Fix an integer $j \leq n$. An application of Proposition 3.5.6 implies

$$\mathbb{P}_{X_j}([a, b]) = \mathbb{P}_{\vec{X}}(\mathbb{R} \times \cdots \times \underbrace{[a, b]}_{j} \times \cdots \times \mathbb{R})$$

$$= \int_{-\infty}^{\infty} \cdots \underbrace{\int_{a}^{b}}_{j} \cdots \int_{-\infty}^{\infty} p(x_1, \ldots x_n) \, dx_n \ldots dx_1$$

$$= \int_{a}^{b} \left[\int_{-\infty}^{\infty} \cdots \int_{-\infty}^{\infty} p(\ldots, x_{j-1}, x_j, x_{j+1}, \ldots) dx_n \ldots dx_{j+1} \, dx_{j-1} \ldots dx_1 \right] dx_j$$

$$= \int_{a}^{b} p_j(x_j) \, dx_j$$

with p_j defined by eq. (3.13). The interchange of the integrals was justified by Fubini's theorem (Proposition A.5.5); note that p is a density, hence it is non-negative. Since the preceding equation holds for all real numbers $a < b$, the function p_j has to be a density of $\mathbb{P}_{X_j}$. This completes the proof. ■

Remark 3.5.18. Another way to formulate Proposition 3.5.17 is as follows: if the function $p : \mathbb{R}^n \to \mathbb{R}$ is a joint density of $X_1, \ldots, X_n$, then $p_1, \ldots, p_n$ defined in eq. (3.13) are densities of the random variables $X_1, \ldots, X_n$, respectively.

Example 3.5.19. Choose by random a point $x = (x_1, x_2, x_3)$ in the unit ball of $\mathbb{R}^3$. How are the coordinates x_1, x_2, and x_3 distributed?

Answer: Let $\vec{X} = (X_1, X_2, X_3)$ be uniformly distributed on the unit ball

$$K = \{(x_1, x_2, x_3) : x_1^2 + x_2^2 + x_3^2 \leq 1\}.$$

Then the joint density is given by[7]

$$p(x) = \begin{cases} \frac{3}{4\pi} & : \quad x \in K \\ 0 & : \quad x \notin K \end{cases}$$

An application of Proposition 3.5.17 leads to $p_1(x_1) = 0$ whenever $|x_1| > 1$ and, if $|x_1| \leq 1$, then it follows that

$$p_1(x_1) = \frac{3}{4\pi} \iint\limits_{x_2^2 + x_3^2 \leq 1 - x_1^2} dx_2 dx_3 = \frac{3}{4\pi}(1 - x_1^2)\pi = \frac{3}{4}(1 - x_1^2).$$

Hence, X_1 has the density

$$p_1(s) = \begin{cases} \frac{3}{4}(1 - s^2) & : \quad -1 \leq s \leq 1 \\ 0 & : \quad \text{otherwise} \end{cases}$$

Of course, by symmetry X_2 and X_3 possess exactly the same distribution densities.

Example 3.5.20. Suppose the two-dimensional random vector (X_1, X_2) has the density p defined by[8]

$$p(x_1, x_2) := \begin{cases} 8 x_1 x_2 & : \quad 0 \leq x_1 \leq x_2 \leq 1 \\ 0 & : \quad \text{otherwise} \end{cases}$$

Then, the density p_1 of X_1 is given by

$$p_1(x_1) = \int_{-\infty}^{\infty} p(x_1, x_2)\, dx_2 = 8 x_1 \int_{x_1}^{1} x_2\, dx_2 = 4(x_1 - x_1^3), \quad 0 \leq x_1 \leq 1,$$

and $p_1(x_1) = 0$ if $x_1 \notin [0, 1]$.

In the case of p_2, the density of X_2, it follows that

$$p_2(x_2) = \int_{-\infty}^{\infty} p(x_1, x_2)\, dx_1 = 8 x_2 \int_{0}^{x_2} x_1\, dx_1 = 4 x_2^3, \quad 0 \leq x_2 \leq 1,$$

and $p_2(x_2) = 0$ if $x_2 \notin [0, 1]$.

---

7 Recall $\mathrm{vol}_3(K) = \frac{4}{3}\pi$.
8 Check that p is indeed a probability density.

3.6 Independence of Random Variables

The central question considered in this section is as follows: when are n given random variables independent? Surely everybody has an intuitive idea about the independence or dependence of random values. But how do we express this property by a mathematical formula? Let us try to approach a solution of this problem with an example.

Example 3.6.1. Roll a fair die twice and define the two random variables X_1 and X_2 as a result of the first and second roll, respectively. These random variables are intuitively independent of each other. But what formulas do these express? Take two subsets $B_1, B_2 \in \{1, \ldots, 6\}$ and look at their preimages $A_1 = X_1^{-1}(B_1)$ and $A_2 = X_2^{-1}(B_2)$. Then A_1 occurs if the first result belongs to B_1 while the same is true for A_2 whenever the second result belongs to B_2. For example, A_1 might be that the first result is an even number while A_2 could occur if the second result equals "4." The basic observation is, no matter how B_1 and B_2 were chosen, the occurrence of their preimages A_1 and A_2 only depends on the first or second roll, respectively. Therefore, they should be independent (as events) in the sense of Definition 2.2.2, that is, the following equation should hold:

$$\mathbb{P}\{X_1 \in B_1, X_2 \in B_2\} = \mathbb{P}\left(X_1^{-1}(B_1) \cap X_2^{-1}(B_2)\right) = \mathbb{P}\left(A_1 \cap A_2\right)$$

$$= \mathbb{P}(A_1) \cdot \mathbb{P}(A_2) = \mathbb{P}\left(X_1^{-1}(B_1)\right) \cdot \mathbb{P}\left(X_2^{-1}(B_2)\right) = \mathbb{P}\{X_1 \in B_1\} \cdot \mathbb{P}\{X_2 \in B_2\}.$$

This observation leads us to the following definition of independence.

Definition 3.6.2. Let $X_1, \ldots, X_n$ be n random variables mapping Ω into $\mathbb{R}$. These variables are said to be (stochastically) **independent** if, for all Borel sets $B_j \subseteq \mathbb{R}$,

$$\mathbb{P}\{X_1 \in B_1, \ldots, X_n \in B_n\} = \mathbb{P}\{X_1 \in B_1\} \cdots \mathbb{P}\{X_n \in B_n\}. \qquad (3.14)$$

Remark 3.6.3. By virtue of Remark 3.5.4, eq. (3.14) may also be written as

$$\mathbb{P}_{(X_1, \ldots, X_n)}(B_1 \times \cdots B_n) = \mathbb{P}_{X_1}(B_1) \cdots \mathbb{P}_{X_n}(B_n), \quad B_j \in \mathcal{B}(\mathbb{R}).$$

Before proceeding further, we shortly recall Corollary 1.9.7.

Corollary 3.6.4. *Given n probability measures $\mathbb{P}_1, \ldots, \mathbb{P}_n$ defined on $\mathcal{B}(\mathbb{R})$, there exists a unique probability measure $\mathbb{P}$ on $\mathcal{B}(\mathbb{R}^n)$, the product measure denoted by $P = \mathbb{P}_1 \otimes \cdots \otimes \mathbb{P}_n$, such that for all Borel sets $B_j \subseteq \mathbb{R}$*

$$\mathbb{P}(B_1 \times \cdots \times B_n) = \mathbb{P}_1(B_1) \cdots \mathbb{P}_n(B_n). \qquad (3.15)$$

Now, we are prepared to state the characterization of independent random variables by properties of their distributions.

Proposition 3.6.5. *The random variables $X_1, \ldots, X_n$ are independent if and only if their joint distribution coincides with the product probability of the marginal distributions. That is, if and only if*

$$\mathbb{P}_{(X_1,\ldots,X_n)} = \mathbb{P}_{X_1} \otimes \cdots \otimes \mathbb{P}_{X_n}.$$

Proof: In view of Corollary 3.6.4, the product probability $\mathbb{P}$ of $\mathbb{P}_{X_1}, \ldots, \mathbb{P}_{X_n}$ is the unique probability measure on $\mathcal{B}(\mathbb{R}^n)$ satisfying

$$\mathbb{P}(B_1 \times \cdots \times B_n) = \mathbb{P}_{X_1}(B_1) \cdots \mathbb{P}_{X_n}(B_n), \quad B_j \in \mathcal{B}(\mathbb{R}).$$

On the other hand, by Remark 3.6.3, the X_js are independent if and only if

$$\mathbb{P}_{(X_1,\ldots,X_n)}(B_1 \times \cdots B_n) = \mathbb{P}_{X_1}(B_1) \cdots \mathbb{P}_{X_n}(B_n), \quad B_j \in \mathcal{B}(\mathbb{R}). \tag{3.16}$$

Consequently, eq. (3.16) holds for all Borel sets B_j if and only if $\mathbb{P}_{(X_1,\ldots,X_n)}$ is the product probability $\mathbb{P}_{X_1} \otimes \cdots \otimes \mathbb{P}_{X_n}$. This completes the proof. ∎

Corollary 3.6.6. *If $X_1, \ldots, X_n$ are independent, the joint distribution $\mathbb{P}_{(X_1,\ldots,X_n)}$ is uniquely determined by its marginal distributions $\mathbb{P}_{X_1}, \ldots, \mathbb{P}_{X_n}$.*

Proof: Proposition 3.6.5 asserts $\mathbb{P}_{(X_1,\ldots,X_n)} = \mathbb{P}_{X_1} \otimes \cdots \otimes \mathbb{P}_{X_n}$. Hence, the joint distribution is uniquely described by the marginal ones. ∎

The next proposition clarifies the relation between the properties "independence of events" and "independence of random variables." At a first glance the assertion looks trivial or self-evident, but it is not at all. The reason is that the definition of independence for more than two events, as given in Definition 2.2.12, is more complicated than in the case of two events.

Proposition 3.6.7. *The random variables $X_1, \ldots, X_n$ are independent if and only if for all Borel sets $B_1, \ldots, B_n$ in $\mathbb{R}$ the events*

$$X_1^{-1}(B_1), \ldots, X_n^{-1}(B_n)$$

are stochastically independent in $(\Omega, \mathcal{A}, \mathbb{P})$.

Proof [9]: When are $X_1^{-1}(B_1), \ldots, X_n^{-1}(B_n)$ independent? According to Definition 2.2.12 this holds if for all subsets $I \subseteq \{1, \ldots, n\}$

$$\mathbb{P}\left(\bigcap_{i\in I} X_i^{-1}(B_i)\right) = \prod_{i\in I} \mathbb{P}(X_i^{-1}(B_i)). \tag{3.17}$$

On the other hand, by Definition 3.6.2, the $X_1, \ldots, X_n$ are independent if

$$\mathbb{P}\left(\bigcap_{i=1}^n X_i^{-1}(B_i)\right) = \mathbb{P}\{X_1 \in B_1, \ldots, X_n \in B_n\}$$

$$= \prod_{i=1}^n \mathbb{P}\{X_i \in B_i\} = \prod_{i=1}^n \mathbb{P}(X_i^{-1}(B_i)). \tag{3.18}$$

Of course, eq. (3.17) implies eq. (3.18); use eq. (3.17) with $I = \{1, \ldots, n\}$. But it is far from clear why, conversely, eq. (3.18) should imply eq. (3.17). As we saw in Example 2.2.10, for fixed sets B_j this is even false. The key observation is that eq. (3.17) has to be valid for **all** Borel sets B_j. This allows us to choose the Borel sets in an appropriate way.

Thus let us assume the validity of eq. (3.18) for all Borel sets in $\mathbb{R}$. Given $B_j \in \mathcal{B}(\mathbb{R})$ and a subset I of $\{1, \ldots, n\}$ we introduce "new" $B_1', \ldots, B_n'$ as follows: $B_i' = B_i$ if $i \in I$ and $B_i' = \mathbb{R}$ if $i \notin I$. This choice of the B_j' implies $X_j^{-1}(B_j') = \Omega$ whenever $i \notin I$. An application of eq. (3.18) to $B_1', \ldots, B_n'$ leads to (recall $X_i^{-1}(B_i') = \Omega$ if $i \notin I$)

$$\mathbb{P}\left(\bigcap_{i\in I} X_i^{-1}(B_i)\right) = \mathbb{P}\left(\bigcap_{i=1}^n X_i^{-1}(B_i')\right) = \prod_{i=1}^n \mathbb{P}(X_i^{-1}(B_i')) = \prod_{i\in I} \mathbb{P}(X_i^{-1}(B_i)).$$

This proves eq. (3.17) for any subset I of $\{1, \ldots, n\}$. Hence, $X_1^{-1}(B_1), \ldots, X_n^{-1}(B_n)$ are independent as asserted. ∎

Remark 3.6.8. To verify the independence of $X_1, \ldots, X_n$, it is not necessary to check eq. (3.14) for all Borel sets B_j. It suffices if this is valid for real intervals $[a_j, b_j]$. In other words, $X_1, \ldots, X_n$ are independent if and only if, for all $a_j < b_j$,

$$\mathbb{P}\{a_1 \leq X_1 \leq b_1, \ldots, a_n \leq X_n \leq b_n\}$$

$$= \mathbb{P}\{a_1 \leq X_1 \leq b_1\} \cdots \mathbb{P}\{a_n \leq X_n \leq b_n\}. \tag{3.19}$$

Furthermore, it also suffices to choose the Borel sets as intervals $(-\infty, t_j]$ for $t_j \in \mathbb{R}$, i.e., $X_1, \ldots, X_n$ are independent if and only if, for all $t_j \in \mathbb{R}$,

$$\mathbb{P}\{X_1 \leq t_1, \ldots, X_n \leq t_n\} = \mathbb{P}\{X_1 \leq t_1\} \cdots \mathbb{P}\{X_n \leq t_n\}.$$

---

[9] The proposition and its proof are not necessarily needed for further reading. But they may be helpful for a better understanding of the independence of events and of random variables.

3.6.1 Independence of Discrete Random Variables

As in Section 3.5.1, we restrict ourselves to the case of two random variables. The extension to more than two variables is straightforward and will be shortly considered at the end of this section. We use the same notation as in Section 3.5.1. That is, the two random variables are denoted by X and Y, and they map Ω into $D = \{x_1, x_2, \ldots\}$ and $E = \{y_1, y_2, \ldots\}$, respectively. The joint distribution (X, Y) as well as the marginal distributions, that is, the distributions of X and Y, are described as in eqs. (3.11) and (3.12) by

$$p_{ij} = \mathbb{P}\{X = x_i, Y = y_j\}, \quad q_i = \mathbb{P}\{X = x_i\} \quad \text{and} \quad r_j = \mathbb{P}\{Y = y_j\}.$$

With these notations the following result is valid.

Proposition 3.6.9. *For the independence of two random variables X and Y, it is necessary and sufficient that*

!

$$p_{ij} = q_i \cdot r_j, \quad 1 \le i, j < \infty$$

Proof: The assertion is an immediate consequence of Propositions 1.9.9 and 3.6.5. But, because of the importance of the result, we give an alternative proof avoiding the direct use of product probabilities; only the techniques are similar.

Let us first show that the condition is necessary. Therefore, choose indices i and j, and put $B_1 := \{x_i\}$ and $B_2 := \{y_j\}$. Then $\{X \in B_1\}$ occurs if and only if $X = x_i$, and, in the same way, the occurrence of $\{Y \in B_2\}$ is equivalent to $Y = y_j$. Since X and Y are assumed to be independent, as claimed,

$$p_{ij} = \mathbb{P}\{X = x_i, Y = y_j\} = \mathbb{P}\{X \in B_1, Y \in B_2\} = \mathbb{P}\{X \in B_1\} \cdot \mathbb{P}\{Y \in B_2\}$$
$$= \mathbb{P}\{X = x_i\} \cdot \mathbb{P}\{Y = y_j\} = q_i \cdot r_j .$$

To prove the converse implication, assume we have $p_{ij} = q_i \cdot r_j$ for all pairs (i, j) of integers. Let B_1 and B_2 be two arbitrary subsets of $\mathbb{R}$. Then it follows

$$\mathbb{P}\{X \in B_1, Y \in B_2\} = \mathbb{P}_{(X,Y)}(B_1 \times B_2) = \sum_{\{(i,j):(x_i,y_j)\in B_1\times B_2\}} p_{ij}$$

$$= \sum_{\{(i,j):x_i\in B_1, y_j\in B_2\}} q_i \cdot r_j = \sum_{\{i:x_i\in B_1\}} \sum_{\{j:y_j\in B_2\}} q_i \cdot r_j$$

$$= \left(\sum_{\{i:x_i\in B_1\}} q_i \right) \cdot \left(\sum_{\{j:y_j\in B_2\}} r_j \right) = \mathbb{P}_X(B_1) \cdot \mathbb{P}_Y(B_2)$$

$$= \mathbb{P}(X \in B_1) \cdot \mathbb{P}(Y \in B_2) .$$

Since B_1 and B_2 were arbitrary, the random variables X and Y are independent. This completes the proof. ∎

Remark 3.6.10. The previous proposition implies again that for (discrete) independent random variables the joint distribution is determined by the marginal ones. Indeed, in order to know the p_{ij}s, it suffices to know the q_is and r_js.

Let us represent the assertion of Proposition 3.6.9 graphically. It asserts that the random variables X and Y are independent if and only if the table describing their joint distribution may be represented as follows:

$Y\backslash X$	x_1	x_2	x_3	·	·	
y_1	$q_1 r_1$	$q_2 r_1$	$q_3 r_1$	·	·	r_1
y_2	$q_1 r_2$	$q_2 r_2$	$q_3 r_2$	·	·	r_2
y_3	$q_1 r_3$	$q_2 r_3$	$q_3 r_3$	·	·	r_3
·	·	·	·	·	·	·
·	·	·	·	·	·	·
	q_1	q_2	q_3	·	·	1

Example 3.6.11. Proposition 3.6.9 lets us conclude that X and Y in Example 3.5.8 (without replacing) are dependent while X' and Y' (with replacement) are independent. Furthermore, by the same argument, the random variables X and Y in Example 3.5.9 (minimum and maximum value when rolling a die twice) are dependent as well.

Example 3.6.12. Let X and Y be two independent Pois_λ-distributed random variables. Then the joint distribution of the vector (X, Y) is determined by

$$\mathbb{P}\{X = k, Y = l\} = \frac{\lambda^{k+l}}{k!\, l!}\, e^{-2\lambda}, \quad (k, l) \in \mathbb{N}_0 \times \mathbb{N}_0 .$$

For example, applying this for $\mathbb{P}_{(X,Y)}(B)$ with $B = \{(k, l) : k = l\}$ leads to

$$\mathbb{P}\{X = Y\} = \sum_{k=0}^{\infty} \mathbb{P}\{X = k, Y = k\} = \sum_{k=0}^{\infty} \frac{\lambda^{2k}}{(k!)^2}\, e^{-2\lambda} .$$

Example 3.6.13. Suppose X and Y are two independent geometric distributed random variables, with parameters p and q, respectively. Evaluate $\mathbb{P}(X \le Y)$.

Solution: By the independence of X and Y,

$$\mathbb{P}(X \leq Y) = \sum_{k=1}^{\infty} \mathbb{P}(X = k, Y \geq k) = \sum_{k=1}^{\infty} \mathbb{P}(X = k) \cdot \mathbb{P}(Y \geq k)$$

$$= \sum_{k=1}^{\infty} p(1-p)^{k-1} \sum_{l=k}^{\infty} q(1-q)^{l-1} = pq \left(\sum_{k=0}^{\infty} (1-p)^k \right) \left(\sum_{l=k+1}^{\infty} (1-q)^{l-1} \right)$$

$$= pq \left(\sum_{k=0}^{\infty} (1-p)^k \right) \left(\sum_{l=k}^{\infty} (1-q)^l \right) = pq \left(\sum_{k=0}^{\infty} (1-p)^k (1-q)^k \right) \left(\sum_{l=0}^{\infty} (1-q)^l \right)$$

$$= \frac{p}{1 - (1-p)(1-q)} = \frac{p}{p + q - pq} .$$

Example of application: Player A rolls a die and, simultaneously, player B tosses two fair coins labeled with "0" and "1." Find the probability that player A observes the number "6" for the first time strictly before player B gets "1" two times.

Answer: Let $\{Y = k\}$ be the event that player A observes his first "6" in trial k. Similarly, $\{X = k\}$ occurs if player B has his first two "ones" in trial k. Then we ask for the probability $\mathbb{P}\{Y < X\}$. Note that X is geometrically distributed with parameter $p = 1/4$, while the success probability for Y is $q = 1/6$. Hence, by the above calculations,

$$\mathbb{P}\{Y < X\} = 1 - \mathbb{P}\{X \leq Y\} = 1 - \frac{1/4}{1/4 + 1/6 - 1/24} = \frac{1}{3} .$$

The next objective is to investigate in which cases two quite special random variables are independent. To this end we need the following notation.

Definition 3.6.14. Let Ω be a set and $A \subseteq \Omega$. Then the **indicator function** $\mathbb{1}_A :$ $\Omega \to \mathbb{R}$ of A is defined by

$$\mathbb{1}_A(\omega) := \begin{cases} 1 : \omega \in A \\ 0 : \omega \notin A \end{cases} \tag{3.20}$$

Let us state some basic properties of indicator functions.

Proposition 3.6.15. *Let $(\Omega, \mathcal{A}, \mathbb{P})$ be a probability space.*
(1) *The indicator function of a set $A \subseteq \Omega$ is a random variable if and only if $A \in \mathcal{A}$.*
(2) *If $A \in \mathcal{A}$, then $\mathbb{1}_A$ is $B_{1,p}$-distributed (binomial) where $p = \mathbb{P}(A)$.*
(3) *If $A, B \in \mathcal{A}$, then the random variables $\mathbb{1}_A$ and $\mathbb{1}_B$ are independent if and only if the events A and B are so.*

Proof: Given $t \in \mathbb{R}$, the event $\{\omega \in \Omega : \mathbb{1}_A(\omega) \leq t\}$ is either empty, A^c or Ω in dependence of $t < 0$, $0 \leq t < 1$ or $t \geq 1$. Consequently, the set $\{\omega \in \Omega : \mathbb{1}_A(\omega) \leq t\}$ is in $\mathcal{A}$ for all $t \in \mathbb{R}$ if and only if $A^c \in \mathcal{A}$. But this happens if and only if $A \in \mathcal{A}$, which proves the first assertion.

To prove the second part we first observe that $\mathbb{1}_A$ attains only the values "0" and "1." Since

$$\mathbb{P}\{\mathbb{1}_A = 1\} = \mathbb{P}\{\omega \in \Omega : \mathbb{1}_A(\omega) = 1\} = \mathbb{P}(A) = p,$$

it is $B_{1,p}$-distributed with $p = \mathbb{P}(A)$ as claimed.

Let us turn to the last assertion. Given $A, B \in \mathcal{A}$ their joint distribution in table form is

$\mathbb{1}_B\backslash\mathbb{1}_A$	0	1	
0	$\mathbb{P}(A^c \cap B^c)$	$\mathbb{P}(A \cap B^c)$	$\mathbb{P}(B^c)$
1	$\mathbb{P}(A^c \cap B)$	$\mathbb{P}(A \cap B)$	$\mathbb{P}(B)$
	$\mathbb{P}(A^c)$	$\mathbb{P}(A)$	

Consequently, by Proposition 3.6.9 the random variables $\mathbb{1}_A$ and $\mathbb{1}_B$ are independent if and only if the following equations are valid:

$$\mathbb{P}(A^c \cap B^c) = \mathbb{P}(A^c) \cdot \mathbb{P}(B^c), \quad \mathbb{P}(A^c \cap B) = \mathbb{P}(A^c) \cdot \mathbb{P}(B)$$
$$\mathbb{P}(A \cap B^c) = \mathbb{P}(A) \cdot \mathbb{P}(B^c), \quad \mathbb{P}(A \cap B) = \mathbb{P}(A) \cdot \mathbb{P}(B).$$

Because of Proposition 2.2.7 these four equations are satisfied if and only if the events A and B are independent. This proves the third assertion. ∎

Finally, let us shortly discuss the independence of more than two discrete random variables. Hereby we use the same notation as in Proposition 3.5.10, that is, the random variables $X_1, \ldots, X_n$ satisfy $X_j : \Omega \to D_j$, where D_j is either finite or countably infinite. Then the following generalization of Proposition 3.6.9 is valid. Its proof is almost identical to that for two variables. Therefore, we omit it.

Proposition 3.6.16. *The random variables $X_1, \ldots, X_n$ are independent if and only if for all $x_j \in D_j$*

$$\mathbb{P}\{X_1 = x_1, \ldots, X_n = x_n\} = \mathbb{P}\{X_1 = x_1\} \cdots \mathbb{P}\{X_n = x_n\}.$$

Example 3.6.17. Let us consider the problem of tossing a biased coin n times. The sample space is $\Omega = \{0, 1\}^n$, and the describing probability measure $\mathbb{P}$ is as in eq. (3.5). The random variables X_j are defined as results of toss j. Then $X_j : \Omega \to D_j$, where

$D_j = \{0, 1\}$. If we choose arbitrary $x_j \in D_j$, then either $x_j = 0$ or $x_j = 1$. Let k be the number of those x_j, which equals 1, that is, $k = x_1 + \cdots + x_n$. Formula (3.5) implies

$$P\{X_1 = x_1, \ldots, X_n = x_n\} = P\{(x_1, \ldots, x_n)\} = p^k(1-p)^{n-k}.$$

On the other hand, as shown in Example 3.2.16, the probability distribution of each X_j satisfies

$$P\{X_j = 0\} = 1 - p \quad \text{and} \quad P\{X_j = 1\} = p.$$

Since exactly k of the x_js are "1" and $n - k$ are "0," this implies

$$P\{X_1 = x_1\} \cdots P\{X_n = x_n\} = p^k(1-p)^{n-k}.$$

Summing up, for all $x_j \in D_j$,

$$P\{X_1 = x_1, \ldots, X_n = x_n\} = p^k(1-p)^{n-k} = P\{X_1 = x_1\} \cdots P\{X_n = x_n\},$$

that is, $X_1, \ldots, X_n$ are independent.

3.6.2 Independence of Continuous Random Variables

We will consider the question in which cases *continuous* random variables are independent. Thus, let $X_1, \ldots, X_n$ be continuous random variables with distribution densities $p_1, \ldots, p_n$, that is, for $1 \le j \le n$ and real numbers $a < b$

$$P_{X_j}([a, b]) = P\{a \le X_j \le b\} = \int_a^b p_j(x)\, dx.$$

With this notation the independence of the X_js may be characterized as follows.

Proposition 3.6.18. *For random variables* $X_1, \ldots, X_n$ *with densities* $p_1, \ldots, p_n$ *we define a function* $p : \mathbb{R}^n \to \mathbb{R}$ *by*

$$p(x_1, \ldots, x_n) := p_1(x_1) \cdots p_n(x_n), \quad (x_1, \ldots, x_n) \in \mathbb{R}^n. \tag{3.21}$$

Then the X_j*s are independent if and only if* p *defined by eq.* (3.21) *is a distribution density of the random vector* $\vec{X} = (X_1, \ldots, X_n)$.

Proof: As in the discrete case, the result follows directly from Propositions 1.9.12 and 3.6.5. Without using product probabilities, we may argue as follows.

First we observe that p defined by eq. (3.21) is a distribution density of $\vec{X}$ if and only if for all $a_j < b_j$

$$\mathbb{P}\{a_1 \le X_1 \le b_1, \ldots, a_n \le X_n \le b_n\}$$

$$= \int_{a_1}^{b_1} \cdots \int_{a_n}^{b_n} p_1(x_1) \cdots p_n(x_n)\, dx_n \ldots dx_1 . \tag{3.22}$$

The right-hand side of eq. (3.22) coincides with

$$\left(\int_{a_1}^{b_1} p_1(x_1)\, dx_1 \right) \cdots \left(\int_{a_n}^{b_n} p_n(x_n)\, dx_n \right)$$

$$= \mathbb{P}\{a_1 \le X_1 \le b_1\} \cdots \mathbb{P}\{a_n \le X_n \le b_n\} .$$

From this we derive that eq. (3.22) is valid for all $a_j < b_j$ if and only if

$$\mathbb{P}\{a_1 \le X_1 \le b_1, \ldots, a_n \le X_n \le b_n\} = \mathbb{P}\{a_1 \le X_1 \le b_1\} \cdots \mathbb{P}\{a_n \le X_n \le b_n\} .$$

By Remark 3.6.8 this is equivalent to the independence of the X_js. This completes the proof. ■

Example 3.6.19. Throw a dart to a target, which is a circle of radius 1. The center of the circle is the point $(0, 0)$ and $(x_1, x_2) \in K$ denotes the point where the dart hits the target. We assume that the point hit is uniformly distributed on K. The question is whether or not the coordinates x_1 and x_2 of the point hit are dependent or independent of each other.

Answer: Let $\mathbb{P}$ be the uniform distribution on K, and define two random variables X_1 and X_2 by $X_1(x_1, x_2) = x_1$ and $X_2(x_1, x_2) = x_2$. In this notation, the above question is whether the random variables X_1 and X_2 are independent. The density p_1 of X_1 was found in eq. (3.7). By symmetry, p_2, the density of X_2, coincides with p_1, that is, we have

$$p_1(x_1) = \begin{cases} \frac{2}{\pi}\sqrt{1 - x_1^2} & : |x_1| \le 1 \\ 0 & : |x_1| > 1 \end{cases} \quad \text{and} \quad p_2(x_2) = \begin{cases} \frac{2}{\pi}\sqrt{1 - x_2^2} & : |x_2| \le 1 \\ 0 & : |x_2| > 1 . \end{cases}$$

But $p_1(x_1) \cdot p_2(x_2)$ cannot be a distribution density of $\mathbb{P}_{(X_1, X_2)}$. Indeed, the vector $\vec{X} = (X_1, X_2)$ is uniformly distributed on K, thus its (correct) density is p with

$$p(x_1, x_2) = \begin{cases} \frac{1}{\pi} & : x_1^2 + x_2^2 \le 1 \\ 0 & : \text{otherwise} \end{cases}$$

Thus, we conclude that X_1 and X_2 are dependent, hence also the coordinates x_1 and x_2 of the point hit.

Example 3.6.20. We suppose now that the dart does not hit a circle but some rectangle set $R := [\alpha_1, \beta_1] \times [\alpha_2, \beta_2]$. Again we assume that the point $(x_1, x_2) \in R$ is uniformly distributed on R. The posed question is the same as in Example 3.6.19, namely whether x_1 and x_2 are independent of each other.

Answer: Define X_1 and X_2 as in the previous example. By assumption, the vector $\vec{X} = (X_1, X_2)$ is uniformly distributed on R, hence its distribution density p is given by

$$p(x_1, x_2) = \begin{cases} \frac{1}{\text{vol}_2(R)} & : \quad (x_1, x_2) \in R \\ 0 & : \quad (x_1, x_2) \notin R. \end{cases}$$

For the density p_1 of X_1 we get

$$p_1(x_1) = \int_{-\infty}^{\infty} p(x_1, x_2)\, dx_2 = \frac{\beta_2 - \alpha_2}{\text{vol}_2(R)} = \frac{1}{\beta_1 - \alpha_1}$$

provided that $\alpha_1 \le x_1 \le \beta_1$. Otherwise, we have $p_1(x_1) = 0$. This tells us that X_1 is uniformly distributed on $[\alpha_1, \beta_1]$. In the same way, we obtain for $x_2 \in [\alpha_2, \beta_2]$ that

$$p_2(x_2) = \frac{1}{\beta_2 - \alpha_2}$$

and $p_2(x_2) = 0$ otherwise. Hence, X_2 is also uniformly distributed, but this time on $[\alpha_2, \beta_2]$. From the equations for p_1 and p_2 it follows that for the joint density p holds

$$p(x_1, x_2) = p_1(x_1) \cdot p(x_2), \quad (x_1, x_2) \in \mathbb{R}^2.$$

Consequently, by Proposition 3.6.18 the random variables X_1 and X_2 are independent, and so are the coordinates x_1 and x_2 of the point hit.

Example 3.6.21. Let us look at Example 3.6.20 from the reversed side. Now we assume that the coordinates are uniformly distributed, not the vector. Thus let $U_1, \dots, U_n$ be independent random variables with U_j uniformly distributed on the interval $[\alpha_j, \beta_j]$, $1 \le j \le n$. Then the random vector $\vec{U} = (U_1, \dots, U_n)$ is (multivariate) uniformly distributed on the box $K = [\alpha_1, \beta_1] \times \cdots \times [\alpha_n, \beta_n]$. This is an immediate consequence of Example 1.9.13 combined with Proposition 3.6.5. A direct proof of this fact, without using product measures, is as follows.

The density of U_j is $p_j = \frac{1}{\beta_j - \alpha_j} \mathbb{1}_{[\alpha_j, \beta_j]}$, hence by Proposition 3.6.18 the joint density p of $\vec{U}$ is given by

$$p(x) = p_1(x_1) \cdots p_n(x_n) = \prod_{j=1}^{n} (\beta_j - \alpha_j)^{-1} = \frac{1}{\text{vol}_n(K)}, \quad x = (x_1, \dots, x_n) \in K,$$

and $p(x) = 0$ if $x \notin K$. Therefore,

$$\mathbb{P}\{\vec{U} \in B\} = \int_B p(x)\,dx = \frac{\mathrm{vol}_n(K \cap B)}{\mathrm{vol}_n(K)}, \quad B \in \mathcal{B}(\mathbb{R}^n),$$

and $\vec{U}$ is uniformly distributed on K as asserted.

Example 3.6.22. Let $X_1, \ldots, X_n$ be independent standard normally distributed. Which joint density does the vector $\vec{X} = (X_1, \ldots, X_n)$ possess?

Answer: The densities p_j of each of the X_js are

$$p_j(x) = \frac{1}{\sqrt{2\pi}} e^{-x^2/2}, \quad x \in \mathbb{R}.$$

Consequently, by the independence of the X_js the joint density p equals

$$p(x) = p_1(x_1) \cdots p_n(x_n) = \frac{1}{(2\pi)^{n/2}} e^{-(x_1^2 + \cdots + x_n^2)/2}$$

$$= \frac{1}{(2\pi)^{n/2}} e^{-|x|^2/2}, \quad x = (x_1, \ldots, x_n).$$

This tells us that $\mathbb{P}_{\vec{X}} = \mathcal{N}(0,1)^{\otimes n}$ (cf. Definition 1.9.16) or, equivalently, $\vec{X}$ is n-dimensional standard normally distributed.

Example 3.6.23. If $X_1, \ldots, X_n$ are independent E_λ-distributed, then

$$p_j(t) = \begin{cases} 0 & : t < 0 \\ \lambda e^{-\lambda t} & : t \geq 0 \end{cases}$$

hence, the random vector $\vec{X} = (X_1, \ldots, X_n)$ has the joint density

$$p(t) = \lambda^n e^{-\lambda(t_1 + \cdots + t_n)}, \quad t = (t_1, \ldots, t_n), \ t_j \geq 0,$$

and $p(t) = 0$ if one of the t_js is negative.

3.7 *Order Statistics

This section is devoted to a quite practical problem. Suppose we execute a random experiment n times so that different trials are independent of each other. The results of these trials are $x_1, \ldots, x_n$. For example, one may think of n different measurements of the same item, and $x_1, \ldots, x_n$ are the observed values. After getting $x_1, \ldots, x_n$ we reorder them by their size. These "new" numbers are denoted by $x_1^* \leq \cdots \leq x_n^*$. In other words, the numbers are the same as before but in nondecreasing order. We now ask for the distribution of the ordered x_k^*s.

The precise mathematical formulation of this problem is as follows: let $X_1, \ldots, X_n$ be n independent identically distributed random variables defined on a sample space Ω. For each fixed $\omega \in \Omega$, we choose a permutation $\pi_\omega \in S_n$ (we use the notation of Section A.3.1), such that

$$X_{\pi_\omega(1)}(\omega) \leq \cdots \leq X_{\pi_\omega(n)}(\omega). \tag{3.23}$$

Of course, it may happen that there exists more than one permutation for which the inequalities (3.23) hold, namely if $X_i(\omega) = X_j(\omega)$ for some $i \neq j$. In this case, we choose any of these permutations. Finally, for each $\omega \in \Omega$ we set

$$X_1^*(\omega) = X_{\pi_\omega(1)}(\omega), \ldots, X_n^*(\omega) = X_{\pi_\omega(n)}(\omega).$$

In this way[10] we obtain random variables X_k^* satisfying $X_1^* \leq \cdots \leq X_n^*$. For example, it holds

$$X_1^* = \min\{X_1, \ldots, X_n\}, \ldots, X_n^* = \max\{X_1, \ldots, X_n\}.$$

Remark 3.7.1. It is worthwhile to mention that the X_k^*s are no longer independent nor identical distributed.

Remark 3.7.2. For a better understanding of the procedure, let us look at the case $n = 3$. There exist $6 = 3!$ possible ways the $X_j(\omega)$s may be ordered. For example, if $X_2(\omega) < X_3(\omega) < X_1(\omega)$, then set $\pi_\omega(1) = 2$, $\pi_\omega(2) = 3$, and $\pi_\omega(3) = 1$ or, equivalently, $\omega \in A_\pi$ where $\pi(1) = 2$, $\pi(2) = 3$, and $\pi(3) = 1$. Hence, in that example we have $X_1^*(\omega) = X_2(\omega)$, $X_2^*(\omega) = X_3(\omega)$, and $X_3^*(\omega) = X_1(\omega)$. At the end, we get 6 subsets A_π of Ω where $\pi_\omega = \pi$ for a given $\pi \in S_3$, that is, on each of these six sets, the same type of reordering is applied.

Definition 3.7.3. The ordered random variables $X_1^*, \ldots, X_n^*$ are called **order statistics** of $X_1, \ldots, X_n$.

Remark 3.7.4. Order statistics play an important role in Mathematical Statistics. For example, suppose at time $t = 0$ we switch on n light bulbs of the same type. Let us record the times $0 < t_1^* < t_2^* < \cdots < t_n^*$, where some of the n bulbs burns out. Then these times are nothing else than the order statistics of the life times $t_1, \ldots, t_n$ of the first, second, and so on light bulb.

---

10 Another way to describe the procedure of reordering is as follows. For each permutation π let $A_\pi \subseteq \Omega$ be the set of those $\omega \in \Omega$ for which $X_{\pi(1)}(\omega) \leq \cdots \leq X_{\pi(n)}(\omega)$, that is, where $\pi = \pi_\omega$. Then it follows $X_j^*(\omega) = X_{\pi(j)}(\omega)$ whenever $\omega \in A_\pi$. Note that there are at most $n!$ different sets A_π.

Before we state and prove the main result of this section, let us recall that the X_js are assumed to be identically distributed. Consequently, all of them possess the same distribution function F. That is, for all $j \le n$, we have

$$F(t) = \mathbb{P}\{X_j \le t\}, \quad t \in \mathbb{R},$$

Proposition 3.7.5. *Let $X_1, \ldots, X_n$ be independent identically distributed random variables with distribution function F. Then for each $k \le n$ we have*

$$\mathbb{P}\{X_k^* \le t\} = \sum_{i=k}^{n} \binom{n}{i} F(t)^i (1 - F(t))^{n-i}, \ t \in \mathbb{R}. \tag{3.24}$$

Proof: Fix $t \in \mathbb{R}$. When does the event $\{X_k^* \le t\}$ occur? To answer this, for $i \le n$ introduce disjoint sets A_i as follows: the event A_i occurs if and only if exactly i of the X_js attain a value in $(-\infty, t]$. More precisely,

$$A_i = \{\omega \in \Omega : \#\{j \le n : X_j(\omega) \le t\} = i\}.$$

Next, observe that the event $\{X_k^* \le t\}$ occurs if and only if at least k of the X_js attain a value in $(-\infty, t]$. For example, it holds $X_1^* \le t$ if at least one of the X_js is less than or equal to t while we have $X_n^* \le t$ if $X_j \le t$ for all $j \le n$. Thus, by the definition of the A_is the event $\{X_k^* \le t\}$ coincides with $\bigcup_{i=k}^{n} A_i$. Consequently, since the A_is are disjoint, it follows that

$$\mathbb{P}\{X_k^* \le t\} = \sum_{i=k}^{n} \mathbb{P}(A_i). \tag{3.25}$$

Let $Y_j = \mathbb{1}_{(-\infty,t]}(X_j)$. Then $Y_j = 1$ if and only if $X_j \le t$ while $Y_j = 0$ otherwise. Hence, the Y_js are binomial distributed with parameters 1 and p, where

$$p = \mathbb{P}\{Y_j = 1\} = \mathbb{P}\{X_j \le t\} = F(t).$$

Since the X_js are independent, so are the Y_js and their sum[11] $Y_1 + \cdots + Y_n$ is binomial distributed with parameters n and $p = F(t)$. Note that the event A_i occurs if and only if $Y_1 + \cdots + Y_n = i$, which implies

$$\mathbb{P}(A_i) = \mathbb{P}\{Y_1 + \cdots + Y_n = i\} = \binom{n}{i} p^i (1-p)^{n-i} = \binom{n}{i} F(t)^i (1 - F(t))^{n-i}. \tag{3.26}$$

Plugging eq. (3.26) into eq. (3.25) proves eq. (3.24). ∎

---

11 Here we already use a result, which will be proved later on in Proposition 4.6.1.

Example 3.7.6. Let us choose independently and according to the uniform distribution n numbers $x_1, \ldots, x_n$ out of $\{1, \ldots, N\}$. Here, the same number may be chosen more than one time. Given integers $m \le N$ and $k \le n$, find the probability that the kth largest number x_k^* equals m.

Answer: The distribution function F of the uniform distribution on $\{1, \ldots, N\}$ satisfies

$$F(m) = \frac{m}{N}, \quad m = 1, \ldots, N.$$

Thus Proposition 3.7.5 implies

$$\mathbb{P}\{x_k^* \le m\} = \sum_{i=k}^{n} \binom{n}{i} \left(\frac{m}{N}\right)^i \left(1 - \frac{m}{N}\right)^{n-i}.$$

Because of $\{x_k^* = m\} = \{x_k^* \le m\} \backslash \{x_k^* \le m - 1\}$ we obtain

$$\mathbb{P}\{x_k^* = m\} = \mathbb{P}\{x_k^* \le m\} - \mathbb{P}\{x_k^* \le m - 1\}$$

$$= \sum_{i=k}^{n} \binom{n}{i} \left[\left(\frac{m}{N}\right)^i \left(1 - \frac{m}{N}\right)^{n-i} - \left(\frac{m-1}{N}\right)^i \left(1 - \frac{m-1}{N}\right)^{n-i} \right].$$

For example, roll a die four times and order the results in nondecreasing order as $x_1^* \le \cdots \le x_4^*$. What is the probability that x_3^* equals 5 ?

Answer: Let us apply the previous formula with $N = 6$, $k = 3$, and $n = 4$. For $m = 1, \ldots, 6$ this implies

$$\mathbb{P}\{x_3^* = m\} = \sum_{i=3}^{4} \binom{4}{i} \left[\left(\frac{m}{6}\right)^i \left(\frac{6-m}{6}\right)^{4-i} - \left(\frac{m-1}{6}\right)^i \left(\frac{6-m+1}{6}\right)^{4-i} \right].$$

The probabilities are

m	$\mathbb{P}\{x_3^* = m\}$
1	0.0162037
2	0.0949074
3	0.201389
4	0.280093
5	0.275463
6	0.131944

thus, $x_3^* = 4$ is most likely.

Let us now turn to the case of *continuous* random variables. That is, we assume that the random variables X_j possess a distribution density p satisfying

$$\mathbb{P}\{X_j \le t\} = \int_{-\infty}^{t} p(x)\,dx, \quad t \in \mathbb{R}.$$

Again we remark that the preceding formula holds for all $j \le n$. Indeed, the X_js are identically distributed, hence they all have the same density. A natural question arises: what distribution density does X_k^* possess?

Proposition 3.7.7. *Suppose p is the common density of the X_js. Let $X_1^* \le \cdots \le X_n^*$ be the order statistics of the X_j. Then the distribution density p_k of X_k^* is given by*

$$p_k(t) = \frac{n!}{(k-1)!(n-k)!}\, p(t)\, F(t)^{k-1}(1 - F(t))^{n-k}.$$

Proof: It holds that

$$p_k(t) = \frac{d}{dt}\mathbb{P}\{X_k^* \le t\} = \frac{d}{dt}\sum_{i=k}^{n}\binom{n}{i}F(t)^{i}(1 - F(t))^{n-i}$$

$$= \sum_{i=k}^{n} i\binom{n}{i}p(t)F(t)^{i-1}(1 - F(t))^{n-i} - \sum_{i=k}^{n}(n-i)\binom{n}{i}p(t)F(t)^{i}(1 - F(t))^{n-i-1}. \quad (3.27)$$

In fact, the index i in the second sum of eq. (3.27) runs only from k to $n-1$. Hence, shifting it by 1, this sum becomes

$$\sum_{i=k+1}^{n}(n - i + 1)\binom{n}{i-1}p(t)\,F(t)^{i-1}(1 - F(t))^{n-i}.$$

Because of

$$i\binom{n}{i} = \frac{n!}{(i-1)!(n-i)!} = (n - i + 1)\binom{n}{i-1}$$

both sums in eq. (3.27) cancel out for $i = k+1, \ldots, n$, and it remains the term for $i = k$. Thus, we obtain

$$p_k(t) = k\binom{n}{k}p(t)\,F(t)^{k-1}(1 - F(t))^{n-k}$$

$$= \frac{n!}{(k-1)!(n-k)!}\, p(t)\,F(t)^{k-1}(1 - F(t))^{n-k}$$

as asserted. ∎

Example 3.7.8. Let us choose independently and according to the uniform distribution on $[0, 1]$ numbers $x_1, \ldots, x_n$. After reordering them, we get $0 \leq x_1^* \leq \cdots \leq x_n^* \leq 1$. Which distribution does x_k^* possess?

Answer: The density p of the uniform distribution on $[0, 1]$ is $\mathbb{1}_{[0,1]}$. Furthermore, its distribution function F is given by $F(t) = t$ for $0 \leq t \leq 1$. Thus, by Proposition 3.7.7, the density p_k coincides with

$$p_k(t) = \frac{n!}{(k-1)!(n-k)!} t^{k-1}(1-t)^{n-k}, \quad 0 \leq t \leq 1.$$

As already mentioned in Example 1.6.31, this is nothing else than the density of a beta distribution with parameters k and $n - k + 1$. Hence, for all $k = 1, \ldots n$ and all $0 \leq a < b \leq 1$, it follows that

$$\mathbb{P}\{a \leq x_k^* \leq b\} = \mathcal{B}_{k,n-k+1}([a, b]) = \frac{n!}{(k-1)!\,(n-k)!} \int\limits_a^b x^{k-1}(1-x)^{n-k}\, dx.$$

Example 3.7.9. Let us investigate here the example that was already mentioned in Remark 3.7.4. At time $t = 0$ we switch on n electric bulbs of the same type. The times $0 < t_1^* \leq \cdots \leq t_n^*$ are those where we observe that some of the bulbs burns out. If we assume that the lifetime of each bulb is exponentially distributed, what can we say about the distribution of the t_k^*s?

Answer: Let $X_1, \ldots, X_n$ be the lifetimes of the n light bulbs. By assumption, they are independent and all exponentially distributed with some parameter $\lambda > 0$. Then the distribution of t_k^* is that of X_k^*. Furthermore, we have $p(t) = \lambda e^{-\lambda t}$ and $F(t) = 1 - e^{-\lambda t}$ for $t \geq 0$. By Proposition 3.7.7, the distribution density p_k of X_k^* equals

$$p_k(t) = \lambda \frac{n!}{(k-1)!(n-k)!} (1 - e^{-\lambda t})^{k-1} e^{-\lambda t(n-k+1)}, \quad t \geq 0.$$

For example, for t_1^*, the time when we observe the first burnout of any of the bulbs, this implies

$$p_1(t) = \lambda n e^{-\lambda t n}, \quad t \geq 0,$$

that is, t_1^* is $E_{\lambda n}$-distributed.

3.8 Problems

Problem 3.1. The joint distribution of a random vector $\vec{X} = (X_1, X_2)$ is described by

$X_2 \backslash X_1$	0	1
0	$\frac{1}{10}$	$\frac{2}{5}$
1	$\frac{2}{5}$	$\frac{1}{10}$

Define another vector $\vec{Y} = (Y_1, Y_2)$ by $Y_1 := \min\{X_1, X_2\}$ and $Y_2 := \max\{X_1, X_2\}$. Find the probability distribution of $\vec{Y} = (Y_1, Y_2)$. Are Y_1 and Y_2 independent?

Problem 3.2. Let $\vec{X} = (X_1, X_2)$ be uniformly distributed on the square in $\mathbb{R}^2$ with corner points $(0, 1)$, $(1, 0)$, $(0, -1)$, and $(-1, 0)$. Find the marginal distributions of $\vec{X}$.

Problem 3.3. In a lottery, six numbers are chosen out of $\{1, \ldots, 49\}$. As usual in lotteries, chosen numbers are not replaced. Let $X_1, \ldots, X_6$ be the chosen numbers as they appeared. That is, X_1 is the number chosen first while X_6 is the number, which appeared last.
1. Determine the joint distribution of the vector $\vec{X} = (X_1, \ldots, X_6)$, as well its marginal distributions.
2. Argue why $X_1, \ldots, X_6$ are **not** independent.
3. Reordering the six chosen numbers leads to the order statistics $X_1^* < \cdots < X_6^*$. Find the joint distribution of the vector $(X_1^*, \ldots, X_6^*)$, as well as its marginal distributions.

Problem 3.4. A random variable X is geometric distributed. Given natural numbers k and n, show that

$$\mathbb{P}\{X = k + n | X > n\} = \mathbb{P}\{X = k\}.$$

Why is this property called "lack of memory property"?

Problem 3.5. A random variable is exponentially distributed. Prove

$$\mathbb{P}(X > s + t | X > s) = \mathbb{P}(X > t)$$

for all $t, s \geq 0$.

Problem 3.6. Two random variables X and Y are independent and geometrically distributed with parameters p and q for some $0 < p, q < 1$. Evaluate $\mathbb{P}\{X \leq Y \leq 2X\}$.

Problem 3.7. Suppose two independent random variables X and Y satisfy

$$\mathbb{P}\{X = k\} = \mathbb{P}\{Y = k\} = \frac{1}{2^k}, \quad k = 1, 2, \ldots.$$

Find the probabilities $\mathbb{P}\{X \leq Y\}$ and $\mathbb{P}\{X = Y\}$.

Problem 3.8. Choose two numbers b and c independently, b according to the uniform distribution on $[-1, 1]$ and c according to the uniform distribution on $[0, 1]$. Find the probability that the equation

$$x^2 + bx + c = 0$$

does **not** possess a real solution.

Problem 3.9. Use Problem 1.31 to prove the following: If X is a random variable, then the number of points $t \in \mathbb{R}$ with $\mathbb{P}\{X = t\} > 0$ is at most countably infinite.

Problem 3.10. Suppose a fair coin is labeled with "0" and "1." Toss the coin n times. Let X be the maximum observed value and Y the sum of the n values. Determine the joint distribution of (X, Y). Argue that X and Y are **not** independent.

Problem 3.11. Suppose a random vector (X, Y) has the joint density function p defined by

$$p(u, v) := \begin{cases} c \cdot uv : u, v \geq 0, \ u + v \leq 1 \\ 0 \ : \quad \text{otherwise} \end{cases}$$

1. Find the value of the constant c so that p becomes a density function.
2. Determine the density functions of X and Y.
3. Evaluate $\mathbb{P}\{X + Y \leq 1/2\}$.
4. Are X and Y independent?

Problem 3.12. Gambler A has a biased coin with "head" having probability p for some $0 < p < 1$, and gambler B's coin is biased with "head" having probability q for some $0 < q < 1$. A and B toss their coins simultaneously. Whoever lands on "head" first wins. If both gamblers observe "head" at the same time, then the game ends in a draw. Evaluate the probability that A wins and the probability that the game ends in a draw.

Problem 3.13. Randomly choose two integers x_1 and x_2 from 1 to 10. Let X be the **minimum** of x_1 and x_2. Determine the distribution and the probability mass functions of X in the two following cases:
- The number chosen first is replaced.
- The first number is not replaced.

Evaluate in both cases $\mathbb{P}\{2 \leq X \leq 3\}$ and $\mathbb{P}\{X \geq 8\}$.

Problem 3.14. There are four balls labeled with "0" and three balls are labeled with "2" in an urn. Choose three balls without replacement. Let X be the sum of the values on the three chosen balls. Find the distribution of X.

4 Operations on Random Variables

4.1 Mappings of Random Variables

This section is devoted to the following problem: let $X : \Omega \to \mathbb{R}$ be a random variable and let $f : \mathbb{R} \to \mathbb{R}$ be some function. Set $Y := f(X)$, that is, for all $\omega \in \Omega$ we have $Y(\omega) = f(X(\omega))$. Suppose the distribution of X is known. Then the following question arises:

Which distribution does $Y = f(X)$ possess?

For example, if $f(t) = t^2$, and we know the distribution of X, then we ask for the probability distribution of X^2. Is it possible to compute this by easy methods?

At the moment it is not clear at all whether $Y = f(X)$ is a random variable. Only if this is valid, the probability distribution $\mathbb{P}_Y$ is well-defined. For arbitrary functions f this need not to be true, they have to satisfy the following additional property.

Definition 4.1.1. A function $f : \mathbb{R} \to \mathbb{R}$ is called **measurable** if for $B \in \mathcal{B}(\mathbb{R})$ the preimage $f^{-1}(B)$ is a Borel set as well.

Remark 4.1.2. This is a purely technical condition for f, which will not play an important role later on. All functions of interest, for example, piecewise continuous, monotone, pointwise limits of continuous functions, and so on, are measurable.

The measurability of f is needed to prove the following result.

Proposition 4.1.3. *Let $X : \Omega \to \mathbb{R}$ be a random variable. If $f : \mathbb{R} \to \mathbb{R}$ is measurable, then $Y = f(X)$ is a random variable as well.*

Proof: Take a Borel set $B \in \mathcal{B}(\mathbb{R})$. Then

$$Y^{-1}(B) = X^{-1}\left(f^{-1}(B)\right) = X^{-1}(B')$$

with $B' := f^{-1}(B)$. We assumed f to be measurable, which implies $B' \in \mathcal{B}(\mathbb{R})$, and hence, since X is a random variable, we conclude $Y^{-1}(B) = X^{-1}(B') \in \mathcal{A}$. The Borel set B was arbitrary, thus, as asserted, Y is a random variable. ∎

There does not exist a general method for the description of $\mathbb{P}_Y$ in terms of $\mathbb{P}_X$. Only for some special functions, for example, for linear functions or for strictly monotone and differentiable, there exist general rules for the computation of $\mathbb{P}_Y$. Nevertheless, quite often we are able to determine $\mathbb{P}_Y$ directly. Mostly the following two approaches turn out to be helpful.

If X is **discrete** with values in $D := \{x_1, x_2, \ldots\}$, then $Y = f(X)$ maps the sample space Ω into $f(D) = \{f(x_1), f(x_2), \ldots\}$. Problems arise if f is not one-to-one. In this case one has to combine those x_js that are mapped onto the same element in $f(D)$. For example, if X is uniformly distributed on $D = \{-2, -1, 0, 1, 2\}$ and $f(x) = x^2$, then $Y = X^2$ has values in $f(D) = \{0, 1, 4\}$. Combining -1 and 1 and -2 and 2 leads to

$$\mathbb{P}\{Y = 0\} = \mathbb{P}\{X = 0\} = \frac{1}{5}, \quad \mathbb{P}\{Y = 1\} = \mathbb{P}\{X = -1\} + \mathbb{P}\{X = 1\} = \frac{2}{5},$$

$$\mathbb{P}\{Y = 4\} = \mathbb{P}\{X = -2\} + \mathbb{P}\{X = 2\} = \frac{2}{5}.$$

The case of one-to-one functions f is easier to handle because then

$$\mathbb{P}\{Y = f(x_j)\} = \mathbb{P}\{X = x_j\}, \quad j = 1, 2, \ldots, ,$$

and the distribution of Y can be directly computed from that of X.

For **continuous** X one tries to determine the distribution function F_Y of Y. Recall that this was defined as

$$F_Y(t) = \mathbb{P}\{Y \leq t\} = \mathbb{P}\{f(X) \leq t\}.$$

If we are able to compute F_Y, then we are almost done because then we get the distribution density q of Y as derivative of F_Y.

For instance, if f is increasing, we get F_Y easily by

$$F_Y(t) = \mathbb{P}\{X \leq f^{-1}(t)\} = F_X(f^{-1}(t))$$

with inverse function f^{-1} (cf. Problem 4.15).

The following examples demonstrate how we compute the distribution of $f(X)$ in some special cases.

Example 4.1.4. Assume the random variable X is $\mathcal{N}(0, 1)$-distributed. Which distribution does $Y := X^2$ possess?

Answer: Of course, $F_Y(t) = \mathbb{P}\{Y \leq t\} = 0$ when $t \leq 0$. Consequently, it suffices to determine $F_Y(t)$ for $t > 0$. Then

$$F_Y(t) = \mathbb{P}\{X^2 \leq t\} = \mathbb{P}\{-\sqrt{t} \leq X \leq \sqrt{t}\} = \frac{1}{\sqrt{2\pi}} \int_{-\sqrt{t}}^{\sqrt{t}} e^{-s^2/2} \, ds$$

$$= \frac{2}{\sqrt{2\pi}} \int_0^{\sqrt{t}} e^{-s^2/2} \, ds = h(\sqrt{t}),$$

where

$$h(u) := \frac{\sqrt{2}}{\sqrt{\pi}} \int_0^u e^{-s^2/2} \, ds, \quad u \geq 0.$$

Differentiating F_Y with respect to t, the chain rule and the fundamental theorem of Calculus lead to

$$q(t) = F_Y'(t) = \frac{d}{dt}(\sqrt{t})\, h'(\sqrt{t}) = \frac{t^{-1/2}}{2} \cdot \frac{\sqrt{2}}{\sqrt{\pi}}\, e^{-t/2}$$

$$= \frac{1}{2^{1/2}\Gamma(1/2)}\, t^{\frac{1}{2}-1} e^{-t/2}, \quad t > 0.$$

Hereby, in the last step, we used $\Gamma(1/2) = \sqrt{\pi}$. Consequently, Y possesses the density function

$$q(t) = \begin{cases} 0 & : t \le 0 \\ \frac{1}{2^{1/2}\Gamma(1/2)}\, t^{-1/2} e^{-t/2} & : t > 0. \end{cases}$$

But this is the density of a $\Gamma_{2,\frac{1}{2}}$-distribution. Therefore, we obtained the following result, which we, because of its importance, state as a separate proposition.

Proposition 4.1.5. *If X is $\mathcal{N}(0, 1)$-distributed, then X^2 is $\Gamma_{2,\frac{1}{2}}$-distributed or, equivalently, distributed according to χ_1^2.*

Example 4.1.6. Let U be uniformly distributed on $[0, 1]$. Which distribution does $Y :=$ $1/U$ possess?

Answer: Again we determine F_Y. From $\mathbb{P}\{X \in (0, 1]\} = 1$ we derive $\mathbb{P}(Y \ge 1) = 1$, thus, $F_Y(t) = 0$ if $t < 1$. Therefore, we only have to regard numbers $t \ge 1$. Here we have

$$F_Y(t) = \mathbb{P}\left\{\frac{1}{U} \le t\right\} = \mathbb{P}\left\{U \ge \frac{1}{t}\right\} = 1 - \frac{1}{t}.$$

Hence, the density function q of Y is given by

$$q(t) = F_Y'(t) = \begin{cases} 0 & : t < 1 \\ \frac{1}{t^2} & : t \ge 1 \end{cases}$$

Example 4.1.7 (Random walk on $\mathbb{Z}$). A particle is located at the point 0 of $\mathbb{Z}$. In a first step it moves either to -1 or to $+1$. In the second step it jumps, independently of the first move, again to the left or to the right. Thus, after two steps it is located either at -2, 0, or 2. Hereby we assume that p is the probability for jumps to the right, hence $1-p$ for jumps to the left. This procedure is repeated arbitrarily often. Let S_n be the position of the particle after n jumps or, equivalently, after n steps.[1] The (random) sequence $(S_n)_{n\ge 0}$ is called a (next-neighbor) **random walk** on $\mathbb{Z}$, where by the construction $\mathbb{P}\{S_0 = 0\} = 1$.

---

[1] S_n can also be viewed as the loss or win after n games, where p is the probability to win one dollar in a single game, while $1 - p$ is the probability to lose one dollar.

After n steps the possible positions of the particle are in

$$D_n = \{-n, -n+2, \ldots, n-2, n\}.$$

In other words, S_n is a random variable with values in D_n. Which distribution does S_n possess? To answer this question define

$$Y_n := \frac{1}{2}(S_n + n).$$

The random variable Y_n attains values in $\{0, 1, \ldots, n\}$ and, moreover, $Y_n = m$ if the position of the particle after n steps is $2m - n$, that is, if it jumped m times to the right and $n - m$ times to the left. To see this, take $m = 0$, hence $S_n = -n$, which can only be achieved if all jumps were to the left. If $m = 1$, then $S_n = -n + 2$, that is, there were $n - 1$ jumps to the left and 1 to the right. The same argument applies for all $m \le n$.

This observation tells us that Y_n is $B_{n,p}$-distributed, that is,

$$\mathbb{P}\{Y_n = m\} = \binom{n}{m} p^m (1-p)^{n-m}, \quad m = 0, \ldots, n.$$

Since $Y_n = \frac{1}{2}(S_n + n)$, if $k \in D_n$, then it follows that

$$\mathbb{P}\{S_n = k\} = \mathbb{P}\left\{Y_n = \frac{1}{2}(k+n)\right\} = \binom{n}{\frac{n+k}{2}} p^{(n+k)/2}(1-p)^{(n-k)/2}. \tag{4.1}$$

For even n we have $0 \in D_n$, thus one may ask for the probability of $S_n = 0$, that is, for the probability that the particle returns to its starting point after n steps. Applying (4.1) with $k = 0$ gives

$$\mathbb{P}\{S_n = 0\} = \binom{n}{\frac{n}{2}} p^{n/2}(1-p)^{n/2},$$

hence, if $p = 1/2$, then

$$\mathbb{P}\{S_n = 0\} = \binom{n}{\frac{n}{2}} 2^{-n} = \frac{n!}{((n/2)!)^2} 2^{-n}.$$

An application of Stirling's formula (1.51) implies

$$\lim_{n\to\infty} n^{1/2}\,\mathbb{P}\{S_n = 0\} = \lim_{n\to\infty} n^{1/2} \frac{\sqrt{2\pi n}\,(n/e)^n}{\left[\sqrt{\pi n}\,(n/2e)^{n/2}\right]^2} 2^{-n} = \sqrt{\frac{2}{\pi}},$$

that is, if $n \to \infty$, then $\mathbb{P}\{S_n = 0\} \sim \sqrt{\frac{2}{\pi}}\,n^{-1/2}$.

Example 4.1.8. Suppose X is $B_{n,p}^-$-distributed, that is,

$$\mathbb{P}\{X = k\} = \binom{k-1}{k-n} p^n (1-p)^{k-n}, \quad k = n, n+1, \dots.$$

Let $Y = X - n$. Which probability distribution does Y possess?
    Answer: An easy transformation (cf. formula (1.34)) leads to

$$\mathbb{P}\{Y = k\} = \mathbb{P}\{X = k+n\} = \binom{n+k-1}{k} p^n (1-p)^k = \binom{-n}{k} p^n (p-1)^k \tag{4.2}$$

for all $k = 0, 1, \dots$
    Additional question: Which random experiment does Y describe?
    Answer: We perform a series of random trials where each time we may obtain either failure or success. Hereby, the success probability is $p \in (0,1)$. Then the event $\{Y = k\}$ occurs if and only if we observe the nth success in trial $k + n$.

We conclude this section with the investigation of the following problem. Suppose $X_1, \dots, X_n$ are independent random variables. Given n measurable functions $f_1, \dots, f_n$ from $\mathbb{R}$ to $\mathbb{R}$, we define "new" random variables $Y_1, \dots, Y_n$ by

$$Y_i := f_i(X_i), \quad 1 \le i \le n.$$

It is intuitively clear that then $Y_1, \dots, Y_n$ are also independent; the values of Y_i only depend on those of X_i, thus the independence should be preserved. For example, if X_1 and X_2 are independent, then this should also be valid for X_1^2 and $2X_2$.
    The next result shows that this is indeed true.

Proposition 4.1.9. *Let $X_1, \dots, X_n$ be independent random variables and let $(f_i)_{i=1}^n$ be measurable functions from $\mathbb{R}$ to $\mathbb{R}$. Then $f_1(X_1), \dots, f_n(X_n)$ are independent as well.*

Proof: Choose arbitrary Borel sets $B_1, \dots, B_n$ in $\mathbb{R}$ and set $A_i := f_i^{-1}(B_i)$, $1 \le i \le n$. With this notation, an $\omega \in \Omega$ satisfies $f_i(X_i(\omega)) \in B_i$ if and only if $X_i(\omega) \in A_i$. Hence, an application of the independence of X_i (use 3.14 with the X_is and the A_is) leads to

$$\mathbb{P}\{f_1(X_1) \in B_1, \dots, f_n(X_n) \in B_n\} = \mathbb{P}\{X_1 \in A_1, \dots, X_n \in A_n\}$$
$$= \mathbb{P}\{X_1 \in A_1\} \cdots \mathbb{P}\{X_n \in A_n\} = \mathbb{P}\{f_1(X_1) \in B_1\} \cdots \mathbb{P}\{f_n(X_n) \in B_n\}.$$

The B_is were chosen arbitrarily, thus the random variables $f_1(X_1), \dots, f_n(X_n)$ are independent as well. ∎

Remark 4.1.10. Without proof we still mention that the independence of random variables is preserved whenever they are put together into disjoint groups. For example, if $X_1, \ldots, X_n$ are independent, then so are $f(X_1, \ldots, X_k)$ and $g(X_{k+1}, \ldots, X_n)$ for suitable functions f and g. Assume we roll a die five times and let $X_1, \ldots, X_5$ be the results. Then these random variables are independent, but so are also the two random variables $\max\{X_1, X_2\}$ and $X_3 + X_4 + X_5$ or the three X_1, $\max\{X_2, X_3\}$ and $\min\{X_4, X_5\}$.

4.2 Linear Transformations

Let a and b real numbers with $a \neq 0$. Given a random variable X set $Y := aX + b$, that is, Y arises from X by a linear transformation. We ask now for the probability distribution of Y.

Proposition 4.2.1. *Define $Y = aX + b$ with $a, b \in \mathbb{R}$ and $a \neq 0$.*

(a) *In respective of $a > 0$ or $a < 0$,*

$$F_Y(t) = F_X\left(\frac{t-b}{a}\right) \quad \text{or} \quad F_Y(t) = 1 - \mathbb{P}\left\{X < \frac{t-b}{a}\right\}.$$

If $a < 0$ and F_X is continuous at $\frac{t-b}{a}$, then $F_Y(t) = 1 - F_X\left(\frac{t-b}{a}\right)$.

(b) *Let X be a continuous random variable with density p. Then Y is also continuous with density q given by*

$$q(t) = \frac{1}{|a|} p\left(\frac{t-b}{a}\right), \quad t \in \mathbb{R}. \tag{4.3}$$

Proof: Let us first treat the case $a > 0$. Then we get

$$F_Y(t) = \mathbb{P}\{aX + b \le t\} = \mathbb{P}\left\{X \le \frac{t-b}{a}\right\} = F_X\left(\frac{t-b}{a}\right)$$

as asserted.

In the case $a < 0$ we conclude as follows:

$$F_Y(t) = \mathbb{P}\{aX + b \le t\} = \mathbb{P}\left\{X \ge \frac{t-b}{a}\right\} = 1 - \mathbb{P}\left\{X < \frac{t-b}{a}\right\}.$$

If F_X is continuous at $\frac{t-b}{a}$, then $\mathbb{P}\left\{X = \frac{t-b}{a}\right\} = 0$, hence

$$1 - \mathbb{P}\left\{X < \frac{t-b}{a}\right\} = 1 - \mathbb{P}\left\{X \le \frac{t-b}{a}\right\} = 1 - F_X\left(\frac{t-b}{a}\right),$$

completing the proof of part (a).

Suppose now that p is a density function of X, that is,

$$F_X(t) = \mathbb{P}\{X \le t\} = \int_{-\infty}^{t} p(x)dx, \quad t \in \mathbb{R}.$$

If $a > 0$, by part (a) and the change of variables $x = \frac{y-b}{a}$, we get

$$F_Y(t) = F_X\left(\frac{t-b}{a}\right) = \int_{-\infty}^{\frac{t-b}{a}} p(x)dx = \int_{-\infty}^{t} \frac{1}{a} p\left(\frac{y-b}{a}\right) dy = \int_{-\infty}^{t} q(y)\,dy.$$

Thus, q is a density of Y.

If $a < 0$, the same change of variables[2] leads to

$$F_Y(t) = 1 - F_X\left(\frac{t-b}{a}\right) = \int_{\frac{t-b}{a}}^{\infty} p(x)dx = -\int_{-\infty}^{t} \frac{1}{a} p\left(\frac{y-b}{a}\right) dy$$

$$= \int_{-\infty}^{t} \frac{1}{-a} p\left(\frac{y-b}{a}\right) dy = \int_{-\infty}^{t} \frac{1}{|a|} p\left(\frac{y-b}{a}\right) dy = \int_{-\infty}^{t} q(y)\,dy.$$

This being true for all $t \in \mathbb{R}$ completes the proof. ∎

Example 4.2.2. Let X be $\mathcal{N}(0, 1)$-distributed. Given $a \neq 0$ and $\mu \in \mathbb{R}$, we ask for the distribution of $Y := aX + \mu$.

Answer: The random variable X is assumed to be continuous with density

$$p(t) = \frac{1}{\sqrt{2\pi}} e^{-t^2/2}.$$

We apply eq. (4.3) with $b = \mu$ to deduce that the density q of Y equals

$$q(t) = \frac{1}{|a|} p\left(\frac{t-\mu}{|a|}\right) = \frac{1}{\sqrt{2\pi}\,|a|} e^{-(t-\mu)^2/2a^2}.$$

That is, the random variable Y is $\mathcal{N}(\mu, |a|^2)$-distributed. In particular, if $\sigma > 0$ and $\mu \in \mathbb{R}$, then $\sigma X + \mu$ is distributed according to $\mathcal{N}(\mu, \sigma^2)$.

Additional question: Suppose Y is $\mathcal{N}(\mu, \sigma^2)$-distributed. Which probability distribution does $X := \frac{Y-\mu}{\sigma}$ possess?

Answer: Formula (4.3) immediately implies that X is standard normally distributed.

Because of the importance of the previous observation, we formulate it as proposition.

Proposition 4.2.3. *Suppose $\mu \in \mathbb{R}$ and $\sigma > 0$. Then the following are equivalent:*

X is $\mathcal{N}(0, 1)$-distributed $\iff$ $\sigma X + \mu$ is distributed according to $\mathcal{N}(\mu, \sigma^2)$. $\boxed{!}$

---

2 Observe that now $a < 0$, hence the order of integration changes and a minus sign appears.

Corollary 4.2.4. *Let Φ be the Gaussian Φ-function introduced in eq. (1.62). For each interval $[a, b]$,*

$$N(\mu, \sigma^2)([a, b]) = \Phi\left(\frac{b - \mu}{\sigma}\right) - \Phi\left(\frac{a - \mu}{\sigma}\right).$$

Proof: This is a direct consequence of Proposition 4.2.3. Indeed, if X is standard normally distributed, then

$$N(\mu, \sigma^2)([a, b]) = \mathbb{P}\{a \le \sigma X + \mu \le b\} = \mathbb{P}\left\{\frac{a - \mu}{\sigma} \le X \le \frac{b - \mu}{\sigma}\right\}$$

$$= \Phi\left(\frac{b - \mu}{\sigma}\right) - \Phi\left(\frac{a - \mu}{\sigma}\right)$$

as asserted. ∎

Let X be an $N(\mu, \sigma^2)$-distributed random variable. The next result shows that X with high probability (more than 99.7%) attains values in $[\mu - 3\sigma, \mu + 3\sigma]$. Therefore, in most cases, one may assume that X maps into $[\mu - 3\sigma, \mu + 3\sigma]$. This observation is usually called 3σ-**rule**.

Corollary 4.2.5 (3σ-rule). *If X is distributed according to $N(\mu, \sigma^2)$, then*

$$\mathbb{P}\{|X - \mu| \le 2\sigma\} \ge 0.954 \quad and \quad \mathbb{P}\{|X - \mu| \le 3\sigma\} \ge 0.997.$$

Proof: By virtue of Corollary 4.2.4, for each $c > 0$

$$\mathbb{P}\{|X - \mu| \le c\sigma\} = \Phi(c) - \Phi(-c),$$

hence the desired estimates follow by

$$\Phi(2) - \Phi(-2) = 0.9545 \quad and \quad \Phi(3) - \Phi(-3) = 0.9973.$$

∎

Example 4.2.6. Let U be uniformly distributed on $[0, 1]$. What is the probability distribution of $aU + b$ if $a \ne 0$ and $b \in \mathbb{R}$?

Answer: The distribution density p of U is given by $p(t) = 1$ if $0 \le t \le 1$ and $p(t) = 0$ otherwise. Therefore, the density q of $aU + b$ equals

$$q(t) = \begin{cases} \frac{1}{|a|} : 0 \le \frac{t-b}{a} \le 1 \\ 0 : \text{otherwise}. \end{cases}$$

Assume first $a > 0$. Then $q(t) = 1/a$ if and only if $b \le t \le a + b$ and $q(t) = 0$ otherwise. Consequently, $aU + b$ is uniformly distributed on $[b, a + b]$.

If, in contrast, $a < 0$, then $q(t) = 1/|a|$ if and only if $a + b \le t \le b$ and $q(t) = 0$ otherwise. Hence, now $aU + b$ is uniformly distributed on $[a + b, b]$.

It is easy to see that the reversed implications are also true. That is, we have

---

$$U \text{ unif. distr. on } [0, 1] \iff aU + b \text{ unif. distr. on } \begin{cases} [b, a + b] : a > 0 \\ [a + b, b] : a < 0 \end{cases}$$

!

Corollary 4.2.7. *A random variable X is uniformly distributed on* $[0, 1]$ *if and only if* $1 - X$ *is so. In particular, if U is uniformly distributed on* $[0, 1]$, *then* $U \stackrel{d}{=} 1 - U$.

Example 4.2.8. Suppose a random variable X is $\Gamma_{\alpha,\beta}$-distributed for some $\alpha, \beta > 0$ and let $a > 0$. Which distribution does aX possess?

Answer: The distribution density p of X satisfies $p(t) = 0$ if $t \le 0$ and, if $t > 0$, then

$$p(t) = \frac{1}{\alpha^\beta \, \Gamma(\beta)} \, t^{\beta-1} e^{-t/\alpha}.$$

An application of eq. (4.3) implies that the density q of aX is given by $q(t) = 0$ if $t \le 0$ and, if $t > 0$, then

$$q(t) = \frac{1}{a} \, p\left(\frac{t}{a}\right) = \frac{1}{a \, \alpha^\beta \, \Gamma(\beta)} \left(\frac{t}{a}\right)^{\beta-1} e^{-t/a\alpha} = \frac{1}{(a\alpha)^\beta \, \Gamma(\beta)} t^{\beta-1} e^{-t/a\alpha}.$$

Thus, aX is $\Gamma_{a\alpha,\beta}$-distributed.

In the case of the exponential distribution $E_\lambda = \Gamma_{1/\lambda,1}$ the previous result implies the following: if $a > 0$, then a random variable X is E_λ-distributed if and only if aX possesses an $E_{\lambda/a}$ distribution.

4.3 Coin Tossing versus Uniform Distribution

4.3.1 Binary Fractions

We start this section with the following statement: each real number $x \in [0, 1)$ may be represented as binary fraction $x = 0.x_1x_2\cdots$, where $x_k \in \{0, 1\}$. This is a shortened way to express that

$$x = \sum_{k=1}^{\infty} \frac{x_k}{2^k}.$$

The representation of x as binary fraction is in general not unique. For example,

$$\frac{1}{2} = 0.10000 \cdots, \quad \text{but also} \quad \frac{1}{2} = 0.01111 \cdots.$$

Check this by computing the infinite sums in both cases.

It is not difficult to prove that exactly those $x \in [0, 1)$ admit two different representations, which may be written as $x = k/2^n$ for some $n \in \mathbb{N}$ and some $k = 1, 3, 5, \ldots, 2^n - 1$.

To make the binary representation unique we declare the following:

Convention 4.1. *If a number $x \in [0, 1)$ admits the representations*

$$x = 0.x_1 \cdots x_{n-1}1000 \cdots \quad \text{and} \quad x = 0.x_1 \cdots x_{n-1}0111 \cdots,$$

then we always choose the former one. In other words, there do not exist numbers $x \in [0, 1)$ whose binary representation consists from a certain point only of 1s.

How do we get the binary fraction for a given $x \in [0, 1)$?

The procedure is not difficult. First, one checks whether $x < \frac{1}{2}$ or $x \geq \frac{1}{2}$. In the former case one takes $x_1 = 0$ and in the latter $x_1 = 1$.

By this choice it follows that $0 \leq x - \frac{x_1}{2} < \frac{1}{2}$. In the next step one asks whether $x - \frac{x_1}{2} < \frac{1}{4}$ or $x - \frac{x_1}{2} \geq \frac{1}{4}$. Depending on this one chooses either $x_2 = 0$ or $x_2 = 1$. This choice implies $0 \leq x - \frac{x_1}{2} - \frac{x_2}{2^2} < \frac{1}{4}$, and if this difference belongs either to $[0, \frac{1}{8})$ or to $[\frac{1}{8}, \frac{1}{4})$, then $x_3 = 0$ or $x_3 = 1$, respectively. Proceeding further in that way leads to the binary fraction representing x.

After that heuristic method we now present a mathematically more exact way. To this end, for each $n \geq 1$, we divide the interval $[0, 1)$ into 2^n intervals of length 2^{-n}.

We start with $n = 1$ and divide $[0, 1)$ into the two intervals

$$I_0 := \left[0, \frac{1}{2} \right) \quad \text{and} \quad I_1 := \left[\frac{1}{2}, 1 \right).$$

In the second step we divide each of the two intervals I_0 and I_1 further into two parts of equal length. In this way we obtain the four intervals

$$I_{00} := \left[0, \frac{1}{4} \right), \quad I_{01} := \left[\frac{1}{4}, \frac{1}{2} \right), \quad I_{10} := \left[\frac{1}{2}, \frac{3}{4} \right) \quad \text{and} \quad I_{11} := \left[\frac{3}{4}, 1 \right).$$

Observe that the left corner point of $I_{a_1 a_2}$ equals $a_1/2 + a_2/4$, that is,

$$I_{a_1 a_2} = \left[\sum_{j=1}^{2} \frac{a_j}{2^j}, \sum_{j=1}^{2} \frac{a_j}{2^j} + \frac{1}{2^2} \right), \quad a_1, a_2 \in \{0, 1\}.$$

It is clear now how to proceed. Given $n \geq 1$ and numbers $a_1, \ldots, a_n \in \{0, 1\}$, set

$$I_{a_1 \cdots a_n} = \left[\sum_{j=1}^{n} \frac{a_j}{2^j}, \sum_{j=1}^{n} \frac{a_j}{2^j} + \frac{1}{2^n} \right). \tag{4.4}$$

In this way, we obtain 2^n disjoint intervals of length 2^{-n} where the left corner points are $0.a_1 a_2 \cdots a_n$.

The following lemma makes the above heuristic method more precise.

Lemma 4.3.1. *For all $a_1, \ldots, a_n \in \{0, 1\}$ the intervals in definition (4.4) are characterized by*

$$I_{a_1 \cdots a_n} = \{x \in [0, 1) : x = 0.a_1 a_2 \cdots a_n \cdots\}.$$

Verbally, a number in $[0, 1)$ belongs to $I_{a_1 \cdots a_n}$ if and only if its first n digits in the binary fraction are $a_1, \ldots, a_n$.

Proof: Assume first $x \in I_{a_1 \cdots a_n}$. If $a := 0.a_1 \cdots a_n$ denotes the left corner point of $I_{a_1, \ldots, a_n}$, by definition $a \le x < a + 1/2^n$ or, equivalently, $0 \le x - a < 1/2^n$. Therefore, the binary fraction of $x - a$ is of the form $0.00 \cdots 0 b_{n+1} \cdots$ with certain numbers $b_{n+1}, b_{n+2}, \ldots \in \{0, 1\}$. This yields

$$x = a + (x - a) = 0.a_1 \cdots a_n b_{n+1} \cdots .$$

Thus, as asserted, the first n digits in the representation of x are $a_1, \ldots, a_n$.

Conversely, if x can be written as $x = 0.x_1 x_2 \cdots$ with $x_1 = a_1, \ldots, x_n = a_n$, then $a \le x$ where, as above, a denotes the left corner point of $I_{a_1 \cdots a_n}$. Moreover, by Convention 4.3.1 at least one of the x_ks, $k > n$, has to be zero. Consequently,

$$x - a = \sum_{k=n+1}^{\infty} \frac{x_k}{2^k} < \sum_{k=n+1}^{\infty} \frac{1}{2^k} = \frac{1}{2^n},$$

that is, we have $a \le x < a + \frac{1}{2^n}$ or, equivalently, $x \in I_{a_1 \cdots a_n}$ as asserted. ∎

A direct consequence of Lemma 4.3.1 is as follows.

Corollary 4.3.2. *For each $n \ge 1$ the 2^n sets $I_{a_1 \cdots a_n}$ form a disjoint partition of $[0, 1)$, that is,*

$$\bigcup_{a_1, \ldots, a_n \in \{0,1\}} I_{a_1 \cdots a_n} = [0, 1) \quad and \quad I_{a_1 \cdots a_n} \cap I_{a'_1 \cdots a'_n} = \varnothing$$

provided that $(a_1, \ldots, a_n) \ne (a'_1, \ldots, a'_n)$. Furthermore,

$$\{x \in [0, 1) : x_k = 0\} = \bigcup_{a_1, \ldots, a_{k-1} \in \{0,1\}} I_{a_1 \cdots a_{k-1} 0} .$$

4.3.2 Binary Fractions of Random Numbers

We saw above each number $x \in [0, 1)$ admits a representation $x = 0.x_1 x_2 \cdots$ with certain $x_k \in \{0, 1\}$. What does happen if we choose a number x randomly, say according to the uniform distribution on $[0, 1]$? Then the x_ks in the binary fraction are also random, with values in $\{0, 1\}$. How are they distributed?

The mathematical formulation of this question is as follows: let $U : \Omega \to \mathbb{R}$ be a random variable uniformly distributed on $[0, 1]$. If $\omega \in \Omega$, write[3]

$$U(\omega) = 0.X_1(\omega)X_2(\omega) \cdots = \sum_{k=1}^{\infty} \frac{X_k(\omega)}{2^k} . \tag{4.5}$$

In this way we obtain infinitely many random variables $X_k : \Omega \to \{0, 1\}$.

Which distribution do these random variables possess? Answer gives the next proposition.

Proposition 4.3.3. *If $k \in \mathbb{N}$, then*

$$\mathbb{P}\{X_k = 0\} = \mathbb{P}\{X_k = 1\} = \frac{1}{2} . \tag{4.6}$$

Furthermore, given $n \geq 1$, the random variables $X_1, \ldots, X_n$ are independent.

Proof: By assumption $\mathbb{P}_U$ is the uniform distribution on $[0, 1]$. Thus, the finite additivity of $\mathbb{P}_U$, Corollary 4.3.2 and eq. (1.45) imply

$$\mathbb{P}\{X_k = 0\} = \mathbb{P}_U \left(\bigcup_{a_1, \ldots, a_{k-1} \in \{0,1\}} I_{a_1 \cdots a_{k-1} 0} \right)$$

$$= \sum_{a_1, \ldots, a_{k-1} \in \{0,1\}} \mathbb{P}_U \left(I_{a_1 \cdots a_{k-1} 0} \right) = \sum_{a_1, \ldots, a_{k-1} \in \{0,1\}} \frac{1}{2^k} = \frac{2^{k-1}}{2^k} = \frac{1}{2} .$$

Since X_k attains only two different values, $\mathbb{P}\{X_k = 1\} = 1/2$ as well, proving the first part.

We want to verify that for all $n \geq 1$ the random variables $X_1, \ldots, X_n$ are independent. Equivalently, according to Proposition 3.6.16, the following has to be proven: if $a_1, \ldots, a_n \in \{0, 1\}$, then

$$\mathbb{P}\{X_1 = a_1, \ldots, X_n = a_n\} = \mathbb{P}\{X_1 = a_1\} \cdots \mathbb{P}\{X_n = a_n\} . \tag{4.7}$$

By eq. (4.6) the right-hand side of eq. (4.7) equals

$$\mathbb{P}\{X_1 = a_1\} \cdots \mathbb{P}\{X_n = a_n\} = \underbrace{\frac{1}{2} \cdots \frac{1}{2}}_{n} = \frac{1}{2^n} .$$

---

3 Note that $\mathbb{P}\{U \in [0, 1)\} = 1$. Thus, without losing generality, we may assume $U(\omega) \in [0, 1)$.

To compute the left-hand side of eq. (4.7), note that Lemma 4.3.1 implies that we have $X_1 = a_1$ up to $X_n = a_n$ if and only if U attains a value in $I_{a_1 \cdots a_n}$. The intervals $I_{a_1 \cdots a_n}$ are of length 2^{-n}, hence by eq. (1.45) (recall that $\mathbb{P}_U$ is the uniform distribution on $[0, 1]$),

$$\mathbb{P}\{X_1 = a_1, \ldots, X_n = a_n\} = \mathbb{P}\{U \in I_{a_1 \cdots a_n}\} = \mathbb{P}_U(I_{a_1 \cdots a_n}) = \frac{1}{2^n}.$$

Thus, for all $a_1, \ldots, a_n \in \{0, 1\}$ eq. (4.7) is valid, and, as asserted, the random variables $X_1, \ldots, X_n$ are independent. ∎

To formulate the previous result in a different way, let us introduce the following notation.

Definition 4.3.4. An infinite sequence $X_1, X_2, \ldots$ of random variables is said to be **independent** provided that any finite collection of the X_ks is independent.

Remark 4.3.5. Since any subcollection of independent random variables is independent as well, the independence of $X_1, X_2, \ldots$ is equivalent to the following. For all $n \geq 1$ the random variables $X_1, \ldots, X_n$ are independent, that is, for all $n \geq 1$ and all Borel sets $B_1, \ldots, B_n$ it follows that

$$\mathbb{P}\{X_1 \in B_1, \ldots, X_n \in B_n\} = \mathbb{P}\{X_1 \in B_1\} \cdots \mathbb{P}\{X_n \in B_n\}.$$

Remark 4.3.6. In view of Definition 4.3.4 the basic observation in Example 3.6.17 may now be formulated in the following way. If we toss a (maybe biased) coin, labeled with "0" and "1," infinitely often and if we let $X_1, X_2, \ldots$ be the results of the single tosses, then this infinite sequence of random variables is independent with $\mathbb{P}\{X_k = 0\} = 1 - p$ and $\mathbb{P}\{X_k = 1\} = p$. In particular, for a fair coin the X_ks possess the following properties:

(a) If $k \in \mathbb{N}$, then $\mathbb{P}\{X_k = 0\} = \mathbb{P}\{X_k = 1\} = 1/2$.
(b) $X_1, X_2, \ldots$ is an infinite sequence of independent random variables.

This observation leads us to the following definition.

Definition 4.3.7. An infinite sequence $X_1, X_2, \ldots$ of **independent** random variables with values in $\{0, 1\}$ satisfying

$$\mathbb{P}\{X_k = 0\} = \mathbb{P}\{X_k = 1\} = 1/2, \quad k = 1, 2, \ldots,$$

is said to be a **model for tossing a fair coin infinitely often.**

Consequently, Proposition 4.3.3 asserts that the random variables $X_1, X_2, \ldots$ defined by eq. (4.5) serve as model for tossing a fair coin infinitely often.

4.3.3 Random Numbers Generated by Coin Tossing

We saw in Proposition 4.3.3 that choosing a random number in $[0, 1]$ leads to a model for tossing a fair coin infinitely often. Our aim is now to investigate the converse question. That is, we are given an infinite random sequence of zeros and ones and we want to construct a uniformly distributed number in $[0, 1]$. The precise mathematical question is as follows: suppose we are given an infinite sequence $(X_k)_{k \geq 1}$ of independent random variables with

$$\mathbb{P}\{X_k = 0\} = \mathbb{P}\{X_k = 1\} = 1/2, \quad k = 1, 2, \ldots. \tag{4.8}$$

Is it possible to construct from these X_ks a uniform distributed U ? The next proposition answers this question to the affirmative.

Proposition 4.3.8. *Let* $X_1, X_2, \ldots$ *be an* **arbitrary** *sequence of independent random variables satisfying eq. (4.8). If U is defined by*

$$U(\omega) := \sum_{k=1}^{\infty} \frac{X_k(\omega)}{2^k}, \quad \omega \in \Omega,$$

then this random variable is uniformly distributed on $[0, 1]$.

Proof: In order to prove that U is uniformly distributed on $[0, 1]$, we have to show that, if $t \in [0, 1)$, then

$$\mathbb{P}\{U \leq t\} = t. \tag{4.9}$$

We start the proof of eq. (4.9) with the following observation: suppose the binary fraction of some $t \in [0, 1)$ is $0.t_1 t_2 \cdots$ for certain $t_i \in \{0, 1\}$. If $s = 0.s_1 s_2 \cdots$, then $s < t$ if and only if there is an $n \in \mathbb{N}$ so that the following is satisfied[4]:

$$s_1 = t_1, \ldots, s_{n-1} = t_{n-1}, \quad s_n = 0 \text{ and } t_n = 1.$$

Fix $t \in [0, 1)$ for a moment and set

$$A_n(t) := \{s \in [0, 1) : s_1 = t_1, \ldots, s_{n-1} = t_{n-1}, s_n < t_n\}.$$

---

4 In the case $n = 1$ this says $s_1 = 0$ and $t_1 = 1$.

Of course, $A_n(t) \cap A_m(t) = \varnothing$ whenever $n \neq m$ and, moreover, $A_n(t) \neq \varnothing$ if and only if $t_n = 1$. Furthermore, by the previous remark

$$[0, t) = \bigcup_{n=1}^{\infty} A_n(t) = \bigcup_{\{n:t_n=1\}} A_n(t).$$

Finally, if $A_n(t) \neq \varnothing$, that is, if $t_n = 1$, then

$$\mathbb{P}\{U \in A_n(t)\} = \mathbb{P}\{X_1 = t_1, \ldots, X_{n-1} = t_{n-1}, X_n = 0\}$$
$$= \mathbb{P}\{X_1 = t_1\} \cdots \mathbb{P}\{X_{n-1} = t_{n-1}\} \cdot \mathbb{P}\{X_n = 0\} = \frac{1}{2^n}.$$

In the last step we used both properties of the X_ks, that is, they are independent and satisfy $\mathbb{P}\{X_k = 0\} = \mathbb{P}\{X = 1\} = 1/2$.

Summing up, we get

$$\mathbb{P}\{U < t\} = \mathbb{P}\left\{U \in \bigcup_{\{n:t_n=1\}} A_n(t)\right\} = \sum_{\{n:t_n=1\}} \mathbb{P}\{U \in A_n(t)\}$$
$$= \sum_{\{n:t_n=1\}} \frac{1}{2^n} = \sum_{n=1}^{\infty} \frac{t_n}{2^n} = t.$$

This "almost" proves eq. (4.9). It remains to show that $\mathbb{P}\{U < t\} = \mathbb{P}\{U \leq t\}$ or, equivalently, $\mathbb{P}\{U = t\} = 0$. To verify this we use the continuity of $\mathbb{P}$ from above. Then

$$\mathbb{P}\{U = t\} = \mathbb{P}\{X_1 = t_1, X_2 = t_2, \ldots\}$$
$$= \lim_{n\to\infty} \mathbb{P}\{X_1 = t_1, \ldots, X_n = t_n\} = \lim_{n\to\infty} \frac{1}{2^n} = 0.$$

Consequently, eq. (4.9) holds for all $t \in [0, 1)$ and, as asserted, U is uniformly distributed on $[0, 1]$. ∎

Remark 4.3.9. Another possibility to write U is as binary fraction

$$U(\omega) = 0.X_1(\omega)X_2(\omega) \cdots, \quad \omega \in \Omega.$$

Consequently, in order to construct a random number u in $[0, 1]$ one may proceed as follows: toss a fair coin with "0" and "1" infinitely often and take the obtained sequence as binary fraction of u. The u obtained in this way is uniformly distributed on $[0, 1]$.

Of course, in practice one tosses a coin not infinitely often. One stops the procedure after N trials for some "large" N. In this way one gets a number u, which is "almost" uniformly distributed on $[0, 1]$.

Then how does one construct n independent numbers $u_1, \ldots, u_n$, all uniformly distributed on $[0, 1]$? The answer is quite obvious. Take n coins and toss them. As functions of independent observations the generated $u_1, \ldots, u_n$ are independent as well and, by the construction, each of these numbers is uniformly distributed on $[0, 1]$. Another way is to toss the same coin n times "infinitely often," thus getting n infinite sequences of zeroes and ones.

4.4 Simulation of Random Variables

Proposition 4.3.8 provides us with a technique to simulate a uniformly distributed random variable U by tossing a fair coin. The aim of this section is to find a suitable function $f : [0, 1] \to \mathbb{R}$, so that the transformed random variable $X = f(U)$ possesses a given probability distribution.

Example 4.4.1. Typical questions of this kind are as follows: find a function f so that $X = f(U)$ is standard normally distributed. Does there exist another function $g : [0, 1] \to \mathbb{R}$ for which $g(U)$ is $B_{n,p}$-distributed?

Suppose for a moment we already found such functions f and g. According to Remark 4.3.9, we construct independent numbers $u_1, \ldots, u_n$, uniformly distributed on $[0, 1]$, and set $x_i = f(u_i)$ and $y_i = g(u_i)$. In this way we get either n standard normally distributed numbers $x_1, \ldots, x_n$ or n binomial distributed numbers $y_1, \ldots, y_n$. Moreover, by Proposition 4.1.9 these numbers are independent. In this way we may simulate independent random numbers possessing a given probability distribution.

We start with simulating **discrete** random variables. Thus suppose we are given real numbers $x_1, x_2, \ldots$ and $p_k \geq 0$ with $\sum_{k=1}^{\infty} p_k = 1$, and we look for a random variable $X = f(U)$ such that

$$\mathbb{P}\{X = x_k\} = p_k, \quad k = 1, 2, \ldots.$$

One possible way to find such a function f is as follows: divide $[0, 1)$ into disjoint intervals $I_1, I_2, \ldots$ of length $|I_k| = p_k$, $k = 1, 2, \ldots$. Since $\sum_{k=1}^{\infty} p_k = 1$, such intervals exist. For example, take $I_1 = [0, p_1)$ and

$$I_k = \left[\sum_{i=1}^{k-1} p_i, \sum_{i=1}^{k} p_i \right), \quad k = 2, 3, \ldots.$$

With these intervals I_k we define $f : [0, 1] \to \mathbb{R}$ by

$$f(x) := x_k \quad \text{if} \quad x \in I_k, \tag{4.10}$$

or, equivalently,

$$f(x) = \sum_{k=1}^{\infty} x_k \, \mathbb{1}_{I_k}(x) \, . \tag{4.11}$$

Then the following is true.

Proposition 4.4.2. *Let U be uniformly distributed on $[0, 1]$, and set $X = f(U)$ with f defined by eq. (4.10) or eq. (4.11). Then*

$$\mathbb{P}\{X = x_k\} = p_k, \quad k = 1, 2, \dots$$

Proof: Using that U is uniformly distributed on $[0, 1]$, this is an easy consequence of eq. (1.45) in view of

$$\mathbb{P}\{X = x_k\} = \mathbb{P}\{f(U) = x_k\} = \mathbb{P}\{U \in I_k\} = |I_k| = p_k \, .$$

∎

Remark 4.4.3. Note that the concrete shape of the intervals[5] I_k is not important at all. They only have to satisfy $|I_k| = p_k$, $k = 1, 2, \dots$. Moreover, these intervals need not necessarily to be disjoint; a "small" overlap does not influence the assertion. Indeed, it suffices that $\mathbb{P}\{U \in I_k \cap I_l\} = 0$ whenever $k \neq l$. For example, if always $\#(I_k \cap I_l) < \infty$, $k \neq l$, then the construction works as well. In particular, we may choose also $I_k = \left[\sum_{i=1}^{k-1} p_i, \sum_{i=1}^{k} p_i\right]$.

Example 4.4.4. We want to simulate a random variable X, which is uniformly distributed on $\{x_1, \dots, x_N\}$. How to proceed?

Answer: Divide the interval $[0, 1)$ into N intervals $I_1, \dots, I_N$ of length $\frac{1}{N}$. For example, choose $I_k := \left[\frac{k-1}{N}, \frac{k}{N}\right)$, $k = 1, \dots, N$. If $f = \sum_{k=1}^{N} x_k \, \mathbb{1}_{I_k}$, then $X = f(U)$ is uniformly distributed on $\{x_1, \dots, x_N\}$.

Example 4.4.5. Suppose we want to simulate a number $k \in \mathbb{N}_0$, which is Pois_λ-distributed. Set

$$I_k := \left[\sum_{j=0}^{k-1} \frac{\lambda^j}{j!} \, e^{-\lambda}, \sum_{j=0}^{k} \frac{\lambda^j}{j!} \, e^{-\lambda}\right), \quad k = 0, 1, \dots,$$

where the left-hand sum is supposed to be zero if $k = 0$. Choose randomly a number $u \in [0, 1]$ and take the k with $u \in I_k$. Then k is the number we are interested in.

---

5 They do not even need to be intervals.

Our next aim is to simulate **continuous** random variables. More precisely, suppose we are given a probability density p. Then we look for a function $f : [0, 1] \to \mathbb{R}$ such that p is the density of $X = f(U)$, that is, that

$$\mathbb{P}\{X \le t\} = \int_{-\infty}^{t} p(x)\,dx, \quad t \in \mathbb{R}, \tag{4.12}$$

To this end set

$$F(t) = \int_{-\infty}^{t} p(x)\,dx, \quad t \in \mathbb{R}. \tag{4.13}$$

Thus, F is the distribution function of the random variable X, which we are going to construct.

Suppose first that F is one-to-one on a finite or infinite interval (a, b), so that $F(x) = 0$ if $x < a$, and $F(x) = 1$ if $x > b$. Since F is continuous, the inverse function F^{-1} exists and maps $(0, 1)$ to (a, b).

Proposition 4.4.6. *Let p be a probability density and define F by eq. (4.12). Suppose F satisfies the above condition. If $X = F^{-1}(U)$, then*

$$\mathbb{P}\{X \le t\} = \int_{-\infty}^{t} p(x)\,dx, \quad t \in \mathbb{R},$$

that is, p is a density of X.

Proof: First note that the assumptions about F imply that F is increasing on (a, b). Hence, if $t \in \mathbb{R}$, then

$$\mathbb{P}\{X \le t\} = \mathbb{P}\{F^{-1}(U) \le t\} = \mathbb{P}\{U \le F(t)\} = F(t) = \int_{-\infty}^{t} p(x)\,dx, \quad t \in \mathbb{R}.$$

Here we used $0 \le F(t) \le 1$ and $\mathbb{P}\{U \le s\} = s$ whenever $0 \le s \le 1$. This completes the proof. ∎

But what do we do if F does not satisfy the above assumption? For example, this happens if $p(x) = 0$ on an interval $I = (\alpha, \beta)$ and $p(x) > 0$ on some left- and right-hand intervals[6] of I. In this case F^{-1} does not exist, and we have to modify the construction.[7]

---

6 Take, for instance, p with $p(x) = \frac{1}{2}$ if $x \in [0, 1]$ and if $x \in [1, 2]$, and $p(x) = 0$ otherwise.

7 All subsequent distribution functions F possess an inverse function on a suitable interval (a, b). Thus, Proposition 4.4.6 applies in almost all cases of interest. Therefore, to whom the statements about pseudo-inverse functions look too complicated, you may skip them.

Definition 4.4.7. Let F be defined by eq. (4.13). Then we set

$$F^-(s) = \inf\{t \in \mathbb{R} : F(t) = s\}, \quad 0 \le s < 1.$$

The function F^-, mapping $[0, 1)$ to $[-\infty, \infty)$, is called the **pseudo-inverse** of F.

Remark 4.4.8. If $0 < s < 1$, then $F^-(s) \in \mathbb{R}$ while $F^-(0) = -\infty$. Moreover, whenever F is increasing on some interval I, then $F^-(s) = F^{-1}(s)$ for $s \in I$.

Lemma 4.4.9. *The pseudo-inverse function F^- possesses the following properties.*
1. *If $s \in (0, 1)$ and $t \in \mathbb{R}$, then*

$$F(F^-(s)) = s \quad and \quad F^-(F(t)) \le t.$$

2. *Given $t \in (0, 1)$ we have*

$$F^-(s) \le t \iff s \le F(t). \tag{4.14}$$

Proof: The equality $F(F^-(s)) = s$ is a direct consequence of the continuity of F. Indeed, if there are $t_n \searrow F^-(s)$ with $F(t_n) = s$, then

$$s = \lim_{n \to \infty} F(t_n) = F(F^-(s)).$$

The second part of the first assertion follows by the definition of F^-.

Now let us come to the proof of property (4.14). If $F^-(s) \le t$, then the monotonicity of F as well as $F(F^-(s)) = s$ lead to $s = F(F^-(s)) \le F(t)$.

Conversely, if $s \le F(t)$, then $F^-(s) \le F^-(F(t)) \le t$ by the first part, thus, property (4.14) is proved. ∎

Now choose a uniform distributed U and set $X = F^-(U)$. Since $\mathbb{P}\{U = 0\} = 0$, we may assume that X attains values in $\mathbb{R}$.

Proposition 4.4.10. *Let p be a probability density, that is, we have $p(x) \ge 0$ and $\int_{-\infty}^{\infty} p(x)dx = 1$. Define F by eq. (4.13) and let F^- be its pseudo-inverse. Take U uniform on $[0, 1]$ and set $X = F^-(U)$. Then p is a distribution density of the random variable X.*

Proof: Using property (4.14) it follows

$$F_X(t) = \mathbb{P}\{X \le t\} = \mathbb{P}\{\omega \in \Omega : F^-(U(\omega)) \le t\} = \mathbb{P}\{\omega \in \Omega : U(\omega) \le F(t)\} = F(t),$$

which completes the proof. ∎

Remark 4.4.11. Since $F^- = F^{-1}$ whenever the inverse function exists, Proposition 4.4.6 is a special case of Proposition 4.4.10.

Example 4.4.12. Let us simulate an $\mathcal{N}(0, 1)$-distributed random variable, that is, we are looking for a function $f : (0, 1) \to \mathbb{R}$ such that for uniformly distributed U

$$\mathbb{P}\{f(U) \le t\} = \frac{1}{\sqrt{2\pi}} \int_{-\infty}^{t} e^{-x^2/2} \, dx, \quad t \in \mathbb{R}.$$

The distribution function

$$\Phi(t) = \frac{1}{\sqrt{2\pi}} \int_{-\infty}^{t} e^{-x^2/2} \, dx$$

is one-to-one from $\mathbb{R} \to (0, 1)$, hence Proposition 4.4.6 applies, and $\Phi^{-1}(U)$ is a standard normal random variable.

How does one get an $\mathcal{N}(\mu, \sigma^2)$-distributed random variable? If X is standard normal, by Proposition 4.2.3 the transformed variable $\sigma X + \mu$ is $\mathcal{N}(\mu, \sigma^2)$-distributed. Consequently, $\sigma\Phi^{-1}(U) + \mu$ possesses the desired distribution.

How do we find n independent $\mathcal{N}(\mu, \sigma^2)$-distributed numbers $x_1, \ldots, x_n$? To achieve this, choose $u_1, \ldots, u_n$ in $[0, 1]$ according to the construction presented in Remark 4.3.9 and set $x_i = \sigma\Phi^{-1}(u_i) + \mu, 1 \le i \le n$.

Example 4.4.13. Our next aim is to simulate an E_λ-distributed (exponentially distributed) random variable. Here

$$F(t) = \begin{cases} 0 & : t \le 0 \\ 1 - e^{-\lambda t} & : t > 0, \end{cases}$$

which satisfies the assumptions of Proposition 4.4.6 on the interval $(0, \infty)$. Its inverse F^{-1} maps $(0, 1)$ to $(0, \infty)$ and equals

$$F^{-1}(s) = -\frac{\ln(1 - s)}{\lambda}, \quad 0 < s < 1.$$

Therefore, if U is uniformly distributed on $[0, 1]$, then $X = -\frac{\ln(1-U)}{\lambda}$ is E_λ distributed. This is true for any uniformly distributed random variable U. By Corollary 4.2.7 the random variable $1 - U$ has the same distribution as U, hence, setting

$$Y = -\frac{\ln(1 - (1 - U))}{\lambda} = -\frac{\ln(U)}{\lambda},$$

Y is E_λ distributed as well.

Example 4.4.14. Let us simulate a random variable with Cauchy distribution (cf. Definition 1.6.33). The distribution function F is given by

$$F(t) = \frac{1}{\pi} \int_{-\infty}^{t} \frac{1}{1+x^2} \, dx = \frac{1}{\pi} \arctan(t) + \frac{1}{2}, \quad t \in \mathbb{R},$$

hence $X := \tan(\pi U - \frac{\pi}{2})$ possesses a Cauchy distribution.

Example 4.4.15. Finally, let us give an example where Proposition 4.4.10 applies and Proposition 4.4.6 does not. Suppose we want to simulate a random variable X with distribution function F defined by

$$F(t) = \begin{cases} 0 & : \ t < 0 \\ \frac{t}{2} & : 0 \leq t < 1 \\ \frac{1}{2} & : 1 \leq t < 2 \\ \frac{1}{2} + \frac{t-2}{2} & : 2 \leq t < 3 \\ 1 & : \ t \geq 3 \end{cases} \tag{4.15}$$

Direct computations imply

$$F^{-}(s) = \begin{cases} 2s & : 0 < s < \frac{1}{2} \\ 1 & : \ s = \frac{1}{2} \\ 2s + 1 : \frac{1}{2} < s \leq 1, \end{cases}$$

hence, if X is defined by

$$X = 2U \, \mathbb{1}_{\{U \leq \frac{1}{2}\}} + (2U + 1) \, \mathbb{1}_{\{U > \frac{1}{2}\}},$$

then $\mathbb{P}\{X \leq t\} = F(t)$ with F defined by eq. (4.15). In other words, X is acting as follows. Choose by random a number $u \in [0, 1]$. If $u \leq \frac{1}{2}$, then $X(u) = 2u$ while for $u > \frac{1}{2}$ we take $X(u) = 2u + 1$.

4.5 Addition of Random Variables

Suppose we are given two random variables X and Y, both mapping from Ω into $\mathbb{R}$. As usual, their sum $X + Y$ is defined by

$$(X + Y)(\omega) := X(\omega) + Y(\omega), \quad \omega \in \Omega.$$

The main question we investigate in this section is as follows: suppose we know the probability distributions of X and Y. Is there a way to compute the distribution of $X+Y$? For example, if we roll a die twice, X is the result of the first roll, Y that of the second, then we know $\mathbb{P}_X$ and $\mathbb{P}_Y$. But how do we get $\mathbb{P}_{X+Y}$?

Before we treat this question, we have to be sure that $X + Y$ is also a random variable. This is not obvious at all. Otherwise, the probability distribution of $X + Y$ is not defined and our question does not make sense.

Proposition 4.5.1. *If X and Y are random variables, then so is $X + Y$.*

Proof: We start the proof with the following observation. For two real numbers a and b holds $a < b$ if and only if there is a rational number $q \in \mathbb{Q}$ such that $a < q$ and $b > q$. Therefore, given $t \in \mathbb{R}$, it follows that

$$\{\omega \in \Omega : X(\omega) + Y(\omega) < t\} = \{\omega \in \Omega : X(\omega) < t - Y(\omega)\}$$
$$= \bigcup_{q \in \mathbb{Q}} \left[\{\omega : X(\omega) < q\} \cap \{\omega : q < t - Y(\omega)\} \right]. \tag{4.16}$$

By assumption, X and Y are random variables. Hence, for each $q \in \mathbb{Q}$,

$$A_q := \{\omega : X(\omega) < q\} \in \mathcal{A} \quad \text{and} \quad B_q := \{\omega : Y(\omega) < t - q\} \in \mathcal{A},$$

which by the properties of σ-fields implies $C_q := A_q \cap B_q \in \mathcal{A}$. With this notation we may write eq. (4.16) as

$$\{\omega \in \Omega : X(\omega) + Y(\omega) < t\} = \bigcup_{q \in \mathbb{Q}} C_q .$$

The σ-field $\mathcal{A}$ is closed under countable union, thus, since $\mathbb{Q}$ is countably infinite and $C_q \in \mathcal{A}$, it follows that $\bigcup_{q \in \mathbb{Q}} C_q \in \mathcal{A}$. Therefore, we have proven that, if $t \in \mathbb{R}$, then

$$\{\omega \in \Omega : X(\omega) + Y(\omega) < t\} \in \mathcal{A}.$$

Proposition 3.1.6 lets us conclude that, as asserted, $X + Y$ is a random variable . ∎

Remark 4.5.2. In view of Proposition 4.5.1 the following *question* makes sense: does there exist a general approach to evaluate $\mathbb{P}_{X+Y}$ by virtue of $\mathbb{P}_X$ and of $\mathbb{P}_Y$?

Answer: Such a general way does not exist. The deeper reason behind this is that, in order to get $\mathbb{P}_{X+Y}$, one has to know the joint distribution of (X, Y). And as we saw in Section 3.5, in general, the knowledge of $\mathbb{P}_X$ and $\mathbb{P}_Y$ does not suffice to determine their joint distribution, hence generally we also do not know $\mathbb{P}_{X+Y}$.

The next example emphasizes the previous remark.

Example 4.5.3. Let X, Y, X', and Y' be as in Example 3.5.8, that is,

$$\mathbb{P}\{X = 0, Y = 0\} = \frac{1}{6}, \quad \mathbb{P}\{X = 0, Y = 1\} = \frac{1}{3}$$

$$\mathbb{P}\{X = 1, Y = 0\} = \frac{1}{3}, \quad \mathbb{P}\{X = 1, Y = 1\} = \frac{1}{6} \quad \text{and}$$

$$\mathbb{P}\{X' = 0, Y' = 0\} = \mathbb{P}\{X' = 0, Y' = 1\} = \mathbb{P}\{X' = 1, Y' = 0\}$$

$$= \mathbb{P}\{X' = 1, Y' = 1\} = \frac{1}{4}.$$

Then $\mathbb{P}_X = \mathbb{P}_{X'}$ and $\mathbb{P}_Y = \mathbb{P}_{Y'}$, but

$$\mathbb{P}\{X + Y = 0\} = \frac{1}{6}, \ \mathbb{P}\{X + Y = 1\} = \frac{2}{3} \quad \text{and} \quad \mathbb{P}\{X + Y = 2\} = \frac{1}{6},$$

$$\mathbb{P}\{X' + Y' = 0\} = \frac{1}{4}, \ \mathbb{P}\{X' + Y' = 1\} = \frac{1}{2} \quad \text{and} \quad \mathbb{P}\{X' + Y' = 2\} = \frac{1}{4}.$$

Thus, X and X' as well as Y and Y' are identically distributed, but the sums $X + Y$ and $X' + Y'$ are not.

On the other hand, as we saw in Proposition 3.6.5, the joint distribution is uniquely determined by the marginal ones, provided the random variables are independent. Therefore, for **independent** random variables X and Y, the distribution of $X + Y$ is determined by those of X and Y. The question remains, how $\mathbb{P}_{X+Y}$ can be computed.

4.5.1 Sums of Discrete Random Variables

We first consider an important special case, namely that X and Y attain values in $\mathbb{Z}$. Here we have

Proposition 4.5.4 (Convolution formula for $\mathbb{Z}$-valued random variables). *Let X and Y be two independent random variables with values in $\mathbb{Z}$. If $k \in \mathbb{Z}$, then*

$$\mathbb{P}\{X + Y = k\} = \sum_{i=-\infty}^{\infty} \mathbb{P}\{X = i\} \cdot \mathbb{P}\{Y = k - i\}.$$

!

Proof: Fix $k \in \mathbb{Z}$ and define $B_k \subseteq \mathbb{Z} \times \mathbb{Z}$ by

$$B_k := \{(i, j) \in \mathbb{Z} \times \mathbb{Z} : i + j = k\}.$$

Then we get

$$\mathbb{P}\{X + Y = k\} = \mathbb{P}\{(X, Y) \in B_k\} = \mathbb{P}_{(X,Y)}(B_k) \tag{4.17}$$

with joint distribution $\mathbb{P}_{(X,Y)}$. Proposition 3.6.9 asserts that for independent X and Y and $B \subseteq \mathbb{Z} \times \mathbb{Z}$,

$$\mathbb{P}_{(X,Y)}(B) = \sum_{(i,j)\in B} \mathbb{P}_X(\{i\}) \cdot \mathbb{P}_Y(\{j\}) = \sum_{(i,j)\in B} \mathbb{P}\{X = i\} \cdot \mathbb{P}\{Y = j\}.$$

We apply this formula with $B = B_k$, and by eq. (4.17) we obtain

$$\mathbb{P}\{X + Y = k\} = \sum_{(i,j)\in B_k} \mathbb{P}\{X = i\} \cdot \mathbb{P}\{Y = j\}$$

$$= \sum_{\{(i,j)\,:\,i+j=k\}} \mathbb{P}\{X = i\} \cdot \mathbb{P}\{Y = j\} = \sum_{i=-\infty}^{\infty} \mathbb{P}\{X = i\} \cdot \mathbb{P}\{Y = k - i\},$$

as asserted. ∎

Example 4.5.5. Two independent random variables X and Y are distributed according to $\mathbb{P}\{X = j\} = \mathbb{P}\{Y = j\} = 1/2^j$, $j = 1, 2, \dots$. Determine the probability distribution of $X - Y$.

Solution: First note that $\mathbb{P}\{X = j\} = \mathbb{P}\{Y = j\} = 0$ for $j \leq 0$. Hence, given $k \in \mathbb{Z}$, an application of Proposition 4.5.4 to X and $-Y$ yields

$$\mathbb{P}\{X - Y = k\} = \sum_{i=-\infty}^{\infty} \mathbb{P}\{X = i\} \cdot \mathbb{P}\{-Y = k - i\} = \sum_{i=1}^{\infty} \mathbb{P}\{X = i\} \cdot \mathbb{P}\{Y = i - k\}.$$

If $k \geq 0$, then $\mathbb{P}\{Y = i - k\} = 0$ for $i \leq k$, thus

$$\mathbb{P}\{X - Y = k\} = \sum_{i=k+1}^{\infty} \mathbb{P}\{X = i\} \cdot \mathbb{P}\{Y = i - k\} = \sum_{i=k+1}^{\infty} \frac{1}{2^i} \cdot \frac{1}{2^{i-k}}$$

$$= 2^k \sum_{i=k+1}^{\infty} \frac{1}{2^{2i}} = 2^k \cdot 2^{-2k-2} \cdot \sum_{i=0}^{\infty} \frac{1}{2^{2i}} = 2^{-k-2} \cdot \frac{4}{3} = \frac{2^{-k}}{3}.$$

For $k < 0$ it follows that

$$\mathbb{P}\{X - Y = k\} = \sum_{i=1}^{\infty} \frac{1}{2^i} \cdot \frac{1}{2^{i-k}} = 2^k \sum_{i=1}^{\infty} \frac{1}{2^{2i}} = 2^k \sum_{i=1}^{\infty} \frac{1}{4^i} = \frac{2^k}{3}.$$

We combine both cases and obtain

$$\mathbb{P}\{X - Y = k\} = \frac{2^{-|k|}}{3}, \quad k \in \mathbb{Z}.$$

Which random experiment does $X - Y$ describe? Suppose player A and B both toss a fair coin. Let X be the number of necessary trials for A to observe the first "head." Similarly, Y describes how often B has to toss his coin to get the first "head." Thus, the value of $X - Y$ tells us how many trials later (or earlier) player A got his first "head" than B got his one.

For example, if B got his first "head" one trial earlier than A, then $X - Y = 1$. The probability that this occurs equals 1/6.

One special case of Proposition 4.5.4 is of particular interest.

Proposition 4.5.6 (Convolution formula for $\mathbb{N}_0$-valued random variables). *Let X and Y be two independent random variables with values in $\mathbb{N}_0$. If $k \in \mathbb{N}_0$, then it follows that*

!

$$\mathbb{P}\{X + Y = k\} = \sum_{i=0}^{k} \mathbb{P}\{X = i\} \cdot \mathbb{P}\{Y = k - i\}.$$

Proof: Regard X and Y as $\mathbb{Z}$-valued random variables with $\mathbb{P}\{X = i\} = \mathbb{P}\{Y = i\} = 0$ for all $i = -1, -2 \ldots$. If $k \in \mathbb{N}_0$, then Proposition 4.5.4 lets us conclude that

$$\mathbb{P}\{X + Y = k\} = \sum_{i=-\infty}^{\infty} \mathbb{P}\{X = i\} \cdot \mathbb{P}\{Y = k - i\} = \sum_{i=0}^{k} \mathbb{P}\{X = i\} \cdot \mathbb{P}\{Y = k - i\}.$$

Here we used $\mathbb{P}\{X = i\} = 0$ for $i < 0$ and $\mathbb{P}\{Y = k - i\} = 0$ if $i > k$. For $k < 0$ it follows that $\mathbb{P}\{X + Y = k\} = 0$ because in this case $\mathbb{P}\{Y = k - i\} = 0$ for all $i \geq 0$. This completes the proof. ∎

Example 4.5.7. Let X and Y be two independent random variables, both uniformly distributed on $\{1, 2, \ldots, N\}$. Which probability distribution does $X + Y$ possess?

Answer: Of course, $X + Y$ attains only values in $\{2, 3, \ldots, 2N\}$. Hence, $\mathbb{P}\{X + Y = k\}$ is only of interest for $2 \leq k \leq 2N$. Here we get

$$\mathbb{P}\{X + Y = k\} = \frac{\#(I_k)}{N^2}, \tag{4.18}$$

where I_k is defined by

$$I_k := \{i \in \{1, \ldots, N\} : 1 \leq k - i \leq N\} = \{i \in \{1, \ldots, N\} : k - N \leq i \leq k - 1\}.$$

To verify eq. (4.18) use that for $i \notin I_k$ either $\mathbb{P}\{X = i\} = 0$ or $\mathbb{P}\{Y = k - i\} = 0$. It is not difficult to prove that

$$\#(I_k) = \begin{cases} k - 1 & : \ 2 \le k \le N + 1 \\ 2N - k + 1 & : \ N + 1 < k \le 2N \end{cases}$$

which leads to

$$\mathbb{P}\{X + Y = k\} = \begin{cases} \frac{k-1}{N^2} & : \ 2 \le k \le N + 1 \\ \frac{2N-k+1}{N^2} & : \ N + 1 < k \le 2N \\ 0 & : \quad \text{otherwise} \end{cases}$$

If $N = 6$, then $X + Y$ may be viewed as the sum of two rolls of a die. Here the above formula leads to the values of $\mathbb{P}\{X + Y = k\}$, $k = 2, \ldots, 12$, which we, by a direct approach, already computed in Example 3.2.15.

Finally, let us shortly discuss the case of two arbitrary independent discrete random variables. Assume that X and Y have values in at most countable infinite sets D and E, respectively. Then $X + Y$ maps into

$$D + E := \{x + y : x \in D, \ y \in E\}.$$

Note that $D + E$ is also at most countably infinite.

Under these assumptions the following is valid.

Proposition 4.5.8. *Suppose X and Y are two independent discrete random variables with values in the (at most) countably infinite sets D and E, respectively. For $z \in D + E$ it follows that*

$$\mathbb{P}\{X + Y = z\} = \sum_{\{(x,y) \in D \times E \, : \, x+y=z\}} \mathbb{P}\{X = x\} \cdot \mathbb{P}\{Y = y\}.$$

Proof: For fixed $z \in D + E$ define $B_z \subseteq D \times E$ by $B_z := \{(x, y) : x + y = z\}$. Using this notation we get

$$\mathbb{P}\{X + Y = z\} = \mathbb{P}\{(X, Y) \in B_z\} = \mathbb{P}_{(X,Y)}(B_z),$$

where again $\mathbb{P}_{(X,Y)}$ denotes the joint distribution of X and Y. Now we may proceed as in the proof of Proposition 4.5.4. The independence of X and Y implies

$$\mathbb{P}_{(X,Y)}(B_z) = \sum_{\{(x,y) \in D \times E : x+y=z\}} \mathbb{P}\{X = x\} \cdot \mathbb{P}\{Y = y\},$$

proving the proposition. ∎

Remark 4.5.9. If $D = E = \mathbb{Z}$, then Proposition 4.5.8 implies Proposition 4.5.4, while for $D = E = \mathbb{N}_0$ we rediscover Proposition 4.5.6.

4.5.2 Sums of Continuous Random Variables

In this section we investigate the following question: let X and Y be two continuous random variables with density functions p and q. Is $X + Y$ continuous as well, and if this is so, how do we compute its density?

To answer this question we need a special type of composing two functions.

Definition 4.5.10. Let f and g be two Riemann integrable functions from $\mathbb{R}$ to $\mathbb{R}$. Their **convolution** $f \star g$ is defined by

$$(f \star g)(x) := \int_{-\infty}^{\infty} f(x - y) g(y) \, dy, \quad x \in \mathbb{R}. \tag{4.19}$$

Remark 4.5.11. The convolution is a commutative operation, that is,

$$f \star g = g \star f.$$

This follows by the change of variables $u = x - y$ in eq. (4.19), thus

$$(f \star g)(x) = \int_{-\infty}^{\infty} f(x - y) g(y) \, dy = \int_{-\infty}^{\infty} f(u) g(x - u) \, du = (g \star f)(x), \quad x \in \mathbb{R}.$$

Remark 4.5.12. For general functions f and g the integral in eq. (4.19) does not always exist for all $x \in \mathbb{R}$. The investigation of this question requires facts and notations[8] from Measure Theory; therefore, we will not treat it here. We only state a special case, which suffices for our later purposes. Moreover, for concrete functions f and g it is mostly easy to check for which $x \in \mathbb{R}$ the value $(f \star g)(x)$ exists.

Proposition 4.5.13. *Let p and q be two probability densities and suppose that at least one of them is bounded. Then $(p \star q)(x)$ exists for all $x \in \mathbb{R}$.*

Proof: Say p is bounded, that is, there is a constant $c \geq 0$ such that $0 \leq p(z) \leq c$ for all $z \in \mathbb{R}$. Since $q(y) \geq 0$, if $x \in \mathbb{R}$, then

$$0 \leq \int_{-\infty}^{\infty} p(x - y) q(y) \, dy \leq c \int_{-\infty}^{\infty} q(y) \, dy = c < \infty.$$

---

8 For example, exists "almost everywhere."

This proves that $(p \star g)(x)$ exists for all $x \in \mathbb{R}$.

Since $p \star q = q \star p$, the same argument applies if q is bounded. ∎

The next result provides us with a formula for the evaluation of the density function of $X + Y$ for independent continuous X and Y.

Proposition 4.5.14 (Convolution formula for continuous random variables). *Let X and Y be two independent random variables with distribution densities p and q. Then X + Y is continuous as well, and its density r may be computed by*

> **!**

$$r(x) = (p \star q)(x) = \int_{-\infty}^{\infty} p(y)\, q(x - y)\, dy$$

Proof: We have to show that $r = p \star q$ satisfies

$$\mathbb{P}\{X + Y \le t\} = \int_{-\infty}^{t} r(x)\, dx, \quad t \in \mathbb{R}. \tag{4.20}$$

Fix $t \in \mathbb{R}$ for a moment and define $B_t \subseteq \mathbb{R}^2$ by

$$B_t := \{(u, y) \in \mathbb{R}^2 : u + y \le t\}.$$

Then we get

$$\mathbb{P}\{X + Y \le t\} = \mathbb{P}\{(X, Y) \in B_t\} = \mathbb{P}_{(X,Y)}(B_t). \tag{4.21}$$

To compute the right-hand side of eq. (4.21) we use Proposition 3.6.18. It asserts that the joint distribution $\mathbb{P}_{(X,Y)}$ of independent X and Y is given by $(u, y) \mapsto p(u)q(y)$, that is, if $B \subseteq \mathbb{R}^2$, then

$$\mathbb{P}_{(X,Y)}(B) = \iint_{B} p(u)q(y)\, dy\, du.$$

Choosing $B = B_t$ in the last formula, eq. (4.21) may now be written as

$$\mathbb{P}\{X + Y \le t\} = \iint_{B_t} p(u)\, q(y)\, dy\, du = \int_{-\infty}^{\infty} \left[\int_{-\infty}^{t-y} p(u)\, du \right] q(y)\, dy. \tag{4.22}$$

Next we change the variables in the inner integral as follows[9]: $u = x - y$, hence $du = dx$. Then the right-hand integrals in eq. (4.22) coincide with

$$\int_{-\infty}^{\infty} \left[\int_{-\infty}^{t} p(x-y) \, dx \right] q(y) \, dy = \int_{-\infty}^{t} \left[\int_{-\infty}^{\infty} p(x-y) \, q(y) \, dy \right] dx$$

$$= \int_{-\infty}^{t} (p \star q)(x) \, dx.$$

Hereby we used that p and q are non-negative, so that we may interchange the integrals by virtue of Proposition A.5.5. Thus, eq. (4.20) is satisfied, which completes the proof. ∎

4.6 Sums of Certain Random Variables

Let us start with the investigation of the sum of independent **binomial distributed** random variables. Here the following is valid.

Proposition 4.6.1. *Let X and Y be two independent random variables, accordingly $B_{n,p}$ and $B_{m,p}$ distributed for some $n, m \geq 1$, and some $p \in [0, 1]$. Then $X + Y$ is $B_{n+m,p}$-distributed.*

Proof: By Proposition 4.5.6 we get that for $0 \leq k \leq m + n$

$$\mathbb{P}\{X + Y = k\} = \sum_{j=0}^{k} \left[\binom{n}{j} p^j (1-p)^{n-j} \right] \cdot \left[\binom{m}{k-j} p^{k-j} (1-p)^{m-(k-j)} \right]$$

$$= p^k (1-p)^{n+m-k} \sum_{j=0}^{k} \binom{n}{j} \binom{m}{k-j}.$$

To evaluate the sum we apply Vandermonde's identity (Proposition A.3.8), which asserts

$$\sum_{j=0}^{k} \binom{n}{j} \binom{m}{k-j} = \binom{n+m}{k}.$$

This leads to

$$\mathbb{P}\{X + Y = k\} = \binom{n+m}{k} p^k (1-p)^{m+n-k},$$

and $X + Y$ is $B_{n+m,p}$-distributed. ∎

---

9 Note that in the inner integral y is a constant.

Interpretation: In a first experiment we toss a biased coin n times and in a second one m times. We combine these two experiments to one and toss the coin now $n + m$ times. Then we observe exactly k times "head" during the $n + m$ trials if there is some $j \leq k$ so that we had j "heads" among the first n trials and $k - j$ among the second m ones. Finally, we have to sum the probabilities of all these events over $j \leq k$.

Corollary 4.6.2. *Let $X_1, \ldots, X_n$ be independent $B_{1,p}$-distributed, that is,*

$$\mathbb{P}\{X_j = 0\} = 1 - p \quad and \quad \mathbb{P}\{X_j = 1\} = p, \quad j = 1, \ldots, n.$$

Then their sum $X_1 + \cdots + X_n$ is $B_{n,p}$-distributed.

Proof: Apply Proposition 4.6.1 successively, first to X_1 and X_2, then to $X_1 + X_2$ and X_3, and so on. ∎

Remark 4.6.3. Observe that

$$X_1 + \cdots + X_n = \#\{j \leq n : X_j = 1\}.$$

Corollary 4.6.2 justifies the interpretation of the binomial distribution given in Section 1.4.3. Indeed, the event $\{X_j = 1\}$ occurs if in trial j we observe success. Thus, $X_1, \ldots, X_n$ equals the number of successes in n independent trials. Hereby, the success probability is $\mathbb{P}\{X_j = 1\} = p$.

In the literature the following notation is common.

Definition 4.6.4. A sequence $X_1, \ldots, X_n$ of independent $B_{1,p}$-distributed random variables is called a **Bernoulli trial** or **Bernoulli process** with success probability $p \in [0, 1]$.

With these notations, Corollary 4.6.2 may now be formulated as follows:
Let $X_1, X_2, \ldots$ be a Bernoulli trial with success probability p. Then for $n \geq 1$,

$$\mathbb{P}\{X_1 + \cdots + X_n = k\} = \binom{n}{k} p^k (1 - p)^{n-k}, \quad k = 0, \ldots, n.$$

Let X and Y be two independent **Poisson distributed** random variables. Which distribution does $X + Y$ possess? The next result answers this question.

Proposition 4.6.5. *Let X and Y be independent $Pois_\lambda$- and $Pois_\mu$-distributed for some $\lambda > 0$ and $\mu > 0$, respectively. Then $X + Y$ is $Pois_{\lambda+\mu}$-distributed.*

Proof: Proposition 4.5.6 and the binomial theorem (cf. Proposition A.3.7) imply

$$\mathbb{P}\{X + Y = k\} = \sum_{j=0}^{k} \left[\frac{\lambda^j}{j!} e^{-\lambda} \right] \left[\frac{\mu^{k-j}}{(k-j)!} e^{-\mu} \right]$$

$$= \frac{e^{-(\lambda+\mu)}}{k!} \sum_{j=0}^{k} \frac{k!}{j!\,(k-j)!} \lambda^j \mu^{k-j}$$

$$= \frac{e^{-(\lambda+\mu)}}{k!} \sum_{j=0}^{k} \binom{k}{j} \lambda^j \mu^{k-j} = \frac{(\lambda+\mu)^k}{k!} e^{-(\lambda+\mu)}.$$

Consequently, as asserted, $X + Y$ is $\text{Pois}_{\lambda+\mu}$-distributed. ∎

Interpretation: The number of phone calls arriving per day at some call centers A and B are Poisson distributed with parameters[10] λ and μ. Suppose that these two centers have different customers, that is, we assume that the number of calls in A and B is independent of each other. Proposition 4.6.5 asserts that the number of calls arriving per day either in A or in B is again Poisson distributed, yet now with parameter $\lambda + \mu$.

Example 4.6.6. This example deals with the distribution of raisins in a set of dough. More precisely, suppose we have N pounds of dough and therein are n raisins uniformly distributed. Choose by random a one-pound piece of dough. Find the probability that there are $k \geq 0$ raisins in the chosen piece.

Approach 1: Since the raisins are uniformly distributed in the dough, the probability that a single raisin is in the chosen piece equals $1/N$. Hence, if X is the number of raisins in that piece, it is $B_{n,p}$-distributed with $p = 1/N$. Assuming that N is big, the random variable X is approximately Pois_λ-distributed with $\lambda = n/N$, that is,

$$\mathbb{P}\{X = k\} = \frac{\lambda^k}{k!} e^{-\lambda}, \quad k = 0, 1, \ldots.$$

Note that $\lambda = n/N$ coincides with the average number of raisins per pound dough.

Approach 2: Assume that we took in the previous model $N \to \infty$, that is, we have an "infinite" amount of dough and "infinitely" many raisins. Which distribution does X, the number of raisins in a one-pound piece, now possess?

First we have to determine what it means that the amount of dough is "infinite" and that the raisins are uniformly distributed[11] therein. This is expressed by the following conditions:

(a) The mass of dough is unbelievably huge, hence, whenever we choose two different pieces, the numbers of raisins in each of these pieces are independent of each other.

---

10 Later on, in Proposition 5.1.16, we will see that λ and μ are the mean values of arriving calls per day.
11 Note that the multivariate uniform distribution only makes sense (cf. Definition 1.8.10) if the underlying set has a finite volume.

(b) The fact that the raisins are uniformly distributed is expressed by the following condition: suppose the number of raisins in a one-pound piece is $n \geq 0$. If this piece is split into two pieces, say K_1 and K_2 of weight α and $1 - \alpha$ pounds, then the probability that a single raisin (of the n) is in K_1 equals α, and that it is in K_2 is $1 - \alpha$.

Fix $0 < \alpha < 1$ and choose in a first step a piece K_1 of α pounds and in a second one another piece K_2 of weight $1 - \alpha$. Let X_1 and X_2 be the number of raisins in each of the two pieces. By condition (a), the random variables X_1 and X_2 are independent. If $X := X_1 + X_2$, then X is the number of raisins in a randomly chosen one-pound piece. Suppose now $X = n$, that is, there are n raisins in the one-pound piece. Then by condition (b), the probability for k raisins in K_1 is described by the binomial distribution $B_{n,\alpha}$. Recall that the success probability for a single raisin is α, thus, $X_1 = k$ means, we have k times success. This may be formulated as follows: for $0 \leq k \leq n$,

$$P\{X_1 = k | X = n\} = B_{n,\alpha}(\{k\}) = \binom{n}{k} \alpha^k (1 - \alpha)^{n-k} . \tag{4.23}$$

Rewriting eq. (4.23) leads to

$$P\{X_1 = k, \ X_2 = n - k\} = P\{X_1 = k, \ X = n\}$$
$$= P\{X_1 = k | X = n\} \cdot P\{X = n\} = P\{X = n\} \cdot \binom{n}{k} \alpha^k (1 - \alpha)^{n-k} . \tag{4.24}$$

Observe that in contrast to eq. (4.23), eq. (4.24) remains valid if $P\{X = n\} = 0$. Indeed, if $P\{X = n\} = 0$, by Proposition 4.5.6, the event $\{X_1 = k, \ X_2 = n - k\}$ has probability zero as well.

The independence of X_1 and X_2 and eq. (4.24) imply that, if $n = 0, 1, \ldots$ and $k = 0, \ldots, n$, then

$$P\{X_1 = k\} \cdot P\{X_2 = n - k\} = P\{X = n\} \cdot \binom{n}{k} \alpha^k (1 - \alpha)^{n-k} .$$

Seting $k = n$, we get

$$P\{X_1 = n\} \cdot P\{X_2 = 0\} = P\{X = n\} \cdot \alpha^n , \tag{4.25}$$

while for $n \geq 1$ and $k = n - 1$ we obtain

$$P\{X_1 = n - 1\} \cdot P\{X_2 = 1\} = P\{X = n\} \cdot n \cdot \alpha^{n-1} (1 - \alpha). \tag{4.26}$$

In particular, from eq. (4.25) follows $P\{X_2 = 0\} > 0$. If this probability would be zero, then this would imply $P\{X = n\} = 0$ for all $n \in \mathbb{N}_0$, which is impossible in view of $P\{X \in \mathbb{N}_0\} = 1$.

In a next step we solve eqs. (4.25) and (4.26) with respect to $\mathbb{P}\{X = n\}$ and make them equal. Doing so, for $n \geq 1$ we get

$$
\begin{aligned}
\mathbb{P}\{X_1 = n\} &= \frac{\alpha}{n}(1-\alpha)^{-1} \cdot \frac{\mathbb{P}\{X_2 = 1\}}{\mathbb{P}\{X_2 = 0\}} \cdot \mathbb{P}\{X_1 = n-1\} \\
&= \frac{\alpha\lambda}{n} \cdot \mathbb{P}\{X_1 = n-1\},
\end{aligned}
\tag{4.27}
$$

where $\lambda \geq 0$ is defined by

$$
\lambda := (1-\alpha)^{-1} \cdot \frac{\mathbb{P}\{X_2 = 1\}}{\mathbb{P}\{X_2 = 0\}}.
\tag{4.28}
$$

Do we have $\lambda > 0$? If $\lambda = 0$, then $\mathbb{P}\{X_2 = 1\} = 0$ and by eq. (4.26) follows $\mathbb{P}\{X = n\} = 0$ for $n \geq 1$. Consequently, $\mathbb{P}\{X = 0\} = 1$, which says that there are no raisins in the dough. We exclude this trivial case, thus it follows that $\lambda > 0$.

Finally, a successive application of eq. (4.27) implies for $n \in \mathbb{N}_0$ [12]

$$
\mathbb{P}\{X_1 = n\} = \frac{(\alpha\lambda)^n}{n!} \cdot \mathbb{P}\{X_1 = 0\},
\tag{4.29}
$$

leading to

$$
1 = \sum_{n=0}^{\infty} \mathbb{P}\{X_1 = n\} = \mathbb{P}\{X_1 = 0\} \cdot \sum_{n=0}^{\infty} \frac{(\alpha\lambda)^n}{n!} = \mathbb{P}\{X_1 = 0\}\, e^{\alpha\lambda},
$$

that is, we have $\mathbb{P}\{X_1 = 0\} = e^{-\alpha\lambda}$. Plugging this into eq. (4.29) gives

$$
\mathbb{P}\{X_1 = n\} = \frac{(\alpha\lambda)^n}{n!}\, e^{-\alpha\lambda},
$$

and X_1 is Poisson distributed with parameter $\alpha\lambda$.

Let us interchange now the roles of X_1 and X_2, hence also of α and $1-\alpha$. An application of the first step to X_2 tells us that it is Poisson distributed, but now with parameter $(1-\alpha)\lambda'$, where in view of eq. (4.28) λ' is given by [13]

$$
\lambda' = \alpha^{-1} \cdot \frac{\mathbb{P}\{X_1 = 1\}}{\mathbb{P}\{X_1 = 0\}} = \alpha^{-1}\frac{\alpha\lambda\, e^{-\alpha\lambda}}{e^{-\alpha\lambda}} = \lambda.
$$

Thus, X_2 is $\mathrm{Pois}_{(1-\alpha)\lambda}$-distributed.

---

[12] If $n = 0$, the equation holds by trivial reason.
[13] Observe that we have to replace X_2 by X_1 and $1 - \alpha$ by α.

Since X_1 and X_2 are independent, Proposition 4.6.5 applies, hence $X = X_1 + X_2$ is Pois$_\lambda$-distributed or, equivalently,

$$\mathbb{P}\{\text{There are } k \text{ raisins in an one pound piece}\} = \frac{\lambda^k}{k!}\, e^{-\lambda}.$$

Remark 4.6.7. Which role does the parameter $\lambda > 0$ play in this model? As already mentioned, Proposition 5.1.16 will tell us that λ is the average number of raisins per pound dough. Thus, if $\rho > 0$ and we ask for the number of raisins in a piece of ρ pounds, then this number is Pois$_{\rho\lambda}$-distributed[14], that is,

$$\mathbb{P}\{k \text{ raisins in } \rho \text{ pounds dough}\} = \frac{(\rho\lambda)^k}{k!}\, e^{-\rho\lambda}.$$

Assume a dough contains on the average 20 raisins per pound. Let X be number of raisins in a bread baked of five pounds dough. Then X is Pois$_{100}$-distributed and

$$\mathbb{P}(\{95 \le X \le 105\}) = 0.4176, \quad \mathbb{P}(\{90 \le X \le 110\}) = 0.7065,$$
$$\mathbb{P}(\{85 \le X \le 115\}) = 0.8793, \quad \mathbb{P}(\{80 \le X \le 120\}) = 0.9599,$$
$$\mathbb{P}(\{75 \le X \le 125\}) = 0.9892, \quad \mathbb{P}(\{70 \le X \le 130\}) = 0.9976.$$

Additional question: Suppose we buy two loaves of bread baked from ρ pounds dough each. What is the probability that one of these two loaves contains at least twice as many raisins than the other one?

Answer: Let X be the number of raisins in the first loaf, and Y is the number of raisins in the second one. By assumption, X and Y are independent, and both are Pois$_{\rho\lambda}$-distributed, where as before $\lambda > 0$ is the average number of raisins per pound dough. The probability we are interested in is

$$\mathbb{P}\{X \ge 2Y \text{ or } Y \ge 2X\} = \mathbb{P}\{X \ge 2Y\} + \mathbb{P}\{Y \ge 2X\} = 2\,\mathbb{P}\{X \ge 2Y\}.$$

It follows

$$2\,\mathbb{P}(X \ge 2Y) = 2\sum_{k=0}^{\infty} \mathbb{P}(Y = k, X \ge 2k) = 2\sum_{k=0}^{\infty} \mathbb{P}(Y = k) \cdot \mathbb{P}(X \ge 2k)$$

$$= 2\sum_{k=0}^{\infty} \mathbb{P}(Y = k) \cdot \sum_{j=2k}^{\infty} \mathbb{P}(X = j) = 2\,e^{-2\rho\lambda} \sum_{k=0}^{\infty} \frac{(\rho\lambda)^k}{k!} \sum_{j=2k}^{\infty} \frac{(\rho\lambda)^j}{j!}.$$

———

14 Because on average there are $\rho\lambda$ raisins in a piece of ρ pounds.

If the average number of raisins per pound is $\lambda = 20$, and if the loaves are baked from $\rho = 5$ pounds dough, then this probability is approximatively

$$\mathbb{P}(X \geq 2Y \text{ or } Y \geq 2X) = 3.17061 \times 10^{-6}.$$

If $\rho = 1$, that is, the loaves are made from one-pound dough each, then

$$\mathbb{P}(X \geq 2Y \text{ or } Y \geq 2X) = 0.0430079.$$

Now we investigate the distribution of the sum of two independent **negative binomial distributed** random variables. Recall that X is $B_{n,p}^-$-distributed if

$$\mathbb{P}\{X = k\} = \binom{k-1}{k-n} p^n (p-1)^{k-n}, \quad k = n, n+1, \dots.$$

Proposition 4.6.8. *Let X and Y be independent accordingly $B_{n,p}^-$ and $B_{m,p}^-$ distributed for some $n, m \geq 1$. Then $X + Y$ is $B_{n+m,p}^-$-distributed.*

Proof: We derive from Example 4.1.8 that, if $k \in \mathbb{N}_0$, then

$$\mathbb{P}\{X - n = k\} = \binom{-n}{k} p^n (p-1)^k \quad \text{and} \quad \mathbb{P}\{Y - m = k\} = \binom{-m}{k} p^m (p-1)^k.$$

An application of Proposition 4.5.6 to $X - n$ and $Y - m$ implies

$$\mathbb{P}\{X + Y - (n+m) = k\} = \sum_{j=0}^{k} \left[\binom{-n}{j} p^n (p-1)^j \right] \left[\binom{-m}{k-j} p^m (p-1)^{k-j} \right]$$

$$= p^{n+m}(p-1)^k \sum_{j=0}^{k} \binom{-n}{j} \binom{-m}{k-j}.$$

To compute the last sum we use Proposition A.5.3, which asserts that

$$\sum_{j=0}^{k} \binom{-n}{j} \binom{-m}{k-j} = \binom{-n-m}{k}.$$

Consequently, for each $k \in \mathbb{N}_0$,

$$\mathbb{P}\{X + Y - (n+m) = k\} = \binom{-n-m}{k} p^{n+m}(p-1)^k.$$

Another application of eq. (4.2) (this time with $n + m$) leads to

$$\mathbb{P}\{X + Y = k\} = \binom{k-1}{k-(n+m)} p^{n+m}(1-p)^{k-(n+m)}, \quad k = n+m, n+m+1, \dots.$$

This completes the proof. ∎

Corollary 4.6.9. *Let $X_1, \ldots, X_n$ be independent G_p-distributed (geometric distributed) random variables. Then their sum $X_1 + \cdots + X_n$ is $B^-_{n,p}$-distributed.*

Proof: Use $G_p = B^-_{1,p}$ and apply Proposition 4.6.8 n times. ∎

Interpretation: The following two experiments are completely equivalent: one is to play the same game as long as one observes success for the nth time. The other experiment is, after each success to start a new game, as long as one observes success in the nth (and last) game. Here we assume that all n games are executed independently and possess the same success probability.

Let U and V be two independent random variables, both **uniformly distributed** on $[0, 1]$. Which distribution density does $X + Y$ possess?

Proposition 4.6.10. *The sum of two independent random variables U and V, uniformly distributed on $[0, 1]$, has the density r defined by*

$$
r(x) = \begin{cases} x & : 0 \le x < 1 \\ 2 - x & : 1 \le x \le 2 \\ 0 & : otherwise. \end{cases} \tag{4.30}
$$

Proof: The distribution densities p and q of U and V are given by $p(x) = q(x) = 1$ if $0 \le x \le 1$ and $p(x) = q(x) = 0$ otherwise. Proposition 4.5.14 asserts that $U + V$ has density $r = p \star q$ computed by

$$
r(x) = \int_{-\infty}^{\infty} p(x - y) q(y)\, dy = \int_0^1 p(x - y)\, dy.
$$

But, $p(x - y) = 1$ if and only if $0 \le x - y \le 1$ or, equivalently, if and only if $x - 1 \le y \le x$. Taking into account the restriction $0 \le y \le 1$, it follows $p(x - y)q(y) = 1$ if and only if $y \in [\max\{x - 1, 0\}, \min\{x, 1\}]$. In particular, $r(x) = 0$ for $x \notin [0, 2]$, and if $0 \le x \le 2$, then

$$
r(x) = \min\{x, 1\} - \max\{x - 1, 0\}.
$$

It is not difficult to see that r may also be written as in eq. (4.30). This completes the proof. ∎

Application: Suppose we choose independently and according to the uniform distribution two numbers u_1 and u_2 in $[0, 1]$. Then the probability that $a \le u_1 + u_2 \le b$ equals $\int_a^b r(x)\, dx$ with r given by eq. (4.30). For example,

$$
\mathbb{P}\left\{\frac{1}{2} \le u_1 + u_2 \le \frac{3}{2}\right\} = \int_{1/2}^1 x\, dx + \int_1^{3/2} (2 - x)\, dx = \frac{3}{4}.
$$

We investigate now the sum of two **gamma distributed** random variables. Recall that the density of a $\Gamma_{\alpha,\beta}$-distributed random variable is given by

$$p_{\alpha,\beta}(x) = \frac{1}{\alpha^\beta \, \Gamma(\beta)} \, x^{\beta-1} \, e^{-x/\alpha}$$

if $x > 0$, while $p_{\alpha,\beta}(x) = 0$ otherwise.

Proposition 4.6.11. *Let X_1 and X_2 be two independent random variables distributed according to Γ_{α,β_1} and Γ_{α,β_2}, respectively. Then $X_1 + X_2$ is $\Gamma_{\alpha,\beta_1+\beta_2}$-distributed.*

Proof: If r denotes the density of $X_1 + X_2$, Proposition 4.5.14 implies

$$r(x) = (p_{\alpha,\beta_1} \star p_{\alpha,\beta_2})(x) = \int_{-\infty}^{\infty} p_{\alpha,\beta_1}(x-y) p_{\alpha,\beta_2}(y) \, dy, \quad x \in \mathbb{R}, \tag{4.31}$$

and we have to show that $r = p_{\alpha,\beta_1+\beta_2}$.

It is easy to see that $r(x) = 0$ if $x \le 0$, hence it suffices to evaluate eq. (4.31) for $x > 0$. Since $p_{\alpha,\beta_2}(x-y) = 0$ if $y > x$,

$$r(x) = \frac{1}{\alpha^{\beta_1+\beta_2}\Gamma(\beta_1)\Gamma(\beta_2)} \int_0^x y^{\beta_1-1}(x-y)^{\beta_2-1} \, e^{-y/\alpha} \, e^{-(x-y)/\alpha} \, dy$$

$$= \frac{1}{\alpha^{\beta_1+\beta_2}\Gamma(\beta_1)\Gamma(\beta_2)} \, x^{\beta_1+\beta_2-2} e^{-x/\alpha} \int_0^x \left(\frac{y}{x}\right)^{\beta_1-1} \left(1-\frac{y}{x}\right)^{\beta_2-1} \, dy.$$

Changing the variable as $u := y/x$, hence $dy = x \, du$, we obtain

$$r(x) = \frac{1}{\alpha^{\beta_1+\beta_2}\Gamma(\beta_1)\Gamma(\beta_2)} \, x^{\beta_1+\beta_2-1} e^{-x/\alpha} \int_0^1 u^{\beta_1-1}(1-u)^{\beta_2-1} du$$

$$= \frac{B(\beta_1,\beta_2)}{\alpha^{\beta_1+\beta_2}\,\Gamma(\beta_1)\,\Gamma(\beta_2)} \cdot x^{\beta_1+\beta_2-1} e^{-x/\alpha}, \tag{4.32}$$

where B denotes the beta function defined by eq. (1.58). Equation (1.59) yields

$$\frac{B(\beta_1,\beta_2)}{\Gamma(\beta_1)\,\Gamma(\beta_2)} = \frac{1}{\Gamma(\beta_1+\beta_2)}, \tag{4.33}$$

hence, if $x > 0$, then by eqs. (4.32) and (4.33) it follows that $r(x) = p_{\alpha,\beta_1+\beta_2}(x)$. This completes the proof. ∎

Recall that the **Erlang distribution** is defined as $E_{\lambda,n} = \Gamma_{\lambda^{-1},n}$. Thus, Proposition 4.6.11 implies the following corollary.

Corollary 4.6.12. *Let X and Y be independent and distributed according to $E_{\lambda,n}$ and $E_{\lambda,m}$, respectively. Then their sum $X + Y$ is $E_{\lambda,n+m}$-distributed.*

Another corollary of Proposition 4.5.14 (or of Corollary 4.6.12) describes the sum of independent **exponentially distributed** random variables.

Corollary 4.6.13. Let $X_1, \ldots, X_n$ be independent E_λ-distributed. Then their sum $X_1 + \cdots + X_n$ is Erlang distributed with parameters λ and n.

Proof: Recall that $E_\lambda = E_{\lambda,1}$. By Corollary 4.6.12 $X_1 + X_2$ is $E_{\lambda,2}$-distributed. Proceeding in this way, every time applying Corollary 4.6.12 leads to the desired result. ∎

Example 4.6.14. The lifetime of light bulbs is assumed to be E_λ-distributed for a certain $\lambda > 0$. At time zero we switch on a first bulb. In the moment it burns out, we replace it by a second one of the same type. If the second burns out, we replace it by a third one, and so on. Let S_n be the moment when the nth light bulb burns out. Which distribution does S_n possess?

Answer: Let $X_1, X_2, \ldots$ be the lifetimes of the first, second, and so on light bulb. Then $S_n = X_1 + \cdots + X_n$. Since the light bulbs are assumed to be of the same type, the random variables X_j are all E_λ-distributed. Furthermore, the different lifetimes do not influence each other, thus, we may assume that the X_js are independent. Now Corollary 4.6.13 lets us conclude that S_n is Erlang distributed with parameters λ and n, hence, if $t > 0$, by Proposition 1.6.25 we get

$$\mathbb{P}\{S_n \le t\} = \frac{\lambda^n}{(n-1)!} \int_0^t x^{n-1} e^{-\lambda x}\, dx = 1 - \sum_{j=0}^{n-1} \frac{(\lambda t)^j}{j!} e^{-\lambda t}. \tag{4.34}$$

Example 4.6.15. We continue the preceding example, but ask now a different question. How often do we have to change light bulbs before some given time $T > 0$?

Answer: Let Y be the number of changes necessary until time T. Then for $n \ge 0$ the event $\{Y = n\}$ occurs if and only if $S_n \le T$, but $S_{n+1} > T$. Hereby, we use the notation of Example 4.6.14. In other words,

$$\mathbb{P}\{Y = n\} = \mathbb{P}\{S_n \le T,\ S_{n+1} > T\} = \mathbb{P}\big(\{S_n \le T\} \setminus \{S_{n+1} \le T\}\big), \quad n = 0, 1, \ldots.$$

Since $\{S_{n+1} \le T\} \subseteq \{S_n \le T\}$, by eq. (4.34) follows that

$$\mathbb{P}\{Y = n\} = \mathbb{P}\{S_n \le T\} - \mathbb{P}\{S_{n+1} \le T\}$$

$$= \left[1 - \sum_{j=0}^{n-1} \frac{(\lambda T)^j}{j!} e^{-\lambda T} \right] - \left[1 - \sum_{j=0}^{n} \frac{(\lambda T)^j}{j!} e^{-\lambda T} \right]$$

$$= \frac{(\lambda T)^n}{n!} e^{-\lambda T} = \mathrm{Pois}_{\lambda T}(\{n\}).$$

Let us still mention an important equivalent random "experiment": customers arrive at the post office randomly. We assume that the times between their arrivals are independent and E_λ-distributed. Then S_n is the time when the nth customer arrives. Hence, under these assumptions, the number of arriving customers until a certain time $T > 0$ is Poisson distributed with parameter λT.

We investigate now the sum of two independent **chi-squared distributed** random variables. Recall Definition 1.6.26: A random variable X is χ_n^2-distributed if it is $\Gamma_{2, \frac{n}{2}}$-distributed. Hence, Proposition 4.6.11 implies the following result.

Proposition 4.6.16. *Suppose that X is χ_n^2-distributed and that Y is χ_m^2-distributed for some $n, m \geq 1$. If X and Y are independent, then $X + Y$ is χ_{n+m}^2-distributed.*

Proof: Because of Proposition 4.6.11, the sum $X + Y$ is $\Gamma_{2, \frac{n}{2} + \frac{m}{2}} = \chi_{n+m}^2$-distributed. This proves the assertion. ∎

Proposition 4.6.16 has the following important consequence.

Proposition 4.6.17. *Let $X_1, \ldots, X_n$ be a sequence of independent $\mathcal{N}(0, 1)$-distributed random variables. Then $X_1^2 + \cdots + X_n^2$ is χ_n^2-distributed.*

Proof: Proposition 4.1.5 asserts that the random variables X_j^2 are χ_1^2-distributed. Furthermore, because of Proposition 4.1.9 they are also independent. Thus a successive application of Proposition 4.6.16 proves the assertion. ∎

Our next and final aim in this section is to investigate the distribution of the sum of two independent **normally distributed** random variables. Here the following important result is valid.

Proposition 4.6.18. *Let X_1 and X_2 be two independent random variables distributed according to $\mathcal{N}(\mu_1, \sigma_1^2)$ and $\mathcal{N}(\mu_2, \sigma_2^2)$. Then $X_1 + X_2$ is $\mathcal{N}(\mu_1 + \mu_2, \sigma_1^2 + \sigma_2^2)$-distributed.*

Proof: In a first step we treat a special case, namely $\mu_1 = \mu_2 = 0$ and $\sigma_1 = 1$. To simplify the notation, set $\lambda = \sigma_2$. Thus we have to prove the following: if X_1 and X_2 are $\mathcal{N}(0, 1)$- and $\mathcal{N}(0, \lambda^2)$-distributed, then $X_1 + X_2$ is $\mathcal{N}(0, 1 + \lambda^2)$-distributed.

Let $p_{0,1}$ and p_{0,λ^2} be the corresponding densities introduced in eq. (1.47). Then we have to prove that

$$p_{0,1} \star p_{0,\lambda^2} = p_{0,1+\lambda^2} . \tag{4.35}$$

To verify this start with

$$(p_{0,1} \star p_{0,\lambda^2})(x) = \frac{1}{2\pi\lambda} \int_{-\infty}^{\infty} e^{-(x-y)^2/2} \, e^{-y^2/2\lambda^2} \, dy$$

$$= \frac{1}{2\pi\lambda} \int_{-\infty}^{\infty} e^{-\frac{1}{2}(x^2 - 2xy + (1+\lambda^{-2})y^2)} \, dy \,. \tag{4.36}$$

We use

$$x^2 - 2xy + (1 + \lambda^{-2}) y^2$$

$$= \left((1+\lambda^{-2})^{1/2} y - (1+\lambda^{-2})^{-1/2} x \right)^2 - x^2 \left(\frac{1}{1+\lambda^{-2}} - 1 \right)$$

$$= \left((1+\lambda^{-2})^{1/2} y - (1+\lambda^{-2})^{-1/2} x \right)^2 + \frac{x^2}{1+\lambda^2}$$

$$= \left(\alpha y - \frac{x}{\alpha} \right)^2 + \frac{x^2}{1+\lambda^2}$$

with $\alpha := (1+\lambda^{-2})^{1/2}$. Plugging this transformation into eq. (4.36) leads to

$$(p_{0,1} \star p_{0,\lambda^2})(x) = \frac{e^{-x^2/2(1+\lambda^2)}}{2\pi\lambda} \int_{-\infty}^{\infty} e^{-(\alpha y - \frac{x}{\alpha})^2/2} \, dy \,. \tag{4.37}$$

Next change the variables by $u := \alpha y - x/\alpha$, thus, $dy = du/\alpha$, and observe that $\alpha\lambda = (1+\lambda^2)^{1/2}$. Then the right-hand side of eq. (4.37) transforms to

$$\frac{e^{-x^2/2(1+\lambda^2)}}{2\pi (1+\lambda^2)^{1/2}} \int_{-\infty}^{\infty} e^{-u^2/2} \, du = p_{0,1+\lambda^2}(x) \,.$$

Hereby, we used Proposition 1.6.6 asserting $\int_{-\infty}^{\infty} e^{-u^2/2} \, du = \sqrt{2\pi}$. This proves the validity of eq. (4.35).

In a second step we treat the general case, that is, X_1 is $\mathcal{N}(\mu_1, \sigma_1^2)$- and X_2 is $N(\mu_2, \sigma_2^2)$-distributed. Set

$$Y_1 := \frac{X_1 - \mu_1}{\sigma_1} \quad \text{and} \quad Y_2 := \frac{X_2 - \mu_2}{\sigma_2} \,.$$

By Proposition 4.2.3, the random variables Y_1 and Y_2 are standard normal and, moreover, because of Proposition 4.1.9, also independent. Thus, the sum $X_1 + X_2$ may be represented as

$$X_1 + X_2 = \mu_1 + \mu_2 + \sigma_1 Y_1 + \sigma_2 Y_2 = \mu_1 + \mu_2 + \sigma_1 Z$$

where $Z = Y_1 + \lambda Y_2$ with $\lambda = \sigma_2/\sigma_1$.

An application of the first step shows that Z is $\mathcal{N}(0, 1 + \lambda^2)$-distributed. Hence, Proposition 4.2.3 implies the existence of a standard normally distributed Z_0 such that $Z = (1 + \lambda^2)^{1/2} Z_0$. Summing up, $X_1 + X_2$ may now be written as

$$X_1 + X_2 = \mu_1 + \mu_2 + \sigma_1 (1 + \lambda^2)^{1/2} Z_0 = \mu_1 + \mu_2 + \left(\sigma_1^2 + \sigma_2^2 \right)^{1/2} Z_0,$$

and another application of Proposition 4.2.3 lets us conclude that, as asserted, the sum $X_1 + X_2$ is $\mathcal{N}(\mu_1 + \mu_2, \sigma_1^2 + \sigma_2^2)$-distributed. ∎

4.7 Products and Quotients of Random Variables

Let X and Y be two random variables mapping a sample space Ω into $\mathbb{R}$. Then their product $X \cdot Y$ and their quotient X/Y (assume $Y(\omega) \neq 0$ for $\omega \in \Omega$) are defined by

$$(X \cdot Y)(\omega) := X(\omega) \cdot Y(\omega) \quad \text{and} \quad \left(\frac{X}{Y} \right)(\omega) := \frac{X(\omega)}{Y(\omega)}, \quad \omega \in \Omega.$$

The aim of this section is to investigate the distribution of such products and quotients. We restrict ourselves to continuous X and Y because, later on, we will only deal with products and quotients of those random variables. Furthermore, we omit the proof of the fact that products and fractions are random variables as well. The proofs of these permanent properties are not complicated and follow the ideas used in the proof of Proposition 4.5.1. Thus, our interest are products $X \cdot Y$ and quotients X/Y for independent X and Y, where, to simplify the computations, we suppose $\mathbb{P}\{Y > 0\} = 1$.

We start with the investigation of **products** of continuous random variables. Thus, let X and Y be two random variables with distribution densities p and q. Since we assumed $\mathbb{P}\{Y > 0\} = 1$, we may choose the density q such that $q(x) = 0$ if $x \leq 0$.

Proposition 4.7.1. *Let X and Y be two independent random variables possessing the stated properties. Then $X \cdot Y$ is continuous as well, and its density r may be calculated by*

$$r(x) = \int_0^\infty p\left(\frac{x}{y} \right) \frac{q(y)}{y} \, dy, \quad x \in \mathbb{R}. \tag{4.38}$$

Proof: For $t \in \mathbb{R}$ we evaluate $\mathbb{P}\{X \cdot Y \leq t\}$. To this end fix $t \in \mathbb{R}$ and set

$$A_t := \{(u, y) \in \mathbb{R} \times (0, \infty) : u \cdot y \leq t\}.$$

As in the proof of Proposition 4.5.14, it follows that

$$\mathbb{P}\{X \cdot Y \leq t\} = \mathbb{P}_{(X,Y)}(A_t) = \int_0^\infty \left[\int_{-\infty}^{t/y} p(u) \, du \right] q(y) \, dy. \tag{4.39}$$

In the inner integral of eq. (4.39) we change the variables by $x := uy$, hence $dx = y\,du$. Notice that in the inner integral y is a constant. After this change of variables the right-hand integral in eq. (4.39) becomes[15]

$$\int\limits_0^\infty \left[\int\limits_{-\infty}^t p\left(\frac{x}{y}\right) dx \right] \frac{q(y)}{y}\, dy = \int\limits_{-\infty}^t \left[\int\limits_0^\infty p\left(\frac{x}{y}\right) \frac{q(y)}{y}\, dy \right] dx = \int\limits_{-\infty}^t r(x)\, dx.$$

This being valid for all $t \in \mathbb{R}$, the function r is a density of $X \cdot Y$. ■

Example 4.7.2. Let U and V be two independent random variables uniformly distributed on $[0, 1]$. Which probability distribution does $U \cdot V$ possess?

Answer: We have $p(y) = q(y) = 1$ if $0 \le y \le 1$, and $p(y) = q(y) = 0$ otherwise. Furthermore, $0 \le U \cdot V \le 1$, hence its density r satisfies $r(x) = 0$ if $x \notin [0, 1]$. For $x \in [0, 1]$ we apply formula (4.38) and obtain

$$r(x) = \int\limits_0^\infty p\left(\frac{x}{y}\right) \frac{q(y)}{y}\, dy = \int\limits_x^1 \frac{1}{y}\, dy = -\ln(x) = \ln\left(\frac{1}{x}\right), \quad 0 < x \le 1.$$

Consequently, if $0 < a < b \le 1$, then

$$\mathbb{P}\{a \le U \cdot V \le b\} = -\int\limits_a^b \ln(x)\, dx = -[x \ln x - x]_a^b = a \ln(a) - b \ln(b) + b - a.$$

In particular, it follows that

$$\mathbb{P}\{U \cdot V \le t\} = t - t \ln t, \quad 0 < t \le 1. \tag{4.40}$$

Our next objective are **quotients** of random variables X and Y. We denote their densities by p and q, thereby assuming $q(x) = 0$ if $x \le 0$. Then we get

Proposition 4.7.3. *Let X and Y be independent with $\mathbb{P}\{Y > 0\} = 1$. Then their quotient X/Y has the density r given by*

$$r(x) = \int\limits_0^\infty y\, p(xy)\, q(y)\, dy, \quad x \in \mathbb{R}.$$

Proof: The proof of Proposition 4.7.3 is quite similar to that of 4.7.1. Therefore, we present only the main steps. Setting

$$A_t := \{(u, y) \in \mathbb{R} \times (0, \infty) : u \le ty\},$$

---

[15] The interchange of the integrals is justified by Proposition A.5.5. Note that p and q are nonnegative.

we obtain

$$\mathbb{P}\{(X/Y) \le t\} = \mathbb{P}_{(X,Y)}(A_t) = \int_0^\infty \left[\int_{-\infty}^{ty} p(u)\,du \right] q(y)\,dy. \qquad (4.41)$$

We change the variables in the inner integral of eq. (4.41) by putting $x = u/y$. After that we interchange the integrals and arrive at

$$\mathbb{P}\{(X/Y) \le t\} = \int_{-\infty}^{t} r(x)\,dx$$

for all $t \in \mathbb{R}$. This proves that r is a density of X/Y. ∎

Example 4.7.4. Let U and V be as in Example 4.7.2. We investigate now their quotient U/V. By Proposition 4.7.3 its density r can be computed by

$$r(x) = \int_0^\infty y\,p(xy)q(y)\,dy = \int_0^1 yp(xy)dy = \int_0^1 y\,dy = \frac{1}{2}$$

in the case $0 \le x \le 1$. If $1 \le x < \infty$, then $p(xy) = 0$ if $y > 1/x$, and it follows that

$$r(x) = \int_0^{1/x} y\,dy = \frac{1}{2x^2}$$

for those x. Combining both cases, the density r of U/V may be written as

$$r(x) = \begin{cases} \frac{1}{2} & : \ 0 \le x \le 1 \\ \frac{1}{2x^2} & : \ 1 < x < \infty \\ 0 & : \ \text{otherwise}. \end{cases}$$

Question: Does there exist an easy geometric explanation for $r(x) = \frac{1}{2}$ in the case $0 \le x \le 1$?

Answer: If $t > 0$, then

$$F_{U/V}(t) = \mathbb{P}\{U/V \le t\} = \mathbb{P}\{U \le t\,V\} = \mathbb{P}_{(U,V)}(A_t),$$

where

$$A_t := \{(u, v) \in [0, 1]^2 : 0 \le u \le vt\}.$$

If $0 < t \le 1$, then A_t is a triangle in $[0, 1]^2$ with area $\mathrm{vol}_2(A_t) = \frac{t}{2}$. The independence of U and V implies (cf. Example 3.6.21) that $\mathbb{P}_{(U,V)}$ is the uniform distribution on $[0, 1]^2$, hence

$$F_{U/V}(t) = \mathbb{P}_{(U,V)}(A_t) = \mathrm{vol}_2(A_t) = \frac{t}{2}, \quad 0 < t \le 1,$$

leading to $r(t) = F'_{U/V}(t) = \frac{1}{2}$ for those t.

4.7.1 Student's t-Distribution

Let us use Proposition 4.7.3 to compute the density of a distribution, which plays a crucial role in Mathematical Statistics.

Proposition 4.7.5. *Let X be $\mathcal{N}(0,1)$-distributed and Y be independent of X and χ_n^2-distributed for some $n \geq 1$. Define the random variable Z as*

$$Z := \frac{X}{\sqrt{Y/n}} \, .$$

Then Z possesses the density r given by

$$r(x) = \frac{\Gamma\left(\frac{n+1}{2}\right)}{\sqrt{n\pi}\,\Gamma\left(\frac{n}{2}\right)} \left(1 + \frac{x^2}{n}\right)^{-n/2-1/2} , \qquad x \in \mathbb{R} . \tag{4.42}$$

Proof: In a first step we determine the density of $\sqrt{Y}$ with Y distributed according to χ_n^2. If $t > 0$, then

$$F_{\sqrt{Y}}(t) = \mathbb{P}\{\sqrt{Y} \leq t\} = \mathbb{P}\{Y \leq t^2\} = \frac{1}{2^{n/2}\,\Gamma\left(\frac{n}{2}\right)} \int_0^{t^2} x^{n/2-1}\, e^{-x/2}\, dx \, .$$

Thus, if $t > 0$, then the density q of $\sqrt{Y}$ equals

$$q(t) = \frac{d}{dt} F_{\sqrt{Y}}(t) = \frac{1}{2^{n/2}\,\Gamma\left(\frac{n}{2}\right)} (2t)\, t^{n-2}\, e^{-t^2/2}$$

$$= \frac{1}{2^{n/2-1}\,\Gamma\left(\frac{n}{2}\right)} t^{n-1}\, e^{-t^2/2} \, . \tag{4.43}$$

Of course, we have $q(t) = 0$ if $t \leq 0$.

In a second step we determine the density $\tilde{r}$ of $\tilde{Z} = Z/\sqrt{n} = X/\sqrt{Y}$. An application of Proposition 4.7.3 for $p(x) = \frac{1}{\sqrt{2\pi}} e^{-x/2}$ and q given in eq. (4.43) leads to

$$\tilde{r}(x) = \int_0^\infty y \left[\frac{1}{\sqrt{2\pi}} e^{-(xy)^2/2} \right] \left[\frac{1}{2^{n/2-1}\,\Gamma\left(\frac{n}{2}\right)} y^{n-1}\, e^{-y^2/2} \right] dy$$

$$= \frac{1}{\sqrt{\pi}\, 2^{n/2-1/2}\,\Gamma\left(\frac{n}{2}\right)} \int_0^\infty y^n\, e^{-(1+x^2)y^2/2}\, dy \, . \tag{4.44}$$

Change the variables in the last integral by $v = \frac{x^2}{2}(1 + x^2)$. Then $y = \frac{\sqrt{2v}}{(1+x^2)^{1/2}}$ and, consequently, $dy = \frac{1}{\sqrt{2}} v^{-1/2} (1 + x^2)^{-1/2}\, dv$. Inserting this into eq. (4.44) shows that

$$\tilde{r}(x) = \frac{1}{\sqrt{\pi}\, 2^{n/2}\, \Gamma\left(\frac{n}{2}\right)} \int_0^\infty \frac{2^{n/2}\, v^{n/2-1/2}\, e^{-v}}{(1+x^2)^{n/2+1/2}}\, dv$$

$$= \frac{\Gamma\left(\frac{n+1}{2}\right)}{\sqrt{\pi}\, \Gamma\left(\frac{n}{2}\right)} (1+x^2)^{-n/2-1/2}. \tag{4.45}$$

In a third step, we finally obtain the density r of Z. Since $Z = \sqrt{n}\, \tilde{Z}$, formula (4.3) applies with $b = 0$ and $a = \sqrt{n}$. Thus, by eq. (4.45) for $\tilde{r}$, as asserted,

$$r(x) = \frac{1}{\sqrt{n}}\, \tilde{r}\left(\frac{x}{\sqrt{n}}\right) = \frac{\Gamma\left(\frac{n+1}{2}\right)}{\sqrt{n\pi}\, \Gamma\left(\frac{n}{2}\right)} \left(1+\frac{x^2}{n}\right)^{-n/2-1/2}. \qquad\blacksquare$$

Definition 4.7.6. The probability measure on $(\mathbb{R}, \mathcal{B}(\mathbb{R}))$ with density r, given by eq. (4.42), is called t_n-**distribution** or **Student's t-distribution** with n degrees of freedom. A random variable Z is said to be t_n-**distributed** or **t-distributed** with n degrees of freedom, provided its probability distribution is a t_n-distribution, that is, for $a < b$

$$\mathbb{P}\{a \le Z \le b\} = \frac{\Gamma\left(\frac{n+1}{2}\right)}{\sqrt{n\pi}\, \Gamma\left(\frac{n}{2}\right)} \int_a^b \left(1+\frac{x^2}{n}\right)^{-n/2-1/2}\, dx.$$

Remark 4.7.7. The t_1-distribution coincides with the Cauchy distribution introduced in 1.6.33. Observe that $\Gamma(1/2) = \sqrt{\pi}$ and $\Gamma(1) = 1$.

In view of Definition 4.7.6, we may now formulate Proposition 4.7.5 as follows.

Proposition 4.7.8. *If X and Y are independent and $\mathcal{N}(0, 1)$- and χ_n^2 distributed, then $\frac{X}{\sqrt{Y/n}}$ is t_n-distributed.*

Proposition 4.6.17 leads still to another version of Proposition 4.7.5:

Proposition 4.7.9. *If $X, X_1, \ldots, X_n$ are independent $\mathcal{N}(0, 1)$-distributed, then*

$$\frac{X}{\sqrt{\frac{1}{n}\sum_{i=1}^n X_i^2}}$$

is t_n-distributed.

Corollary 4.7.10. *If X and Y are independent and $\mathcal{N}(0, 1)$-distributed, then $X/|Y|$ possesses a Cauchy distribution.*

Proof: An application of Proposition 4.7.9 with $n = 1$ and $X_1 = Y$ implies that $X/|Y|$ is t_1-distributed. We saw in Remark 4.7.7 the t_1 and the Cauchy distribution coincide, thus, $X/|Y|$ is also Cauchy distributed. ∎

4.7.2 F-Distribution

We present now another important class of probability measures or probability distributions playing a central role in Mathematical Statistics.

Proposition 4.7.11. *For two natural numbers m and n let X and Y be independent and χ_m^2- and χ_n^2-distributed. Then $Z := \frac{X/m}{Y/n}$ has the distribution density r defined as*

$$r(x) = \begin{cases} 0 & : x \le 0 \\ m^{m/2} \, n^{n/2} \cdot \dfrac{\Gamma(\frac{m+n}{2})}{\Gamma(\frac{m}{2})\Gamma(\frac{n}{2})} \cdot \dfrac{x^{m/2-1}}{(mx+n)^{(m+n)/2}} & : x > 0 \end{cases} \tag{4.46}$$

Proof: We first evaluate the density $\tilde{r}$ of $\tilde{Z} = X/Y$. To this end we apply Proposition 4.7.3 with functions p and q given by

$$p(x) = \frac{1}{2^{m/2}\,\Gamma(m/2)}\, x^{m/2-1}\, e^{-x/2} \quad \text{and} \quad q(y) = \frac{1}{2^{n/2}\,\Gamma(n/2)}\, y^{n/2-1}\, e^{-y/2}$$

whenever $x, y > 0$. Then we get

$$\tilde{r}(x) = \frac{1}{2^{(m+n)/2}\,\Gamma(m/2)\,\Gamma(n/2)} \int_0^\infty y\,(xy)^{m/2-1}\, y^{n/2-1}\, e^{-xy/2}\, e^{-y/2}\, dy$$

$$= \frac{x^{m/2-1}}{2^{(m+n)/2}\,\Gamma(m/2)\,\Gamma(n/2)} \int_0^\infty y^{(m+n)/2-1}\, e^{-y(1+x)/2}\, dy. \tag{4.47}$$

We replace in eq. (4.47) the variable y by $u = y(1+x)/2$, thus, $dy = \frac{2}{1+x}\, du$. Inserting this into eq. (4.47), the last expression transforms to

$$\frac{x^{m/2-1}}{\Gamma(m/2)\,\Gamma(n/2)}\,(1+x)^{-(n+m)/2} \int_0^\infty u^{(m+n)/2-1}\, e^{-u}\, du$$

$$= \frac{\Gamma\left(\frac{m+n}{2}\right)}{\Gamma\left(\frac{m}{2}\right)\Gamma\left(\frac{n}{2}\right)} \cdot \frac{x^{m/2-1}}{(1+x)^{(m+n)/2}}.$$

Because of $Z = \frac{n}{m} \cdot \tilde{Z}$, we obtain the density r of Z by Proposition 1.7.17. Indeed, then

$$r(x) = \frac{m}{n} \tilde{r}\left(\frac{mx}{n}\right) = m^{m/2} n^{n/2} \cdot \frac{\Gamma\left(\frac{m+n}{2}\right)}{\Gamma\left(\frac{m}{2}\right)\Gamma\left(\frac{n}{2}\right)} \cdot \frac{x^{m/2-1}}{(mx+n)^{(m+n)/2}}$$

as asserted. ■

Remark 4.7.12. Using relation (1.59) between the beta and the gamma functions, the density r of Z may also be written as

$$r(x) = \frac{m^{m/2} n^{n/2}}{B\left(\frac{m}{2}, \frac{n}{2}\right)} \cdot \frac{x^{m/2-1}}{(mx+n)^{(m+n)/2}}, \quad x > 0.$$

Definition 4.7.13. The probability measure on $(\mathbb{R}, \mathcal{B}(\mathbb{R}))$ with density r defined by eq. (4.46) is called **Fisher–Snecedor distribution** or **F-distribution** (with m and n degrees of freedom).

A random variable Z is **F-distributed** (with m and n degrees of freedom), provided its probability distribution is an F-distribution. Equivalently, if $0 \le a < b$, then

$$\mathbb{P}\{a \le Z \le b\} = m^{m/2} n^{n/2} \cdot \frac{\Gamma\left(\frac{m+n}{2}\right)}{\Gamma\left(\frac{m}{2}\right)\Gamma\left(\frac{n}{2}\right)} \int_a^b \frac{x^{m/2-1}}{(mx+n)^{(m+n)/2}}\, dx.$$

The random variable Z is also said to be **$F_{m,n}$-distributed**.

With this notation, Proposition 4.7.11 may now be formulated as follows:

Proposition 4.7.14. *If two independent random variables X and Y are χ_m^2- and χ_n^2-distributed, then $\frac{X/m}{Y/n}$ is $F_{m,n}$-distributed.*

Finally, Proposition 4.6.17 implies the following version of the previous result.

Proposition 4.7.15. *Let $X_1, \ldots, X_m, Y_1, \ldots, Y_n$ be independent $\mathcal{N}(0,1)$-distributed. Then*

$$\frac{\frac{1}{m}\sum_{i=1}^m X_i^2}{\frac{1}{n}\sum_{j=1}^n Y_j^2}$$

is $F_{m,n}$-distributed.

Corollary 4.7.16. *If a random variable Z is $F_{m,n}$-distributed, then $1/Z$ possesses an $F_{n,m}$ distribution.*

Proof: This is an immediate consequence of Proposition 4.7.11. ■

4.8 Problems

Problem 4.1. Let U be uniformly distributed on $[0, 1]$. Which distributions do the following random variables possess

$$\min\{U, 1 - U\}, \quad \max\{U, 1 - U\}, \quad |2U - 1| \text{ and } \left| U - \frac{1}{3} \right| ?$$

Problem 4.2 (Generating functions). Let X be a random variable with values in $\mathbb{N}_0$. For $k \in \mathbb{N}_0$ let $p_k = \mathbb{P}\{X = k\}$. Then its **generating function** φ_X is defined by

$$\varphi_X(t) = \sum_{k=0}^{\infty} p_k t^k .$$

1. Show that $\varphi_X(t)$ exists if $|t| \leq 1$.
2. Let X and Y be two independent random variables with values in $\mathbb{N}_0$. Prove that then

$$\varphi_{X+Y} = \varphi_X \cdot \varphi_Y .$$

3. Compute φ_X in each of the following cases:
   (a) X is uniformly distributed on $\{1, \ldots, N\}$ for some $N \geq 1$.
   (b) X is $B_{n,p}$-distributed for some $n \geq 1$ and $p \in [0, 1]$.
   (c) X is Pois_λ-distributed for some $\lambda > 0$.
   (d) X is G_p-distributed for a certain $0 < p < 1$.
   (e) X is $B_{n,p}^-$-distributed.

Problem 4.3. Roll two dice simultaneously. Let X be result of the first die and Y that of the second one. Is it possible to falsify these two dice in such a way so that $X + Y$ is uniformly distributed on $\{2, \ldots, 12\}$? It is not assumed that both dice are falsified in the same way.

   Hint: One possible way to answer this question is as follows: investigate the generating functions of X and Y and compare their product with the generating function of the uniform distribution on $\{2, \ldots, 12\}$.

Problem 4.4. Let $X_1, \ldots, X_n$ be a sequence of independent identically distributed random variables with common distribution function F and distribution density p, that is,

$$\mathbb{P}\{X_j \leq t\} = F(t) = \int_{-\infty}^{t} p(x) \, dx, \quad j = 1, \ldots, n .$$

Define random variables X_* and X^* by

$$X_* := \min\{X_1, \ldots, X_n\} \quad \text{and} \quad X^* := \max\{X_1, \ldots, X_n\}.$$

1. Determine the distribution functions and densities of X_* and X^*.
2. Describe the distribution of the random variable X_* in the case that the X_js are exponentially distributed with parameter $\lambda > 0$.
3. Suppose now the X_js are uniformly distributed on $[0, 1]$. Describe the distribution of X_* and X^* in this case.

Problem 4.5. Find a function f from $(0, 1)$ to $\mathbb{R}$ such that

$$\mathbb{P}\{f(U) = k\} = \frac{1}{2^k}, \quad k = 1, 2, \ldots$$

for U uniformly distributed on $[0, 1]$.

Problem 4.6. Let U be uniform distributed on $[0, 1]$. Find functions f and g such that $X = f(U)$ and $Y = g(U)$ have the distribution densities p and q with

$$p(x) := \begin{cases} 0 & : x \notin (0, 1] \\ \frac{x^{-1/2}}{2} & : x \in (0, 1] \end{cases} \quad \text{and} \quad q(x) := \begin{cases} 0 & : \quad |x| > 1 \\ x + 1 & : -1 \le x \le 0 \\ 1 - x & : \quad 0 < x \le 1 \end{cases}.$$

Problem 4.7. Let X and Y be independent random variables with

$$\mathbb{P}\{X = k\} = \mathbb{P}\{Y = k\} = \frac{1}{2^k}, \quad k = 1, 2, \ldots.$$

How is $X + Y$ distributed?

Problem 4.8. The number of customers visiting a shop per day is Poisson distributed with parameter $\lambda > 0$. The probability that a single customer buys something equals p for a given $0 \le p \le 1$. Let X be the number of customers per day buying some goods. Determine the distribution of X.

Remark: We assume that the decision whether or not a single customer buys something is independent of the number of daily visitors.

A different way to formulate the above question is as follows: let $X_0, X_1, \ldots$ be independent random variables with $\mathbb{P}\{X_0 = 0\} = 1$,

$$\mathbb{P}\{X_j = 1\} = p \quad \text{and} \quad \mathbb{P}\{X_j = 0\} = 1 - p, \quad j = 1, 2, \ldots,$$

for a certain $p \in [0, 1]$. Furthermore, let Y be a Poisson-distributed random variable with parameter $\lambda > 0$, independent of the X_j. Determine the distribution of

$$X := \sum_{j=0}^{Y} X_j .$$

Hint: Use the "infinite" version of the law of total probability as stated in Problem 2.4.

Problem 4.9. Suppose X and Y are independent and exponentially distributed with parameter $\lambda > 0$. Find the distribution densities of $X - Y$ and X/Y.

Problem 4.10. Two random variables U and V are independent and uniformly distributed on $[0, 1]$. Given $n \in \mathbb{N}$, find the distribution density of $U + nV$.

Problem 4.11. Let X and Y be independent random variable distributed according to Pois_λ and Pois_μ, respectively. Given $n \in \mathbb{N}_0$ and some $k \in \{0, \ldots, n\}$, prove

$$\mathbb{P}\{X = k \mid X + Y = n\} = \binom{n}{k} \left(\frac{\lambda}{\lambda + \mu} \right)^k \left(\frac{\mu}{\lambda + \mu} \right)^{n-k} = B_{n,p}(\{k\})$$

with $p = \frac{\lambda}{\lambda + \mu}$.

Reformulation of the preceding problem: An owner of two stores, say store A and store B, observes that the number of customers in each of these stores is independent and Pois_λ and Pois_μ distributed. One day he was told that there were n customers in both stores together. What is the probability that k of the n customers were in store A, hence $n - k$ in store B ?

Problem 4.12. Let X and Y be independent standard normal variables. Show that X/Y is Cauchy distributed.

Hint: Use Corollary 4.7.10 and the fact that the vectors (X, Y), $(-X, Y)$, $(X, -Y)$, and $(-X, -Y)$ are identically distributed. Note that the probability distribution of each of these two-dimensional vectors is the (two-dimensional) standard normal distribution.

Problem 4.13. Let X and Y be independent G_p-distributed. Find the probability distribution of $X - Y$.

Hint: Compare Example 4.5.5. There we evaluated the distribution of $X - Y$ if $p = \frac{1}{2}$.

Problem 4.14. Let U and V be as in Example 4.7.2. Find an analytic (or geometric) explanation for

$$\mathbb{P}\{U \cdot V \le t\} = t - t \ln t, \quad 0 < t \le 1,$$

proved in (4.40).

Hint: Use that the vector (U, V) is uniformly distributed on $[0, 1]^2$.

Problem 4.15. Suppose X is a random variable with values in $(a, b) \subseteq \mathbb{R}$ and with density p. Let f from $(a, b) \to \mathbb{R}$ be (strictly) monotone and differentiable. Give a formula for q, the density of $f(X)$.

Hint: Investigate the cases of decreasing and increasing functions f separately.

Use this formula to evaluate the density of e^X and of e^{-X} for a $\mathcal{N}(0, 1)$-distributed X.

5 Expected Value, Variance, and Covariance

5.1 Expected Value

5.1.1 Expected Value of Discrete Random Variables

What is an expected value (also called mean value or expectation) of a random variable? How is it defined? Which property of the random variable does it describe and how it can be computed? Does every random variable possess an expected value?

To approach the solution of these questions, let us start with an example.

Example 5.1.1. Suppose N students attend a certain exam. The number of possible points is 100. Given $j = 0, 1, \ldots, 100$, let n_j be the number of students who achieved j points. Now choose randomly, according to the uniform distribution (a single student is chosen with probability $1/N$), one student. Name him or her ω, and define $X(\omega)$ as the number of points that the chosen student achieved. Then X is a random variable with values in $D = \{0, 1, \ldots, 100\}$. How is X distributed? Since X has values in D, its distribution is described by the probabilities

$$p_j = \mathbb{P}\{X = j\} = \frac{n_j}{N}, \quad j = 0, 1, \ldots, 100. \tag{5.1}$$

As expected value of X we take the average number A of points in this exam. How is A evaluated? The easiest way to do this is

$$A = \frac{1}{N} \sum_{j=0}^{100} j \cdot n_j = \sum_{j=0}^{100} j \cdot \frac{n_j}{N} = \sum_{j=0}^{100} j \cdot p_j,$$

where the p_js are defined by eq. (5.1). If we write $\mathbb{E}X$ for the expected value (or mean value) of X, and if we assume that this value coincides with A, then the preceding equation says

$$\mathbb{E}X = \sum_{j=0}^{100} j\, p_j = \sum_{j=0}^{100} j\, \mathbb{P}\{X = j\} = \sum_{j=0}^{100} x_j\, \mathbb{P}\{X = x_j\},$$

where the $x_j = j, j = 0, \ldots, 100$ denote the possible values of X.

In view of this example, the following definition for the expected value of a discrete random variable X looks feasible. Suppose X has values in $D = \{x_1, x_2, \ldots\}$, and let $p_j = \mathbb{P}\{X = x_j\}, j = 1, 2, \ldots$. Then the expected value $\mathbb{E}X$ of X is given by

$$\mathbb{E}X = \sum_{j=1}^{\infty} x_j\, p_j = \sum_{j=1}^{\infty} x_j\, \mathbb{P}\{X = x_j\}. \tag{5.2}$$

Unfortunately, the sum in eq. (5.2) does not always exist. In order to overcome this difficulty, let us recall some basic facts about infinite series of real numbers.

A sequence $(a_j)_{j\geq 1}$ of real numbers is called **summable**, provided its sequence of partial sums $(s_n)_{n\geq 1}$ with $s_n = \sum_{j=1}^{n} a_j$ converges in $\mathbb{R}$. Then one defines

$$\sum_{j=1}^{\infty} a_j = \lim_{n\to\infty} s_n.$$

If the sequence of partial sums diverges, nevertheless, in some cases we may assign to the infinite series a limit. If either $\lim_{n\to\infty} s_n = -\infty$ or $\lim_{n\to\infty} s_n = \infty$, then we write $\sum_{j=1}^{\infty} a_j = -\infty$ or $\sum_{j=1}^{\infty} a_j = \infty$, respectively. In particular, if $a_j \geq 0$ for $j \in \mathbb{N}$, then the sequence of partial sums is nondecreasing, which implies that only two different cases may occur: Either $\sum_{j=1}^{\infty} a_j < \infty$ (in this case the sequence is summable) or $\sum_{j=1}^{\infty} a_j = \infty$.

Let $(a_j)_{j\geq 1}$ be an arbitrary sequence of real numbers. If $\sum_{j=1}^{\infty} |a_j| < \infty$, then it is called **absolutely summable**. Note that each absolutely summable sequence is summable. This is a direct consequence of Cauchy's convergence criterion. The converse implication is wrong, as can be seen by considering $((-1)^n/n)_{n\geq 1}$.

Now we are prepared to define the expected value of a non-negative random variable.

Definition 5.1.2. Let X be a discrete random variable with values in $\{x_1, x_2, \ldots\}$ for some $x_j \geq 0$. Equivalently, the random variable X is discrete with $X \geq 0$. Then the **expected value** of X is defined by

$$\mathbb{E}X := \sum_{j=1}^{\infty} x_j\, \mathbb{P}\{X = x_j\}. \tag{5.3}$$

Remark 5.1.3. Since $x_j\, \mathbb{P}\{X = x_j\} \geq 0$ for non-negative X, for those random variables the sum in eq. (5.3) is always well-defined, but may be infinite. That is, each non-negative discrete random variable X possesses an expected value $\mathbb{E}X \in [0, \infty]$.

Let us now turn to the case of arbitrary (not necessarily non-negative) random variables. The next example shows which problems may arise.

Example 5.1.4. We consider the probability measure introduced in Example 1.3.6 and choose a random variable X with values in $\mathbb{Z}$ distributed according to the probability measure in this example. In other words,

$$\mathbb{P}\{X = k\} = \frac{3}{\pi^2}\frac{1}{k^2}, \quad k \in \mathbb{Z}\setminus\{0\}.$$

If we try to evaluate the expected value of X by formula (5.2), then this leads to the undetermined expression

$$\frac{3}{\pi^2} \sum_{\substack{k=-\infty \\ k \neq 0}}^{\infty} \frac{k}{k^2} = \frac{3}{\pi^2} \lim_{\substack{n \to \infty \\ m \to \infty}} \sum_{k=-m}^{n} \frac{1}{k} = \frac{3}{\pi^2} \left[\lim_{n \to \infty} \sum_{k=1}^{n} \frac{1}{k} + \lim_{m \to \infty} \sum_{k=-m}^{1} \frac{1}{k} \right]$$

$$= \frac{3}{\pi^2} \left[\lim_{n \to \infty} \sum_{k=1}^{n} \frac{1}{k} - \lim_{m \to \infty} \sum_{k=1}^{m} \frac{1}{k} \right] = \infty - \infty .$$

To exclude phenomenons as in Example 5.1.4, we suppose that a random variable has to meet the following condition.

Definition 5.1.5. Let X be discrete with values in $\{x_1, x_2, \ldots\} \subset \mathbb{R}$. Then the expected value of X **exists**, provided that

$$\mathbb{E}|X| = \sum_{j=1}^{\infty} |x_j| \, \mathbb{P}\{X = x_j\} < \infty . \qquad (5.4)$$

We mentioned above that an absolutely summable sequence is summable. Hence, under assumption (5.4), the sum in the subsequent definition is a well-defined real number.

Definition 5.1.6. Let X be a discrete random variable satisfying $\mathbb{E}|X| < \infty$. Then its **expected value** is defined as

$$\mathbb{E}X = \sum_{j=1}^{\infty} x_j \, \mathbb{P}\{X = x_j\} .$$

As before, $x_1, x_2, \ldots$ are the possible values of X.

Example 5.1.7. We start with an easy example that demonstrates how to compute the expected value in concrete cases. If the distribution of a random variable X is defined as $\mathbb{P}\{X = -1\} = 1/6$, $\mathbb{P}\{X = 0\} = 1/8$, $\mathbb{P}\{X = 1\} = 3/8$, and $\mathbb{P}\{X = 2\} = 1/3$, then its expected value equals

$$\mathbb{E}X = (-1) \cdot \mathbb{P}\{X = -1\} + 0 \cdot \mathbb{P}\{X = 0\} + 1 \cdot \mathbb{P}\{X = 1\} + 2 \cdot \mathbb{P}\{X = 2\}$$
$$= -\frac{1}{6} + \frac{3}{8} + \frac{2}{3} = \frac{7}{8} .$$

Example 5.1.8. The next example shows that $\mathbb{E}X = \infty$ may occur even for quite natural random variables. Thus, let us come back to the model presented in Example 1.4.39. There we developed a strategy how to win always one dollar in a series of games. The basic idea was, after losing a game, next time one doubles the amount in the pool. As in Example 1.4.39, let $X(k)$ be the amount of money needed when winning for the first time in game k. We obtained

$$\mathbb{P}\{X = 2^k - 1\} = p(1-p)^{k-1}, \quad k = 1, 2 \ldots .$$

Recall that $0 < p < 1$ is the probability to win a single game. We ask for the expected value of money needed to apply this strategy. It follows

$$\mathbb{E}X = \sum_{k=1}^{\infty}(2^k - 1)\mathbb{P}\{X = 2^k - 1\} = p \sum_{k=1}^{\infty}(2^k - 1)(1-p)^{k-1}. \tag{5.5}$$

If the game is fair, that is, if $p = 1/2$, then this leads to

$$\mathbb{E}X = \sum_{k=1}^{\infty} \frac{2^k - 1}{2^k} = \infty,$$

because of $(2^k - 1)/2^k \to 1$ as $k \to \infty$. This yields $\mathbb{E}X = \infty$ for all[1] $p \le 1/2$.

Let us sum up: if $p \le 1/2$ (which is the case in all provided games), the obtained result tells us that the average amount of money needed, to use this strategy, is arbitrarily large. The owners of gambling casinos know this strategy as well. Therefore, they limit the possible amount of money in the pool. For example, if the largest possible stakes is N dollars, then the strategy breaks down as soon as one loses n games for some n with $2^n > N$. And, as our calculations show, on average this always happens.

Remark 5.1.9. If $p > 1/2$, then the average amount of money needed is finite, and it can be calculated by

$$\mathbb{E}X = p \sum_{k=1}^{\infty}(2^k - 1)(1-p)^{k-1} = 2p \sum_{k=0}^{\infty}(2 - 2p)^k - p \sum_{k=0}^{\infty}(1-p)^k$$

$$= \frac{2p}{1 - (2 - 2p)} - \frac{p}{1 - (1 - p)} = \frac{2p}{2p - 1} - 1 = \frac{1}{2p - 1}.$$

5.1.2 Expected Value of Certain Discrete Random Variables

The aim of this section is to compute the expected value of the most interesting discrete random variables. We start with uniformly distributed ones.

---

[1] If $p \le 1/2$ then $1 - p \ge 1/2$, hence the sum in eq. (5.5) becomes bigger and, therefore, it also diverges.

Proposition 5.1.10. *Let X be uniformly distributed on the set $\{x_1, \ldots, x_N\}$ of real numbers. Then it follows that*

$$\mathbb{E}X = \frac{1}{N} \sum_{j=1}^{N} x_j. \tag{5.6}$$

That is, $\mathbb{E}X$ is the arithmetic mean of the x_js.

Proof: This is an immediate consequence of $\mathbb{P}\{X = x_j\} = 1/N$, implying

$$\mathbb{E}X = \sum_{j=1}^{N} x_j \cdot \mathbb{P}\{X = x_j\} = \sum_{j=1}^{N} x_j \cdot \frac{1}{N}. \qquad \blacksquare$$

Remark 5.1.11. For general discrete random variables X with values $x_1, x_2, \ldots$, their expected value $\mathbb{E}X$ may be regarded as a weighted (the weights are the p_js) mean of the x_js.

Example 5.1.12. Let X be uniformly distributed on $\{1, \ldots, 6\}$. Then X is a model for rolling a fair die. Its expected value is, as is well known,

$$\mathbb{E}X = \frac{1 + \cdots + 6}{6} = \frac{21}{6} = \frac{7}{2}.$$

Next we determine the expected value of a binomial distributed random variable.

Proposition 5.1.13. *Let X be binomial distributed with parameters n and p. Then we get*

$$\mathbb{E}X = np. \tag{5.7}$$

Proof: The possible values of X are $0, \ldots, n$. Thus, it follows that

$$\mathbb{E}X = \sum_{k=0}^{n} k \cdot \mathbb{P}\{X = k\} = \sum_{k=1}^{n} k \cdot \binom{n}{k} p^k (1 - p)^{n-k}$$

$$= \sum_{k=1}^{n} \frac{n!}{(k-1)!\,(n-k)!} p^k (1 - p)^{n-k}$$

$$= np \sum_{k=1}^{n} \frac{(n-1)!}{(k-1)!(n-k)!} p^{k-1} (1 - p)^{n-k}.$$

Shifting the index from $k - 1$ to k in the last sum implies

$$\mathbb{E}X = np \sum_{k=0}^{n-1} \frac{(n-1)!}{k!\,(n-1-k)!}\, p^k (1-p)^{n-1-k}$$

$$= np \sum_{k=0}^{n-1} \binom{n-1}{k} p^k (1-p)^{n-1-k}$$

$$= np\left[p + (1-p)\right]^{n-1} = np\,.$$

This completes the proof. ∎

Remark 5.1.14. The previous result tells us the following: if we perform n independent trials of an experiment with success probability p, then on the average we will observe np times success.

Example 5.1.15. One kilogram of a radioactive material consists of N atoms. The atoms decay independently of each other and, moreover, the lifetime of each of the atoms is exponentially distributed with some parameter $\lambda > 0$. We ask for the time $T_0 > 0$, at which, on the average, half of the atoms are decayed. T_0 is usually called radioactive half-life.

Answer: If $T > 0$, then the probability that a single atom decays before time T is given by

$$p(T) = E_\lambda([0, T]) = 1 - e^{\lambda T}.$$

Since the atoms decay independently, the number of atoms decaying before time T is $B_{N,p(T)}$-distributed. Therefore, by Proposition 5.1.13, the expected value of decayed atoms equals $N \cdot p(T) = N(1 - e^{\lambda T})$. Hence, T_0 has to satisfy

$$N(1 - e^{-\lambda T_0}) = \frac{N}{2},$$

leading to $T_0 = \ln 2/\lambda$. Conversely, if we know T_0 and want to determine λ, then $\lambda = \ln 2/T_0$. Consequently, the probability that a single atom decays before time $T > 0$ can also be described by

$$E_\lambda([0, T]) = 1 - e^{-T \ln 2/T_0} = 1 - 2^{-T/T_0}\,.$$

Next, we determine the expected value of *Poisson distributed* random variables.

Proposition 5.1.16. *For some $\lambda > 0$, let X be distributed according to Pois_λ. Then it follows that $\mathbb{E}X = \lambda$.*

Proof: The possible values of X are $0, 1, \ldots$. Hence, the expected value is given by

$$\mathbb{E}X = \sum_{k=0}^{\infty} k \cdot \mathbb{P}\{X = k\} = \sum_{k=1}^{\infty} k \frac{\lambda^k}{k!} e^{-\lambda} = \lambda \sum_{k=1}^{\infty} \frac{\lambda^{k-1}}{(k-1)!} e^{-\lambda},$$

which transforms by a shift of the index to

$$\lambda \left[\sum_{k=0}^{\infty} \frac{\lambda^k}{k!} \right] e^{-\lambda} = \lambda [e^{\lambda}] e^{-\lambda} = \lambda.$$

This proves the assertion. ∎

Interpretation: Proposition 5.1.16 explains the role of the parameter λ in the definition of the Poisson distribution. Whenever certain numbers are Poisson distributed, then $\lambda > 0$ is the average of the observed values. For example, if the number of accidents per week is known to be Pois_{λ}-distributed, then the parameter λ is determined by the average number of accidents per week in the past. Or, as we already mentioned in Example 4.6.6, the number of raisins in a piece of ρ pounds of dough is $\text{Pois}_{\lambda\rho}$-distributed, where λ is the proportion of raisins per pound dough, hence $\lambda\rho$ is the average number of raisins per ρ pounds.

Example 5.1.17. Let us once more take a look at Example 4.6.15. There we considered light bulbs with E_{λ}-distributed lifetime. Every time a bulb burned out, we replaced it by a new one of the same type. It turned out that the number of necessary replacements until time $T > 0$ was $\text{Pois}_{\lambda T}$-distributed. Consequently, by Proposition 5.1.16, on average, until time T we have to change the light bulbs λT times.

Finally, we compute the expected value of a negative binomial distributed random variable. According to Definition 1.4.41, a random variable X is $B_{n,p}^-$-distributed if

$$\mathbb{P}\{X = k\} = \binom{k-1}{k-n} p^n (1-p)^{k-n}, \quad k = n, n+1, \ldots$$

or, equivalently, if

$$\mathbb{P}\{X = k + n\} = \binom{-n}{k} p^n (p-1)^k, \quad k = 0, 1, \ldots \tag{5.8}$$

Proposition 5.1.18. *Suppose X is $B_{n,p}^-$-distributed for some $n \geq 1$ and $p \in (0, 1)$. Then*

$$\mathbb{E}X = \frac{n}{p}.$$

Proof: Using eq. (5.8), the expected value of X is computed as

$$\mathbb{E}X = \sum_{k=n}^{\infty} k\,\mathbb{P}\{X = k\} = \sum_{k=0}^{\infty} (k + n)\,\mathbb{P}\{X = k + n\}$$

$$= p^n \sum_{k=1}^{\infty} k \binom{-n}{k} (p - 1)^k + n p^n \sum_{k=0}^{\infty} \binom{-n}{k} (p - 1)^k. \tag{5.9}$$

To evaluate the two sums in eq. (5.9) we use Proposition A.5.2, which asserts

$$\frac{1}{(1 + x)^n} = \sum_{k=0}^{\infty} \binom{-n}{k} x^k \tag{5.10}$$

for $|x| < 1$. Applying this with $x = p - 1$ (recall $0 < p < 1$),

$$n p^n \sum_{k=0}^{\infty} \binom{-n}{k} (p - 1)^k = n p^n \frac{1}{(1 + (p - 1))^n} = n. \tag{5.11}$$

Next we differentiate eq. (5.10) with respect to x and obtain

$$\frac{-n}{(1 + x)^{n+1}} = \sum_{k=1}^{\infty} k \binom{-n}{k} x^{k-1},$$

which, multiplying both sides by x, gives

$$\frac{-nx}{(1 + x)^{n+1}} = \sum_{k=1}^{\infty} k \binom{-n}{k} x^k. \tag{5.12}$$

Letting $x = p - 1$ in eq. (5.12), the first sum in eq. (5.9) becomes

$$p^n \sum_{k=1}^{\infty} k \binom{-n}{k} (p - 1)^k = p^n \frac{-n(p - 1)}{(1 + (p - 1))^{n+1}} = \frac{n(1 - p)}{p}. \tag{5.13}$$

Finally, we combine eqs. (5.9), (5.11), and (5.13) and obtain

$$\mathbb{E}X = \frac{n(1 - p)}{p} + n = \frac{n}{p}$$

as claimed. ∎

Remark 5.1.19. Proposition 5.1.18 asserts that on average the nth success occurs in trial n/p. For example, rolling a die, on average, the first appearance of number "6" will be in trial 6, the second in trial 12, and so on.

Corollary 5.1.20. *If X is geometric distributed with parameter p, then*

$$\mathbb{E}X = \frac{1}{p}.$$
(5.14)

Proof: Recall that $G_p = B^-_{1,p}$, hence X is $B^-_{1,p}$-distributed, and $\mathbb{E}X = \frac{1}{p}$ by Proposition 5.1.18. ∎

Alternative proof of Corollary 5.1.20: Suppose X is G_p-distributed. Then we write

$$\mathbb{E}X = p \sum_{k=1}^{\infty} k(1-p)^{k-1} = p \sum_{k=0}^{\infty} (k+1)(1-p)^k$$

$$= (1-p) \sum_{k=0}^{\infty} kp(1-p)^{k-1} + p \sum_{k=0}^{\infty} (1-p)^k$$

$$= (1-p)\mathbb{E}X + 1.$$

Solving this equation with respect to $\mathbb{E}X$ proves eq. (5.14) as asserted. Observe that this alternative proof is based upon the knowledge of $\mathbb{E}X < \infty$. Otherwise, we could not solve the equation with respect to $\mathbb{E}X$. But, because of $0 < p < 1$, this fact is an easy consequence of

$$\mathbb{E}X = p \sum_{k=1}^{\infty} k(1-p)^{k-1} < \infty.$$

5.1.3 Expected Value of Continuous Random Variables

Let X be a continuous random variable with distribution density p, that is, if $t \in \mathbb{R}$, then

$$\mathbb{P}\{X \le t\} = \int_{-\infty}^{t} p(x)\,dx.$$

How to define $\mathbb{E}X$ in this case?

To answer this question, let us present formula (5.3) in an equivalent way. Suppose X maps Ω into a set $D \subset \mathbb{R}$, which is either finite or countably infinite. Let $p : \mathbb{R} \to [0,1]$ be the probability mass function of X introduced in eq. (3.4). Then the expected value of X may also be written as

$$\mathbb{E}X = \sum_{x \in \mathbb{R}} x\, p(x).$$

In this form, the preceding formula suggests that in the continuous case the sum should be replaced by an integral. This can made more precise by approximating continuous random variables by discrete ones. But this is only a heuristic explanation; for a precise approach, deeper convergence theorems for random variables are needed. Therefore, we do not give more details here, we simply replace sums by integrals.

Doing so, for continuous random variables the following approach for the definition of $\mathbb{E}X$ might be taken. If $p : \mathbb{R} \to [0, \infty)$ is the distribution density of X, set

$$\mathbb{E}X := \int_{-\infty}^{\infty} x\, p(x)\, dx.$$

(5.15)

However, here we have a similar problem as in the discrete case, namely that the integral in eq. (5.15) need not exist. Therefore, let us give a short digression about the integrability of real functions.

Let $f : \mathbb{R} \to \mathbb{R}$ be a function such that for all $a < b$ the integral $\int_a^b f(x)dx$ is a well-defined real number. Then

$$\int_{-\infty}^{\infty} f(x)\, dx := \lim_{\substack{a \to -\infty \\ b \to \infty}} \int_a^b f(x)\, dx,$$

(5.16)

provided **both** limits on the right-hand side of eq. (5.16) exist. In this case we call f **integrable** (in the Riemann sense) on $\mathbb{R}$.

If $f(x) \geq 0$, $x \in \mathbb{R}$, then the limit $\lim_{\substack{a \to -\infty \\ b \to \infty}} \int_a^b f(x)\, dx$ always exists in a generalized sense, that is, it may be finite (then f is integrable) or infinite, then this is expressed by $\int_{-\infty}^{\infty} f(x)\, dx = \infty$.

If $\int_{-\infty}^{\infty} |f(x)|\, dx < \infty$, then f is said to be **absolutely integrable**, and as in the case of infinite series, absolutely integrable function are integrable. Note that $x \mapsto \sin x / x$ is integrable, but not absolutely integrable.

After this preparation, we come back to the definition of the expected value for continuous random variables.

Definition 5.1.21. Let X be a random variable with distribution density p. If $p(x) = 0$ for $x < 0$, or, equivalently, $\mathbb{P}\{X \geq 0\} = 1$, then the **expected value** of X is defined by

$$\mathbb{E}X := \int_0^{\infty} x\, p(x)\, dx.$$

(5.17)

Observe that under these conditions upon p or X, we have $x\, p(x) \geq 0$. Therefore, the integral in eq. (5.17) is always well-defined, but might be infinite. In this case we write $\mathbb{E}X = \infty$.

Let us turn now to the case of $\mathbb{R}$-valued random variables. The following example shows that the integral in eq. (5.15) may not exist, hence, in general, without an additional assumption the expected value cannot be defined by eq. (5.15).

Example 5.1.22. A random variable X is supposed to possess the density (check that this is indeed a density function)

$$p(x) = \begin{cases} 0 & : -1 < x < 1 \\ \frac{1}{2x^2} & : |x| \geq 1 \end{cases}$$

If we try to evaluate $\mathbb{E}X$ by virtue of eq. (5.15), then, because of

$$\int_{-\infty}^{\infty} x p(x) dx = \lim_{\substack{a \to -\infty \\ b \to \infty}} \int_{a}^{b} x p(x) \, dx = \frac{1}{2} \left[\lim_{b \to \infty} \int_{1}^{b} \frac{dx}{x} + \lim_{a \to -\infty} \int_{a}^{-1} \frac{dx}{x} \right]$$

$$= \frac{1}{2} \left[\lim_{b \to \infty} \int_{1}^{b} \frac{dx}{x} - \lim_{a \to \infty} \int_{1}^{a} \frac{dx}{x} \right] = \infty - \infty,$$

we observe an undetermined expression. Thus, there is no meaningful way to introduce an expected value for X.

We enforce the existence of the integral by the following condition.

Definition 5.1.23. Let X be a (real-valued) random variable with distribution density p. We say the expected value of X **exists**, provided p satisfies the following integrability condition[2]:

$$\mathbb{E}|X| := \int_{-\infty}^{\infty} |x| \, p(x) \, dx < \infty. \tag{5.18}$$

Condition (5.18) says nothing but that $f(x) := x \, p(x)$ is absolutely integrable. Hence, as mentioned above, f is integrable, and the integral in the following definition is well-defined.

Definition 5.1.24. Suppose condition (5.18) is satisfied. Then the **expected value** of X is defined by

$$\mathbb{E}X := \int_{-\infty}^{\infty} x \, p(x) \, dx.$$

---

2 At this point it is not clear that the right-hand integral is indeed the expected value of $|X|$. This will follow later on by Proposition 5.1.36. Nevertheless, we use this notation before giving a proof.

5.1.4 Expected Value of Certain Continuous Random Variables

We start with computing the expected value of a *uniformly distributed* (continuous) random variable.

Proposition 5.1.25. *Let X be uniformly distributed on the finite interval I =* $[\alpha, \beta]$. *Then*

$$\mathbb{E}X = \frac{\alpha + \beta}{2},$$

that is, the expected value is the midpoint of the interval I.

Proof: The distribution density of X is the function p defined as $p(x) = (\beta - \alpha)^{-1}$ if $x \in I$, and $p(x) = 0$ if $x \notin I$. Of course, X possesses an expected value,[3] which can be evaluated by

$$\mathbb{E}X = \int_{-\infty}^{\infty} xp(x)\,dx = \int_{\alpha}^{\beta} \frac{x}{\beta - \alpha}\,dx = \frac{1}{\beta - \alpha}\left[\frac{x^2}{2}\right]_{\alpha}^{\beta} = \frac{1}{2}\cdot\frac{\beta^2 - \alpha^2}{\beta - \alpha} = \frac{\alpha + \beta}{2}.$$

This proves the proposition. ∎

Next we determine the expected value of a *gamma distributed* random variable.

Proposition 5.1.26. *Suppose X is $\Gamma_{\alpha,\beta}$-distributed with $\alpha, \beta > 0$. Then its expected value is*

$$\mathbb{E}X = \alpha\beta.$$

Proof: Because of $\mathbb{P}\{X \geq 0\} = 1$, its expected value is well-defined and computed by

$$\mathbb{E}X = \int_{0}^{\infty} xp(x)\,dx = \frac{1}{\alpha^\beta\, \Gamma(\beta)} \int_{0}^{\infty} x \cdot x^{\beta-1}\, e^{-x/\alpha}\,dx$$

$$= \frac{1}{\alpha^\beta\, \Gamma(\beta)} \int_{0}^{\infty} x^\beta\, e^{-x/\alpha}\,dx. \tag{5.19}$$

The change of variables $u := x/\alpha$ transforms eq. (5.19) into

$$\mathbb{E}X = \frac{\alpha^{\beta+1}}{\alpha^\beta\, \Gamma(\beta)} \int_{0}^{\infty} u^\beta\, e^{-u}\,du = \frac{\alpha^{\beta+1}}{\alpha^\beta\, \Gamma(\beta)} \cdot \Gamma(\beta + 1) = \alpha\beta,$$

where we used eq. (1.48) in the last step. This completes the proof. ∎

---

3 $|x|p(x)$ is bounded and nonzero only on a finite interval.

Corollary 5.1.27. *Let X be E_λ-distributed for a certain $\lambda > 0$. Then*

$$\mathbb{E}X = \frac{1}{\lambda}.$$

Proof: Note that $E_\lambda = \Gamma_{\lambda^{-1},1}$. ∎

Example 5.1.28. The lifetime of a special type of light bulbs is exponentially distributed. Suppose the average lifetime constitutes 100 units of time. This implies $\lambda = 1/100$, hence, if X describes the lifetime, then

$$\mathbb{P}\{X \le t\} = 1 - e^{-t/100}, \quad t \ge 0.$$

For example, the probability that the light bulb burns longer than 200 time units equals

$$\mathbb{P}\{X \ge 200\} = e^{-200/100} = e^{-2} = 0.135335 \cdots$$

Remark 5.1.29. If we evaluate in the previous example

$$\mathbb{P}\{X \ge \mathbb{E}X\} = \mathbb{P}\{X \ge 100\} = e^{-1},$$

then we see that in general $\mathbb{P}\{X \ge \mathbb{E}X\} \ne 1/2$. Thus, in this case, the expected value is different from the **median** of X defined as a real number M satisfying $\mathbb{P}\{X \ge M\} \ge 1/2$ and $\mathbb{P}\{X \le M\} \ge 1/2$. In particular, if F_X satisfies the condition of Proposition 4.4.6, then the median is uniquely determined by $M = F_X^{-1}(1/2)$, i.e., by $\mathbb{P}\{X \le M\} = 1/2$. It is easy to see that the above phenomenon appears for all exponentially distributed random variables. Indeed, if X is E_λ-distributed, then $M = \ln 2/\lambda$ while, as we saw, $\mathbb{E}X = 1/\lambda$.

Corollary 5.1.30. *If X is χ_n^2-distributed, then*

$$\mathbb{E}X = n.$$

Proof: Since $\chi_n^2 = \Gamma_{2,n/2}$, by Proposition 5.1.26 follows that $\mathbb{E}X = 2 \cdot n/2 = n$. ∎

Which expected value does a beta distributed random variable possess? The next proposition answers this question.

Proposition 5.1.31. *Let X be $B_{\alpha,\beta}$-distributed for certain $\alpha, \beta > 0$. Then*

$$\mathbb{E}X = \frac{\alpha}{\alpha + \beta}.$$

Proof: Using eq. (1.60), by eq. (5.17) we obtain, as asserted,

$$\mathbb{E}X = \frac{1}{B(\alpha, \beta)} \int_0^1 x \cdot x^{\alpha-1}(1-x)^{\beta-1}\, dx$$

$$= \frac{1}{B(\alpha, \beta)} \int_0^1 x^\alpha (1-x)^{\beta-1}\, dx = \frac{B(\alpha+1, \beta)}{B(\alpha, \beta)} = \frac{\alpha}{\alpha+\beta}.$$ ∎

Example 5.1.32. Suppose we choose independently n numbers $x_1, \ldots, x_n$ uniformly distributed on $[0, 1]$ and order them by their size. Then we get the order statistics $0 \le x_1^* \le \cdots \le x_n^* \le 1$. According to Example 3.7.8, if $1 \le k \le n$, then the number x_k^* is $B_{k, n-k+1}$-distributed. Thus Proposition 5.1.31 implies that the average value of x_k^*, that is, of the kth largest number, equals

$$\frac{k}{k + (n - k + 1)} = \frac{k}{n + 1}.$$

In particular, the expected value of the smallest number is $\frac{1}{n+1}$ while that of the largest one is $\frac{n}{n+1}$.

Does a Cauchy distributed random variable possess an expected value? Here we obtain the following.

Proposition 5.1.33. *If X Cauchy distributed, then $\mathbb{E}X$ does not exist.*

Proof: First observe that we may not use Definition 5.17. The distribution density of X is given by $p(x) = \frac{1}{\pi} \cdot \frac{1}{1+x^2}$, hence, it does not satisfy $p(x) = 0$ for $x < 0$. Consequently, we have to check whether condition (5.18) is satisfied. Here we get

$$\mathbb{E}|X| = \frac{1}{\pi} \int_{-\infty}^{\infty} \frac{|x|}{1 + x^2}\, dx = \frac{2}{\pi} \int_0^{\infty} \frac{x}{1 + x^2}\, dx = \frac{1}{\pi} \left[\ln(1 + x^2)\right]_0^{\infty} = \infty.$$

Thus, $\mathbb{E}|X| = \infty$, that is, X does not possess an expected value. ∎

Finally, we determine the expected value of normally distributed random variables.

Proposition 5.1.34. *If X is $\mathcal{N}(\mu, \sigma^2)$-distributed, then*

$$\mathbb{E}X = \mu.$$

Proof: First, we check whether the expected value exists. The density of X is given by eq. (1.47), hence

$$\int_{-\infty}^{\infty} |x|\, p_{\mu,\sigma}(x)\, dx = \frac{1}{\sqrt{2\pi}\,\sigma} \int_{-\infty}^{\infty} |x|\, e^{-(x-\mu)^2/2\sigma^2}\, dx$$

$$= \frac{1}{\sqrt{\pi}} \int_{-\infty}^{\infty} |\sqrt{2}\sigma u + \mu|\, e^{-u^2}\, du$$

$$\leq \sigma \frac{2\sqrt{2}}{\sqrt{\pi}} \int_{0}^{\infty} u\, e^{-u^2}\, du + |\mu|\, \frac{2}{\sqrt{\pi}} \int_{0}^{\infty} e^{-u^2}\, du < \infty,$$

where we used the well-known fact[4] that for all $k \in \mathbb{N}_0$

$$\int_{0}^{\infty} u^k\, e^{-u^2}\, du < \infty.$$

The expected value $\mathbb{E}X$ is now evaluated in a similar way by

$$\mathbb{E}X = \int_{-\infty}^{\infty} x\, p_{\mu,\sigma}(x)\, dx = \frac{1}{\sqrt{2\pi}\,\sigma} \int_{-\infty}^{\infty} x\, e^{-(x-\mu)^2/2\sigma^2}\, dx$$

$$= \frac{1}{\sqrt{2\pi}} \int_{-\infty}^{\infty} (\sigma v + \mu)\, e^{-v^2/2}\, dv$$

$$= \sigma\, \frac{1}{\sqrt{2\pi}} \int_{-\infty}^{\infty} v\, e^{-v^2/2}\, dv + \mu\, \frac{1}{\sqrt{2\pi}} \int_{-\infty}^{\infty} e^{-v^2/2}\, dv. \qquad (5.20)$$

The function $f(v) := v\, e^{-v^2/2}$ is odd, that is, $f(-v) = -f(v)$, thus $\int_{-\infty}^{\infty} f(v)\, dv = 0$, and the first integral in eq. (5.20) vanishes. To compute the second integral use Proposition 1.6.6 and obtain

$$\mu\, \frac{1}{\sqrt{2\pi}} \int_{-\infty}^{\infty} e^{-v^2/2}\, dv = \mu\, \frac{1}{\sqrt{2\pi}} \sqrt{2\pi} = \mu.$$

This completes the proof. ∎

Remark 5.1.35. Proposition 5.1.34 justifies the notation "expected value" for the parameter μ in the definition of the probability measure $\mathcal{N}(\mu, \sigma^2)$.

---

4 See either [Spi08] or use that for all $k \geq 1$ one has $\sup_{u>0} u^k\, e^{-u} < \infty$.

5.1.5 Properties of the Expected Value

In this section we summarize the main properties of the expected value. They are valid for both discrete and continuous random variables. But, unfortunately, within the framework of this book it is not possible to prove most of them in full generality. To do so one needs an integral (Lebesgue integral) $\int_\Omega f d\mathbb{P}$ of functions $f : \Omega \to \mathbb{R}$ for some probability space $(\Omega, \mathcal{A}, \mathbb{P})$. Then $\mathbb{E}X = \int_\Omega X d\mathbb{P}$, and all subsequent properties of $X \mapsto \mathbb{E}X$ follow from those of the (Lebesgue) integral.

Proposition 5.1.36. *The expected value of random variables owns the following properties:*
(1) *The expected value of X only depends on its probability distribution $\mathbb{P}_X$, not on the way how X is defined. That is, if $X \overset{d}{=} Y$ for two random variables X and Y, then $\mathbb{E}X = \mathbb{E}Y$.*
(2) *If X is with probability 1 constant, that is, there is some $c \in \mathbb{R}$ with $\mathbb{P}(X = c) = 1$, then $\mathbb{E}X = c$.*
(3) *The expected value is linear: let X and Y be two random variables possessing an expected value and let $a, b \in \mathbb{R}$. Then $\mathbb{E}(aX + bY)$ exists as well and, moreover,*

$$\mathbb{E}(aX + bY) = a\,\mathbb{E}X + b\,\mathbb{E}Y\,.$$

(4) *Suppose X is a discrete random variable with values in $x_1, x_2, \ldots$. Given a function f from $\mathbb{R}$ to $\mathbb{R}$, the expected value $\mathbb{E}f(X)$ exists if and only if*

$$\sum_{i=1}^\infty |f(x_i)|\, \mathbb{P}(X = x_i) < \infty\,,$$

and, moreover, then

$$\mathbb{E}f(X) = \sum_{i=1}^\infty f(x_i)\, \mathbb{P}(X = x_i)\,. \tag{5.21}$$

(5) *If X is continuous with density p, then for any measurable function $f : \mathbb{R} \to \mathbb{R}$ the expected value $\mathbb{E}f(X)$ exists if and only if*

$$\int_{-\infty}^\infty |f(x)|\, p(x)\, dx < \infty\,.$$

In this case it follows that

$$\mathbb{E}f(X) = \int_{-\infty}^\infty f(x)\, p(x)\, dx\,. \tag{5.22}$$

(6) For **independent** X and Y possessing an expected value, the expected value of $X \cdot Y$ exists as well and, moreover,

$$\mathbb{E}[X\,Y] = \mathbb{E}X \cdot \mathbb{E}Y.$$

(7) Write $X \leq Y$ provided that $X(\omega) \leq Y(\omega)$ for all $\omega \in \Omega$. If in this sense $|X| \leq Y$ for some Y with $\mathbb{E}Y < \infty$, then $\mathbb{E}|X| < \infty$ and, hence, $\mathbb{E}X$ exists.

(8) Suppose $\mathbb{E}X$ and $\mathbb{E}Y$ exist. Then $X \leq Y$ implies $\mathbb{E}X \leq \mathbb{E}Y$. In particular, if $X \geq 0$, then $\mathbb{E}X \geq 0$.

Proof: We only prove properties (1), (2), (4), and (8). Some of the other properties are not difficult to verify in the case of discrete random variables, for example, (3), but because the proofs are incomplete, we do not present them here. We refer to [Bil12], [Dur10] or [Kho07] for the proofs of the remaining properties.

We begin with the proof of (1). If X and Y are identically distributed, then either both are discrete or both are continuous. If they are discrete, and $\mathbb{P}_X(D) = 1$ for an at most countably infinite set D, then $X \overset{d}{=} Y$ implies $\mathbb{P}_Y(D) = 1$. Moreover, by the same argument $\mathbb{P}_X(\{x\}) = \mathbb{P}_Y(\{x\})$ for any $x \in D$. Hence, in view of Definition 5.1.2, $\mathbb{E}X$ exists if and only if $\mathbb{E}Y$ does so. Moreover, if this is valid, then $\mathbb{E}X = \mathbb{E}Y$ by the same argument.

In the continuous case we argue as follows. Let p be a density of X. By $X \overset{d}{=} Y$ it follows that

$$\int_{-\infty}^{t} p(x)\,\mathrm{d}x = \mathbb{P}_X((-\infty, t]) = \mathbb{P}_Y((-\infty, t]), \quad t \in \mathbb{R}.$$

Thus, p is also a distribution density of Y and, consequently, in view of Definition 5.1.23, the expected value of X exists if and only if this is the case for Y. Moreover, by Definition 5.1.24 we get $\mathbb{E}X = \mathbb{E}Y$.

Next we show that (2) is valid. Thus, suppose $\mathbb{P}\{X = c\} = 1$ for some $c \in \mathbb{R}$. Then X is discrete with $\mathbb{P}_X(D) = 1$ where $D = \{c\}$, and by Definition 5.1.2 we obtain

$$\mathbb{E}X = c \cdot \mathbb{P}\{X = c\} = c \cdot 1 = c$$

as asserted.

To prove (4) we assume that X has values in $D = \{x_1, x_2, \dots\}$. Then $Y = f(X)$ maps into $f(D) = \{y_1, y_2, \dots\}$. Given $j \in \mathbb{N}$ let $D_j = \{x_i : f(x_i) = y_j\}$. Thus,

$$\mathbb{P}\{Y = y_j\} = \mathbb{P}\{X \in D_j\} = \sum_{x_i \in D_j} \mathbb{P}\{X = x_i\}.$$

Consequently, since $D_j \cap D_{j'} = \emptyset$ if $j \neq j'$, by $\bigcup_{j=1}^{\infty} D_j = D$ we get

$$
\mathbb{E}|Y| = \sum_{j=1}^{\infty} |y_j| \mathbb{P}\{Y = y_j\} = \sum_{j=1}^{\infty} \sum_{x_i \in D_j} |y_j| \mathbb{P}\{X = x_i\}
$$

$$
= \sum_{j=1}^{\infty} \sum_{x_i \in D_j} |f(x_i)| \mathbb{P}\{X = x_i\} = \sum_{i=1}^{\infty} |f(x_i)| \mathbb{P}\{X = x_i\} = \mathbb{E}|X| .
$$

This proves the first part of (4). The second part follows by exactly the same arguments (replace $|y_j|$ by y_j). Therefore, we omit its proof.

We finally prove (8). To this end we first show the second part, that is, $\mathbb{E}X \geq 0$ for $X \geq 0$. If X is discrete, then X attains values in D, where D consists only of non-negative real numbers. Hence, $x_j \mathbb{P}\{X = x_j\} \geq 0$, which implies $\mathbb{E}X \geq 0$. If X is continuous, in view of $X \geq 0$, we may choose its density p such that $p(x) = 0$ if $x < 0$. Then $\mathbb{E}X = \int_0^{\infty} p(x)dx \geq 0$.

Suppose now $X \leq Y$. Setting $Z = Y - X$, by the first step follows $\mathbb{E}Z \geq 0$. But, property (3) implies $\mathbb{E}Z = \mathbb{E}Y - \mathbb{E}X$, from which we derive $\mathbb{E}X \leq \mathbb{E}Y$ as asserted. Note that by assumption $\mathbb{E}X$ and $\mathbb{E}Y$ are real numbers, so that $\mathbb{E}Y - \mathbb{E}X$ is not an undetermined expression. ∎

Remark 5.1.37. Properties (4) and (5) of the previous proposition, applied with $f(x) = |x|$, lead to

$$
\mathbb{E}|X| = \sum_{j=1}^{\infty} |x_j| \mathbb{P}\{X = x_j\} \quad \text{or} \quad \mathbb{E}|X| = \int_{-\infty}^{\infty} |x| \, p(x) \, dx ,
$$

as we already stated in conditions (5.4) and (5.18).

Corollary 5.1.38. *If $\mathbb{E}X$ exists, then shifting X by $\mu = \mathbb{E}X$, it becomes centralized (the expected value is zero).*

Proof: If $Y = X - \mu$, then properties (2) and (3) of Proposition 5.1.36 imply

$$
\mathbb{E}Y = \mathbb{E}(X - \mu) = \mathbb{E}X - \mathbb{E}\mu = \mu - \mu = 0 ,
$$

as asserted. ∎

An important consequence of (8) in Proposition 5.1.36 reads as follows.

Corollary 5.1.39. *If $\mathbb{E}X$ exists, then*

$$
|\mathbb{E}X| \leq \mathbb{E}|X| .
$$

Proof: For all $\omega \in \Omega$ follows that

$$-|X(\omega)| \le X(\omega) \le |X(\omega)|,$$

that is, we have $-|X| \le X \le |X|$. We apply now (3) and (8) of Proposition 5.1.36 and conclude that

$$- \mathbb{E}|X| = \mathbb{E}(-|X|) \le \mathbb{E}X \le \mathbb{E}|X|. \tag{5.23}$$

Since $|a| \le c$ for $a, c \in \mathbb{R}$ is equivalent to $-c \le a \le c$, the desired estimate is a consequence of inequalities (5.23) with $a = \mathbb{E}X$ and $c = \mathbb{E}|X|$. $\blacksquare$

We now present some examples that show how Proposition 5.1.36 may be used to evaluate certain expected values.

Example 5.1.40. Suppose we roll n fair dice. Let S_n be the sum of the observed values. What is the expected value of S_n ?

Answer: If X_j denotes the value of die j, then $X_1, \ldots, X_n$ are uniformly distributed on $\{1, \ldots, 6\}$ with $\mathbb{E}X_j = 7/2$ and, moreover, $S_n = X_1 + \cdots + X_n$. Thus, property (3) lets us conclude that

$$\mathbb{E}S_n = \mathbb{E}(X_1 + \cdots + X_n) = \mathbb{E}X_1 + \cdots + \mathbb{E}X_n = \frac{7n}{2}.$$

Example 5.1.41. In Example 4.1.7 we investigated the random walk of a particle on $\mathbb{Z}$. Each time it jumped with probability p either one step to the right or with probability $1 - p$ one step to the left. S_n denoted the position of the particle after n steps. What is the expected position after n steps?

Answer: We proved that $S_n = 2Y_n - n$ with a $B_{n,p}$-distributed random variable Y_n. Proposition 5.1.13 implies $\mathbb{E}Y_n = np$, hence the linearity of the expected value leads to

$$\mathbb{E}S_n = 2\mathbb{E}Y_n - n = 2np - n = n(2p - 1).$$

For $p = 1/2$ we obtain the (not very surprising) result $\mathbb{E}S_n = 0$.

Alternative approach: If X_j is the size of jump j, then $\mathbb{P}\{X_j = -1\} = 1 - p$ and $\mathbb{P}\{X_j = +1\} = p$. Hence, $\mathbb{E}X_j = (-1)(1 - p) + 1 \cdot p = 2p - 1$, and because of $S_n = X_1 + \cdots + X_n$ we get $\mathbb{E}S_n = n \mathbb{E}X_1 = n(2p - 1)$ as before.

The next example demonstrates how property (4) of Proposition 5.1.36 may be used.

Example 5.1.42. Let X be Pois_λ-distributed. Find $\mathbb{E}X^2$.

Solution: Property (4) of Proposition 5.1.36 implies

$$\mathbb{E}X^2 = \sum_{k=0}^{\infty} k^2 \, \mathbb{P}\{X = k\} = \sum_{k=1}^{\infty} k^2 \frac{\lambda^k}{k!} \, e^{-\lambda} = \lambda \sum_{k=1}^{\infty} k \frac{\lambda^{k-1}}{(k-1)!} \, e^{-\lambda}.$$

We shift the index of summation in the right-hand sum by 1 and get

$$\lambda \sum_{k=0}^{\infty} (k+1) \frac{\lambda^k}{k!} e^{-\lambda} = \lambda \sum_{k=0}^{\infty} k \frac{\lambda^k}{k!} e^{-\lambda} + \lambda \sum_{k=0}^{\infty} \frac{\lambda^k}{k!} e^{-\lambda}.$$

By Proposition 5.1.16, the first sum coincides with $\lambda \mathbb{E} X = \lambda^2$, while the second one gives $\lambda \operatorname{Pois}_\lambda(\mathbb{N}_0) = \lambda \cdot 1 = \lambda$. Adding both values leads to

$$\mathbb{E} X^2 = \lambda^2 + \lambda.$$

The next example rests upon an application of properties (3), (4) and (6) in Proposition 5.1.36.

Example 5.1.43. Compute $\mathbb{E} X^2$ for X being $B_{n,p}$-distributed.

Solution: Let $X_1, \ldots, X_n$ be independent $B_{1,p}$-distributed random variables. Then Corollary 4.6.2 asserts that $X = X_1 + \cdots + X_n$ is $B_{n,p}$-distributed. Therefore, it suffices to evaluate $\mathbb{E} X^2$ with $X = X_1 + \cdots + X_n$. Thus, property (3) of Proposition 5.1.36 implies

$$\mathbb{E} X^2 = \mathbb{E}(X_1 + \cdots + X_n)^2 = \sum_{i=1}^{n} \sum_{j=1}^{n} \mathbb{E} X_i X_j.$$

If $i \neq j$, then X_i and X_j are independent, hence property (6) applies and yields

$$\mathbb{E} X_i X_j = \mathbb{E} X_i \cdot \mathbb{E} X_j = p \cdot p = p^2.$$

For $i = j$, property (4) gives

$$\mathbb{E} X_j^2 = 0^2 \cdot \mathbb{P}\{X_j = 0\} + 1^2 \cdot \mathbb{P}\{X_j = 1\} = p.$$

Combining both cases leads to

$$\mathbb{E} X^2 = \sum_{i \neq j} \mathbb{E} X_i \cdot \mathbb{E} X_j + \sum_{j=1}^{n} \mathbb{E} X_j^2 = n(n-1) p^2 + n p = n^2 p^2 + n p(1-p).$$

Example 5.1.44. Let X be G_p-distributed. Compute $\mathbb{E} X^2$.

Solution: We claim that

$$\mathbb{E} X^2 = \frac{2-p}{p^2}. \tag{5.24}$$

To prove this, let us start with

$$\mathbb{E} X^2 = \sum_{k=1}^{\infty} k^2 \, \mathbb{P}\{X = k\} = p \sum_{k=1}^{\infty} k^2 (1-p)^{k-1}. \tag{5.25}$$

We evaluate the right-hand sum by the following approach. If $|x| < 1$, then

$$\frac{1}{1-x} = \sum_{k=0}^{\infty} x^k .$$

Differentiating both sides of this equation leads to

$$\frac{1}{(1-x)^2} = \sum_{k=1}^{\infty} k x^{k-1} .$$

Next we multiply this equation by x and arrive at

$$\frac{x}{(1-x)^2} = \sum_{k=1}^{\infty} k x^k .$$

Another time differentiating of both functions on $\{x \in \mathbb{R} : |x| < 1\}$ implies

$$\frac{1}{(1-x)^2} + \frac{2x}{(1-x)^3} = \sum_{k=1}^{\infty} k^2 x^{k-1} .$$

If we use the last equation with $x = 1 - p$, then by eq. (5.25)

$$\mathbb{E}X^2 = p\left[\frac{1}{(1-(1-p))^2} + \frac{2(1-p)}{(1-(1-p))^3}\right] = \frac{2-p}{p^2} ,$$

as we claimed in eq. (5.24).

In the next example we us property (5) of Proposition 5.1.36.

Example 5.1.45. Let U be uniformly distributed on $[0, 1]$. Which expected value does $\sqrt{U}$ possess?

Solution: By property (5) it follows that

$$\mathbb{E}\sqrt{U} = \int_0^1 \sqrt{x}\, p(x)\, dx = \int_0^1 \sqrt{x}\, dx = \frac{2}{3}\left[x^{3/2}\right]_0^1 = \frac{2}{3} .$$

Here $p = \mathbb{1}_{[0,1]}$ denotes the distribution density of U.

Another approach is as follows. Because of

$$F_{\sqrt{U}}(t) = \mathbb{P}\{\sqrt{U} \le t\} = \mathbb{P}\{U \le t^2\} = t^2$$

for $0 \le t \le 1$, the density q of $\sqrt{U}$ is given by $q(x) = 2x$, $0 \le x \le 1$, and $q(x) = 0$, otherwise. Thus,

$$\mathbb{E}\sqrt{U} = \int_0^1 x\, 2x\, dx = 2\left[\frac{x^3}{3}\right]_0^1 = \frac{2}{3}.$$

Let us present now an interesting example called **Coupon collector's problem**. It was first mentioned in 1708 by A. De Moivre. We formulate it in a present-day version.

Example 5.1.46. A company produces cornflakes. Each pack contains a picture. We assume that there are n different pictures and that they are equally likely. That is, when buying a pack, the probability to get a certain fixed picture is $1/n$. How many packs of cornflakes have to be bought on the average before one gets all possible n pictures?

An equivalent formulation of the problem is as follows. In an urn are n balls numbered from 1 to n. One chooses balls out of the urn with replacement. How many balls have to be chosen on average before one observes all n numbers?

Answer: Assume we already have k different pictures for some $k = 0, 1, \ldots, n-1$. Let X_k be the number of necessary purchases to obtain a new picture, that is, to get one which we do not have. Since each pack contains a picture,

$$\mathbb{P}\{X_0 = 1\} = 1.$$

If $k \ge 1$, then there are still $n - k$ pictures that one does not possess. Hence, X_k is geometric distributed with success probability $p_k = (n-k)/n$. If $S_n = X_0 + \cdots + X_{n-1}$, then S_n is the totality of necessary purchases. By Corollary 5.1.20 we obtain

$$\mathbb{E}X_k = \frac{1}{p_k} = \frac{n}{n-k}, \quad k = 0, \ldots, n-1.$$

Note that $\mathbb{E}X_0 = 1$, thus the previous formula also holds in this case. Then the linearity of the expected value implies

$$\mathbb{E}S_n = 1 + \mathbb{E}X_1 + \cdots + \mathbb{E}X_{n-1} = 1 + \frac{1}{p_1} + \cdots + \frac{1}{p_{n-1}}$$

$$= 1 + \frac{n}{n-1} + \frac{n}{n-2} + \cdots + \frac{n}{1} = n \sum_{k=1}^n \frac{1}{k}.$$

Consequently, on average, one needs $n \sum_{k=1}^{n} \frac{1}{k}$ purchases to obtain a complete collection of all pictures.

For example, if $n = 50$, on average, we have to buy 225 packs, if $n = 100$, then 519, for $n = 200$, on average, there are 1176 purchases necessary, if $n = 300$, then 1885, if $n = 400$ we have to buy 2628 packs, and, finally, if $n = 500$ we need to buy 3397 ones.

Remark 5.1.47. As $n \to \infty$, the harmonic series $\sum_{k=1}^{n} \frac{1}{k}$ behaves like $\ln n$. More precisely (cf. [Lag13] or [Spi08], Problem 12, Chapter 22)

$$\lim_{n \to \infty} \left[\sum_{k=1}^{n} \frac{1}{k} - \ln n \right] = y, \tag{5.26}$$

where y denotes Euler's constant, which is approximately 0.57721. Therefore, for large n, the average number of necessary purchases is approximately $n [\ln n + y]$. For example, if $n = 300$, then the approximative value is 1884.29, leading also to 1885 necessary purchases.

5.2 Variance

5.2.1 Higher Moments of Random Variables

Definition 5.2.1. Let $n \geq 1$ be some integer. A random variable X possesses an **nth moment**, provided that $\mathbb{E}|X|^{n} < \infty$. We also say X has a finite **absolute nth moment**. If this is so, then $\mathbb{E}X^{n}$ exists, and it is called **nth moment** of X.

Remark 5.2.2. Because of $|X|^{n} = |X^{n}|$, the assumption $\mathbb{E}|X|^{n} < \infty$ implies the existence of the nth moment $\mathbb{E}X^{n}$.

Note that a random variable X has a first moment if and only if the expected value of X exists, cf. conditions (5.4) and (5.18). Moreover, then the first moment coincides with $\mathbb{E}X$.

Proposition 5.2.3. *Let X be either a discrete random variable with values in $\{x_1, x_2, \ldots\}$ and with $p_j = \mathbb{P}\{X = x_j\}$, or let X be continuous with density p. If $n \geq 1$, then*

$$\mathbb{E}|X|^{n} = \sum_{j=1}^{\infty} |x_j|^{n} \cdot p_j \quad or \quad \mathbb{E}|X|^{n} = \int_{-\infty}^{\infty} |x|^{n} p(x)\, dx. \tag{5.27}$$

Consequently, X possesses a finite absolute nth moment if and only if either the sum or the integral in eq. (5.27) are finite. If this is satisfied, then

$$\mathbb{E}X^n = \sum_{j=1}^{\infty} x_j^n \cdot p_j \quad or \quad \mathbb{E}X^n = \int_{-\infty}^{\infty} x^n p(x)\, dx.$$

Proof: Apply properties (4) and (5) in Proposition 5.1.36 with $f(x) = |x|^n$ or $f(x) = x^n$, respectively. ∎

Example 5.2.4. Let U be uniformly distributed on $[0, 1]$. Then

$$\mathbb{E}|U|^n = \mathbb{E}\, U^n = \int_0^1 x^n\, dx = \frac{1}{n+1}.$$

For the subsequent investigations, we need the following elementary lemma.

Lemma 5.2.5. *If $0 < \alpha < \beta$, then for all $x \geq 0$*

$$x^\alpha \leq x^\beta + 1.$$

Proof: If $0 \leq x \leq 1$, by $x^\beta \geq 0$ follows that

$$x^\alpha \leq 1 \leq x^\beta + 1,$$

and the inequality is valid.

If $x > 1$, then $\alpha < \beta$ implies $x^\alpha < x^\beta$, hence also for those x we arrive at

$$x^\alpha < x^\beta \leq x^\beta + 1,$$

which proves the lemma. ∎

Proposition 5.2.6. *Suppose a random variable X has a finite absolute nth moment. Then X possesses all mth moments with $m < n$.*

Proof: Suppose $\mathbb{E}|X|^n < \infty$ and choose an $m < n$. For fixed $\omega \in \Omega$ we apply Lemma 5.2.5 with $\alpha = m$, $\beta = n$ and $x = |X(\omega)|$. Doing so, we obtain

$$|X(\omega)|^m \leq |X(\omega)|^n + 1,$$

and this being true for all $\omega \in \Omega$ implies $|X|^m \leq |X|^n + 1$. Hence, property (7) of Proposition 5.1.36 yields

$$\mathbb{E}|X|^m \leq \mathbb{E}(|X|^n + 1) = \mathbb{E}|X|^n + 1 < \infty.$$

Consequently, as asserted, X possesses also an absolute mth moment.

Remark 5.2.7. There exist much stronger estimates between different absolute moments of X. For example, Hölder's inequality asserts that for any $0 < \alpha \le \beta$

$$\left[\mathbb{E}|X|^\alpha\right]^{1/\alpha} \le \left[\mathbb{E}|X|^\beta\right]^{1/\beta} .$$

∎

The case $n = 2$ and $m = 1$ in Proposition 5.2.6 is of special interest. Here we get the following useful result.

Corollary 5.2.8. *If X possesses a finite second moment, then $\mathbb{E}|X| < \infty$, that is, its expected value exists.*

Let us state another important consequence of Proposition 5.2.6.

Corollary 5.2.9. *Suppose X has a finite absolute nth moment. Then for any $b \in \mathbb{R}$ we also have $\mathbb{E}|X + b|^n < \infty$.*

Proof: An application of the binomial theorem (Proposition A.3.7) implies

$$|X + b|^n \le (|X| + |b|)^n = \sum_{k=0}^{n} \binom{n}{k} |X|^k |b|^{n-k} .$$

Hence, using properties (3) and (7) of Proposition 5.1.36, we obtain

$$\mathbb{E}|X + b|^n \le \sum_{k=0}^{n} \binom{n}{k} |b|^{n-k} \mathbb{E}|X|^k < \infty .$$

Note that Proposition 5.2.6 implies $\mathbb{E}|X|^k < \infty$ for all $k < n$. This ends the proof. ∎

Example 5.2.10. Let X be $\Gamma_{\alpha,\beta}$-distributed with parameters $\alpha, \beta > 0$. Which moments does X possess, and how can they be computed?
Answer: In view of $X \ge 0$ it suffices to investigate $\mathbb{E}X^n$. For all $n \ge 1$ it follows that

$$\mathbb{E}X^n = \frac{1}{\alpha^\beta \Gamma(\beta)} \int_0^\infty x^{n+\beta-1} e^{-x/\alpha} \, dx = \frac{\alpha^{n+\beta}}{\alpha^\beta \Gamma(\beta)} \int_0^\infty y^{n+\beta-1} e^{-y} \, dy$$

$$= \alpha^n \frac{\Gamma(\beta + n)}{\Gamma(\beta)} = \alpha^n \, (\beta + n - 1)(\beta + n - 2) \cdots (\beta + 1)\beta .$$

In particular, X has moments of any order $n \ge 1$.

In the case of an E_λ-distributed random variable X we have $\alpha = 1/\lambda$ and $\beta = 1$, hence

$$\mathbb{E}X^n = \frac{n!}{\lambda^n}.$$

Example 5.2.11. Suppose a random variable is t_n-distributed. Which moments does X possess?

Answer: We already know that a t_1-distributed random variable does not possess a first moment. Recall that X is t_1-distributed if it is Cauchy distributed. And in Proposition 5.1.33 we proved $\mathbb{E}|X| = \infty$ for Cauchy distributed random variables.

But what can be said if $n \geq 2$?

According to Definition 4.7.6, the random variable X has the density p with

$$p(x) = \frac{\Gamma\left(\frac{n+1}{2}\right)}{\sqrt{n\pi}\,\Gamma\left(\frac{n}{2}\right)}\left(1+\frac{x^2}{n}\right)^{-n/2-1/2}, \quad x \in \mathbb{R}.$$

If $m \in \mathbb{N}$, then

$$\mathbb{E}|X|^m = \frac{\Gamma\left(\frac{n+1}{2}\right)}{\sqrt{n\pi}\,\Gamma\left(\frac{n}{2}\right)}\int_{-\infty}^{\infty}|x|^m\left(1+\frac{x^2}{n}\right)^{-n/2-1/2}dx.$$

Hence, X has an mth moment, if and only if the integral

$$\int_{-\infty}^{\infty}|x|^m\left(1+\frac{x^2}{n}\right)^{-n/2-1/2}dx = 2\int_{0}^{\infty}x^m\left(1+\frac{x^2}{n}\right)^{-n/2-1/2}dx \qquad (5.28)$$

is finite. Note that

$$\lim_{x\to\infty}x^{n+1}\left(1+\frac{x^2}{n}\right)^{-n/2-1/2} = \lim_{x\to\infty}\left(x^{-2}+\frac{1}{n}\right)^{-n/2-1/2} = n^{n/2+1/2},$$

thus, there are constants $0 < c_1 < c_2$ (depending on n, but not on x) such that

$$\frac{c_1}{x^{n-m+1}} \leq x^m\left(1+\frac{x^2}{n}\right)^{-n/2-1/2} \leq \frac{c_2}{x^{n-m+1}} \qquad (5.29)$$

for large x, that is, if $x > x_0$ for a suitable $x_0 \in \mathbb{R}$.

Recall that $\int_1^{\infty}x^{-\alpha}\,dx < \infty$ if and only if $\alpha > 1$. Having this in mind, in view of eq. (5.28) and by the estimates in (5.29) we get $\mathbb{E}|X|^m < \infty$ if and only if $n - m + 1 > 1$, that is, if and only if $m < n$.

Summing up, a t_n-distributed random variable has moments of order $1, \ldots, n-1$, but no moments of order greater than or equal n.

Finally, let us investigate the moments of normally distributed random variables.

Example 5.2.12. How do we calculate $\mathbb{E}X^n$ for an $\mathcal{N}(0, 1)$-distributed random variable?

Answer: Well-known properties of the exponential function imply

$$\mathbb{E}|X|^n = \frac{1}{\sqrt{2\pi}} \int_{-\infty}^{\infty} |x|^n \, e^{-x^2/2} \, dx = \frac{2}{\sqrt{2\pi}} \int_{0}^{\infty} x^n \, e^{-x^2/2} \, dx < \infty$$

for all $n \in \mathbb{N}$. Thus, a normally distributed random variable possesses moments of any order. These moments are evaluated by

$$\mathbb{E}X^n = \int_{-\infty}^{\infty} x^n \, p_{0,1}(x) \, dx = \frac{1}{\sqrt{2\pi}} \int_{-\infty}^{\infty} x^n \, e^{-x^2/2} \, dx .$$

If n is an odd integer, then $x \mapsto x^n \, e^{-x^2/2}$ is an odd function, hence $\mathbb{E}X^n = 0$ for these n. Therefore, it suffices to investigate even $n = 2m$ with $m \in \mathbb{N}$. Here we get

$$\mathbb{E}X^{2m} = 2 \cdot \frac{1}{\sqrt{2\pi}} \int_{0}^{\infty} x^{2m} \, e^{-x^2/2} \, dx ,$$

which, by the change of variables $y := x^2/2$, thus $x = \sqrt{2y}$ with $dx = \frac{1}{\sqrt{2}} y^{-1/2} \, dy$, transforms into

$$\mathbb{E}X^{2m} = \frac{1}{\sqrt{\pi}} 2^m \int_{0}^{\infty} y^{m-1/2} \, e^{-y} \, dy = \frac{2^m}{\sqrt{\pi}} \Gamma\left(m + \frac{1}{2}\right) .$$

By $\Gamma(1/2) = \sqrt{\pi}$ and an application of eq. (1.48) we finally obtain

$$\begin{aligned}
\mathbb{E}X^{2m} &= \frac{2^m}{\sqrt{\pi}} \Gamma\left(m + \frac{1}{2}\right) = \frac{2^m}{\sqrt{\pi}} \left(m - \frac{1}{2}\right) \Gamma\left(m - \frac{1}{2}\right) \\
&= \frac{2^m}{\sqrt{\pi}} \left(m - \frac{1}{2}\right)\left(m - \frac{3}{2}\right) \Gamma\left(m - \frac{3}{2}\right) \\
&= \frac{2^m \Gamma(1/2) \cdot 1/2 \cdot 3/2 \cdots (m - 1/2)}{\sqrt{\pi}} \\
&= (2m - 1)(2m - 3) \cdots 3 \cdot 1 := (2m - 1)!! .
\end{aligned}$$

5.2.2 Variance of Random Variables

Let X be a random variable with finite second moment. As we saw in Corollary 5.2.8, then its expected value $\mu := \mathbb{E}X$ exists. Furthermore, letting $b = -\mu$, by Corollary 5.2.9, we also have $\mathbb{E}|X - \mu|^2 < \infty$. After this preparation we can introduce the variance of a random variable.

Definition 5.2.13. Let X be a random variable possessing a finite second moment. If $\mu := \mathbb{E}X$, then its **variance** is defined as

$$\mathbb{V}X := \mathbb{E}|X - \mu|^2 = \mathbb{E}|X - \mathbb{E}X|^2 .$$

Interpretation: The expected value μ of a random variable is its main characteristic. It tells us around which value the observations of X have to be expected. But it does not tell us how far away from μ these observations will be on average. Are they concentrated around μ or are they widely dispersed? This behavior is described by the variance. It is defined as the average quadratic distance of X to its mean value. If $\mathbb{V}X$ is small, then we will observe realizations of X quite near to its mean. Otherwise, if $\mathbb{V}X$ is large, then it is very likely to observe values of X far away from its expected value.

How do we evaluate the variance in concrete cases? We answer this question for discrete and continuous random variables separately.

Proposition 5.2.14. *Let X be a random variable with finite second moment and let $\mu \in \mathbb{R}$ be its expected value. Then it follows that*

$$\mathbb{V}X = \sum_{j=1}^{\infty}(x_j - \mu)^2 \cdot p_j \quad and \quad \mathbb{V}X = \int_{-\infty}^{\infty} (x - \mu)^2 p(x)\, dx \tag{5.30}$$

in the discrete and continuous case, respectively. Hereby, $x_1, x_2, \ldots$ are the possible values of X and $p_j = \mathbb{P}\{X = x_j\}$ in the discrete case, while p denotes the density of X in the continuous case.

Proof: The assertion follows directly by an application of properties (4) and (5) of Proposition 5.1.36 to $f(x) = (x - \mu)^2$. ∎

Before we present concrete examples, let us state and prove certain properties of the variance, which will simplify the calculations later on.

Proposition 5.2.15. *Assume X and Y are random variables with finite second moment. Then the following are valid.*

(i) *We have*

$$\mathbb{V}X = \mathbb{E}X^2 - (\mathbb{E}X)^2 . \tag{5.31}$$

(ii)   If $\mathbb{P}\{X = c\} = 1$ for some $c \in \mathbb{R}$, then[5] $\mathbb{V}X = 0$.

(iii)   For $a, b \in \mathbb{R}$ follows that

$$\mathbb{V}(aX + b) = a^2\,\mathbb{V}X\,.$$

(iv)   In the case of **independent** X and Y one has

$$\mathbb{V}(X + Y) = \mathbb{V}X + \mathbb{V}Y\,.$$

Proof: Let us begin with the proof of (i). With $\mu = \mathbb{E}X$ we obtain

$$\mathbb{V}X = \mathbb{E}(X - \mu)^2 = \mathbb{E}\left[X^2 - 2\mu X + \mu^2\right] = \mathbb{E}X^2 - 2\mu\mathbb{E}X + \mu^2$$
$$= \mathbb{E}X^2 - 2\mu^2 + \mu^2 = \mathbb{E}X^2 - \mu^2\,.$$

This proves (i).

To verify (ii) we use property (2) in Proposition 5.1.36. Then $\mu = \mathbb{E}X = c$, hence $\mathbb{P}\{X - \mu = 0\} = 1$. Another application of property (2) leads to

$$\mathbb{V}X = \mathbb{E}(X - \mu)^2 = 0$$

as asserted.

Next we prove (iii). If $a, b \in \mathbb{R}$, then $\mathbb{E}(aX + b) = a\mathbb{E}X + b$ by the linearity of the expected value. Consequently,

$$\mathbb{V}(aX + b) = \mathbb{E}\left[aX + b - (a\mathbb{E}X + b)\right]^2 = a^2\,\mathbb{E}(X - \mathbb{E}X)^2 = a^2\,\mathbb{V}X\,.$$

Thus (iii) is valid.

To prove (iv) observe that, if $\mu := \mathbb{E}X$ and $v := \mathbb{E}Y$, then $\mathbb{E}(X + Y) = \mu + v$, and hence

$$\begin{aligned}
\mathbb{V}(X + Y) &= \mathbb{E}\left[(X - \mu) + (Y - v)\right]^2 \\
&= \mathbb{E}(X - \mu)^2 + 2\,\mathbb{E}[(X - \mu)(Y - v)] + \mathbb{E}(Y - v)^2 \\
&= \mathbb{V}X + 2\,\mathbb{E}[(X - \mu)(Y - v)] + \mathbb{V}Y\,. \tag{5.32}
\end{aligned}$$

By Proposition 4.1.9 the independence of X and Y implies that of $X - \mu$ and $Y - v$. Therefore, from property (6) in Proposition 5.1.36 we derive

$$\mathbb{E}[(X - \mu)(Y - v)] = \mathbb{E}(X - \mu) \cdot \mathbb{E}(Y - v) = (\mathbb{E}X - \mu) \cdot (\mathbb{E}Y - v) = 0 \cdot 0 = 0\,.$$

Plugging this into eq. (5.32) completes the proof of (iv).   ∎

---

5 The converse implication is also true. If $\mathbb{V}X = 0$, then X is constant with probability 1.

5.2.3 Variance of Certain Random Variables

Our first objective is to describe the variance of a random variable uniformly distributed on a finite set.

Proposition 5.2.16. *If X is uniformly distributed on $\{x_1, \ldots, x_N\}$, then*

$$\mathbb{V}X = \frac{1}{N} \sum_{j=1}^{N} (x_j - \mu)^2,$$

where μ is given by $\mu = \frac{1}{N} \sum_{j=1}^{N} x_j$.

Proof: Because of $p_j = \frac{1}{N}$, $1 \le j \le N$, this is a direct consequence of eq. (5.30). Recall that μ was computed in eq. (5.6). ∎

Example 5.2.17. Suppose X is uniformly distributed on $\{1, \ldots, 6\}$. Then $\mathbb{E}X = 7/2$, and we get

$$\mathbb{V}X = \frac{\left(1 - \frac{7}{2}\right)^2 + \left(2 - \frac{7}{2}\right)^2 + \left(3 - \frac{7}{2}\right)^2 + \left(4 - \frac{7}{2}\right)^2 + \left(5 - \frac{7}{2}\right)^2 + \left(6 - \frac{7}{2}\right)^2}{6}$$

$$= \frac{\frac{25}{4} + \frac{9}{4} + \frac{1}{4} + \frac{1}{4} + \frac{9}{4} + \frac{25}{4}}{6} = \frac{35}{12}.$$

Thus, when rolling a die once, the variance is given by $\frac{35}{12}$.

Now assume that we roll the die n times. Let $X_1, \ldots, X_n$ be the results of the single rolls. The X_js are independent, hence, if $S_n = X_1 + \cdots + X_n$ denotes the sum of the n trials, then by (iv) in Proposition 5.2.15 it follows that

$$\mathbb{V}S_n = \mathbb{V}(X_1 + \cdots + X_n) = \mathbb{V}X_1 + \cdots \mathbb{V}X_n = \frac{35\,n}{12}.$$

The next proposition examines the variance of binomial distributed random variables.

Proposition 5.2.18. *If X is $B_{n,p}$-distributed, then*

$$\mathbb{V}X = np(1 - p). \tag{5.33}$$

Proof: Let X be $B_{n,p}$-distributed. In Example 5.1.44 we found $\mathbb{E}X^2 = n^2 p^2 + np(1 - p)$. Moreover, $\mathbb{E}X = np$ by Proposition 5.1.13. Thus, from formula (5.31) we derive

$$\mathbb{V}X = \mathbb{E}X^2 - (\mathbb{E}X)^2 = n^2 p^2 + np(1 - p) - (np)^2 = np(1 - p)$$

as asserted. ∎

Corollary 5.2.19. *Binomial distributed random variables have maximal variance (with n fixed) if p = 1/2.*

Proof: The function $p \mapsto np(1 - p)$ becomes maximal for $p = \frac{1}{2}$. In the extreme cases $p = 0$ and $p = 1$ the variance is zero. ∎

Next we determine the variance of Poisson distributed random variables.

Proposition 5.2.20. *Let X be Pois_λ-distributed for some $\lambda > 0$. Then*

$$\mathbb{V}X = \lambda.$$

Proof: In Example 5.1.42 we computed $\mathbb{E}X^2 = \lambda^2 + \lambda$. Furthermore, by Proposition 5.1.16 we know that $\mathbb{E}X = \lambda$. Thus, by eq. (5.31) we obtain, as asserted,

$$\mathbb{V}X = \mathbb{E}X^2 - (\mathbb{E}X)^2 = \lambda^2 + \lambda - \lambda^2 = \lambda.$$ ∎

Next, we compute the variance of a geometric distributed random variable.

Proposition 5.2.21. *Let X be G_p-distributed for some $0 < p < 1$. Then its variance equals*

$$\mathbb{V}X = \frac{1-p}{p^2}.$$

Proof: In Example 5.1.44 we found $\mathbb{E}X^2 = \frac{2-p}{p^2}$, and by eq. (5.14) we have $\mathbb{E}X = \frac{1}{p}$. Consequently, formula (5.31) implies

$$\mathbb{V}X = \frac{2-p}{p^2} - \left(\frac{1}{p}\right)^2 = \frac{1-p}{p^2},$$

as asserted. ∎

Corollary 5.2.22. *If X is $B_{n,p}^-$-distributed, then*

$$\mathbb{V}X = n\frac{1-p}{p^2}$$

Proof: Let $X_1, \ldots, X_n$ be independent G_p-distributed random variables. By Corollary 4.6.9 their sum $X := X_1 + \cdots + X_n$ is $B_{n,p}^-$-distributed, hence property (iv) in Proposition 5.2.15 lets us conclude that

$$\mathbb{V}X = \mathbb{V}(X_1 + \cdots + X_n) = \mathbb{V}X_1 + \cdots + \mathbb{V}X_n = n\mathbb{V}X_1 = n\frac{1-p}{p^2}.$$ ∎

Interpretation: The smaller p becomes the bigger is the variance of a geometrically or negative binomially distributed random variable (for n fixed). This is not surprising, because the smaller p is, the larger is the expected value, and the values of X may be very far from $1/p$ (success is very unlikely).

We consider now variances of continuous random variables. Let us begin with uniformly distributed ones.

Proposition 5.2.23. *Let X be uniformly distributed on an interval $[\alpha, \beta]$. Then it follows that*

$$\mathbb{V}X = \frac{(\beta - \alpha)^2}{12}.$$

Proof: We know by Proposition 5.1.25 that $\mathbb{E}X = (\alpha + \beta)/2$. In order to apply formula (5.31), we still have to compute the second moment $\mathbb{E}X^2$. Here we obtain

$$\mathbb{E}X^2 = \frac{1}{\beta - \alpha} \int_\alpha^\beta x^2 \, dx = \frac{1}{3} \cdot \frac{\beta^3 - \alpha^3}{\beta - \alpha} = \frac{\beta^2 + \alpha\beta + \alpha^2}{3}.$$

Consequently, formula (5.31) lets us conclude that

$$\mathbb{V}X = \mathbb{E}X^2 - (\mathbb{E}X)^2 = \frac{\beta^2 + \alpha\beta + \alpha^2}{3} - \left(\frac{\alpha + \beta}{2}\right)^2$$
$$= \frac{\beta^2 + \alpha\beta + \alpha^2}{3} - \frac{\alpha^2 + 2\alpha\beta + \beta^2}{4}$$
$$= \frac{\alpha^2 - 2\alpha\beta + \beta^2}{12} = \frac{(\beta - \alpha)^2}{12}.$$

This completes the proof. ∎

In the case of gamma distributed random variables, the following is valid.

Proposition 5.2.24. *If X is $\Gamma_{\alpha,\beta}$-distributed, then*

$$\mathbb{V}X = \alpha^2\beta.$$

Proof: Recall that $\mathbb{E}X = \alpha\beta$ by Proposition 5.1.26. Furthermore, in Example 5.2.10 we evaluated $\mathbb{E}X^n$ for a gamma distributed X. Taking $n = 2$ implies

$$\mathbb{E}X^2 = \alpha^2 (\beta + 1)\beta,$$

and, hence, by eq. (5.31),

$$\mathbb{V}X = \mathbb{E}X^2 - (\mathbb{E}X)^2 = \alpha^2 (\beta + 1)\beta - (\alpha\beta)^2 = \alpha^2\beta$$

as asserted. ∎

Corollary 5.2.25. *If X is E_λ-distributed, then*

$$\mathbb{V}X = \frac{1}{\lambda^2}\,.$$

Proof: Because of $E_\lambda = \Gamma_{\frac{1}{\lambda},1}$, this directly follows from Proposition 5.2.24. ∎

Corollary 5.2.26. *For a χ_n^2-distributed X holds*

$$\mathbb{V}X = 2n\,.$$

Proof: Let us give two alternative proofs of the assertion. The first one uses Proposition 5.2.24 and $\chi_n^2 = \Gamma_{2,\frac{n}{2}}$.

The second proof is longer, but maybe more interesting. Let $X_1, \ldots, X_n$ be independent $\mathcal{N}(0,1)$-distributed random variables. Proposition 4.6.17 implies that $X_1^2 + \cdots + X_n^2$ is χ_n^2-distributed, thus property (iv) of Proposition 5.2.15 applies and leads to

$$\mathbb{V}X = \mathbb{V}X_1^2 + \cdots + \mathbb{V}X_n^2 = n\mathbb{V}X_1^2\,.$$

In Example 5.2.12 we evaluated the moments of an $\mathcal{N}(0,1)$-distributed random variable. In particular, $\mathbb{E}X_1^2 = 1$ and, $\mathbb{E}(X_1^2)^2 = \mathbb{E}X_1^4 = 3!! = 3$, hence

$$\mathbb{V}X = n\,\mathbb{V}X_1^2 = n\left(\mathbb{E}X_1^4 - (\mathbb{E}X_1^2)^2\right) = (3-1)n = 2n$$

as claimed. ∎

Finally we determine the variance of a normal random variable.

Proposition 5.2.27. *If X is $\mathcal{N}(\mu, \sigma^2)$-distributed, then it follows that*

$$\mathbb{V}X = \sigma^2\,.$$

Proof: Of course, this could be proven by computing the integral

$$\mathbb{V}X = \int_{-\infty}^{\infty} (x - \mu)^2 p_{\mu,\sigma}(x)dx\,.$$

We prefer a different approach that avoids the calculation of integrals. Because of Proposition 4.2.3, the random variable X may be represented as $X = \sigma X_0 + \mu$ for a standard normal X_0. Applying (iii) in Proposition 5.2.15 gives

$$\mathbb{V}X = \sigma^2\,\mathbb{V}X_0\,. \tag{5.34}$$

But $\mathbb{E}X_0 = 0$, and by Example 5.2.12 we have $\mathbb{E}X_0^2 = 1$, thus

$$\mathbb{V}X_0 = 1 - 0 = 1.$$

Plugging this into eq. (5.34) proves $\mathbb{V}X = \sigma^2$. ∎

Remark 5.2.28. The previous result explains why the parameter σ^2 of an $\mathcal{N}(\mu, \sigma^2)$-distribution is called "variance."

5.3 Covariance and Correlation

5.3.1 Covariance

Suppose we know or we conjecture that two given random variables X and Y are dependent. The aim of this section is to introduce a quantity that measures their degree of dependence. Such a quantity should tell us whether the random variables are strongly or only weakly dependent. Furthermore, we want to know what kind of dependence we observe. Do larger values of X trigger larger values of Y or is it the other way round? To illustrate these questions let us come back to the experiment presented in Example 2.2.5.

Example 5.3.1. In an urn are n balls labeled with "0" and another n balls labeled with "1." Choose two balls out of the urn **without** replacement Let X be the number appearing on the first ball and Y that on the second. Then X and Y are dependent (check this), but it is intuitively clear that if n becomes larger, then their dependence diminishes. We ask for a quantity that tells us their degree of dependence. This measure should decrease as n increases and it should tend to zero as $n \to \infty$.

Moreover, if $X = 1$ occurred, then there remained in the urn more balls with "0" than with "1," and the probability of the event $\{Y = 0\}$ increases. Thus, larger values of X make smaller values of Y more likely.

Before we are able to introduce such a "measure of dependence," we need some preparation.

Proposition 5.3.2. *If two random variables X and Y possess a finite second moment, then the expected value of their product $X Y$ exists.*

Proof: We use the elementary estimate $|ab| \le \frac{a^2+b^2}{2}$ valid for $a, b \in \mathbb{R}$. Thus, if $\omega \in \Omega$, then

$$|X(\omega)Y(\omega)| \le \frac{X(\omega)^2}{2} + \frac{Y(\omega)^2}{2},$$

that is, we have

$$|XY| \le \frac{X^2}{2} + \frac{Y^2}{2}.$$ (5.35)

By assumption

$$\mathbb{E}\left[\frac{X^2}{2} + \frac{Y^2}{2}\right] = \frac{1}{2}\left[\mathbb{E}X^2 + \mathbb{E}Y^2\right] < \infty,$$

consequently, because of estimate (5.35), property (7) in Proposition 5.1.36 applies and tells us that $\mathbb{E}|XY| < \infty$. Thus, $\mathbb{E}[XY]$ exists as asserted. ■

How do we compute $\mathbb{E}[XY]$ for given X and Y ? In Section 4.5 we observed that the distribution of $X + Y$ does not only depend on the distributions of X and Y. We have to know their **joint** distribution, that is, the distribution of the vector (X, Y). And the same is true for products and the expected value of the product.

Example 5.3.3. Let us again investigate the random variables X, Y, X', and Y' introduced in Example 3.5.8. Recall that they satisfied

$$\mathbb{P}\{X = 0, Y = 0\} = \frac{1}{6}, \quad \mathbb{P}\{X = 0, Y = 1\} = \frac{1}{3},$$

$$\mathbb{P}\{X = 1, Y = 0\} = \frac{1}{3}, \quad \mathbb{P}\{X = 1, Y = 1\} = \frac{1}{6},$$

$$\mathbb{P}\{X' = 0, Y' = 0\} = \frac{1}{4}, \quad \mathbb{P}\{X' = 0, Y' = 1\} = \frac{1}{4},$$

$$\mathbb{P}\{X' = 1, Y' = 0\} = \frac{1}{4}, \quad \mathbb{P}\{X' = 1, Y' = 1\} = \frac{1}{4}.$$

Then $\mathbb{P}_X = \mathbb{P}_{X'}$ as well as $\mathbb{P}_Y = \mathbb{P}_{Y'}$, but

$$\mathbb{E}[XY] = \frac{1}{6}(0 \cdot 0) + \frac{1}{3}(1 \cdot 0) + \frac{1}{3}(0 \cdot 1) + \frac{1}{6}(1 \cdot 1) = \frac{1}{6} \quad \text{and}$$

$$\mathbb{E}[X'Y'] = \frac{1}{4}(0 \cdot 0) + \frac{1}{4}(1 \cdot 0) + \frac{1}{4}(0 \cdot 1) + \frac{1}{4}(1 \cdot 1) = \frac{1}{4}.$$

This example tells us that we have to know the joint distribution in order to compute $\mathbb{E}[XY]$. The knowledge of the marginal distributions does not suffice.

To evaluate $\mathbb{E}[XY]$ we need the following two-dimensional generalization of formulas (5.21) and (5.22).

Proposition 5.3.4. *Let X and Y be two random variables and let $f : \mathbb{R}^2 \to \mathbb{R}$ be some function.*

1. *Suppose X and Y are discrete with values in $\{x_1, x_2, \ldots\}$ and in $\{y_1, y_2, \ldots\}$. Set $p_{ij} = \mathbb{P}\{X = x_i, Y = y_j\}$. If*

$$\mathbb{E}|f(X, Y)| = \sum_{i,j=1}^{\infty} |f(x_i, y_j)|\, p_{ij} < \infty, \tag{5.36}$$

then $Ef(X, Y)$ exists and can be computed by

$$\mathbb{E}f(X, Y) = \sum_{i,j=1}^{\infty} f(x_i, y_j)\, p_{ij}.$$

2. *Let $f : \mathbb{R}^2 \to \mathbb{R}$ be continuous[6]. If $p : \mathbb{R}^2 \to \mathbb{R}$ is the joint density of (X, Y) (recall Definition 3.5.15), then*

$$\mathbb{E}|f(X, Y)| = \int_{-\infty}^{\infty} \int_{-\infty}^{\infty} |f(x, y)|\, p(x, y)\, dxdy < \infty \tag{5.37}$$

implies the existence of $\mathbb{E}f(X, Y)$, which can be evaluated by

$$\mathbb{E}f(X, Y) = \int_{-\infty}^{\infty} \int_{-\infty}^{\infty} f(x, y)\, p(x, y)\, dxdy. \tag{5.38}$$

Remark 5.3.5. The previous formulas extend easily to higher dimensions. That is, if $\vec{X} = (X_1, \ldots, X_n)$ is an n-dimensional random vector with (joint) distribution density $p : \mathbb{R}^n \to \mathbb{R}$, then for continuous[7] $f : \mathbb{R}^n \to \mathbb{R}$ one has

$$\mathbb{E}f(\vec{X}) = \mathbb{E}f(X_1, \ldots, X_n) = \int_{\mathbb{R}^n} f(x_1, \ldots, x_n)\, p(x_1, \ldots, x_n)\, dx_n \cdots dx_1$$

provided the integral exists. The case of discrete $X_1, \ldots, X_n$ is treated in a similar way. If $\vec{X}$ maps into the finite or uncountably infinite set $D \subset \mathbb{R}^n$, then

$$\mathbb{E}f(\vec{X}) = \mathbb{E}f(X_1, \ldots, X_n) = \sum_{x \in D} f(x)\mathbb{P}\{\vec{X} = x\}.$$

---

6 In fact we need only a measurability in the sense of 4.1.1, but this time for functions f from $\mathbb{R}^2$ to $\mathbb{R}$. For our purposes "continuity" of f suffices.

7 cf. the remark for $n = 2$.

If we apply Proposition 5.3.4 with $f : (x, y) \mapsto x \cdot y$, then we obtain the following formulas for the evaluation of $\mathbb{E}[XY]$. Hereby, we assume that conditions (5.36) or (5.37) are satisfied.

Corollary 5.3.6. *In the notation of Proposition 5.3.4 the following are valid:*

$$\mathbb{E}[XY] = \sum_{i,j=1}^{\infty} (x_i \cdot y_j) p_{ij} \quad \text{and} \quad \mathbb{E}[XY] = \int_{-\infty}^{\infty} \int_{-\infty}^{\infty} (x \cdot y) p(x, y) \, dx dy$$

in the discrete and in the continuous case, respectively.

After all these preparations we are now in position to introduce the covariance of two random variables.

> **Definition 5.3.7.** Let X and Y be two random variables with finite second moments. Setting $\mu = \mathbb{E}X$ and $v = \mathbb{E}Y$, the **covariance** of X and Y is defined as
>
> $$\mathrm{Cov}(X, Y) = \mathbb{E}[(X - \mu)(Y - v)].$$

Remark 5.3.8. Apply Corollary 5.2.9 and Proposition 5.3.2 to see that the covariance is well-defined for random variables with finite second moment. Furthermore, in view of Proposition 5.3.4, the covariance may be computed as

$$\mathrm{Cov}(X, Y) = \sum_{i,j=1}^{\infty} (x_i - \mu)(y_j - v) p_{ij}$$

in the discrete case (recall that $p_{ij} = \mathbb{P}\{X = x_i, Y = y_j\}$), and as

$$\mathrm{Cov}(X, Y) = \int_{-\infty}^{\infty} \int_{-\infty}^{\infty} (x - \mu)(y - v) p(x, y) \, dx dy$$

in the continuous one.

Example 5.3.9. Let us once more consider the random variables X, Y, X', and Y' in Example 3.5.8 or Example 5.3.3, respectively. Each of the four random variables has the expected value 1/2. Therefore, we obtain

$$\mathrm{Cov}(X, Y) = \frac{1}{6} \left(0 - \frac{1}{2}\right) \cdot \left(0 - \frac{1}{2}\right) + \frac{1}{3} \left(1 - \frac{1}{2}\right) \cdot \left(0 - \frac{1}{2}\right)$$
$$+ \frac{1}{3} \left(0 - \frac{1}{2}\right) \cdot \left(1 - \frac{1}{2}\right) + \frac{1}{6} \left(1 - \frac{1}{2}\right) \cdot \left(1 - \frac{1}{2}\right) = -\frac{1}{12},$$

while

$$\text{Cov}(X', Y') = \frac{1}{4}\left(0 - \frac{1}{2}\right)\cdot\left(0 - \frac{1}{2}\right) + \frac{1}{4}\left(1 - \frac{1}{2}\right)\cdot\left(0 - \frac{1}{2}\right)$$
$$+ \frac{1}{4}\left(0 - \frac{1}{2}\right)\cdot\left(1 - \frac{1}{2}\right) + \frac{1}{4}\left(1 - \frac{1}{2}\right)\cdot\left(1 - \frac{1}{2}\right) = 0.$$

The following proposition summarizes the main properties of the covariance.

Proposition 5.3.10. *Let X and Y be random variables with finite second moments. Then the following are valid.*
(1) $\text{Cov}(X, Y) = \text{Cov}(Y, X)$.
(2) $\text{Cov}(X, X) = \mathbb{V}X$.
(3) *The covariance is bilinear, that is, for X_1, X_2 and real numbers a_1 and a_2*

$$\text{Cov}(a_1 X_1 + a_2 X_2, Y) = a_1\text{Cov}(X_1, Y) + a_2\text{Cov}(X_2, Y)$$

and, analogously,

$$\text{Cov}(X, b_1 Y_1 + b_2 Y_2) = b_1\text{Cov}(X, Y_1) + b_2\text{Cov}(X, Y_2)$$

for random variables Y_1, Y_2 and real numbers b_1, b_2.
(4) *The covariance may also evaluated by*

$$\text{Cov}(X, Y) = \mathbb{E}[XY] - (\mathbb{E}X)(\mathbb{E}Y). \tag{5.39}$$

(5) $\text{Cov}(X, Y) = 0$ *for independent X and Y.*

Proof: Properties (1) and (2) follow directly from the definition of the covariance.
Let us verify (3). Setting $\mu_1 = \mathbb{E}X_1$ and $\mu_2 = \mathbb{E}X_2$, the linearity of the expected value implies

$$\mathbb{E}(a_1 X_1 + a_2 X_2) = a_1\mu_1 + a_2\mu_2.$$

Hence, if $v = \mathbb{E}Y$, then

$$\begin{aligned}\text{Cov}(a_1 X_1 + a_2 X_2, Y) &= \mathbb{E}\left[(a_1(X_1 - \mu_1) + a_2(X_2 - \mu_2))(Y - v)\right] \\ &= a_1\mathbb{E}\left[(X_1 - \mu_1)(Y - v)\right] + a_2\mathbb{E}\left[(X_2 - \mu_2)(Y - v)\right] \\ &= a_1\text{Cov}(X_1, Y) + a_2\text{Cov}(X_2, Y).\end{aligned}$$

This proves the first part of (3). The second part can be proven in the same way or one uses $\text{Cov}(X, Y) = \text{Cov}(Y, X)$ and the first part of (3).

Next we prove eq. (5.39). With $\mu = \mathbb{E}X$ and $v = \mathbb{E}Y$ by

$$(X - \mu)(Y - v) = XY - \mu Y - vX + \mu v,$$

we get that

$$\mathrm{Cov}(X, Y) = \mathbb{E}\left[XY - \mu Y - vX + \mu v\right] = \mathbb{E}[XY] - \mu\mathbb{E}Y - v\mathbb{E}X + \mu v$$
$$= \mathbb{E}[XY] - \mu v.$$

This proves (4) by the definition of μ and v.

Finally, we verify (5). If X and Y are independent, then by Proposition 4.1.9 this is also true for $X - \mu$ and $Y - v$. Thus, property (6) of Proposition 5.1.36 applies and leads to

$$\mathrm{Cov}(X, Y) = \mathbb{E}\left[(X - \mu)(Y - v)\right] = \mathbb{E}(X - \mu)\,\mathbb{E}(Y - v) = [\mathbb{E}X - \mu][\mathbb{E}Y - v] = 0.$$

Therefore, the proof is completed. ∎

Remark 5.3.11. Quite often the computation of $\mathrm{Cov}(X, Y)$ can be simplified by the use of eq. (5.39). For example, consider X and Y in Example 3.5.8. In Example 5.3.3 we found $\mathbb{E}[XY] = 1/6$. Since $\mathbb{E}X = \mathbb{E}Y = 1/2$, by eq. (5.39) we immediately get

$$\mathrm{Cov}(X, Y) = \frac{1}{6} - \frac{1}{4} = -\frac{1}{12}.$$

We obtained the same result in Example 5.3.9 with slightly more efforts.

Property (5) in Proposition 5.3.10 is of special interest. It asserts $\mathrm{Cov}(X, Y) = 0$ for independent X and Y. One may ask now whether this characterizes independent random variables. More precisely, are the random variables X and Y independent if and only if $\mathrm{Cov}(X, Y) = 0$?

The answer is negative as the next example shows.

Example 5.3.12. The joint distribution of X and Y is given by the following table:

$Y\backslash X$	-1	0	1	
-1	$\frac{1}{10}$	$\frac{1}{10}$	$\frac{1}{10}$	$\frac{3}{10}$
0	$\frac{1}{10}$	$\frac{2}{10}$	$\frac{1}{10}$	$\frac{2}{5}$
1	$\frac{1}{10}$	$\frac{1}{10}$	$\frac{1}{10}$	$\frac{3}{10}$
	$\frac{3}{10}$	$\frac{2}{5}$	$\frac{3}{10}$	

Of course, $\mathbb{E}X = \mathbb{E}Y = 0$ and, moreover,

$$\mathbb{E}[XY] = \frac{1}{10}\left((-1)(-1) + (-1)(+1) + (+1)(-1) + (+1)(+1)\right) = 0,$$

which by eq. (5.39) implies $\text{Cov}(X, Y) = 0$. On the other hand, Proposition 3.6.9 tells us that X and Y are **not** independent. For example, $\mathbb{P}\{X = 0, Y = 0\} = \frac{1}{5}$ while $\mathbb{P}\{X = 0\}\mathbb{P}\{Y = 0\} = \frac{4}{25}$.

Example 5.3.12 shows that $\text{Cov}(X, Y) = 0$ is in general weaker than the independence of X and Y. Therefore, the following definition makes sense.

> **Definition 5.3.13.** Two random variables X and Y satisfying $\text{Cov}(X, Y) = 0$ are said to be **uncorrelated**. Otherwise, if $\text{Cov}(X, Y) \neq 0$, then X and Y are **correlated**.
> More generally, a sequence $X_1, \dots, X_n$ of random variables is called (pairwise) **uncorrelated**, if $\text{Cov}(X_i, X_j) = 0$ whenever $i \neq j$.

Using this notation, property (5) in Proposition 5.3.10 may now be formulated in the following way:

$$X \text{ and } Y \text{ independent} \quad \overset{\Longrightarrow}{\not\Longleftarrow} \quad X \text{ and } Y \text{ uncorrelated}$$

!

Example 5.3.14. Let $A, B \in \mathcal{A}$ be two events in a probability space $(\Omega, \mathcal{A}, \mathbb{P})$ and let $\mathbb{1}_A$ and $\mathbb{1}_B$ be their indicator functions as introduced in Definition 3.6.14. How can we compute $\text{Cov}(\mathbb{1}_A, \mathbb{1}_B)$?

Answer: Since $\mathbb{E}\mathbb{1}_A = \mathbb{P}(A)$, we get

$$\text{Cov}(\mathbb{1}_A, \mathbb{1}_B) = \mathbb{E}[\mathbb{1}_A \mathbb{1}_B] - (\mathbb{E}\mathbb{1}_A)(\mathbb{E}\mathbb{1}_B) = \mathbb{E}\mathbb{1}_{A \cap B} - \mathbb{P}(A)\mathbb{P}(B)$$
$$= \mathbb{P}(A \cap B) - \mathbb{P}(A)\mathbb{P}(B).$$

This tells us that $\mathbb{1}_A$ and $\mathbb{1}_B$ are uncorrelated if and only if the events A and B are independent. But as we saw in Proposition 3.6.15, this happens if and only if the random variables $\mathbb{1}_A$ and $\mathbb{1}_B$ are independent. In other words, two indicator functions are independent if and only if they are uncorrelated.

Finally we consider the covariance of two continuous random variables.

Example 5.3.15. Suppose a random vector (X, Y) is uniformly distributed on the unit ball of $\mathbb{R}^2$. Then the joint density of (X, Y) is given by

$$p(x, y) = \begin{cases} \frac{1}{\pi} & : x^2 + y^2 \leq 1 \\ 0 & : x^2 + y^2 > 1. \end{cases}$$

We proved in Example 3.5.19 that X and Y possess the distribution densities

$$q(x) = \begin{cases} \frac{2}{\pi}\sqrt{1-x^2} : |x| \le 1 \\ 0 \quad : |x| > 1 \end{cases} \quad \text{and} \quad r(y) = \begin{cases} \frac{2}{\pi}\sqrt{1-y^2} : |y| \le 1 \\ 0 \quad : |y| > 1 \end{cases}$$

The function $y \mapsto y(1-y^2)^{1/2}$ is odd. Consequently, because we integrate over an interval symmetric around the origin,

$$\mathbb{E}X = \mathbb{E}Y = \frac{2}{\pi}\int_{-1}^{1} y(1-y^2)^{1/2}\, dy = 0\,.$$

By the same argument we obtain

$$\mathbb{E}[XY] = \int_{-\infty}^{\infty}\int_{-\infty}^{\infty}(x \cdot y)\, p(x, y)\, dx\, dy = \frac{1}{\pi}\int_{-1}^{1} y\left[\int_{-\sqrt{1-y^2}}^{\sqrt{1-y^2}} x\, dx\right] dy = 0\,,$$

and these two assertions imply $\mathrm{Cov}(X, Y) = 0$. Hence, X and Y are uncorrelated, but as we already observed in Example 3.6.19, they are not independent.

5.3.2 Correlation Coefficient

The question arises whether or not the covariance is the quantity that we are looking for, that is, which measures the degree of dependence. The answer is only partially affirmative. Why? Suppose X and Y are dependent. If a is a nonzero real number, then a natural demand is that the degree of dependence between X and Y should be the same as that between aX and Y. But

$$\mathrm{Cov}(aX, Y) = a\,\mathrm{Cov}(X, Y)\,,$$

thus, if $a \ne 1$, then the measure of dependence would increase or decrease. To overcome this drawback, we normalize the covariance in the following way.

> **Definition 5.3.16.** Let X and Y be random variables with finite second moments. Furthermore, we assume that neither X nor Y are constant with probability 1, that is, we have $\mathbb{V}X > 0$ and $\mathbb{V}Y > 0$. Then the quotient
>
> $$\rho(X, Y) := \frac{\mathrm{Cov}(X, Y)}{(\mathbb{V}X)^{1/2}(\mathbb{V}Y)^{1/2}} \tag{5.40}$$
>
> is called **correlation coefficient** of X and Y.

To verify a crucial property of the correlation coefficient we need the following version of the Cauchy–Schwarz inequality.

Proposition 5.3.17 (Cauchy–Schwarz inequality). *For any two random variables X and Y with finite second moments it follows that*

$$|\mathbb{E}(XY)| \le \left(\mathbb{E}X^2\right)^{1/2} \left(\mathbb{E}Y^2\right)^{1/2}. \tag{5.41}$$

Proof: By property (8) of Proposition 5.1.36 we have

$$0 \le \mathbb{E}(|X| - \lambda|Y|)^2 = \mathbb{E}X^2 - 2\lambda\mathbb{E}|XY| + \lambda^2\,\mathbb{E}Y^2 \tag{5.42}$$

for any $\lambda \in \mathbb{R}$. To proceed further, we have to assume[8] $\mathbb{E}X^2 > 0$ and $\mathbb{E}Y^2 > 0$. The latter assumption allows us to choose λ as

$$\lambda := \frac{(\mathbb{E}X^2)^{1/2}}{(\mathbb{E}Y^2)^{1/2}}.$$

If we apply inequality (5.42) with this λ, then we obtain

$$0 \le EX^2 - 2\frac{(\mathbb{E}X^2)^{1/2}}{(\mathbb{E}Y^2)^{1/2}}\,\mathbb{E}|XY| + \mathbb{E}X^2 = 2\,EX^2 - 2\frac{(\mathbb{E}X^2)^{1/2}}{(\mathbb{E}Y^2)^{1/2}}\,\mathbb{E}|XY|,$$

which easily implies (recall that we assumed $\mathbb{E}X^2 > 0$)

$$\mathbb{E}|XY| \le \left(\mathbb{E}X^2\right)^{1/2} \left(\mathbb{E}Y^2\right)^{1/2}.$$

To complete the proof, we use Corollary 5.1.39 and get

$$|\mathbb{E}(XY)| \le \mathbb{E}|XY| \le \left(\mathbb{E}X^2\right)^{1/2} \left(\mathbb{E}Y^2\right)^{1/2}$$

as asserted. ∎

Corollary 5.3.18. *The correlation coefficient satisfies*

$$-1 \le \rho(X, Y) \le 1.$$

Proof: Let as before $\mu = \mathbb{E}X$ and $v = \mathbb{E}Y$. Applying inequality (5.41) to $X - \mu$ and $Y - v$ leads to

$$|\mathrm{Cov}(X, Y)| = |\mathbb{E}(X - \mu)(Y - v)| \le \left(\mathbb{E}(X - \mu)^2\right)^{1/2} \left(\mathbb{E}(Y - v)^2\right)^{1/2}$$
$$= (\mathbb{V}X)^{1/2}\,(\mathbb{V}Y)^{1/2},$$

---

8 The Cauchy–Schwarz inequality remains valid for $\mathbb{E}X^2 = 0$ or $\mathbb{E}Y^2 = 0$. In this case follows $\mathbb{P}\{X = 0\} = 1$ or $\mathbb{P}\{Y = 0\} = 1$, hence $\mathbb{P}\{XY = 0\} = 1$ and $\mathbb{E}[XY] = 0$.

or, equivalently,

$$- (\mathbb{V}X)^{1/2} \, (\mathbb{V}Y)^{1/2} \leq \mathrm{Cov}(X, Y) \leq (\mathbb{V}X)^{1/2} \, (\mathbb{V}Y)^{1/2} \, .$$

By the definition of $\rho(X, Y)$ given in eq. (5.40), this implies $-1 \leq \rho(X, Y) \leq 1$ as asserted.

∎

Interpretation: For uncorrelated X and Y we have $\rho(X, Y) = 0$. In particular, this is valid if X and Y are independent. On the contrary, $\rho(X, Y) \neq 0$ tells us that X and Y are dependent. Thereby, values near to zero correspond to weak dependence, while $\rho(X, Y)$ near to 1 or -1 indicate a strong dependence. The strongest possible dependence is when $Y = aX$ for some $a \neq 0$. Then $\rho(X, Y) = 1$ if $a > 0$ while $\rho(X, Y) = -1$ for $a < 0$.

> **Definition 5.3.19.** Two random variables X and Y are said to be **positively correlated** if $\rho(X, Y) > 0$. In the case that $\rho(X, Y) < 0$, they are said to be **negatively correlated**.

Interpretation: X and Y are positively correlated, provided that larger (or smaller) values of X make larger (or smaller) values of Y more likely. This does **not** mean that a larger X-value always implies a larger Y-value. Only that the probability for those larger values increases. And in the same way, if X and Y are negatively correlated, then larger values of X make smaller Y-values more likely. Let us explain this with two typical examples. Choose by random a person ω in the audience. Let $X(\omega)$ be his height and $Y(\omega)$ his weight. Then X and Y will surely be positively correlated. But this does not necessarily mean that each taller person has a bigger weight. Another example of negatively correlated random variables could be as follows: X is the average number of cigarettes that a randomly chosen person smokes per day and Y is his lifetime.

Example 5.3.20. Let us come back to Example 5.3.1: in an urn are n balls labeled with "0" and n labeled with "1." One chooses two balls without replacement. Then X is the value of the first ball, Y that of the second. How does the correlation coefficient of X and Y depend on n?

Answer: The joint distribution of X and Y is given by the following table:

$Y\backslash X$	0	1	
0	$\frac{n-1}{4n-2}$	$\frac{n}{4n-2}$	$\frac{1}{2}$
1	$\frac{n}{4n-2}$	$\frac{n-1}{4n-2}$	$\frac{1}{2}$
	$\frac{1}{2}$	$\frac{1}{2}$	

Direct computations show $\mathbb{E}X = \mathbb{E}Y = 1/2$ and $\mathbb{V}X = \mathbb{V}Y = 1/4$. Moreover, it easily follows $\mathbb{E}[XY] = \frac{n-1}{4n-2}$, hence

$$\text{Cov}(X, Y) = \frac{n-1}{4n-2} - \frac{1}{4} = \frac{-1}{8n-4},$$

and the correlation coefficient equals

$$\rho(X, Y) = \frac{\frac{-1}{8n-4}}{\sqrt{\frac{1}{4}}\sqrt{\frac{1}{4}}} = \frac{-1}{2n-1}.$$

If $n \to \infty$, then $\rho(X, Y)$ is of order $\frac{-1}{2n}$. Hence, if n is large, then the random variables X and Y are "almost" uncorrelated.

Since $\rho(X, Y) < 0$, the two random variables are negatively correlated. Why? This was already explained in Example 5.3.1: an occurrence of $X = 1$ makes $Y = 0$ more likely, while the occurrence of $X = 0$ increases the likelihood of $Y = 1$. Some word about the case $n = 1$. Here Y is completely determined by the value of X, expressed by $\rho(X, Y) = -1$.

5.4 Problems

Problem 5.1.

1. Put successively and independently of each other n particles into N boxes. Thereby, each box is equally likely. How many boxes remain empty on average?

    Hint: Define random variables $X_1, \ldots, X_N$ as follows: set $X_i = 1$ if box i remains empty and $X_i = 0$, otherwise.

2. Fifty persons write randomly (according to the uniform distribution), and independently of each other, one of the 26 letters in the alphabet on a sheet of paper. On average, how many different letters appear?

Problem 5.2. Let $(\Omega, \mathcal{A}, \mathbb{P})$ be a probability space. Given (not necessarily disjoint) events $A_1, \ldots, A_n$ in $\mathcal{A}$ and real numbers $\alpha_1, \ldots, \alpha_n$, define $X : \Omega \to \mathbb{R}$ by[9].

$$X := \sum_{j=1}^{n} \alpha_j \mathbb{1}_{A_j} .$$

1. Why is X a random variable?
2. Prove

$$\mathbb{E}X = \sum_{j=1}^{n} \alpha_j \mathbb{P}(A_j) \quad \text{and} \quad \mathbb{V}X = \sum_{i,j=1}^{n} \alpha_i \alpha_j \left[\mathbb{P}(A_i \cap A_j) - \mathbb{P}(A_i)\mathbb{P}(A_j) \right] .$$

How does $\mathbb{V}X$ simplify for independent events $A_1, \ldots, A_n$?

---

9 For the definition indicator functions $\mathbb{1}_{A_i}$ see eq. (3.20).

Problem 5.3. Suppose a fair "die" has k faces labeled by the numbers from 1 to k.
1. How often one has to roll the die on the average before the first "1" shows up?
2. Suppose one rolls the die exactly k times. Let p_k be the probability that "1" appears exactly once and q_k is the probability that "1" shows up at least once. Compute p_k and q_k and determine their behavior as $k \to \infty$, that is, find $\lim_{k \to \infty} p_k$ and $\lim_{k \to \infty} q_k$.

Problem 5.4.
1. Let X be a random variable with values in $\mathbb{N}_0 = \{0, 1, 2, \dots\}$. Prove that

$$\mathbb{E}X = \sum_{k=1}^{\infty} \mathbb{P}\{X \geq k\}.$$

2. Suppose now that X is continuous with $\mathbb{P}\{X \geq 0\} = 1$. Verify

$$\sum_{k=1}^{\infty} \mathbb{P}\{X \geq k\} \leq \mathbb{E}X \leq 1 + \sum_{k=1}^{\infty} \mathbb{P}\{X \geq k\}.$$

Problem 5.5. Let X be an $\mathbb{N}_0$-valued random variable with

$$\mathbb{P}\{X = k\} = q^{-k}, \quad k = 1, 2, \dots$$

for some $q \geq 2$.
(a) Why we have to suppose $q \geq 2$, although $\sum_{k=1}^{\infty} q^{-k} < \infty$ for $q > 1$?
(b) Determine $\mathbb{P}\{X = 0\}$?
(c) Compute $\mathbb{E}X$ by the formula in Problem 5.4.
(d) Compute $\mathbb{E}X$ directly by $\mathbb{E}X = \sum_{k=0}^{\infty} k \mathbb{P}\{X = k\}$.

Problem 5.6. Two independent random variables X and Y with third moment satisfy $\mathbb{E}X = \mathbb{E}Y = 0$. Prove that then

$$\mathbb{E}(X + Y)^3 = \mathbb{E}X^3 + \mathbb{E}Y^3.$$

Problem 5.7. A random variable X is Pois_λ-distributed for some $\lambda > 0$. Evaluate

$$\mathbb{E}\left(\frac{1}{1+X}\right) \quad \text{and} \quad \mathbb{E}\left(\frac{X}{1+X}\right).$$

Problem 5.8. In a lottery are randomly chosen 6 of 49 numbers. Let X be the largest number of the 6 ones. Show that

$$\mathbb{E}X = \frac{6 \cdot 43!}{49!} \sum_{k=6}^{49} k(k-1)(k-2)(k-3)(k-4)(k-5) = 42.8571.$$

Evaluate $\mathbb{E}X$ if X is the smallest number of the 6 chosen.

Hint: Either one modifies the calculations for the maximal value suitably or one reduces the second problem to the first one by an easy algebraic operation.

Problem 5.9. A fair coin is labeled by "0" on one side and with "1" on the other one. Toss it four times. Let X be the sum of the two first tosses and Y be the sum of all four ones. Determine the joint distribution of X and Y. Evaluate $\mathrm{Cov}(X, Y)$ as well as $\rho(X, Y)$.

Problem 5.10. In an urn are five balls, two labeled by "0" and three by "1." Choose two balls without replacement. Let X be the number on the first ball and Y that on the second.
1. Determine the distribution of the random vector (X, Y) as well as its marginal distributions.
2. Compute $\rho(X, Y)$.
3. Which distribution does $X + Y$ possess?

Problem 5.11. Among 40 students are 30 men and 10 women. Also, 25 of the 30 men and 8 of the 10 women passed an exam successfully. Choose randomly, according to the uniform distribution, one of the 40 students. Let $X = 0$ if the chosen person is a man, and $X = 1$ if it is a woman. Furthermore, set $Y = 0$ if the person failed the exam, and $Y = 1$ if she or he passed.
1. Find the joint distribution of X and Y.
2. Are X and Y independent? If not, evaluate $\mathrm{Cov}(X, Y)$.
3. Are X and Y negatively or positively correlated? What does it express, when X and Y are positively or negatively correlated?

Problem 5.12. Let $(\Omega, \mathcal{A}, \mathbb{P})$ be a probability space. Prove for any two events A and B in $\mathcal{A}$ the estimate

$$|\mathbb{P}(A \cap B) - \mathbb{P}(A)\,\mathbb{P}(B)| \le \frac{1}{4}\,.$$

Is it possible to improve the upper bound $\frac{1}{4}$?

Problem 5.13. (Problem of Luca Pacioli in 1494; the first correct solution was found by Blaise Pascal in 1654) Two players, say A and B, are playing a fair game consisting of several rounds. The first player who wins six rounds wins the game and the stakes of 20 Taler that have been bet throughout the game. However, one day the game is interrupted and must be stopped. If player A has won five rounds and player B has won three rounds, how should the stakes be divided fairly among the players?

Problem 5.14. In Example 5.1.46 we computed the average number of necessary purchases to get all n pictures. Let m be an integer with $1 \le m < n$. How many purchases are necessary on average to possess m of the n pictures?

For n even choose $m = n/2$ and for n odd take $m = (n-1)/2$. Let M_n be the average number of purchases to get m pictures, that is, to get half of the pictures. Determine

$$\lim_{n \to \infty} \frac{M_n}{n} .$$

Hint: Use eq. (5.26).

Problem 5.15. Compute $\mathbb{E}|X|^{2n+1}$ for a standard normal distributed X and $n = 0, 1, \ldots$.

Problem 5.16. Suppose X has the density

$$p(x) = \begin{cases} 0 & : \quad x < 1 \\ c_\alpha x^\alpha & : \quad x \ge 1 \end{cases}$$

for some $\alpha < -1$.
1. Determine c_α such that p is a density.
2. For which $n \ge 1$ does X possess an nth moment?

Problem 5.17. Let U be uniform distributed on an interval $[\alpha, \beta]$. Show that for $n \ge 1$

$$\mathbb{E}U^n = \frac{\beta^n + \alpha\beta^{n-1} + \cdots \alpha^{n-1}\beta + \beta^n}{n+1} .$$

Problem 5.18. Let $X_1, \ldots, X_n$ be random variables with finite second moment and with $\mathbb{E}X_j = 0$. Show that

$$\mathbb{E}\left[X_1 + \cdots + X_n\right]^2 = \sum_{i,j=1}^{n} \mathrm{Cov}(X_i, X_j) = \sum_{j=1}^{n} \mathbb{V}X_j + 2 \sum_{1 \le i < j \le n} \mathrm{Cov}(X_i, X_j) .$$

Problem 5.19. Show

$$\mathbb{E}X = \frac{nM}{N}$$

for a hypergeometric distributed random variable X with

$$\mathbb{P}\{X = m\} = \frac{\binom{M}{m}\binom{N-M}{n-m}}{\binom{N}{n}}, \quad m = 0, \ldots, n .$$

Problem 5.20. Let X be $\mathcal{N}(0, 1)$-distributed. Determine $\mathbb{V}X^3$ and $\mathbb{V}X^4$.

Problem 5.21. Given a non-negative random variable X, define φ_X from $[0, \infty)$ to $[0, \infty]$ by $\varphi_X(t) = \mathbb{E}\, t^X$. Then φ_X is called **generating function** of X (see [GS01b], Section 5.1).

(1) Suppose X has values in $\mathbb{N}_0$. Show that, if $t \geq 0$, then this "new" definition of the generating function coincides with the one given in Problem 4.2.

(2) Let $X_1, \ldots, X_n$ be independent and non-negative. For $\alpha_j \geq 0, 1 \leq j \leq n$, let

$$X = \alpha_1 X_1 + \cdots + \alpha_n X_n .$$

Prove

$$\varphi_X(t) = \varphi_{X_1}(t^{\alpha_1}) \cdots \varphi_{X_n}(t^{\alpha_n}) .$$

(3) Find φ_X for an exponentially distributed X.

6 Normally Distributed Random Vectors

6.1 Representation and Density

In Example 3.4.3 we considered a two-dimensional random vector (X_1, X_2), where X_1 was the height of a randomly chosen person and X_2 was his weight. From experience and in view of the central limit theorem (cf. Section 7), it is quite reasonable to assume that X_1 and X_2 are normally distributed. Suppose we are able to determine their expected values and their variances. However, this is not sufficient to describe the experiment. Why? The random variables X_1 and X_2 are surely dependent, and the most interesting problem is to describe their degree of dependence. This cannot be done based only on the knowledge of their distributions. What we really need to know is their joint distribution. Therefore, we not only have to suppose X_1 and X_2 to be normal, but the generated vector (X_1, X_2) has to be as well.

But what does it mean that a random vector is normally distributed? This section is devoted to answer this and related questions.

Let us first recall the univariate case, investigated in Example 4.2.2 and in the subsequent Proposition 4.2.3. The main observation was that a random variable Y is normally distributed if and only if it may be written as

$$Y = aX + \mu \tag{6.1}$$

for some $a \neq 0$, $\mu \in \mathbb{R}$, and a standard normal random variable X.

Let now $\vec{Y} = (Y_1, \ldots, Y_n)$ be an n-dimensional random vector. We want to represent it in the same way as Y in eq. (6.1). Consequently, we have to replace X by a multivariate standard normal vector and the function $x \mapsto ax + \mu$ by a suitable mapping from $\mathbb{R}^n$ to $\mathbb{R}^n$. But which kind of mapping this should be and what is an n-dimensional standard normal vector?

Let us begin by answering the second question. Therefore, recall the definition of the multivariate standard normal distribution $\mathcal{N}(0, 1)^{\otimes n}$ introduced in Definition 1.9.16. This probability measure on $(\mathbb{R}^n, \mathcal{B}(\mathbb{R}^n))$ was given by

$$\mathcal{N}(0, 1)^{\otimes n}(B) = \frac{1}{(2\pi)^{n/2}} \int_B e^{-|x|^2/2} \, dx$$

$$= \frac{1}{(2\pi)^{n/2}} \underbrace{\int \cdots \int}_{B} e^{-(x_1^2 + \cdots + x_n^2)/2} \, dx_n \cdots dx_1$$

with $B \in \mathcal{B}(\mathbb{R}^n)$. Thus, a random vector $\vec{X}$ should be standard normal distributed whenever its probability distribution is $\mathcal{N}(0, 1)^{\otimes n}$. Let us formulate this as a definition.

> **Definition 6.1.1.** A random vector $\vec{X} = (X_1, \ldots, X_n)$ is **standard normally** (distributed) if its probability distribution satisfies $\mathbb{P}_{\vec{X}} = \mathcal{N}(0,1)^{\otimes n}$.

To make this definition more descriptive, let us state some equivalent properties.

Proposition 6.1.2. *For a random vector $\vec{X} = (X_1, \ldots, X_n)$ the following are equivalent:*
1. *$\vec{X}$ is standard normal.*
2. *If $B \in \mathcal{B}(\mathbb{R}^n)$, then*

$$\mathbb{P}\{\vec{X} \in B\} = \frac{1}{(2\pi)^{n/2}} \int_B e^{-|x|^2/2} \, dx.$$

3. *The coordinate mappings $X_1, \ldots, X_n$ are (univariate) standard normal distributed and independent. That is, for all $t_j \in \mathbb{R}, 1 \le j \le n$,*

$$\mathbb{P}\{X_1 \le t_1, \ldots, X_n \le t_n\} = \mathbb{P}\{X_1 \le t_1\} \cdots \mathbb{P}\{X_n \le t_n\}$$

$$= \left(\frac{1}{\sqrt{2\pi}} \int_{-\infty}^{t_1} e^{-x_1^2/2} dx_1 \right) \cdots \left(\frac{1}{\sqrt{2\pi}} \int_{-\infty}^{t_n} e^{-x_n^2/2} dx_n \right).$$

Proof: Taking into account the definition of $\mathcal{N}(0,1)^{\otimes n}$, this is an immediate consequence of Propositions 3.6.5 and 3.6.18. Compare also the considerations in Example 3.6.22. ∎

An adequate substitute for $x \mapsto ax + \mu$ in representation (6.1) is still undetermined. Which mappings in $\mathbb{R}^n$ should be considered?

Observe that $x \mapsto ax + \mu$ is affine linear from $\mathbb{R}$ to $\mathbb{R}$. The counterpart in $\mathbb{R}^n$ is of the form $x \mapsto Ax + \mu$, where A is a linear mapping in $\mathbb{R}^n$ and $\mu \in \mathbb{R}^n$. Linear mappings in $\mathbb{R}^n$ are described by $n \times n$ matrices $A = (a_{ij})_{i,j=1}^n$ and act as follows:

$$Ax = \left(\sum_{j=1}^n a_{1j}x_j, \ldots, \sum_{j=1}^n a_{nj}x_j \right), \quad x = (x_1, \ldots, x_n) \in \mathbb{R}^n.$$

Consequently, the suitable generalization of $x \mapsto ax + \mu$ is the mapping $x \mapsto Ax + \mu$ with an $n \times n$ matrix A and $\mu \in \mathbb{R}^n$. The condition $a \ne 0$ transfers to $\det(A) \ne 0$ or, equivalently, A has to be **regular**, that is, the generated mapping is one-to-one from $\mathbb{R}^n$ onto $\mathbb{R}^n$. Here and in the sequel we will use results and notations as presented in Section A.4.

Now we are in position to define normally (distributed) random vectors.

Definition 6.1.3. A random vector $\vec{Y}$ is said to be **normally distributed** (or simply, normal) provided there exists a regular $n \times n$ matrix A and a vector $\mu \in \mathbb{R}^n$ such that

$$\vec{Y} = A\vec{X} + \mu \tag{6.2}$$

for some standard normal $\vec{X}$.

Remark 6.1.4. Let us reformulate Definition 6.1.3 due to its importance. A random vector $\vec{Y} = (Y_1, \ldots, Y_n)$ is normal if and only if there exists a regular matrix $A = (\alpha_{ij})_{i,j=1}^n$ and a vector $\mu = (\mu_1, \ldots, \mu_n)$ such that

$$Y_i = \sum_{j=1}^n \alpha_{ij} X_j + \mu_i, \quad 1 \le i \le n,$$

with $X_1, \ldots, X_n$ independent $\mathcal{N}(0, 1)$-distributed.

Example 6.1.5. Suppose the three-dimensional random vector $\vec{Y} = (Y_1, Y_2, Y_3)$ is defined by

$$Y_1 = 2X_1 + X_2 - X_3 + 4, \quad Y_2 = X_1 - 2X_2 + X_3 - 2 \quad \text{and}$$
$$Y_3 = X_1 - 2X_3 + 5$$

with $\mathcal{N}(0, 1)$-distributed independent X_1, X_2, X_3. Then $\vec{Y}$ is normally distributed. Observe that it may be represented in the form of eq. (6.2) with A given by

$$A = \begin{pmatrix} 2 & 1 & -1 \\ 1 & -2 & 1 \\ 1 & 0 & -2 \end{pmatrix}$$

and with $\mu = (4, -2, 5)$. Moreover, we have $\det(A) = 9$, hence A is regular.

Given a normal vector $\vec{Y}$, how do we get the standard normal $\vec{X}$ in representation (6.2)? The next proposition answers this question.

Proposition 6.1.6. *A random vector $\vec{Y} = (Y_1, \ldots, Y_n)$ is normal if and only if there exists a regular $n \times n$ matrix $B = (\beta_{ij})_{i,j=1}^n$ and a vector $v = (v_1, \ldots, v_n) \in \mathbb{R}^n$ such that the random variables X_i, defined by*

$$X_i := \sum_{j=1}^n \beta_{ij} Y_j + v_i, \quad 1 \le i \le n,$$

are independent standard normal.

Proof: This is a direct consequence of the following observation. One has $\vec{Y} = A\vec{X} + \mu$ if and only if $\vec{X}$ may be represented as $\vec{X} = A^{-1}\vec{Y} - A^{-1}\mu$. Therefore, the assertion follows by choosing B and v such that $B = A^{-1}$ and $v = -A^{-1}\mu$. ∎

Example 6.1.7. For the random vector $\vec{Y}$ investigated in Example 6.1.5, the generated independent standard normal random variables X_1, X_2, and X_3 may be represented as follows:

$$X_1 = \frac{1}{9}(4Y_1 + 2Y_2 - Y_3 + 7) \quad X_2 = \frac{1}{9}(Y_1 - Y_2 - Y_3 + 1)$$

$$X_3 = \frac{1}{9}(2Y_1 + Y_2 - 5Y_3 - 19) \ .$$

Suppose $\vec{Y} = A\vec{X} + \mu$ is a normal vector. How can we evaluate its distribution density? To answer this question, we introduce the following function. Let $R > 0$ be an $n \times n$-matrix and $\mu \in \mathbb{R}^n$. The inverse matrix of R is R^{-1}, and to simplify the notation, set $|R| = \det(R)$. Observe that $R > 0$ implies $|R| > 0$. With these notations we define a function $p_{\mu,R}$ from $\mathbb{R}^n$ to $\mathbb{R}$ by

$$p_{\mu,R}(x) := \frac{1}{(2\pi)^{n/2}|R|^{1/2}} e^{-\langle R^{-1}(x-\mu),(x-\mu)\rangle/2}, \quad x \in \mathbb{R}^n . \tag{6.3}$$

Now we are prepared to answer the above question about the density of $\vec{Y}$.

Proposition 6.1.8. *Suppose the normal vector $\vec{Y}$ is represented as in eq. (6.2) with regular A and $\mu \in \mathbb{R}^n$. Define the positive matrix R by $R = AA^T$. Then $p_{\mu,R}$, as given in eq. (6.3), is the distribution density of $\vec{Y}$. In other words, if $B \in \mathcal{B}(\mathbb{R}^n)$, then*

$$\mathbb{P}\{\vec{Y} \in B\} = \frac{1}{(2\pi)^{n/2}|R|^{1/2}} \int_B e^{-\langle R^{-1}(x-\mu),(x-\mu)\rangle/2} dx .$$

Proof: Because $\vec{Y} = A\vec{X} + \mu$ with $\vec{X}$ standard normal, Proposition 6.1.2 implies

$$\mathbb{P}\{\vec{Y} \in B\} = \mathbb{P}\{A\vec{X} + \mu \in B\} = \mathbb{P}\{\vec{X} \in A^{-1}(B - \mu)\}$$

$$= \frac{1}{(2\pi)^{n/2}} \int_{A^{-1}(B-\mu)} e^{-|y|^2/2} dy$$

for any Borel set $B \subseteq \mathbb{R}^n$. Hereby, $B - \mu$ denotes the set $\{b - \mu : b \in B\}$.

In the next step we change the variables by setting $x = Ay + \mu$. Then $dx = |\det(A)| dy$, where by assumption $\det(A) \neq 0$ and, moreover, we have $y \in A^{-1}(B - \mu)$ if and only if $x \in B$. Therefore, the last integral transforms to

$$\mathbb{P}\{\vec{Y} \in B\} = \frac{1}{(2\pi)^{n/2}} |\det(A)|^{-1} \int_B e^{-|A^{-1}(x-\mu)|^2/2} \, dx. \tag{6.4}$$

Proposition A.4.1 implies $R > 0$ and, moreover,

$$|R| = \det(R) = \det(AA^T) = \det(A) \cdot \det(A^T) = \det(A)^2.$$

Since $|R| = \det(R) > 0$, this leads to $|R|^{1/2} = |\det(A)|$, that is, to

$$|\det(A)|^{-1} = |R|^{-1/2}. \tag{6.5}$$

Note that

$$|A^{-1}(x - \mu)|^2 = \langle A^{-1}(x - \mu), A^{-1}(x - \mu)\rangle = \langle (A^{-1})^T A^{-1}(x - \mu), (x - \mu)\rangle,$$

which by

$$\left(A^{-1}\right)^T \circ A^{-1} = \left(A^T\right)^{-1} \circ A^{-1} = \left(A \circ A^T\right)^{-1} = R^{-1}$$

implies

$$|A^{-1}(x - \mu)|^2 = \langle R^{-1}(x - \mu), (x - \mu)\rangle. \tag{6.6}$$

Plugging eqs. (6.5) and (6.6) into eq. (6.4), we get

$$\mathbb{P}\{\vec{Y} \in B\} = \int_B p_{\mu,R}(x) \, dx$$

with $p_{\mu,R}$ as in eq. (6.3). This completes the proof. ∎

Remark 6.1.9. How does Proposition 6.1.8 look like for $n = 1$? Here $Y = aX + \mu$, that is, $A = (a)$, and since A has to be regular, this implies $a \neq 0$. Hence we get $R = AA^T = (a^2)$, $R^{-1} = (a^{-2})$ and $|R|^{1/2} = |a|$. Thus, the density of Y is given by

$$p_{\mu,R}(x) = \frac{1}{(2\pi)^{1/2} |R|^{1/2}} e^{-\left(R^{-1}(x-\mu), x-\mu\right)/2} = \frac{1}{(2\pi)^{1/2} |a|} e^{-(x-\mu)^2/2a^2}, \quad x \in \mathbb{R}.$$

This coincides with the result obtained in Example 4.2.2.

In view of Proposition 6.1.8 we will use the following notation.

Definition 6.1.10. A normal vector $\vec{Y}$ is said to be $\mathcal{N}(\mu, R)$-distributed if $p_{\mu,R}$ is its density, that is, if

$$\mathbb{P}\{\vec{Y} \in B\} = \frac{1}{(2\pi)^{n/2}|R|^{1/2}} \int_B e^{-\langle R^{-1}(x-\mu),(x-\mu)\rangle/2} \, dx \,.$$

Remark 6.1.11. It follows from Proposition A.4.2 that, given **any** $\mu \in \mathbb{R}^n$ and **any** $R > 0$, there exists a normal vector $\vec{Y}$ that is $\mathcal{N}(\mu, R)$-distributed. Indeed, write $R > 0$ as $R = AA^T$ and set $\vec{Y} = A\vec{X} + \mu$ with $\vec{X}$ standard normal. Then $\vec{Y}$ is $\mathcal{N}(\mu, R)$-distributed by Proposition 6.1.8.

$$\{\text{Distributions of } \mathbb{R}^n\text{-valued normal vectors}\} \quad \Longleftrightarrow \quad \{\mu \in \mathbb{R}^n, \; R > 0\}$$

Example 6.1.12. Assume

$$Y_1 = X_1 - X_2 + 3 \quad \text{and} \quad Y_2 = 2X_1 + X_2 - 2$$

for X_1, X_2 independent $\mathcal{N}(0, 1)$-distributed. Then we get

$$\mu = (3, -2) \quad \text{and} \quad A = \begin{pmatrix} 1 & -1 \\ 2 & 1 \end{pmatrix},$$

which implies

$$R = AA^T = \begin{pmatrix} 1 & -1 \\ 2 & 1 \end{pmatrix} \cdot \begin{pmatrix} 1 & 2 \\ -1 & 1 \end{pmatrix} = \begin{pmatrix} 2 & 1 \\ 1 & 5 \end{pmatrix}. \tag{6.7}$$

Thus, $\vec{Y}$ is $\mathcal{N}(\mu, R)$-distributed with $\mu = (3, -2)$ and R as in eq. (6.7).

Which density does $\vec{Y}$ possess? To answer this, we have to compute $\det(R)$ and R^{-1}. One easily gets $\det(R) = 9$. The inverse matrix of R equals

$$R^{-1} = \frac{1}{9} \begin{pmatrix} 5 & -1 \\ -1 & 2 \end{pmatrix}.$$

Therefore, the distribution density $p_{\mu,R}$ of $\vec{Y} = (Y_1, Y_2)$ is given by

$$p_{\mu,R}(x_1, x_2) = \frac{1}{6\pi} \exp\left(-\frac{1}{2}\langle R^{-1}(x_1 - 3, x_2 + 2), (x_1 - 3, x_2 + 2)\rangle\right)$$

$$= \frac{1}{6\pi} \exp\left(-\frac{1}{18}\left[5(x_1 - 3)^2 - 2(x_1 - 3)(x_2 + 2) + 2(x_2 + 2)^2\right]\right). \tag{6.8}$$

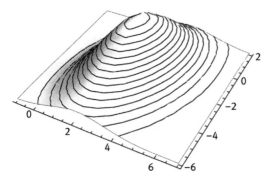

Figure 6.1: The density given by eq. (6.8).

For later purposes we have to name the probability measures on $(\mathbb{R}^n, \mathcal{B}(\mathbb{R}^n))$ appearing as distributions of normal vectors.

Definition 6.1.13. Given $\mu \in \mathbb{R}^n$ and $R > 0$, the probability measure $\mathcal{N}(\mu, R)$ on $(\mathbb{R}^n, \mathcal{B}(\mathbb{R}^n))$ is defined by

$$\mathcal{N}(\mu, R)(B) = \int_B p_{\mu,R}(x)\,dx = \frac{1}{(2\pi)^{n/2}|R|^{1/2}} \int_B e^{-\frac{1}{2}\langle R^{-1}(x-\mu),(x-\mu)\rangle}\,dx.$$

$\mathcal{N}(\mu, R)$ is called a **multivariate normal distribution** with[1] expected value μ and covariance matrix R.

According to Definition 6.1.13, we may now formulate Proposition 6.1.8 as follows:

Proposition 6.1.14. Let $\vec{Y}$ be a random vector. Then the following are equivalent.
1. $\vec{Y}$ is $\mathcal{N}(\mu, R)$-distributed.
2. $\mathbb{P}_{\vec{Y}} = \mathcal{N}(\mu, R)$.
3. There is a regular $n \times n$ matrix A with $R = AA^T$ such that $\vec{Y} = A\vec{X} + \mu$.

Remark 6.1.15. The case $R = I_n$ (as in Section A.4 we denote the identity matrix in $\mathbb{R}^n$ by I_n) and $\mu = 0$ is of special interest. Because $I_n^{-1} = I_n$ and $\det(I_n) = 1$, we get

$$p_{0,I_n}(x) = \frac{1}{(2\pi)^{n/2}} e^{-|x|^2/2}, \quad x \in \mathbb{R}^n.$$

---

1 Why they are named in this way will become clear in the next section.

This tells us that $\mathcal{N}(0, I_n)$ is nothing else as the multivariate standard normal distribution introduced in Definition 1.9.16. Written as formula, this means

$$\mathcal{N}(0, 1)^{\otimes n} = \mathcal{N}(0, I_n).$$

More generally, in view of eq. (1.75) it follows that

$$\mathcal{N}(\mu, \sigma^2)^{\otimes n} = \mathcal{N}(\vec{\mu}, \sigma^2 I_n)$$

where $\vec{\mu} = (\mu, \ldots, \mu) \in \mathbb{R}^n$ and $\sigma > 0$. In other words,

$$\mathcal{N}(\mu, \sigma^2)^{\otimes n}(B) = \mathcal{N}(\vec{\mu}, \sigma^2 I_n)(B) = \frac{1}{(2\pi)^{n/2}\sigma^n} \int_B e^{-|x-\vec{\mu}|^2/2\sigma^2} \, dx. \tag{6.9}$$

For later purposes, the next result is of importance.

Proposition 6.1.16. *Suppose a normal vector* $\vec{Y} = (Y_1, \ldots, Y_n)$ *may be written as*

$$\vec{Y} = U\vec{X}$$

with an $\mathcal{N}(0, I_n)$-*distributed (standard normal)* $\vec{X}$ *and a unitary matrix U. Then its coordinate mappings* $Y_1, \ldots, Y_n$ *are independent standard normal random variables.*

Proof: The random vector $\vec{Y}$ is $\mathcal{N}(0, UU^T)$-distributed. But U is unitary, hence, $UU^T = I_n$ and $\vec{Y}$ is $\mathcal{N}(0, I_n)$ or, equivalently, standard normally distributed. Then the assertion follows by Proposition 6.1.2. ∎

Example 6.1.17. For $\theta \in [0, 2\pi)$ define the 2×2 matrix U by

$$U = \begin{pmatrix} \cos\theta & \sin\theta \\ -\sin\theta & \cos\theta \end{pmatrix}.$$

The matrix U is unitary and by Proposition 6.1.16 the vector $\vec{Y} = U\vec{X}$ is standard normal. In other words, given independent standard normal X_1 and X_2, for each $\theta \in [0, 2\pi)$ the random variables

$$Y_1 := \cos\theta X_1 + \sin\theta X_2 \quad \text{and} \quad Y_2 = -\sin\theta X_1 + \cos\theta X_2$$

are independent and standard normally distributed as well.

6.2 Expected Value and Covariance Matrix

We start with the following definition.

Definition 6.2.1. Let $\vec{Y} = (Y_1, \ldots, Y_n)$ be a random vector such that $\mathbb{E}|Y_j| < \infty$ for all $1 \le j \le n$. Then the vector

$$\mathbb{E}\vec{Y} := (\mathbb{E}Y_1, \ldots, \mathbb{E}Y_n) = (\mu_1, \ldots, \mu_n)$$

is called the (multivariate) **expected value** of $\vec{Y}$.
   If $\mathbb{E}Y_j^2 < \infty$, $1 \le j \le n$, then the matrix

$$\mathrm{Cov}_{\vec{Y}} := \left(\mathrm{Cov}(Y_i, Y_j) \right)_{i,j=1}^n = \left(\mathbb{E}(Y_i - \mu_i)(Y_j - \mu_j) \right)_{i,j=1}^n$$

is said to be the **covariance matrix** of $\vec{Y}$.

Remark 6.2.2. It is important to notice that both $\mathbb{E}\vec{Y}$ and the covariance matrix $\mathrm{Cov}_{\vec{Y}}$ depend only on the distribution of $\vec{Y}$. That is, whenever $\mathbb{P}_{\vec{Y}_1} = \mathbb{P}_{\vec{Y}_2}$, then

$$\mathbb{E}\vec{Y}_1 = \mathbb{E}\vec{Y}_2 \quad \text{and} \quad \mathrm{Cov}_{\vec{Y}_1} = \mathrm{Cov}_{\vec{Y}_2}.$$

The next proposition describes the (multivariate) expected value and the covariance matrix of a normally distributed vector.

Proposition 6.2.3. *Assume* $\vec{Y} = A\vec{X} + \mu$ *for some regular matrix A and $\mu \in \mathbb{R}^n$. Define $R = (r_{ij})_{i,j=1}^n$ as $R = AA^T$. Then the following is valid.*

(1) *We have* $\mathbb{E}\vec{Y} = \mu$ *and* $\mathrm{Cov}_{\vec{Y}} = \left(\mathrm{Cov}(Y_i, Y_j) \right)_{i,j=1}^n = R$.

(2) *Given $a \in \mathbb{R}^n$, $a \ne 0$, then $\langle \vec{Y}, a \rangle$ is a normal random variable with expected value $\langle \mu, a \rangle$ and variance $\langle Ra, a \rangle$.*

(3) *The coordinate mappings Y_i are $\mathcal{N}(\mu_i, r_{ii})$-distributed, $1 \le i \le n$, that is, the marginal distributions of $\vec{Y}$ are the probability measures $\mathcal{N}(\mu_i, r_{ii})$.*

Proof: By assumption

$$Y_i = \sum_{j=1}^n \alpha_{ij} X_j + \mu_i, \quad i = 1, \ldots, n, \tag{6.10}$$

hence, the linearity of the expected value and $\mathbb{E}X_j = 0$ imply

$$\mathbb{E}Y_i = \sum_{j=1}^n \alpha_{ij} \mathbb{E}X_j + \mu_i = \mu_i, \quad 1 \le i \le n.$$

This proves $\mathbb{E}\vec{Y} = (\mathbb{E}Y_1, \ldots, \mathbb{E}Y_n) = \mu$.

Let us now verify the second part of property (1). Using $\mu_j = \mathbb{E}Y_j$, by representation (6.10) we get

$$\text{Cov}(Y_i, Y_j) = \mathbb{E}[(Y_i - \mu_i)(Y_j - \mu_j)] = \mathbb{E}\left(\sum_{k=1}^{n} \alpha_{ik}X_k\right)\left(\sum_{l=1}^{n} \alpha_{jl}X_l\right)$$

$$= \sum_{k,l=1}^{n} \alpha_{ik}\alpha_{jl}\,\mathbb{E}X_kX_l\,.$$

The X_js are independent $\mathcal{N}(0, 1)$-distributed, hence

$$\mathbb{E}X_kX_l = \begin{cases} 1 : k = l \\ 0 : k \neq l \end{cases},$$

leading to

$$\text{Cov}(Y_i, Y_j) = \sum_{k=1}^{n} \alpha_{ik}\alpha_{jk} = r_{ij}\,.$$

To see this, recall that $R = AA^T$, hence $r_{ij} = \sum_{k=1}^{n} \alpha_{ik}\alpha_{jk}$. This proves $\text{Cov}_{\vec{Y}} = R$ as asserted.

To verify property (2) we first treat a special case, namely that the vector is standard normally distributed. So suppose that $\vec{X}$ is $\mathcal{N}(0, I_n)$-distributed. In this case, property (2) asserts the following. For any $b \in \mathbb{R}^n$, $b \neq 0$,

$$\langle\vec{X}, b\rangle \text{ is distributed according to } \mathcal{N}(0, |b|^2)\,. \tag{6.11}$$

If $b = (b_1, \ldots, b_n)$, then

$$\langle\vec{X}, b\rangle = \sum_{j=1}^{n} b_j X_j = \sum_{j=1}^{n} Z_j$$

with $Z_j = b_j X_j$. The random variables $Z_1, \ldots, Z_n$ are independent and, moreover, by Proposition 4.2.3, the Z_js are $\mathcal{N}(0, b_j^2)$-distributed. Proposition 4.6.18 implies that

$$\sum_{j=1}^{n} Z_j \text{ is distributed according to } \mathcal{N}\left(0, \sum_{j=1}^{n} b_j^2\right).$$

In view of $\sum_{j=1}^{n} b_j^2 = |b|^2$ this proves assertion (6.11).

Let us now turn to the general case. Recall that

$$\vec{Y} = A\vec{X} + \mu$$

and $R = AA^T$. If $a \in \mathbb{R}^n$ is a nonzero vector, then we take the scalar product with respect to a on both sides of the last equation and obtain

$$\left\langle \vec{Y}, a \right\rangle = \left\langle A\vec{X}, a \right\rangle + \langle \mu, a \rangle = \left\langle \vec{X}, A^T a \right\rangle + \langle \mu, a \rangle .$$

An application of statement (6.11) with $b = A^T a$ lets us conclude that $\left\langle \vec{X}, A^T a \right\rangle$ is $\mathcal{N}(0, |A^T a|^2)$-distributed, that is, $\left\langle \vec{Y}, a \right\rangle$ is $\mathcal{N}(\langle \mu, a \rangle , |A^T a|^2)$-distributed. Here we used that A, hence also A^T, are regular, so that $a \neq 0$ yields $b = A^T a \neq 0$, and statement (6.11) applies. Assertion (2) follows now by

$$|A^T a|^2 = \left\langle A^T a, A^T a \right\rangle = \left\langle AA^T a, a \right\rangle = \langle Ra, a \rangle .$$

Property (3) is an immediate consequence of the second one. An application of property (2) to the ith unit vector $e_i = (0, \dots, 0, \underbrace{1}_{i}, 0, \dots, 0)$ in $\mathbb{R}^n$ leads on one side to

$$\left\langle \vec{Y}, e_i \right\rangle = Y_i , \quad 1 \leq i \leq n,$$

and on the other side to

$$\langle Re_i, e_i \rangle = r_{ii} \quad \text{and} \quad \langle \mu, e_i \rangle = \mu_i , \quad 1 \leq i \leq n.$$

Thus, by property (2), for each $i \leq n$ the random variable Y_is is $\mathcal{N}(\mu_i, r_{ii})$-distributed. This completes the proof. ∎

Corollary 6.2.4. *If $\vec{Y}$ is $\mathcal{N}(\mu, R)$-distributed, then $\mathbb{E}\vec{Y} = \mu$ and $Cov_{\vec{Y}} = R$.*

Proof: Choose any regular $n \times n$ matrix $\tilde{A}$ such that $R = \tilde{A}\tilde{A}^T$. The existence of such an $\tilde{A}$ is proved in Proposition A.4.2. Set $\vec{Z} = \tilde{A}\vec{X} + \mu$ for some standard normal vector $\vec{X}$. Then $\vec{Y}$ as well as $\vec{Z}$ are both $\mathcal{N}(\mu, R)$-distributed, hence $\vec{Z} \overset{d}{=} \vec{Y}$. Proposition 6.2.3 implies $\mathbb{E}\vec{Z} = \mu$ and $Cov_{\vec{Z}} = R$. Consequently, by Remark 6.2.2 follows

$$\mathbb{E}\vec{Y} = \mathbb{E}\vec{Z} = \mu \quad \text{and} \quad Cov_{\vec{Y}} = Cov_{\vec{Z}} = R,$$

which completes the proof. ∎

In view of property Corollary 6.2.4 we will use the following notation.

Definition 6.2.5. If $\vec{Y}$ is $\mathcal{N}(\mu, R)$, distributed, then the parameters μ and R are called the (multivariate) **expected value** and the **covariance matrix** of $\vec{Y}$, respectively.

Remark 6.2.6. We proved above that for any normal vector $\vec{Y}$ the coordinate mappings $Y_i = \langle \vec{Y}, e_i \rangle$ are normal as well. The converse is not valid. There are random vectors $\vec{Y}$ with all random variables $\langle \vec{Y}, e_i \rangle$ normal, $1 \leq i \leq n$, but $\vec{Y}$ is not so.

In contrast to this remark, the following is valid.

Proposition 6.2.7. If $\langle \vec{Y}, a \rangle$ is normal **for all** nonzero $a \in \mathbb{R}^n$, then $\vec{Y}$ is normal as well.

Idea of the proof. By assumption, for each $a \neq 0$ there are real numbers μ_a and $\sigma_a > 0$ such that $\langle \vec{Y}, a \rangle$ is $\mathcal{N}(\mu_a, \sigma_a^2)$-distributed. In order to prove the proposition, one has to show that there are a $\mu \in \mathbb{R}^n$ with $\mu_a = \langle \mu, a \rangle$ and an $R > 0$ such that $\sigma_a^2 = \langle Ra, a \rangle$, $a \in \mathbb{R}^n$. The existence of the vector μ easily follows from

$$\mu_{\alpha a + \beta b} = \mathbb{E} \langle \vec{Y}, \alpha a + \beta b \rangle = \alpha \mathbb{E} \langle \vec{Y}, a \rangle + \beta \langle \vec{Y}, b \rangle = \alpha \mu_a + \beta \mu_b \,,$$

using the fact that each linear mapping from $\mathbb{R}^n$ to $\mathbb{R}$ is of the form $a \mapsto \langle a, \mu \rangle$ for a suitable $\mu \in \mathbb{R}^n$.

The existence of an $R > 0$ with $\sigma_a^2 = \langle Ra, a \rangle$ is consequence of a representation theorem for positive quadratic forms on $\mathbb{R}^n$. To this end, one has to show that $a \mapsto \sigma_a^2$ is a positive quadratic form, which follows by using $\sigma_a^2 = \mathbb{E} \langle \vec{Y}, a \rangle^2$.

As we saw above (cf. Proposition 5.3.10), independent random variables are uncorrelated. On the other hand, Examples 5.3.12 and 5.3.15 showed the existence of uncorrelated variables that are not independent. Thus, in general, the property of being uncorrelated is weaker than that of being independent.

One of the basic features of normal vectors is that for them uncorrelated coordinate mappings are already independent. This somehow explains why in the common speech these properties are synonymies.

Proposition 6.2.8. Let $\vec{Y} = (Y_1, \ldots, Y_n)$ be a normally distributed vector. Then the following are equivalent.
(1) $Y_1, \ldots, Y_n$ are independent.
(2) $Y_1, \ldots, Y_n$ are uncorrelated.
(3) The covariance matrix $\mathrm{Cov}_{\vec{Y}}$ is a diagonal matrix.

Proof: The implication (1) $\Rightarrow$ (2) follows by Proposition 5.3.10. If the Y_js are uncorrelated, then this tells us that $\mathrm{Cov}(Y_i, Y_j) = 0$ whenever $i \neq j$. Thus, $\mathrm{Cov}_{\vec{Y}}$ is a diagonal matrix, which proves (2) $\Rightarrow$ (3).

It remains to verify (3) $\Rightarrow$ (1). Thus assume that $\vec{Y}$ is $\mathcal{N}(\mu, R)$ distributed, where $R > 0$ is a diagonal matrix. Let $r_{11}, \ldots, r_{nn}$ be the entries of R at the diagonal. Define A as diagonal matrix with $r_{11}^{1/2}, \ldots, r_{nn}^{1/2}$ on the diagonal. Note that $R > 0$ implies $r_{ii} > 0$, hence A is well-defined. Of course, then $AA^T = R$, hence $\vec{Y}$ has the same distribution as the the vector $(Z_1, \ldots, Z_n)$ with

$$Z_i = r_{ii}^{1/2} X_i + \mu_i, \quad 1 \leq i \leq n,$$

where $X_1, \ldots, X_n$ are independent standard normal. Proposition 4.1.9 lets us conclude that $Z_1, \ldots, Z_n$ are independent normal random variables. But since $\vec{Y} \overset{d}{=} \vec{Z}$, the random variables $Y_1, \ldots, Y_n$ are independent as well[2]. ∎

Remark 6.2.9. Another property, being equivalent to those in Proposition 6.2.8, is as follows. The density function of $\vec{Y}$ is

$$p_{\mu,R}(x) = \frac{1}{(2\pi)^{n/2}|R|^{1/2}} e^{-\sum_{j=1}^{n}(x_j-\mu_j)^2/2r_{jj}}, \quad x = (x_1, \ldots, x_n).$$

Note that $|R| = \det(R) = r_{11} \cdots r_{nn}$.

Finally, we investigate the case of two-dimensional normal vectors more thoroughly. Thus assume $\vec{Y} = (Y_1, Y_2)$ is a normal vector. Then the covariance matrix R is given by

$$R = \begin{pmatrix} \mathbb{V}Y_1 & \text{Cov}(Y_1, Y_2) \\ \text{Cov}(Y_1, Y_2) & \mathbb{V}Y_2 \end{pmatrix}$$

Let σ_1^2 and σ_2^2 be the variance of Y_1 and Y_2, respectively, and let $\rho := \rho(Y_1, Y_2)$ be their correlation coefficient. Because of

$$\text{Cov}(Y_1, Y_2) = (\mathbb{V}Y_1)^{1/2}(\mathbb{V}Y_2)^{1/2}\rho(Y_1, Y_2) = \sigma_1\sigma_2\rho$$

we may rewrite R as

$$R = \begin{pmatrix} \sigma_1^2 & \rho\sigma_1\sigma_2 \\ \rho\sigma_1\sigma_2 & \sigma_2^2 \end{pmatrix}.$$

This implies $\det(R) = \sigma_1^2\sigma_2^2(1 - \rho^2)$. Since $\sigma_1^2 > 0$, the matrix R is positive if and only if $|\rho| < 1$. The inverse matrix R^{-1} can be computed by Cramer's rule as

$$R^{-1} = \frac{1}{\sigma_1^2\sigma_2^2(1-\rho^2)} \begin{pmatrix} \sigma_2^2 & -\rho\sigma_1\sigma_2 \\ -\rho\sigma_1\sigma_2 & \sigma_1^2 \end{pmatrix} = \frac{1}{1-\rho^2} \begin{pmatrix} \frac{1}{\sigma_1^2} & \frac{-\rho}{\sigma_1\sigma_2} \\ \frac{-\rho}{\sigma_1\sigma_2} & \frac{1}{\sigma_2^2} \end{pmatrix}.$$

Consequently,

$$\langle R^{-1}x, x \rangle = \frac{1}{1-\rho^2} \left(\frac{x_1^2}{\sigma_1^2} - \frac{2\rho x_1 x_2}{\sigma_1\sigma_2} + \frac{x_2^2}{\sigma_2^2} \right), \quad x = (x_1, x_2) \in \mathbb{R}^2.$$

---

2 Indeed, use the characterization of independent random variables given in Proposition 3.6.5. The condition stated there depends only on the joint distribution.

If $\mu = (\mu_1, \mu_2) = (\mathbb{E}Y_1, \mathbb{E}Y_2)$ denotes the expected value of $\vec{Y}$, then for $a_1 < b_1$ and $a_2 < b_2$,

$$\mathbb{P}\{a_1 \leq Y_1 \leq b_1, a_2 \leq Y_2 \leq b_2\} = \frac{1}{2\pi(1-\rho^2)^{1/2}\sigma_1\sigma_2} \times$$

$$\times \int_{a_1}^{b_1} \int_{a_2}^{b_2} \exp\left(-\frac{1}{2(1-\rho^2)}\left[\frac{(x_1-\mu_1)^2}{\sigma_1^2} - \frac{2\rho(x_1-\mu_1)(x_2-\mu_2)}{\sigma_1\sigma_2}\right.\right.$$

$$\left.\left.+\frac{(x_2-\mu_2)^2}{\sigma_2^2}\right]\right) dx_2\, dx_1 . \qquad (6.12)$$

Compare this with the case of **independent** Y_1 and Y_2. Here it follows that

$$\mathbb{P}\{a_1 \leq Y_1 \leq b_1, a_2 \leq Y_2 \leq b_2\}$$

$$= \frac{1}{2\pi\sigma_1\sigma_2} \int_{a_1}^{b_1} \int_{a_2}^{b_2} \exp\left(-\frac{1}{2}\left[\frac{(x_1-\mu_1)^2}{\sigma_1^2} + \frac{(x_2-\mu_2)^2}{\sigma_2^2}\right]\right) dx_2\, dx_1 . \qquad (6.13)$$

It is worthwhile to mention that in both cases (dependent and independent) the marginal distributions are the same, namely $\mathcal{N}(\mu_1, \sigma_1^2)$ and $\mathcal{N}(\mu_2, \sigma_2^2)$. A comparison of eqs. (6.12) and (6.13) shows clearly the influence of the correlation coefficient to the density (Fig. 6.2).

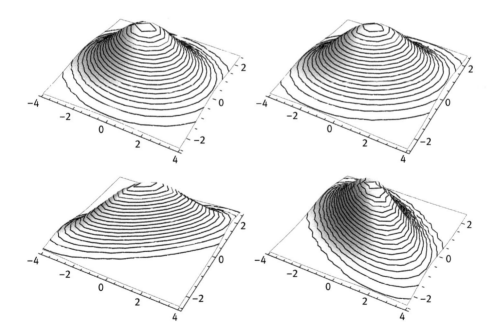

Figure 6.2: $\mu_1 = \mu_2 = 0$, $\sigma_1 = 2$, $\sigma_2 = 1$ and $\rho = 0$, $\rho = 1/4$, $\rho = 3/4$, and $\rho = -1/2$ from top left to bottom right.

6.3 Problems

Problem 6.1. Let $\vec{Y} = (Y_1, \ldots, Y_n)$ be an arbitrary (not necessarily normal) random vector.

1. Show that $\mathbb{E}|Y_j| < \infty$, $1 \leq j \leq n$, if and only if $\mathbb{E}|\vec{Y}| < \infty$. Here $|\vec{Y}|$ denotes the Euclidean distance of $\vec{Y}$.
2. Let A be an arbitrary $n \times n$ matrix. Prove that

$$\mathbb{E}(A\vec{Y}) = A(\mathbb{E}\vec{Y})$$

provided that $\mathbb{E}|\vec{Y}| < \infty$.

3. Show that $\mathbb{E}|Y_j|^2 < \infty$, $1 \leq j \leq n$, if and only if $\mathbb{E}|\vec{Y}|^2 < \infty$.
4. Suppose $\mathbb{E}|\vec{Y}|^2 < \infty$. Let $\mathrm{Cov}_{\vec{Y}}$ be the covariance matrix of $\vec{Y}$. Prove that $\mathrm{Cov}_{\vec{Y}}$ is non-negative definite, that is,

$$\left\langle \mathrm{Cov}_{\vec{Y}} x, x \right\rangle \geq 0, \quad x \in \mathbb{R}^n.$$

Problem 6.2. Let $\vec{X} = (X_1, X_2)$ be a two-dimensional standard normal vector. Compute

$$\mathbb{P}\{|X| \leq 1\} = \mathbb{P}\{X_1^2 + X_2^2 \leq 1\}.$$

Hint: Compare the proof of Proposition 1.6.6.

Problem 6.3. Let $X_1, \ldots, X_{n+m}$ be a sequence of independent standard normal random variables. For an $n \times n$ matrix $A = (\alpha_{ij})_{i,j=1}^n$ and an $m \times m$ matrix $B = (\beta_{kl})_{k,l=1}^m$ define two normal vectors $\vec{Y}$ and $\vec{Z}$ by

$$Y_i = \sum_{j=1}^n \alpha_{ij} X_j \quad \text{and} \quad Z_k = \sum_{l=1}^m \beta_{kl} X_{l+n},$$

$1 \leq i \leq n$ and $1 \leq k \leq m$. Let $(\vec{Y}, \vec{Z})$ be the $(n + m)$-dimensional vector

$$(\vec{Y}, \vec{Z}) = (Y_1, \ldots, Y_n, Z_1, \ldots, Z_m).$$

Why is $(\vec{Y}, \vec{Z})$ normal? Show that the covariance matrix $\mathrm{Cov}_{(\vec{Y}, \vec{Z})}$ is given by

$$\mathrm{Cov}_{(\vec{Y}, \vec{Z})} = \begin{pmatrix} \mathrm{Cov}_{\vec{Y}} & 0 \\ 0 & \mathrm{Cov}_{\vec{Z}} \end{pmatrix}.$$

Problem 6.4. Let X_1, X_2, and X_3 be three standard normal independent random variables. Define the random vector $\vec{Y}$ by

$$\vec{Y} := (X_1 - 1, X_1 + X_2 - 1, X_1 + X_2 + X_3 - 1).$$

1. Argue why $\vec{Y}$ is normal. Determine its expected value, its covariance matrix, and the correlation coefficients $\rho(Y_i, Y_j), 1 \le i < j \le 3$.
2. Determine the distribution density of $\vec{Y}$.

Problem 6.5. The random vector $\vec{Y} = (Y_1, \ldots, Y_n)$ is $\mathcal{N}(\mu, R)$-distributed for some $\mu \in \mathbb{R}^n$ and $R > 0$. Determine the distribution of $Y_1 + \cdots + Y_n$.

Problem 6.6. Prove the following assertion: If $\vec{Y}$ is $\mathcal{N}(0, R)$-distributed, then there exist an orthonormal basis $(f_j)_{j=1}^n$ in $\mathbb{R}^n$, positive numbers $\lambda_1, \ldots, \lambda_n$ and independent $\mathcal{N}(0, 1)$-distributed $\xi_1, \ldots, \xi_n$ such that

$$\vec{Y} = \sum_{j=1}^{n} \lambda_j \xi_j f_j \ . \tag{6.14}$$

Hint: Use the principal axis transformation for symmetric matrices and the fact that unitary matrices map an orthonormal basis onto an orthonormal basis.

Conclude from eq. (6.14) the following: If $\vec{Y}$ is $\mathcal{N}(0, R)$-distributed, then there are $a_1, \ldots, a_n$ in $\mathbb{R}^n$ such that $\left(\vec{Y}, a_1\right), \ldots, \left(\vec{Y}, a_n\right)$ is a sequence of independent standard normal random variables.

Problem 6.7. The n-dimensional vector $\vec{Y}$ is distributed according to $N(\mu, R)$. For some regular $n \times n$ matrix S define $\vec{Z}$ by $\vec{Z} := S\vec{Y}$. Is $\vec{Z}$ normal? If this is so, determine the expected value and the covariance matrix of $\vec{Z}$.

Problem 6.8. Let $\vec{X} = (X_1, X_2)$ be standard normal. Define random variables Y_1 and Y_2 by

$$Y_1 := \frac{1}{\sqrt{2}}(X_1 + X_2) \quad \text{and} \quad Y_2 := \frac{1}{\sqrt{2}}(X_1 - X_2).$$

Why are Y_1 and Y_2 also independent and standard normal?

7 Limit Theorems

Probability Theory does not have the ability to predict the occurrence or nonoccurrence of a single event in a random experiment; besides, this event occurs either with probability one or with probability zero. For example, Probability Theory does not give any information about the next result when rolling a die, it does not predict the numbers appearing next week on the lottery nor is it able to foresee the lifetime of a component in a machine. Such statements are impossible within the theory. The theory is only able to say that some events are more likely and others are less likely. For instance, when rolling a die twice, it is more likely that the sum of both rolls will be "7" than "2." Nevertheless, next when we roll the die the sum may be "2," not "7." The event "the sum is 2" is not impossible, only less likely.

In contrast, Probability Theory provides us with very precise and far-reaching information about the behavior of the results when we execute "many" identical random experiments. As already said, we cannot tell anything about the expected number on a die when we roll it once, but we are able to say a lot about the frequency of the number "6" when rolling a die many times, namely that, on average, this number will appear in one of six cases (provided the die is fair). In this example, certain laws of Probability Theory, which we will present in this section, are operating. These laws are only applicable in the case of many experiments, not in that of a single one.

Limit theorems in Probability Theory belong to the most beautiful and most important assertions within this theory. They are always the highlight of a lecture about advanced Probability Theory. However, their proofs require a longer comprehensive mathematical explanation, which is impossible to give here within the frame of this book. For those who are interested in knowing more about this topic, they may look into one of the more advanced books, such as [Bil12], [Dur10], or [Kho07]. Although the proofs of the limit theorems are mostly quite complicated, they are very important, and their consequences influence our daily lives. Moreover, great parts of Mathematical Statistics are based on these results. Therefore, we decided to state here the crucial assertions without proving most of them. Thus, our main focus is to present the most important limit theorems, to explain them in detail, and to give examples that show how they apply. If possible, we give some hint as to how the results are derived, but mostly we must resign to prove them.

7.1 Laws of Large Numbers

7.1.1 Chebyshev's Inequality

Our first objective is to prove Chebyshev's inequality. To do so, we need the following lemma.

Lemma 7.1.1. *Let Y be a non-negative random variable. Then for each $\lambda > 0$ it follows that*

$$\mathbb{P}\{Y \geq \lambda\} \leq \frac{\mathbb{E}Y}{\lambda}. \tag{7.1}$$

Proof: Let us first treat the case that Y is discrete. Since $Y \geq 0$, its possible values $y_1, y_2, \ldots$ are nonnegative real numbers. Therefore, we get

$$\mathbb{E}Y = \sum_{j=1}^{\infty} y_j \, \mathbb{P}\{Y = y_j\} \geq \sum_{y_j \geq \lambda} y_j \, \mathbb{P}\{Y = y_j\}$$
$$\geq \lambda \sum_{y_j \geq \lambda} \mathbb{P}\{Y = y_j\} = \lambda \mathbb{P}\{Y \geq \lambda\}.$$

Solving the inequality for $\mathbb{P}\{Y \geq \lambda\}$ proves inequality (7.1).

The proof of estimate (7.1) for continuous Y uses similar methods. If q denotes the distribution density of Y, by $Y \geq 0$ we may suppose $q(y) = 0$ for $y < 0$. Then, as in the discrete case, we conclude that

$$\mathbb{E}Y = \int_0^{\infty} yq(y)\,dy \geq \int_{\lambda}^{\infty} yq(y)\,dy \geq \lambda \int_{\lambda}^{\infty} q(y)\,dy = \lambda \mathbb{P}\{Y \geq \lambda\}.$$

From this inequality (7.1) follows directly. ∎

Remark 7.1.2. Sometimes it is useful to apply inequality (7.1) in a slightly modified way. For example, if $Y \geq 0$ and $\alpha > 0$, then one derives

$$\mathbb{P}\{Y \geq \lambda\} = \mathbb{P}\{Y^{\alpha} \geq \lambda^{\alpha}\} \leq \frac{\mathbb{E}Y^{\alpha}}{\lambda^{\alpha}}.$$

Or, if Y is real valued, then for $\lambda \in \mathbb{R}$ we obtain

$$\mathbb{P}\{Y \leq \lambda\} = \mathbb{P}\left\{e^{-Y} \geq e^{-\lambda}\right\} \leq \frac{\mathbb{E}\,e^{-Y}}{e^{-\lambda}} = e^{\lambda}\,\mathbb{E}\,e^{-Y}.$$

Now we are in a position to state and to prove Chebyshev's inequality.

Proposition 7.1.3 (Chebyshev's inequality). *Let X be a random variable with finite second moment. Then, if $c > 0$, it follows that*

$$\mathbb{P}\{|X - \mathbb{E}X| \geq c\} \leq \frac{\mathbb{V}X}{c^2}. \tag{7.2}$$

Proof: Setting $Y := |X - \mathbb{E}X|^2$, we have $Y \geq 0$ and $\mathbb{E}Y = \mathbb{V}X$. Now apply inequality (7.1) to Y with $\lambda = c^2$. This leads to

$$\mathbb{P}\{|X - \mathbb{E}X| \geq c\} = \mathbb{P}\{|X - \mathbb{E}X|^2 \geq c^2\} = \mathbb{P}\{Y \geq c^2\} \leq \frac{\mathbb{E}Y}{c^2} = \frac{\mathbb{V}X}{c^2},$$

and estimate (7.2) is proven. ∎

Interpretation: Inequality (7.2) quantifies the interpretation of $\mathbb{V}X$ as a measure for the dispersion of X. The smaller $\mathbb{V}X$, the less the values of X vary around its expected value $\mathbb{E}X$.

Remark 7.1.4. Another way to formulate inequality (7.2) is as follows. If $\kappa > 0$, then

$$\mathbb{P}\{|X - \mathbb{E}X| \geq \kappa\,(\mathbb{V}X)^{1/2}\} \leq \frac{1}{\kappa^2}.$$

To see this, apply inequality (7.2) with $c = \kappa\,(\mathbb{V}X)^{1/2}$.

Example 7.1.5. Roll a fair die n times. We are interested in the relative frequency of the occurrence of the event $A := \{6\}$. Recall that this frequency was defined in eq. (1.1). Moreover, we claimed in this section that $\lim_{n\to\infty} r_n(A) = \mathbb{P}(A) = \frac{1}{6}$. Is it possible to estimate the probability for $|r_n(A) - \frac{1}{6}|$ being bigger than some given $c > 0$?

Answer: Define the random variable X as the absolute frequency of the occurrence of A, that is, we have $X = k$ for some $k = 0, \dots, n$ provided that A occurred exactly k times. Then X is binomial distributed with parameters n and $p = 1/6$. To see this, define "success" as appearance of "6." Consequently, the relative frequency can be represented as $r_n(A) = \frac{X}{n}$. An application of eqs. (5.7) and (5.33) gives

$$\mathbb{E}\,r_n(A) = \frac{1}{n}\,\mathbb{E}X = \frac{np}{n} = p = \frac{1}{6} \quad \text{and} \quad \mathbb{V}\,r_n(A) = \frac{np(1-p)}{n^2} = \frac{5}{36\,n}.$$

Thus, inequality (7.2) leads to

$$\mathbb{P}\left\{\left|r_n(A) - \frac{1}{6}\right| \geq c\right\} \leq \frac{5}{36\,c^2\,n}.$$

If, for example, $n = 10^3$, and if we choose $c = 1/36$, then Chebyshev's inequality yields

$$\mathbb{P}\left\{\frac{5}{36} < r_{10^3}(A) < \frac{7}{36}\right\} \geq 1 - \frac{9}{50} = 0.82.$$

For the absolute frequency this means

$$\mathbb{P}\{139 \leq a_{10^3} \leq 194\} \geq 0.82.$$

Let us interpret the result. Suppose we roll a fair die 1000 times. Then, with a probability of at least 82%, the frequency of "6" will be between 139 and 194.

Let us present a second quite similar example.

Example 7.1.6. Roll a fair die n times and let S_n be the sum of the n results. Then $S_n = X_1 + \cdots + X_n$, where $X_1, \ldots, X_n$ are uniformly distributed on $\{1, \ldots, 6\}$ and independent. By Example 5.2.17 we know that

$$\mathbb{E}S_n = \mathbb{E}X_1 + \cdots + \mathbb{E}X_n = \frac{7n}{2} \quad \text{and} \quad \mathbb{V}S_n = \mathbb{V}X_1 + \cdots + \mathbb{V}X_n = \frac{35n}{12},$$

hence

$$\mathbb{E}\left(\frac{S_n}{n}\right) = \frac{7}{2} \quad \text{and} \quad \mathbb{V}\left(\frac{S_n}{n}\right) = \frac{35}{12n}.$$

An application of inequality (7.2) leads then to

$$\mathbb{P}\left\{\left|\frac{S_n}{n} - \frac{7}{2}\right| \geq c\right\} \leq \frac{35}{12nc^2}.$$

For example, if $n = 10^3$ and c is chosen as $c = 0.1$, then

$$\mathbb{P}\left\{3.4 < \frac{S_{10^3}}{10^3} < 3.6\right\} \geq 0.709.$$

The interpretation of this result is as in the previous example. With a probability larger than 70% the sum of 1000 rolls of a fair die will be a number between 3400 and 3600.

7.1.2 *Infinite Sequences of Independent Random Variables

Whenever one wants to describe the limit behavior of random variables or random events, one needs a model for the infinite performance of random experiments. Otherwise, we cannot investigate limits or other related quantities. This is comparable with similar investigations in Calculus. In order to analyze limits, infinite sequences are necessary, not finite ones. Thus, for the examination of limits of random variables we need an infinite sequence $X_1, X_2, \ldots$ of random variables, which are, on one hand, independent in the sense of Definition 4.3.4 and, on the other hand, possess some given probability distributions.

Example 7.1.7. In order to describe the infinite tossing of a fair coin we need independent random variables $X_1, X_2, \ldots$ such that $\mathbb{P}\{X_j = 0\} = \mathbb{P}\{X_j = 1\} = \frac{1}{2}$. Or, similarly, for a model of rolling a die infinitely often we need infinitely many independent random variables all uniformly distributed on $\{1, \ldots, 6\}$.

In Proposition 4.3.3 we presented the construction of independent $(X_j)_{j=1}^{\infty}$ distributed according to $B_{1,1/2}$. This technique can be extended to more general sequences of random variables, but it is quite complicated. Another, much smarter way is to use so-called infinite product measures. Their existence follows by a deep theorem due to A. N. Kolmogorov. As a consequence one gets the following result, which cannot be proven within the framework of this book. We refer to [Kho07], Chapter 5, §2, for further reading.

Proposition 7.1.8. *Let $\mathbb{P}_1, \mathbb{P}_2, \ldots$ be arbitrary probability measures on $(\mathbb{R}, \mathcal{B}(\mathbb{R}))$. Then there are a probability space $(\Omega, \mathcal{A}, \mathbb{P})$ and an infinite sequence of random variables $X_j : \Omega \rightarrow \mathbb{R}$ such that the following holds.*
1. *The probability distribution of X_j is $\mathbb{P}_j$, $j = 1, 2, \ldots$. That is, for all $j \geq 1$ and all $B \in \mathcal{B}(\mathbb{R})$ it follows that*

$$\mathbb{P}\{X_j \in B\} = P_j(B).$$

2. *The random variables $X_1, X_2, \ldots$ are independent in the sense of Definition 4.3.4. This says, for all $n \geq 1$ and all $B_j \in \mathcal{B}(\mathbb{R})$ it follows that*

$$\mathbb{P}\{X_1 \in B_1, \ldots, X_n \in B_n\} = \mathbb{P}\{X_1 \in B_1\} \cdots \mathbb{P}\{X_n \in B_n\} = \mathbb{P}_1(B_1) \cdots \mathbb{P}_n(B_n).$$

Of special interest is the case $\mathbb{P}_1 = \mathbb{P}_2 = \cdots = \mathbb{P}_0$ for a certain probability measure $\mathbb{P}_0$ on $\mathbb{R}$. Then the previous proposition implies the following.

Corollary 7.1.9. *Given an arbitrary probability measure $\mathbb{P}_0$ on $\mathcal{B}(\mathbb{R})$, there are random variables $X_1, X_2, \ldots$ such that for all $n \geq 1$ and all $B_j \in \mathcal{B}(\mathbb{R})$*

$$\mathbb{P}\{X_1 \in B_1, \ldots, X_n \in B_n\} = \mathbb{P}_0(B_1) \cdots \mathbb{P}_0(B_n).$$

Example 7.1.10. Choosing as $\mathbb{P}_0$ the uniform distribution on $[0, 1]$, the previous corollary ensures the existence of (independent) random variables $X_1, X_2, \ldots$ such that for all $n \geq 1$ and all $0 \leq a_j < b_j \leq 1$

$$\mathbb{P}\{a_1 \leq X_1 \leq b_1, \ldots, a_n \leq X_n \leq b_n\} = \prod_{j=1}^{n}(b_j - a_j).$$

The sequence $X_1, X_2, \ldots$ models the independent choosing of infinitely many numbers uniformly distributed in $[0, 1]$.

Remark 7.1.11. One may ask whether the kind of independence in Definition 4.3.4 suffices for later purposes. Recall, we only require $X_1, \ldots, X_n$ to be independent for all (finite) $n \geq 1$. Maybe one would expect a condition that involves the whole infinite sequence, not only a finite part of it. The answer is that such a condition for the whole sequence is a consequence of Definition 4.3.4. Namely, if $B_1, B_2, \ldots$ are Borel sets in $\mathbb{R}$, then, by the continuity of probability measures from above, it follows that

$$\mathbb{P}\{X_1 \in B_1, X_2 \in B_2, \ldots\} = \lim_{n \to \infty} \mathbb{P}\{X_1 \in B_1, \ldots, X_n \in B_n\}$$

$$= \lim_{n \to \infty} \mathbb{P}\{X_1 \in B_1\} \cdots \mathbb{P}\{X_n \in B_n\} = \lim_{n \to \infty} \prod_{j=1}^{n} \mathbb{P}\{X_j \in B_j\} = \prod_{j=1}^{\infty} \mathbb{P}\{X_j \in B_j\}.$$

In particular, this implies

$$\mathbb{P}\{a_1 \le X_1 \le b_1, a_2 \le X_2 \le b_2, \ldots\} = \prod_{j=1}^{\infty} \mathbb{P}\{a_j \le X_j \le b_j\}. \tag{7.3}$$

Example 7.1.12. Let $X_1, X_2, \ldots$ be a sequence of independent E_λ-distributed random variables for some $\lambda > 0$. Given real numbers $\alpha_j > 0$, we ask for the probability of

$$\mathbb{P}\{X_1 \le \alpha_1, X_2 \le \alpha_2, \ldots\}.$$

Answer: If we apply eq. (7.3) with $a_j = 0$ and with $b_j = \alpha_j$, then we get

$$\mathbb{P}\{X_1 \le \alpha_1, X_2 \le \alpha_2, \ldots\} = \prod_{j=1}^{\infty} \mathbb{P}\{X_j \le \alpha_j\} = \prod_{j=1}^{\infty} \left[1 - e^{-\lambda \alpha_j} \right].$$

Of special interest are sequences $(\alpha_j)_{j \ge 1}$ such that the infinite product converges, that is, for these sequences $(\alpha_j)_{j \ge 1}$ we have $\prod_{j=1}^{\infty} \left[1 - e^{-\lambda \alpha_j} \right] > 0$. This happens if and only if

$$\ln \left(\prod_{j=1}^{\infty} \left[1 - e^{-\lambda \alpha_j} \right] \right) = \sum_{j=1}^{\infty} \ln[1 - e^{-\lambda \alpha_j}] > -\infty. \tag{7.4}$$

Because of

$$\lim_{x \to 0} \frac{\ln(1 - x)}{-x} = 1,$$

by the limit comparison test for infinite series, condition (7.4) holds if and only if

$$\sum_{j=1}^{\infty} e^{-\lambda \alpha_j} < \infty.$$

If, for example, $\alpha_j = c \cdot \ln(j + 1)$ for some $c > 0$, then

$$\sum_{j=1}^{\infty} e^{-\lambda \alpha_j} = \sum_{j=1}^{\infty} \frac{1}{(j + 1)^{\lambda c}}.$$

This sum is known to be finite if and only if $\lambda c > 1$, that is, if $c > 1/\lambda$.

Another way to formulate this observation is as follows. It holds

$$\mathbb{P}\left\{\sup_{j\geq1}\frac{X_j}{\ln(j+1)}\leq c\right\}=\mathbb{P}\left\{X_j\leq c\ln(j+1),\ \forall j\geq1\right\}=\prod_{j=1}^{\infty}\left(1-\frac{1}{(j+1)^{\lambda c}}\right),$$

and this probability is positive if and only if $c > 1/\lambda$.

7.1.3 * Borel–Cantelli Lemma

The aim of this section is to present one of the most useful tools for the investigation of the limit behavior of infinite sequences of random variables and events. Let $(\Omega, \mathcal{A}, \mathbb{P})$ be a probability space and let $A_1, A_2, \ldots$ be a sequence of events in $\mathcal{A}$. Then two typical questions arise. What is the probability that there exists some $n \in \mathbb{N}$ such that all events A_m with $m \geq n$ occur? The other related question asks for the probability that infinitely many of the events A_n occur.

To explain why these questions are of interest, let us once more regard Example 4.1.7 of the random walk. Here S_n denotes the integer where the particle is located after n random jumps. For example, letting $A_n := \{\omega \in \Omega : S_n(\omega) > 0\}$, then the existence of an $n \in \mathbb{N}$, such that A_m occurs for all $m \geq n$, says that the particle from a certain (random) moment attains only positive numbers and never goes back to the negative ones. Or, if we investigate the events $B_n := \{\omega \in \Omega : S_n(\omega) = 0\}$, then the B_ns occur infinitely often if and only if the particle returns to zero infinitely often. Equivalently, there are (random) $n_1 < n_2 < \cdots$ with $S_{n_j}(\omega) = 0$.

To formulate the two previous questions more precisely, let us introduce the following two events.

Definition 7.1.13. Let $A_1, A_2, \ldots$ be subsets of Ω. Then

$$\liminf_{n\to\infty} A_n := \bigcup_{n=1}^{\infty}\bigcap_{m=n}^{\infty} A_m \quad\text{and}\quad \limsup_{n\to\infty} A_n := \bigcap_{n=1}^{\infty}\bigcup_{m=n}^{\infty} A_m$$

are called the **lower** and the **upper limit** of the A_ns.

Remark 7.1.14. Let us characterize when the lower and the upper limit occur.

1. An element $\omega \in \Omega$ belongs to $\liminf_{n\to\infty} A_n$ if and only if there is an $n \in \mathbb{N}$ such that $\omega \in \bigcap_{m=n}^{\infty} A_m$, that is, if it is an element of A_m for $m \geq n$. In other words, the lower limit occurs if there is an[1] $n \in \mathbb{N}$ such that after n the events A_m always occur.

---

1 Note that this n is random, that is, it may depend on the chosen $\omega \in \Omega$.

Therefore, we say that $\liminf_{n \to \infty} A_n$ occurs if the A_ns **finally always** occur. Thus,

$$\mathbb{P}\{\omega \in \Omega : \exists\, n \text{ s.t. } \omega \in A_m, m \geq n\} = \mathbb{P}\left(\liminf_{n \to \infty} A_n\right).$$

2. An element $\omega \in \Omega$ belongs to $\limsup_{n \to \infty} A_n$ if and only if for each $n \in \mathbb{N}$ there is an $m \geq n$ such that $\omega \in A_m$. But this is nothing else as to say that the number of A_ns with $\omega \in A_n$ is infinite. Therefore, the upper limit consists of those elements for which we have **infinitely often** $\omega \in A_n$. Note that also these events may be different for different ωs. Thus,

$$\mathbb{P}\{\omega \in \Omega : \omega \in A_n \text{ for infinitely many } n\} = \mathbb{P}\left(\limsup_{n \to \infty} A_n\right).$$

Example 7.1.15. Suppose a fair coin is labeled on one side with "0" and on the other side with "1." We toss it infinitely often. Let A_n occur if the nth toss is "1." Then $\liminf_{n \to \infty} A_n$ occurs if after a certain number of tosses "1" always shows up. On the other hand, $\limsup_{n \to \infty} A_n$ occurs if and only if the number "1" appears infinitely often.

Let us formulate and prove some easy properties of the lower and upper limit.

Proposition 7.1.16. *If $A_1, A_2, \ldots$ are subsets of Ω, then*

(1) $\displaystyle \liminf_{n \to \infty} A_n \subseteq \limsup_{n \to \infty} A_n$,

(2) $\displaystyle \left(\limsup_{n \to \infty} A_n\right)^c = \liminf_{n \to \infty} A_n^c$ *and* $\displaystyle \left(\liminf_{n \to \infty} A_n\right)^c = \limsup_{n \to \infty} A_n^c$.

Proof: We prove these properties in the interpretation of the lower and upper limit given in Remark 7.1.14.

Suppose that $\omega \in \liminf_{n \to \infty} A_n$. Then for some $n \geq 1$ it follows that $\omega \in A_m$, $m \geq n$. Of course, then the number of events with $\omega \in A_n$ is infinite, which implies $\omega \in \limsup_{n \to \infty} A_n$. This proves (1).

Observe that we have $\omega \notin \limsup_{n \to \infty} A_n$ if and only if $\omega \in A_n$ for only finitely many $n \in \mathbb{N}$. Equivalently, there is an $n \geq 1$ such that whenever $m \geq n$, then $\omega \notin A_m$, or, that $\omega \in A_m^c$. In other words, this happens if and only if $\omega \in \liminf_{n \to \infty} A_n^c$. This proves the left-hand identity in (2). The second one follows by the same arguments. One may also prove this by applying the left-hand identity with A_n^c. ∎

Before we can formulate the main result in this section, we have to define when an infinite sequence of events is independent.

> **Definition 7.1.17.** A sequence of events $A_1, A_2, \ldots$ in $\mathcal{A}$ is said to be **independent** provided that for all $n \geq 1$ the events $A_1, \ldots, A_n$ are independent in the sense of Definition 2.2.12.

Remark 7.1.18. Using the method for the proof of eq. (7.3) one may deduce the following "infinite" version of independence. For independent $A_1, A_2, \ldots$ follows that

$$\mathbb{P}\left(\bigcap_{n=1}^{\infty} A_n\right) = \prod_{n=1}^{\infty} \mathbb{P}(A_n).$$

Remark 7.1.19. According to Proposition 3.6.7, the independence of random variables and events are linked as follows.

The random variables $X_1, X_2, \ldots$ are independent in the sense of Definition 4.3.4 if and only if for all Borel sets $B_1, B_2, \ldots$ in $\mathbb{R}$ the preimages $X_1^{-1}(B_1), X_2^{-1}(B_2), \ldots$ are independent events as introduced in Definition 7.1.17.

Now we are in the position to state and prove the main result of this section.

Proposition 7.1.20 (Borel–Cantelli lemma). *Let $(\Omega, \mathcal{A}, \mathbb{P})$ be a probability space and let $A_n \in \mathcal{A}, n = 1, 2, \ldots$.*

1. *If $\sum_{n=1}^{\infty} \mathbb{P}(A_n) < \infty$, then this implies*

$$\mathbb{P}(\limsup_{n \to \infty} A_n) = 0. \tag{7.5}$$

2. *For **independent** $A_1, A_2, \ldots$ the following is valid. If $\sum_{n=1}^{\infty} \mathbb{P}(A_n) = \infty$, then*

$$\mathbb{P}(\limsup_{n \to \infty} A_n) = 1.$$

Proof: We start with proving the first assertion. Thus, take arbitrary subsets $A_n \in \mathcal{A}$ satisfying $\sum_{n=1}^{\infty} \mathbb{P}(A_n) < \infty$. Write

$$\limsup_{n \to \infty} A_n = \bigcap_{n=1}^{\infty} B_n$$

with $B_n := \bigcup_{m=n}^{\infty} A_m$. Since $B_1 \supseteq B_2 \supseteq \cdots$, property (7) in Proposition 1.2.1 applies, and together with (5) in the same proposition this leads to

$$\mathbb{P}(\limsup_{n \to \infty} A_n) = \lim_{n \to \infty} \mathbb{P}(B_n) \leq \liminf_{n \to \infty} \sum_{m=n}^{\infty} \mathbb{P}(A_m). \tag{7.6}$$

If $\alpha_1, \alpha_2, \ldots$ are non-negative numbers with $\sum_{n=1}^{\infty} \alpha_n < \infty$, then it is known that $\sum_{m=n}^{\infty} \alpha_m \to 0$ as $n \to \infty$. Applying this observation to $\alpha_n = \mathbb{P}(A_n)$, assertion (7.5) is a direct consequence of estimate (7.6). Thus, the first part is proven.

To prove the second assertion we investigate the probability of the complementary event. Here we have

$$(\limsup_{n \to \infty} A_n)^c = \bigcup_{n=1}^{\infty} \bigcap_{m=n}^{\infty} A_m^c .$$

An application of (5) in Proposition 1.2.1 implies

$$\mathbb{P}\left((\limsup_{n \to \infty} A_n)^c\right) \le \sum_{n=1}^{\infty} \mathbb{P}\left(\bigcap_{m=n}^{\infty} A_m^c\right). \tag{7.7}$$

Fix $n \in \mathbb{N}$ and for $k \ge n$ set $B_k := \bigcap_{m=n}^{k} A_m^c$. Then $B_n \supseteq B_{n+1} \supseteq \cdots$, hence by property (7) in Proposition 1.2.1 it follows that

$$\mathbb{P}\left(\bigcap_{m=n}^{\infty} A_m^c\right) = \mathbb{P}\left(\bigcap_{k=n}^{\infty} B_k\right) = \lim_{k \to \infty} \mathbb{P}(B_k) = \lim_{k \to \infty} \prod_{m=n}^{k} (1 - \mathbb{P}(A_m)) .$$

Here we used in the last step that, according to Proposition 2.2.15, the events $A_1^c, A_2^c, \ldots$ are independent as well. Next we apply the elementary inequality

$$1 - x \le e^{-x}, \quad 0 \le x \le 1,$$

for $x = \mathbb{P}(A_m)$, and because of $\sum_{m=n}^{\infty} \mathbb{P}(A_m) = \infty$ we arrive at

$$\mathbb{P}\left(\bigcap_{m=n}^{\infty} A_m^c\right) \le \limsup_{k \to \infty} \exp\left(-\sum_{m=n}^{k} \mathbb{P}(A_m)\right) = 0 .$$

Plugging this into estimate (7.7) finally implies

$$\mathbb{P}\left((\limsup_{n \to \infty} A_n)^c\right) = 0, \quad \text{hence} \quad \mathbb{P}(\limsup_{n \to \infty} A_n) = 1$$

as asserted. ∎

Remark 7.1.21. The second assertion in Proposition 7.1.20 remains valid under the weaker condition of pairwise independence. But then the proof becomes more complicated.

Corollary 7.1.22. *Let $A_n \in \mathcal{A}$ be **independent** events. Then the following are equivalent.*

$$\mathbb{P}(\limsup_{n\to\infty} A_n) = 0 \iff \sum_{n=1}^{\infty} \mathbb{P}(A_n) < \infty$$

$$\mathbb{P}(\limsup_{n\to\infty} A_n) = 1 \iff \sum_{n=1}^{\infty} \mathbb{P}(A_n) = \infty.$$

Example 7.1.23. Let $(U_n)_{n\geq 1}$ be a sequence of independent random variables, uniformly distributed on $[0, 1]$. Given positive real numbers $(a_n)_{n\geq 1}$, we define events A_n by setting $A_n := \{U_n \leq a_n\}$. Since the U_ns are independent, so are the events A_n, and Corollary 7.1.22 applies. Because of $\mathbb{P}(A_n) = a_n$ this leads to

$$\mathbb{P}\{U_n \leq a_n \text{ infinitely often}\} = \begin{cases} 0 : \sum_{n=1}^{\infty} a_n < \infty \\ 1 : \sum_{n=1}^{\infty} a_n = \infty \end{cases}$$

or, equivalently, to

$$\mathbb{P}\{U_n > a_n \text{ finally always}\} = \begin{cases} 0 : \sum_{n=1}^{\infty} a_n = \infty \\ 1 : \sum_{n=1}^{\infty} a_n < \infty. \end{cases}$$

For example, we have

$$\mathbb{P}\{U_n \leq 1/n \text{ infinitely often}\} = 1 \quad \text{and} \quad \mathbb{P}\{U_n \leq 1/n^2 \text{ infinitely often}\} = 0.$$

Example 7.1.24. Let $(X_n)_{n\geq 1}$ be a sequence of independent $\mathcal{N}(0, 1)$-distributed random variables and let $c_n > 0$. What probability does the event, to observe infinitely often $\{|X_n| \geq c_n\}$, possess?

Answer: It holds

$$\sum_{n=1}^{\infty} \mathbb{P}\{|X_n| \geq c_n\} = \frac{2}{\sqrt{2\pi}} \sum_{n=1}^{\infty} \int_{c_n}^{\infty} e^{-x^2/2}\, dx = \frac{2}{\sqrt{2\pi}} \sum_{n=1}^{\infty} \varphi(c_n),$$

where

$$\varphi(t) := \int_{t}^{\infty} e^{-x^2/2}\, dx, \quad t \in \mathbb{R}.$$

Setting $\psi(t) := t^{-1} e^{-t^2/2}$, $t > 0$, then

$$\varphi'(t) = -e^{-t^2/2} \quad \text{and} \quad \psi'(t) = -\left(1 + \frac{1}{t^2}\right) e^{-t^2/2},$$

hence l'Hôpital's rule implies

$$\lim_{t \to \infty} \frac{\varphi'(t)}{\psi'(t)} = 1, \quad \text{thus} \quad \lim_{t \to \infty} \frac{\varphi(t)}{\psi(t)} = 1.$$

The limit comparison test for infinite series tells us that $\sum_{n=1}^{\infty} \varphi(c_n) < \infty$ if and only if $\sum_{n=1}^{\infty} \psi(c_n) < \infty$. Thus, by the definition of ψ the following are equivalent.

$$\sum_{n=1}^{\infty} \mathbb{P}\{|X_n| \geq c_n\} < \infty \quad \Longleftrightarrow \quad \sum_{n=1}^{\infty} \frac{e^{-c_n^2/2}}{c_n} < \infty.$$

In other words, we have

$$\mathbb{P}\{|X_n| \geq c_n \text{ infinitely often}\} = \begin{cases} 0 \\ 1 \end{cases} \quad \Longleftrightarrow \quad \sum_{n=1}^{\infty} \frac{e^{-c_n^2/2}}{c_n} \begin{array}{l} < \infty \\ = \infty \end{array}$$

For example, if $c_n = c\sqrt{\ln n}$ for some $c > 0$, then

$$\sum_{n=1}^{\infty} \frac{e^{-c_n^2/2}}{c_n} = \frac{1}{c} \sum_{n=1}^{\infty} \frac{1}{n^{c^2/2} \sqrt{\ln n}} < \infty$$

if and only if $c > \sqrt{2}$. In particular, this yields the following interesting fact:

$$\mathbb{P}\{|X_n| \geq \sqrt{2 \ln n} \text{ infinitely often}\} = 1,$$

while for each $c > 2$

$$\mathbb{P}\{|X_n| \geq \sqrt{c \ln n} \text{ infinitely often}\} = 0.$$

From this we derive

$$\mathbb{P}\left\{\omega \in \Omega : \limsup_{n \to \infty} \frac{|X_n(\omega)|}{\sqrt{\ln n}} = \sqrt{2}\right\} = 1.$$

Example 7.1.25. In a lottery 6 of 49 numbers are randomly chosen. Find the probability to have infinitely often the six chosen numbers on your lottery ticket.

Answer: Let A_n be the event to have in the nth drawing the six chosen numbers on the ticket. We saw (compare Example 1.4.3) that

$$\mathbb{P}(A_n) = \frac{1}{\binom{49}{6}} := \delta > 0.$$

Consequently, it follows $\sum_{n=1}^{\infty} \mathbb{P}(A_n) = \infty$, and since the A_ns are independent, Proposition 7.1.20 implies

$$\mathbb{P}\{A_n \text{ infinitely often}\} = 1.$$

Therefore, the event to win infinitely often has probability 1. One does only not play long enough!

Remark 7.1.26. Corollary 7.1.22 shows particularly that for independent A_ns either

$$\mathbb{P}(\limsup_{n\to\infty} A_n) = 0 \quad \text{or} \quad \mathbb{P}(\limsup_{n\to\infty} A_n) = 1.$$

Because of Proposition 7.1.16 the same is valid for the lower limit. Here operate so-called 0-1 laws, which, roughly spoken, assert the following. Whenever the occurrence or nonoccurrence of an event is independent of the first finitely many results, then such events occur either with probability 0 or 1. For example, the occurrence or nonoccurrence of the lower or upper limit is completely independent of what had happened during the first n results, $n \geq 1$.

7.1.4 Weak Law of Large Numbers

Given random variables $X_1, X_2, \ldots$ let

$$S_n := X_1 + \cdots + X_n \tag{7.8}$$

be the sum of the first n values. One of the most important questions in Probability Theory is that about the behavior of S_n as $n \to \infty$. Suppose we play a series of games and X_j denotes the loss or the gain in game $j \geq 1$. Then S_n is nothing else than the total loss or gain after n games. Also recall the random walk presented in Example 4.1.7. Set $X_j = -1$ if in step j the particle jumps to the left and $X_j = 1$, otherwise. Then S_n is the point in $\mathbb{Z}$ where the particle is located after n jumps.

Let us come back to the general case. We are given arbitrary independent and identical distributed random variables $X_1, X_2, \ldots$. Recall that "identically distributed" says that they possess all the same probability distribution. Set $S_n = X_1 + \cdots + X_n$. A first result gives some information about the behavior of the arithmetic mean S_n/n as $n \to \infty$.

Proposition 7.1.27 (Weak law of large numbers). *Let $X_1, X_2, \ldots$ be independent identically distributed random variables with (common) expected value $\mu \in \mathbb{R}$. If $\varepsilon > 0$, then it follows that*

$$\lim_{n\to\infty} \mathbb{P}\left\{ \left| \frac{S_n}{n} - \mu \right| \geq \varepsilon \right\} = 0.$$

Proof: We prove the result with only an additional condition, namely that X_1, and hence all X_j, possess a finite second moment. The result remains true without this condition, but then its proof becomes significantly more complicated.

From (3) in Proposition 5.1.36 we derive

$$\mathbb{E}\left(\frac{S_n}{n}\right) = \frac{\mathbb{E}S_n}{n} = \frac{\mathbb{E}(X_1 + \cdots + X_n)}{n} = \frac{\mathbb{E}X_1 + \cdots + \mathbb{E}X_n}{n} = \frac{n\mu}{n} = \mu.$$

Furthermore, by the independence of the X_js, property (iv) in Proposition 5.2.15 also gives

$$\mathbb{V}\left(\frac{S_n}{n}\right) = \frac{\mathbb{V}S_n}{n^2} = \frac{\mathbb{V}X_1 + \cdots + \mathbb{V}X_n}{n^2} = \frac{\mathbb{V}X_1}{n}.$$

Consequently, inequality (7.2) implies

$$\mathbb{P}\left\{\left|\frac{S_n}{n} - \mu\right| \geq \varepsilon\right\} \leq \frac{\mathbb{V}(S_n/n)}{\varepsilon^2} = \frac{\mathbb{V}X_1}{n\varepsilon^2},$$

and the desired assertion follows by

$$\limsup_{n\to\infty} \mathbb{P}\left\{\left|\frac{S_n}{n} - \mu\right| \geq \varepsilon\right\} \leq \lim_{n\to\infty} \frac{\mathbb{V}X_1}{n\varepsilon^2} = 0.$$

∎

Remark 7.1.28. The type of convergence appearing in Proposition 7.1.27 is usually called **convergence in probability**. More precisely, given random variables $Y_1, Y_2, \ldots$, they converge in probability to some random variable Y provided that for each $\varepsilon > 0$

$$\lim_{n\to\infty} \mathbb{P}\{|Y_n - Y| \geq \varepsilon\} = 0.$$

Hence, in this language the weak law of large numbers asserts that S_n/n converges in probability to a random variable Y, which is the constant μ.

Interpretation of Proposition 7.1.27 : Fix $\varepsilon > 0$ and define events A_n, $n \geq 1$, by

$$A_n := \left\{\omega \in \Omega : \left|\frac{S_n(\omega)}{n} - \mu\right| < \varepsilon\right\}.$$

Then Proposition 7.1.27 implies $\lim_{n\to\infty} \mathbb{P}(A_n) = 1$. Hence, given $\delta > 0$, then there is an $n_0 = n_0(\varepsilon, \delta)$ such that $\mathbb{P}(A_n) \geq 1-\delta$ whenever $n \geq n_0$. In other words, if n is sufficiently large, then with high probability (recall, μ is the expected value of the X_js)

$$\mu - \varepsilon \leq \frac{1}{n}\sum_{j=1}^{n} X_j \leq \mu + \varepsilon.$$

This confirms once more the interpretation of the expected value as (approximative) arithmetic mean of the observed values, provided that we execute the same experiment arbitrarily often and the results do not depend on each other.

7.1.5 Strong Law of Large Numbers

Proposition 7.1.27 does not imply $S_n/n \to \mu$ in the usual sense. It only asserts the convergence of S_n/n in probability, which, in general, does not imply a pointwise convergence. The following theorem shows that, nevertheless, a strong type of convergence takes place. The proof of this result is much more complicated than that of Proposition 7.1.27. Therefore, we cannot present it in the scope of this book, and we refer to [Dur10], Section 2.4, for a proof.

Proposition 7.1.29 (Strong law of large numbers). *Let $X_1, X_2, \ldots$ be a sequence of independent identically distributed random variables with expected value $\mu = \mathbb{E}X_1$. If S_n is defined by eq. (7.8), then*

$$\mathbb{P}\left\{\omega \in \Omega : \lim_{n\to\infty} \frac{S_n(\omega)}{n} = \mu\right\} = 1.$$

Remark 7.1.30. Given random variables $Y_1, Y_2, \ldots$ and Y, one says that the Y_ns converge to Y **almost surely**, if

$$\mathbb{P}\left\{\lim_{n\to\infty} Y_n = Y\right\} = \mathbb{P}\left\{\omega \in \Omega : \lim_{n\to\infty} Y_n(\omega) = Y(\omega)\right\} = 1.$$

Thus, Proposition 7.1.29 asserts that S_n/n converges almost surely to a random variable Y, which is constant μ.

Remark 7.1.31. Proposition 7.1.29 allows the following interpretation. There exists a subset Ω_0 in the sample space Ω with $\mathbb{P}(\Omega_0) = 1$ such that for all $\omega \in \Omega_0$ and all $\varepsilon > 0$, there is an $n_0 = n_0(\varepsilon, \omega)$ with

$$\left|\frac{S_n(\omega)}{n} - \mu\right| < \varepsilon$$

whenever $n \geq n_0$.

In other words, with probability one the following happens: given $\varepsilon > 0$, then there is a certain n_0 depending on ω, hence being random, such that for $n \geq n_0$ the arithmetic mean S_n/n is in an ε-neighborhood of μ and never leaves it again.

Let us emphasize once more that S_n/n is random, hence S_n/n may attain different values for a different series of experiments. Nevertheless, starting from a certain point, which may be different for different experiments, the arithmetic mean of the first n results will be in $(\mu - \varepsilon, \mu + \varepsilon)$.

When we introduced probability measures in Section 1.1.3, we claimed that the number $\mathbb{P}(A)$ may be regarded as limit of the relative frequencies of the occurrence of A. As a first consequence of Proposition 7.1.29 we show that this is indeed true.

Proposition 7.1.32. *Suppose a random experiment is described by a probability space* $(\Omega, \mathcal{A}, \mathbb{P})$. *Execute this experiment arbitrarily often. Given an event* $A \in \mathcal{A}$, *let* $r_n(A)$ *be the relative frequency of* A *in* n *trials as defined in eq. (1.1). Then almost surely*

$$\lim_{n \to \infty} r_n(A) = \mathbb{P}(A).$$

Proof: Define random variables $X_1, X_2, \ldots$ as follows. Set $X_j = 1$ if A occurs in trial j, while $X_j = 0$ otherwise. Since the experiments are executed independently of each other, the X_js are independent as well. Moreover, we execute every time exactly the same experiment, hence the X_js are also identically distributed.

By the definition of the X_js,

$$\frac{S_n}{n} = r_n(A).$$

Thus, it remains to evaluate $\mu = \mathbb{E}X_j$. To this end observe that the X_js are $B_{1,p}$-distributed with success probability $p = \mathbb{P}(A)$. Recall that $X_j = 1$ if and only if A occurs in experiment j, and since the experiment is described by $(\Omega, \mathcal{A}, \mathbb{P})$, the probability for X_j being one is $\mathbb{P}(A)$. Consequently, $\mathbb{E}X_j = \mathbb{P}(A)$.

Proposition 7.1.29 now implies that almost surely

$$\lim_{n \to \infty} r_n(A) = \lim_{n \to \infty} \frac{S_n}{n} = \mathbb{E}X_1 = \mathbb{P}(A).$$

This completes the proof. ∎

What does happen in the case that the X_js do not possess an expected value? Does then S_n/n converge nevertheless? If this is so, could we take this limit as a "generalized" expected value? The next proposition shows that such an approach does not work.

Proposition 7.1.33. *Let* $X_1, X_2, \ldots$ *be independent and identically distributed with* $\mathbb{E}|X_1| = \infty$. *Then it follows that*

$$\mathbb{P}\left\{ \omega \in \Omega : \frac{S_n(\omega)}{n} \text{ diverges} \right\} = 1.$$

For example, if we take an independent sequence $(X_j)_{j \geq 1}$ of Cauchy distributed random variables, then their arithmetic means S_n/n will diverge almost surely.

Remark 7.1.34. Why does one need a **weak** law of large numbers when there exists a **strong** one? This question is justified, and in fact, in the situation described in this book the weak law is a consequence of the strong one, thus, it is not necessarily needed.

The situation is different if one investigates independent, but not necessarily identically distributed, random variables. Then there are sequences $X_1, X_2, \ldots$ satisfying the weak law but not the strong one.[2]

Let us state two applications of Proposition 7.1.29, one taken from Numerical Mathematics, the other from Number Theory.

Example 7.1.35 (Monte Carlo method for integrals). Suppose we are given a quite "complicated" function $f : [0, 1]^n \to \mathbb{R}$. The task is to find the numerical value of

$$\int_{[0,1]^n} f(x)\, dx = \int_0^1 \cdots \int_0^1 f(x_1, \ldots, x_n)\, dx_n \cdots dx_1.$$

For large n this can be a highly nontrivial problem. One way to overcome this difficulty is to use a probabilistic approach that is based on the strong law of large numbers.

To this end, choose an independent sequence $\vec{U}_1, \vec{U}_2, \ldots$ of random vectors uniformly distributed on $[0, 1]^n$. For example, such a sequence can be constructed as follows. Take independent $U_1, U_2, \ldots$ uniformly distributed on[3] $[0, 1]$ and build random vectors by $\vec{U}_1 = (U_1, \ldots, U_n)$, $\vec{U}_2 = (U_{n+1}, \ldots, U_{2n})$, and so on.

Proposition 7.1.36. *As above, let* $\vec{U}_1, \vec{U}_2, \ldots$ *be independent random vectors uniformly distributed on* $[0, 1]^n$. *Given an integrable function* $f : [0, 1]^n \to \mathbb{R}$, *then, with probability one,*

$$\lim_{N \to \infty} \frac{1}{N} \sum_{j=1}^{N} f(\vec{U}_j) = \int_{[0,1]^n} f(x)\, dx.$$

Proof: Set $X_j := f(\vec{U}_j)$, $j = 1, 2, \ldots$. By construction, the X_js are independent and identically distributed random variables. Proposition 3.6.18 implies (compare also Example 3.6.21) that the distribution densities of the random vectors $\vec{U}_j$ are given by

$$p(x) = \begin{cases} 1 : x \in [0, 1]^n \\ 0 : x \notin [0, 1]^n \end{cases}.$$

---

2 In the case of nonidentically distributed X_js one investigates if $\frac{1}{n} \sum_{j=1}^{n} (X_j - \mathbb{E}X_j)$ converges to zero either in probability (weak law) or almost surely (strong law).

3 Use the methods developed in Section 4.4 to construct such U_js.

As already mentioned in Remark 5.3.5, formula (5.38), stated for a function of two variables, also holds for functions of n variables, $n \geq 1$ arbitrary. This implies

$$\mathbb{E}X_1 = \mathbb{E}f(\vec{U}_1) = \int_{\mathbb{R}^n} f(x)\,p(x)\,dx = \int_{[0,1]^n} f(x)\,dx.$$

Thus, Proposition 7.1.29 applies and leads to

$$\mathbb{P}\left\{\lim_{N\to\infty} \frac{1}{N}\sum_{j=1}^{N} f(\vec{U}_j) = \int_{[0,1]^n} f(x)\,dx\right\} = \mathbb{P}\left\{\lim_{N\to\infty} \frac{1}{N}\sum_{j=1}^{N} X_j = \mathbb{E}X_1\right\} = 1$$

as asserted. ∎

Remark 7.1.37. The numerical application of the preceding proposition is as follows. Choose independent numbers $u_i^{(j)}$, $1 \leq i \leq n$, $1 \leq j \leq N$, uniformly distributed on $[0,1]$ and set

$$R_N(f) := \frac{1}{N}\sum_{j=1}^{N} f\bigl(u_1^{(j)}, \ldots, u_n^{(j)}\bigr).$$

Proposition 7.1.36 asserts that $R_N(f)$ converges almost surely to $\int_{[0,1]^n} f(x)\,dx$. Thus, if $N \geq 1$ is large, then $R_N(f)$ may be taken as approximative value for $\int_{[0,1]^n} f(x)\,dx$.

If we apply Proposition 7.1.36 to the indicator function of a Borel set $B \subseteq [0,1]^n$, that is, we choose $f = \mathbb{1}_B$ with $\mathbb{1}_B$ as in Definition 3.6.14, then with probability 1 it follows that

$$\mathrm{vol}_n(B) = \int_{[0,1]^n} \mathbb{1}_B(x)\,dx = \lim_{N\to\infty} \frac{1}{N}\sum_{j=1}^{N} \mathbb{1}_B(\vec{U}_j) = \lim_{N\to\infty} \frac{\#\{j \leq N : \vec{U}_j \in B\}}{N}.$$

This provides us with a method to determine the volume $\mathrm{vol}_n(B)$, even for quite "complicated" Borel sets $B \subseteq \mathbb{R}^n$.

Example 7.1.38 (Normal numbers). As we saw in Section 4.3.1, each $x \in [0,1)$ admits a representation as binary fraction $x = 0.x_1x_2\cdots$ with $x_j \in \{0,1\}$. Take some fixed $x \in [0,1)$ with binary representation $x = 0.x_1x_2\cdots$. Then one may ask whether in the binary representation of x one of the numbers 0 or 1 occurs more frequently than the other one. Or do both numbers possess the same frequency, at least on average?

To investigate this question, for $n \in \mathbb{N}$ set

$$a_n^0(x) := \#\{k \leq n : x_k = 0\} \quad \text{and} \quad a_n^1(x) := \#\{k \leq n : x_k = 1\}, \quad x = 0.x_1x_2\cdots$$

Thus, $a_n^0(x)$ is the frequency of the number 0 among the first n positions in the representation of x.

Definition 7.1.39. A number $x \in [0, 1)$ is said to be **normal** (with respect to base 2) if

$$\lim_{n \to \infty} \frac{a_n^0(x)}{n} = \lim_{n \to \infty} \frac{a_n^1(x)}{n} = \frac{1}{2}.$$

In other words, a number $x \in [0, 1)$ is normal with respect to base 2 if, on average, in its binary representation the frequency of 0, and hence also of 1, equals 1/2. Are there many normal numbers as, for example, $x = 0.0101010 \cdots$ or maybe only a few ones? Answer gives the next proposition.

Proposition 7.1.40. *Let $\mathbb{P}$ be the uniform distribution on $[0, 1]$. Then there is a subset $M \subseteq [0, 1)$ with $\mathbb{P}(M) = 1$ such that all $x \in M$ are normal with respect to base 2.*

Proof: Define random variables $X_k : [0, 1) \to \mathbb{R}$, $k = 0, 1, \ldots$, by $X_k(x) := x_k$ whenever $x = 0.x_1 x_2 \cdots$. Proposition 4.3.3 tells us that the X_ks are independent with $\mathbb{P}\{X_k = 0\} = 1/2$ and $\mathbb{P}\{X_k = 1\} = 1/2$. Recall that the underlying probability measure $\mathbb{P}$ on $[0, 1]$ is the uniform distribution. By the definition of the X_ks it follows that

$$S_n(x) := X_1(x) + \cdots + X_n(x) = \#\{k \le n : X_k(x) = 1\} = a_n^1(x).$$

Since $\mathbb{E}X_1 = 1/2$, Proposition 7.1.29 implies the existence of a subset $M \subseteq [0, 1)$ with $\mathbb{P}(M) = 1$ such that for $x \in M$ it follows that

$$\lim_{n \to \infty} \frac{a_n^1(x)}{n} = \lim_{n \to \infty} \frac{S_n(x)}{n} = \mathbb{E}X_1 = \frac{1}{2}.$$

Since $a_n^0(x) = n - a_n^1(x)$, this completes the proof. ∎

Remark 7.1.41. The previous considerations do not depend on the fact that the base of the representation was 2. It extends easily to representations with respect to any base $b \ge 2$. Here, the definition of normal numbers has to be extended slightly. Fix $b \ge 2$. Each $x \in [0, 1)$ admits the representation $x = 0.x_1 x_2 \cdots$ where $x_j \in \{0, \ldots, b-1\}$ provided that $x = \sum_{k=1}^{\infty} \frac{x_k}{b^k}$. To make this representation unique we do not allow representations $x = 0.x_1 x_2 \cdots$ where for some $k_0 \in \mathbb{N}$ we have $x_k = b - 1$ whenever $k \ge k_0$.

Then a number x is said to be normal with respect to the base $b \ge 2$ if for all $\ell = 0, \ldots b - 1$

$$\lim_{n \to \infty} \frac{\#\{j \le n : x_j = \ell\}}{n} = \frac{1}{b}, \quad x = 0.x_1 x_2 \cdots .$$

Similar methods as used in the proof of Proposition 7.1.40 show that there is a set $M_b \subset [0, 1]$ with $\mathbb{P}(M_b) = 1$ such that all $x \in M_b$ are normal with respect to base b.

Letting $M = \bigcap_{b=2}^{\infty} M_b$, then property (5) (Boole's inequality) in Proposition 1.2.1 easily gives $\mathbb{P}(M) = 1$. Numbers $x \in M$ are **completely normal**, which says that they are normal for any base $b \geq 2$. Again we see that with respect to the uniform distribution on $[0, 1]$ almost all numbers are completely normal.

7.2 Central Limit Theorem

Why does the normal distribution play such an important role in Probability Theory and why are so many observed random phenomenons normally distributed? The reason for this is the central limit theorem, which we are going to present in this section.

Regard a sequence of independent and identically distributed random variables $(X_j)_{j \geq 1}$ with finite second moment. As in eq. (7.8) let S_n be the sum of $X_1, \ldots, X_n$. For example, if X_j is the loss or gain in the jth game, then S_n is the total loss or gain after n games. Which probability distribution does S_n possess? Theoretically, this can be evaluated by the convolution formulas stated in Section 4.5. But practically, this is mostly impossible; imagine, we want to determine the distribution of the sum of 100 rolls with a fair die. Therefore, one is very interested in asymptotic statements about the distribution of S_n.

To get a clue about possible asymptotic distributions of S_n, take independent $B_{1,p}$-distributed X_js. In this case, the distribution of S_n is known to be $B_{n,p}$.

For example, if $p = 0.4$ and $n = 30$, then $\mathbb{P}\{S_n = k\} = B_{n,p}(\{k\})$, $k = 0, \ldots, 30$, may be described in Fig. 7.1.

The summit of the diagram occurs at $k = 12$, which is the expected value of S_{30}. Enlarging the number of trials leads to a shift of the summit to the right. At the same time, the height of the summit becomes smaller.

The shape of the diagram in Figure 7.1 lets us suggest that sums of independent, identically distributed random variables "almost" normally distributed. If this is so, which expected value and which variance would the approximating normal distribution possess?

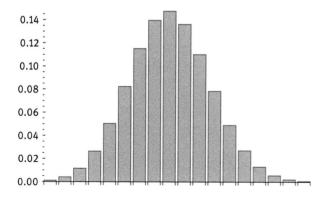

Figure 7.1: Probability mass function of $B_{n,p}$, $n = 30$ and $p = 0.4$.

Let us investigate this question in the general setting. Thus, we are given a sequence $(X_j)_{j\geq 1}$ of independent identically distributed random variables with finite second moment and with $\mu = \mathbb{E}X_1$ and $\sigma^2 = \mathbb{V}X_1 > 0$. If, as before, $S_n = X_1 + \cdots + X_n$, then

$$\mathbb{E}S_n = n\mu \quad \text{and} \quad \mathbb{V}S_n = n\sigma^2.$$

Consequently, if we conjecture that S_n is "approximative" normally distributed, then the normalized sum $(S_n - n\mu)/\sigma\sqrt{n}$ should be "approximative" $\mathcal{N}(0, 1)$-distributed. Recall that Propositions 5.1.36 and 5.2.15 imply

$$\mathbb{E}\left(\frac{S_n - n\mu}{\sigma\sqrt{n}}\right) = 0 \quad \text{and} \quad \mathbb{V}\left(\frac{S_n - n\mu}{\sigma\sqrt{n}}\right) = 1.$$

The question about the possible limit of the normalized sums $(S_n - n\mu)/\sigma\sqrt{n}$ remained open for long time. In 1718 Abraham de Moivre investigated the limit behavior for a special case of binomial distributed random variables. As limit he found some infinite series, not a concrete function. In 1808 the American scientist and mathematician Robert Adrain published a paper where for the first time the normal distribution occurred. A year later, independently of the former work, Carl Friedrich Gauß used the normal distribution for error estimates. In 1812 Pierre-Simon Laplace proved that the normalized sums of independent binomial distributed random variables approximate the normal distribution. Later on, Andrei Andreyevich Markov, Aleksandr Mikhailovich Lyapunov, Jarl Waldemar Lindeberg, Paul Lévy, and other mathematicians continued the work of De Moivre and Laplace. In particular, they showed that the normal distribution occurs **always** as a limit, not only for binomial distributed random variables. The only assumption is that the random variables possess a finite second moment. We refer to the very interesting book [Fis11] for further reading about the history of normal approximation.

It remains the question in which sense does $(S_n - n\mu)/\sigma\sqrt{n}$ converge to the standard normal distribution. To answer this we have to introduce the concept of the **convergence in distribution**.

Definition 7.2.1. Let $Y_1, Y_2, \ldots$ and Y be random variables with distribution functions[4] $F_1, F_2, \ldots$ and F. The sequence $(Y_n)_{n\geq 1}$ **converges** to Y **in distribution** provided that

$$\lim_{n\to\infty} F_n(t) = F(t) \quad \text{for all } t \in \mathbb{R} \text{ at which } F \text{ is continuous.} \tag{7.9}$$

In this case one writes $Y_n \overset{\mathcal{D}}{\longrightarrow} Y$.

---

4 $F_n(t) = \mathbb{P}\{Y_n \leq t\}, t \in \mathbb{R}$.

Remark 7.2.2. An alternative way to formulate property (7.9) is as follows:

$$\lim_{n\to\infty} \mathbb{P}\{Y_n \le t\} = \mathbb{P}\{Y \le t\} \text{ for all } t \in \mathbb{R} \text{ with } \mathbb{P}\{Y = t\} = 0.$$

Without proof we state two other characterizations of convergence in distribution.

Proposition 7.2.3. *One has $Y_n \xrightarrow{D} Y$ if and only if for all bounded continuous functions* $f : \mathbb{R} \to \mathbb{R}$

$$\lim_{n\to\infty} \mathbb{E}f(Y_n) = \mathbb{E}f(Y).$$

Furthermore, this is also equivalent to

$$\limsup_{n\to\infty} \mathbb{P}\{Y_n \in A\} \le \mathbb{P}\{Y \in A\}$$

for all closed subsets $A \subseteq \mathbb{R}$.

If the distribution function of Y is continuous, that is, we have $\mathbb{P}\{Y = t\} = 0$ for all $t \in \mathbb{R}$, then $Y_n \xrightarrow{D} Y$ is equivalent to $\lim_{n\to\infty} F_n(t) = F(t)$ for **all** $t \in \mathbb{R}$. Besides, in this case, the type of convergence is stronger as the next proposition shows.

Proposition 7.2.4. *Let $Y_1, Y_2, \ldots$ and Y be random variables with $\mathbb{P}\{Y = t\} = 0$ for all $t \in \mathbb{R}$. Then $Y_n \xrightarrow{D} Y$ implies that the distribution functions converge uniformly, that is,*

$$\lim_{n\to\infty} \sup_{t\in\mathbb{R}} |\mathbb{P}\{Y_n \le t\} - \mathbb{P}\{Y \le t\}| = 0, \quad \text{hence also}$$
$$\lim_{n\to\infty} \sup_{a<b} |\mathbb{P}\{a \le Y_n \le b\} - \mathbb{P}\{a \le Y \le b\}| = 0.$$

We have now all notations and definitions that are necessary to formulate the central limit theorem. Mostly, this theorem is proved via properties of so-called characteristic functions (see Chapter 3 of [Dur10] for such a proof). For alternative proofs using properties of moment generating functions we refer to [Ros14] and [Gha05].

Proposition 7.2.5 (Central limit theorem). *Let $(X_j)_{j\ge1}$ be a sequence of independent identically distributed random variables with finite second moment. Let μ be the expected value of the X_js and let $\sigma^2 > 0$ be their variance. Then for the sums $S_n = X_1 + \cdots + X_n$ it follows that*

$$\frac{S_n - n\mu}{\sigma\sqrt{n}} \overset{\mathcal{D}}{\longrightarrow} Z. \tag{7.10}$$

Here Z is an $\mathcal{N}(0, 1)$-distributed random variable.

Since the limit Z in statement (7.10) is a continuous random variable, Proposition 7.2.4 applies, and the central limit theorem may also be formulated as follows.

Proposition 7.2.6. *Suppose $(X_j)_{j\geq1}$ and S_n are as in Proposition 7.2.5. Then it follows that*

$$\lim_{n\to\infty} \sup_{t\in\mathbb{R}} \left| \mathbb{P}\left\{ \frac{S_n - n\mu}{\sigma\sqrt{n}} \leq t \right\} - \frac{1}{\sqrt{2\pi}} \int_{-\infty}^{t} e^{-x^2/2}\, dx \right| = 0 \quad and \tag{7.11}$$

$$\lim_{n\to\infty} \sup_{a<b} \left| \mathbb{P}\left\{ a \leq \frac{S_n - n\mu}{\sigma\sqrt{n}} \leq b \right\} - \frac{1}{\sqrt{2\pi}} \int_{a}^{b} e^{-x^2/2}\, dx \right| = 0. \tag{7.12}$$

Remark 7.2.7. Recall that Φ denotes the distribution function of the standard normal distribution as introduced in eq. (1.62). Thus, another way to write eq. (7.11) is as follows:

$$\lim_{n\to\infty} \sup_{t\in\mathbb{R}} \left| \mathbb{P}\left\{ \frac{S_n - n\mu}{\sigma\sqrt{n}} \leq t \right\} - \Phi(t) \right| = 0.$$

Our next objective is another reformulation of eq. (7.12). If we set $a' = a\sigma\sqrt{n} + n\mu$ and $b' = b\sigma\sqrt{n} + n\mu$, then these numbers depend on $n \in \mathbb{N}$. But since the convergence in eq. (7.12) is uniform, we may replace a and b by a' and b', respectively and obtain

$$\lim_{n\to\infty} \sup_{a'<b'} \left| \mathbb{P}\left\{ a' \leq S_n \leq b' \right\} - \mathbb{P}\left\{ \frac{a' - n\mu}{\sigma\sqrt{n}} \leq Z \leq \frac{b' - n\mu}{\sigma\sqrt{n}} \right\} \right| = 0. \tag{7.13}$$

Here, as before, Z denotes a standard normally distributed random variable. For a final reformulation set

$$Z_n := \sigma\sqrt{n}\, Z + n\mu.$$

Then eq. (7.13) is equivalent to

$$\lim_{n\to\infty} \sup_{a'<b'} \left| \mathbb{P}\left\{ a' \leq S_n \leq b' \right\} - \mathbb{P}\left\{ a' \leq Z_n \leq b' \right\} \right| = 0. \tag{7.14}$$

By Proposition 4.2.3, the random variables Z_n are $\mathcal{N}(n\mu, n\sigma^2)$-distributed, which allows us to interpret eq. (7.13), or eq. (7.14), as follows. If $\mu = \mathbb{E}X_j$ and $\sigma^2 = \mathbb{V}X_j$, then for large n, the sum S_n is "approximative" $\mathcal{N}(n\mu, n\sigma^2)$-distributed.

In other words, for $-\infty \le a < b \le \infty$ it follows that

$$P\{a \le S_n \le b\} \approx \Phi\left(\frac{b - n\mu}{\sigma\sqrt{n}}\right) - \Phi\left(\frac{a - n\mu}{\sigma\sqrt{n}}\right).$$

!

Interpretation: We emphasize once more that the central limit theorem is valid for **all** sequences of independent identically distributed random variables possessing a second moment. For example, it is true for X_js that are binomial distributed, for X_js being exponentially distributed, and so on. Thus, no matter how the random variables with second moment are distributed, all their normalized sums possess the same limit, the normal distribution. This explains the outstanding role of the normal distribution.

The deeper reason for this phenomenon is that S_n may be viewed as the superposition of many "small" independent errors or perturbations, all of the same kind.[5] Although each perturbation is distributed according to $\mathbb{P}_{X_1}$, the independent superposition of the perturbations leads to the fact that the final result is approximative normally distributed. This explains why so many random phenomena may be described by normally distributed random variables.

Remark 7.2.8 (Continuity correction). A slight technical problem arises in the case of discrete random variables X_j. Then the S_ns are discrete as well, hence their distribution functions F_n have jumps. If these noncontinuous F_ns approximate the continuous function Φ, then there occur certain errors at the points where the jumps of F_n are. To understand the problem, assume that the X_js possess values in $\mathbb{Z}$, then S_n is also $\mathbb{Z}$-valued, hence for any $0 \le h < 1$, and all integers $k < l$, it follows that

$$\mathbb{P}\{k \le S_n \le l\} = \mathbb{P}\{k - h \le S_n \le l + h\}.$$

Consequently, for each such number h, the value

$$\Phi\left(\frac{l + h - n\mu}{\sigma\sqrt{n}}\right) - \Phi\left(\frac{k - h - n\mu}{\sigma\sqrt{n}}\right)$$

may be taken as normal approximation of the above probability. Which number $h < 1$ should be chosen?

To answer this question observe the following. If $l < m < k$, then

$$\mathbb{P}\{k \le S_n \le l\} = \mathbb{P}\{k \le S_n \le m\} + \mathbb{P}\{m + 1 \le S_n \le l\},$$

---

5 The central limit theorem also holds for not necessarily identically distributed random variables provided that all "errors" become uniformly small. That is, one has to exclude that certain errors are dominating the other ones.

which is approximated by

$$\Phi\left(\frac{l+h-n\mu}{\sigma\sqrt{n}}\right) - \Phi\left(\frac{m+1-h-n\mu}{\sigma\sqrt{n}}\right) + \Phi\left(\frac{m+h-n\mu}{\sigma\sqrt{n}}\right) - \Phi\left(\frac{k-h-n\mu}{\sigma\sqrt{n}}\right).$$

Thus, in order to get neither an overlap nor a gap between $m+1-h-n\mu$ and $m+h-n\mu$, it is customary to choose $h = 0.5$. This leads to the following definition.

Definition 7.2.9. Suppose $X_1, X_2, \ldots$ are independent identically distributed with values in $\mathbb{Z}$. Then the corrected normal approximation is given by

$$\mathbb{P}\{k \le S_n \le l\} \approx \Phi\left(\frac{l+0.5-n\mu}{\sigma\sqrt{n}}\right) - \Phi\left(\frac{k-0.5-n\mu}{\sigma\sqrt{n}}\right).$$

It is called **continuity correction** or **histogram correction** for the normal approximation. In a similar way, one corrects the approximation for infinite intervals by

$$\mathbb{P}\{S_n \le l\} \approx \Phi\left(\frac{l+0.5-n\mu}{\sigma\sqrt{n}}\right)$$

and by

$$\mathbb{P}\{S_n \ge k\} \approx 1 - \Phi\left(\frac{k-0.5-n\mu}{\sigma\sqrt{n}}\right) = \Phi\left(\frac{n\mu-k+0.5}{\sigma\sqrt{n}}\right). \tag{7.15}$$

The next result tells us that the continuity correction is only needed for small ns.

Proposition 7.2.10. *For all $x \in \mathbb{R}$ and $h \in \mathbb{R}$ it follows that*

$$\left|\Phi\left(\frac{x+h-n\mu}{\sigma\sqrt{n}}\right) - \Phi\left(\frac{x-n\mu}{\sigma\sqrt{n}}\right)\right| \le \frac{|h|}{\sigma\sqrt{2\pi n}}.$$

Proof: The mean value theorem of Calculus implies the existence of an intermediate value ξ in $\left(\frac{x-|h|-n\mu}{\sigma\sqrt{n}}, \frac{x+|h|-n\mu}{\sigma\sqrt{n}}\right)$ such that

$$\left|\Phi\left(\frac{x+h-n\mu}{\sigma\sqrt{n}}\right) - \Phi\left(\frac{x-n\mu}{\sigma\sqrt{n}}\right)\right| = |h|\frac{\Phi'(\xi)}{\sigma\sqrt{n}}.$$

Using

$$\Phi'(\xi) = \frac{1}{\sqrt{2\pi}}e^{-\xi^2/2} \le \frac{1}{\sqrt{2\pi}},$$

this proves the asserted estimate. ∎

Remark 7.2.11. An application of Proposition 7.2.10 with $x = k$ or $x = l$, and with $h = \pm 0.5$, shows that the improvement by the continuity correction is at most of order $n^{-1/2}$. Thus, it is no longer needed for large n.

Example 7.2.12. Roll a fair die n times. Let S_n be the sum of the n rolls. In view of eq. (7.14), this sum S_n is approximately $\mathcal{N}\left(\frac{7n}{2}, \frac{35n}{12}\right)$ distributed. In other words, it follows that

$$\lim_{n\to\infty} \mathbb{P}\left\{a \le \frac{S_n - 7n/2}{\sqrt{35n/12}} \le b\right\} = \frac{1}{\sqrt{2\pi}} \int_a^b e^{-x^2/2}\, dx = \Phi(b) - \Phi(a).$$

Moreover, this convergence takes place uniformly for all $a < b$. Therefore, at least for large n, the right-hand side of the last equation may be taken as approximative value of the left-hand one.

At first, we consider an example with a small number of trials. We roll a die three times and ask for the probability of the event $\{7 \le S_3 \le 8\}$. Let us compare the exact value

$$\mathbb{P}\{7 \le S_3 \le 8\} = \frac{1}{6} = 0.16667$$

with the one we get by applying the central limit theorem. Without continuity correction, the approximative value is

$$\Phi\left(\frac{8 - 21/2}{\sqrt{3 \cdot 35/12}}\right) - \Phi\left(\frac{7 - 21/2}{\sqrt{3 \cdot 35/12}}\right) = 0.08065,$$

while an application of the continuity correction leads to

$$\Phi\left(\frac{8 + 0.5 - 21/2}{\sqrt{3 \cdot 35/12}}\right) - \Phi\left(\frac{7 - 0.5 - 21/2}{\sqrt{3 \cdot 35/12}}\right) = 0.16133.$$

We see the improvement using the continuity correction.

Next we treat an example with large n. Let us investigate once more Example 7.1.6, but this time from the point of view of the central limit theorem. Choose again $n = 10^3$, $a = -\frac{100\sqrt{12}}{\sqrt{35000}}$ and $b = \frac{100\sqrt{12}}{\sqrt{35000}}$. Then it follows that

$$\mathbb{P}\{3400 \le S_n \le 3600\} \approx \frac{1}{\sqrt{2\pi}} \int_a^b e^{-x^2/2}\, dx \approx 0.93592.$$

As we see, the use of the central limit theorem improves considerably the bound $0,709$ obtained by Chebyshev's inequality.

Example 7.2.13. We investigate once more the model of a random walk on $\mathbb{Z}$ as presented in Example 4.1.7. What does the central limit theorem imply in this case? To simplify the calculations let us assume $p = 1/2$, that is, jumps to the left and to the right are equally likely. Thus, if $X_j = -1$ if the particle jumps to the left in step j and $X_j = 1$ otherwise, then $\mathbb{P}\{X_j = -1\} = \mathbb{P}\{X_j = 1\} = 1/2$ and $S_n = X_1 + \cdots + X_n$ is the position of the particle after n jumps. Moreover, by assumption the X_js are independent and, of course, identically distributed. Thus, the central limit theorem applies with $\mu = \mathbb{E}X_1 = 0$ and with $\sigma^2 = \mathbb{V}X_1 = 1$. Consequently, S_n is approximately $\mathcal{N}(0, n)$-distributed. More precisely, we have

$$\lim_{n \to \infty} \mathbb{P}\{a\sqrt{n} \le S_n \le b\sqrt{n}\} = \frac{1}{\sqrt{2\pi}} \int_a^b e^{-x^2/2}\, dx\,.$$

For instance, if $a = -2$ and $b = 2$, then it follows that

$$\lim_{n \to \infty} \mathbb{P}\{-2\sqrt{n} \le S_n \le 2\sqrt{n}\} = \frac{1}{\sqrt{2\pi}} \int_{-2}^{2} e^{-x^2/2}\, dx \approx 0.9544997\,.$$

Keep in mind that the possible values of S_n are between $-n$ and n. But in realness, with probability greater than 0.95, the position of the particle will be in the much smaller interval $[-2\sqrt{n}, 2\sqrt{n}]$.

Example 7.2.14 (Round-off errors). Many calculations in a bank, for instance of interest, lead to amounts that are not integral in cents. In this case the bank rounds the calculated value either up or down, whether the remainder is larger or smaller than 0.5 cents. For example, if the calculations lead to \$12.837, then the bank transfers \$12.84. Thus, in this case, the bank loses 0.3 cents. This seems to be a small amount, but if, for example, the bank performs 10^6 calculations per day, the total loss or gain could sum up to an amount of \$5000.00. But does this really happen?

 Answer: Theoretically, the rounding procedure could lead to huge losses or gains of the bank. But, as the central limit theorem shows, in reality such a scenario is extremely unlikely. To make this more precise, we use the following model. Let X_j be the loss or gain (in cents) of the bank in calculation j. Then the X_j are independent and uniformly distributed on $[-0.5, 0.5]$. Thus, the total loss or gain after n calculations equals $S_n = X_1 + \cdots + X_n$. By Propositions 5.1.25 and 5.2.23 we know that

$$\mu = \mathbb{E}X_1 = 0 \quad \text{and} \quad \sigma^2 = \mathbb{V}X_1 = \frac{1}{12}\,,$$

hence, if $a < b$, the central limit theorem implies

$$\lim_{n \to \infty} \mathbb{P}\left\{\frac{a\sqrt{n}}{\sqrt{12}} \le S_n \le \frac{b\sqrt{n}}{\sqrt{12}}\right\} = \frac{1}{\sqrt{2\pi}} \int_a^b e^{-x^2/2}\, dx\,.$$

For example, if $n = 10^6$, then taking $a = \sqrt{12}$ and $b = \infty$, this leads to

$$\mathbb{P}\{S_n \geq \$10\} = \mathbb{P}\{S_n \geq 10^3 \text{ cents}\} \approx \frac{1}{\sqrt{2\pi}} \int\limits_{\sqrt{12}}^{\infty} e^{-x^2/2} \, dx \approx 0.00026603 \,,$$

which is an extremely small probability. By symmetry, it also follows that

$$\mathbb{P}\{S_n \leq -\$10\} \approx \frac{1}{\sqrt{2\pi}} \int\limits_{-\infty}^{-\sqrt{12}} e^{-x^2/2} \, dx \approx 0.00026603 \,.$$

In a similar way one obtains

$$\mathbb{P}\{S_n \geq \$1\} \approx 0.364517 \,, \qquad \mathbb{P}\{S_n \geq \$2\} \approx 0.244211$$
$$\mathbb{P}\{S_n \geq \$5\} \approx 0.0416323 \quad \text{and} \quad \mathbb{P}\{S_n \geq \$20\} \approx 2,1311 \times 10^{-12} \,.$$

This shows that even for many calculations, in our case 10^6 ones, the probability for a loss or gain of more than $5 is very unlikely. Recall that theoretically an amount of $5000.00 would be possible.

Special Cases of the Central Limit Theorem:
Binomial distributed random variables: In 1738 De Moivre, and later on in 1812 Laplace, investigated the normal approximation of binomial distributed[6] random variables. This was the starting point for the investigation of general central limit theorems. Let us state their result.

Proposition 7.2.15 (De Moivre–Laplace theorem). *Let X_j be independent $B_{1,p}$-distributed random variables. Then their sums $S_n = X_1 + \cdots + X_n$ satisfy*

$$\lim_{n\to\infty} \mathbb{P}\left\{a \leq \frac{S_n - np}{\sqrt{np(1-p)}} \leq b\right\} = \frac{1}{\sqrt{2\pi}} \int\limits_a^b e^{-x^2/2} \, dx \,. \tag{7.16}$$

Proof: Recall that for a $B_{1,p}$-distributed random variable X we have $\mu = \mathbb{E}X = p$ and $\sigma^2 = \mathbb{V}X = p(1-p)$. Consequently, Proposition 7.2.6 applies and leads to eq. (7.16). ∎

Remark 7.2.16. By Corollary 4.6.2 we know that $S_n = X_1 + \cdots + X_n$ is $B_{n,p}$-distributed. Consequently, eq. (7.16) may also be written as

$$\lim_{n\to\infty} \sum_{k\in I_{n,a,b}} \binom{n}{k} p^k (1-p)^{n-k} = \frac{1}{\sqrt{2\pi}} \int\limits_a^b e^{-x^2/2} \, dx,$$

---

6 De Moivre investigated sums of $B_{1,1/2}$-distributed random variables while Laplace treated $B_{1,p}$-distributed ones for general $0 \leq p \leq 1$.

where

$$I_{n,a,b} := \left\{ k \geq 0 : a \leq \frac{k - np}{\sqrt{np(1-p)}} \leq b \right\}.$$

Another way to formulate the De Moivre–Laplace theorem is as follows. For "large" n, S_n is approximative $\mathcal{N}(np, np(1-p))$-distributed. That is, if $0 \leq l < m \leq n$, then

$$\sum_{k=l}^{m} \binom{n}{k} p^k (1-p)^{n-k} \approx \Phi\left(\frac{m - np}{\sqrt{np(1-p)}}\right) - \Phi\left(\frac{l - np}{\sqrt{np(1-p)}}\right). \tag{7.17}$$

Since the sums S_n are integer-valued, the continuity correction should be applied for small ns, that is, on the right-hand side of eq. (7.17) the numbers m and l should be replaced by $m + 0.5$ and l by $l - 0.5$, respectively.

Example 7.2.17. Play a series of games with success probability $0 < p < 1$. Let $\alpha \in (0, 1)$ be a given security probability, and $m \in \mathbb{N}$ is some integer. How many games one has to play in order to have with probability greater than or equal to $1 - \alpha$ at least m times success?

Answer: Define random variables X_j by setting $X_j = 1$ when winning game j, while $X_j = 0$ in the case of losing it. Then the X_js are independent and $B_{1,p}$-distributed. Hence, if $S_n = X_1 + \cdots X_n$, then the above question may be formulated as follows. What is the smallest $n \in \mathbb{N}$ for which

$$\mathbb{P}\{S_n \geq m\} \geq 1 - \alpha ? \tag{7.18}$$

By Corollary 4.6.2, the sum S_n is $B_{n,p}$-distributed and, therefore, estimate (7.18) transforms to

$$\sum_{k=m}^{n} \binom{n}{k} p^k (1-p)^{n-k} \geq 1 - \alpha. \tag{7.19}$$

Thus, the "exact" answer to the above question is as follows. Choose the minimal $n \geq 1$ for which estimate (7.19) is valid.

Remark 7.2.18. For large m it may be a difficult task to determine the minimal n satisfying estimate (7.19). Therefore, one looks for an "approximative" approach via Proposition 7.2.15. Rewriting estimate (7.18) as

$$\mathbb{P}\left\{ \frac{S_n - np}{\sqrt{np(1-p)}} \geq \frac{m - np}{\sqrt{np(1-p)}} \right\} \geq 1 - \alpha,$$

an "approximative" condition for n is

$$1 - \alpha \leq 1 - \Phi\left(\frac{m - np}{\sqrt{np(1-p)}}\right) = \Phi\left(\frac{np - m}{\sqrt{np(1-p)}}\right).$$

Given $\beta \in (0,1)$, let us define[7] u_β by $\Phi(u_\beta) = \beta$. Consequently, the approximative solution of the above question is to choose the minimal $n \geq 1$ satisfying

$$\frac{np - m}{\sqrt{np(1-p)}} \geq u_{1-\alpha}. \tag{7.20}$$

For "small" n we have to modify the previous approach slightly. Here we have to use the continuity correction. In view of eq. (7.15) the condition is now

$$1 - \alpha \leq \Phi\left(\frac{np - m + 0.5}{\sqrt{np(1-p)}}\right),$$

leading to

$$\frac{np - m + 0.5}{\sqrt{np(1-p)}} \geq u_{1-\alpha}. \tag{7.21}$$

Let us explain Remark 7.2.18 with the help of a concrete example.

Example 7.2.19. Find the minimal $n \geq 1$ such that, rolling a fair die n times, one observes with probability greater than or equal to 0.9 at least 100 times the number 6 ?
The "exact" answer is, choose the minimal $n \geq 1$ satisfying

$$\sum_{k=100}^{n} \binom{n}{k} \left(\frac{1}{6}\right)^k \left(\frac{5}{6}\right)^{n-k} \geq 0.9.$$

Numerical calculations give that the left-hand side equals 0.897721 if $n = 670$, and it is 0.900691 if $n = 671$. Thus, in order to observe, with probability greater than 0.9, the number "6" at least 100 times, one has to roll the die at least 671 times.

Let us compare this result with the one that we obtain by the approximative approach. First we approximate S_n directly, that is, without applying the continuity correction. Here estimate (7.20) says that we have to look for the minimal $n \geq 1$ satisfying

$$\frac{n - 600}{\sqrt{5n}} \geq u_{0.9} = 1.28155. \tag{7.22}$$

---

[7] Later on, in Section 8.4.3, these numbers u_β will play an important role; compare also Definition 8.4.5.

Since

$$\frac{665 - 600}{\sqrt{5 \cdot 665}} = 1.12724 \quad \text{and} \quad \frac{666 - 600}{\sqrt{5 \cdot 666}} = 1.4373,$$

the smallest n satisfying estimate (7.22) is 666.

Applying the continuity correction, by estimate (7.21), condition (7.22) has to be replaced by

$$\frac{n - 600 + 3}{\sqrt{5n}} \geq u_{0.9} = 1.28155.$$

The left-hand side equals 1.27757 for $n = 671$ and 1.29387 if $n = 672$. Consequently, this type of approximation gives the (more precise) value $n = 672$ for the minimal number of necessary rolls of the die.

Poisson distributed random variables: Let $X_1, X_2, \ldots$ be independent and Pois_λ-distributed. By Propositions 5.1.16 and 5.2.20 we know

$$\mathbb{E}X_1 = \lambda \quad \text{and} \quad \mathbb{V}X_1 = \lambda.$$

Thus, in this case Proposition 7.2.6 reads as follows.

Proposition 7.2.20. *Let $(X_j)_{j \geq 1}$ be independent Pois_λ-distributed. Then the sums $S_n = X_1 + \cdots + X_n$ satisfy*

$$\lim_{n \to \infty} \mathbb{P}\left\{a \leq \frac{S_n - n\lambda}{\sqrt{n\lambda}} \leq b\right\} = \frac{1}{\sqrt{2\pi}} \int_a^b e^{-x^2/2} \, dx. \tag{7.23}$$

Remark 7.2.21. By Proposition 4.6.5, the sum S_n is $\text{Pois}_{\lambda n}$-distributed, hence eq. (7.23) transforms to

$$\lim_{n \to \infty} \sum_{k \in J_{n,a,b}} \frac{(\lambda n)^k}{k!} e^{-\lambda n} = \frac{1}{\sqrt{2\pi}} \int_a^b e^{-x^2/2} \, dx, \tag{7.24}$$

where

$$J_{n,a,b} := \left\{k \in \mathbb{N}_0 : a \leq \frac{k - n\lambda}{\sqrt{n\lambda}} \leq b\right\}.$$

Another way to express this is as follows. If $0 \leq l < m < \infty$, then

$$\sum_{k=l}^m \frac{(\lambda n)^k}{k!} e^{-\lambda n} \approx \Phi\left(\frac{m - n\lambda}{\sqrt{n\lambda}}\right) - \Phi\left(\frac{l - n\lambda}{\sqrt{n\lambda}}\right).$$

Remark 7.2.22. Choosing in eq. (7.24) the numbers as $a = -\infty$, $b = 0$ and $\lambda = 1$, we get

$$\lim_{n \to \infty} e^{-n} \sum_{k=0}^{n} \frac{n^k}{k!} = \frac{1}{2},$$

which is interesting in its own right. Taking $a = -\infty$ and $b_n = \sqrt{n}$ yields

$$\lim_{n \to \infty} \left| e^{-n} \sum_{k=0}^{2n} \frac{n^k}{k!} - \frac{1}{\sqrt{2\pi}} \int_{-\infty}^{b_n} e^{-x^2/2} \, dx \right| = 0,$$

hence, because of

$$\lim_{n \to \infty} \frac{1}{\sqrt{2\pi}} \int_{-\infty}^{\sqrt{n}} e^{-x^2/2} \, dx = 1,$$

we obtain

$$\lim_{n \to \infty} e^{-n} \sum_{k=0}^{2n} \frac{n^k}{k!} = 1.$$

Gamma distributed random variables: Finally, we investigate sums of gamma distributed random variables. Here the central limit theorem leads to the following result.

Proposition 7.2.23. *Let $X_1, X_2, \ldots$ be independent $\Gamma_{\alpha,\beta}$-distributed random variables. Then their sums $S_n = X_1 + \cdots + X_n$ satisfy*

$$\lim_{n \to \infty} \mathbb{P} \left\{ a \le \frac{S_n - n\alpha\beta}{\alpha\sqrt{n\beta}} \le b \right\} = \frac{1}{\sqrt{2\pi}} \int_{a}^{b} e^{-x^2/2} \, dx. \tag{7.25}$$

Proof: Propositions 5.1.26 and 5.2.24 tell us that the expected value and the variance of the X_js are given by $\mu = \mathbb{E}X_1 = \alpha\beta$ and $\sigma^2 = \mathbb{V}X_1 = \alpha^2\beta$. Therefore, eq. (7.25) follows by an application of Proposition 7.2.6. ∎

Remark 7.2.24. Note that Proposition 4.6.11 implies that S_n is $\Gamma_{\alpha,n\beta}$-distributed. Thus, setting

$$I_{n,a,b} := \left\{ x \ge 0 : a \le \frac{x - n\alpha\beta}{\alpha\sqrt{n\beta}} \le b \right\},$$

eq. (7.25) leads to

$$\lim_{n \to \infty} \frac{1}{\alpha^{n\beta}\Gamma(n\beta)} \int_{I_{n,a,b}} x^{n\beta-1} e^{-x/\alpha} \, dx = \frac{1}{\sqrt{2\pi}} \int_{a}^{b} e^{-x^2/2} \, dx.$$

Another way to express this is as follows. If $0 \le a < b$, then

$$\frac{1}{a^{n\beta}\Gamma(n\beta)} \int_a^b x^{n\beta-1} e^{-x/a} \, dx \approx \Phi\left(\frac{b - n\alpha\beta}{a\sqrt{n\beta}}\right) - \Phi\left(\frac{a - n\alpha\beta}{a\sqrt{n\beta}}\right).$$

Two cases of Proposition 7.2.23, or of Remark 7.2.24, are of special interest.

(a) For $n \ge 1$ let S_n be a χ_n^2-distributed random variable. Then it follows that

$$\lim_{n \to \infty} \mathbb{P}\left\{a \le \frac{S_n - n}{\sqrt{2n}} \le b\right\} = \frac{1}{\sqrt{2\pi}} \int_a^b e^{-x^2/2} \, dx.$$

(b) If S_n is distributed according to the Erlang distribution $E_{\lambda,n}$, then we get

$$\lim_{n \to \infty} \mathbb{P}\left\{a \le \frac{\lambda S_n - n}{\sqrt{n}} \le b\right\} = \frac{1}{\sqrt{2\pi}} \int_a^b e^{-x^2/2} \, dx.$$

For $\lambda = 1$ this implies (set $a = -\infty$ and $b = 0$) that

$$\lim_{n \to \infty} \frac{1}{\Gamma(n)} \int_0^n x^{n-1} e^{-x} \, dx = \frac{1}{\sqrt{2\pi}} \int_{-\infty}^0 e^{-x^2/2} \, dx = \frac{1}{2}.$$

Additional Remarks:

(1) We play a series of the same game. Suppose in each game we may lose or win a certain amount of money. A natural condition for this game (among friends) is it should be fair. But **what does it mean that a game is fair?** Is this the case if

(i)   the average loss or gain in each single game is zero or is it fair if

(ii) the probability that, after n games, the total loss or gain is positive, tends to 1/2 as n tends to infinity?

The mathematical formulation of the previous question is as follows. Let $X_1, X_2, \ldots$ denote the win or loss in the first game, the second one, and so on. Then the X_js are independent identically distributed random variables. The above question reads now as follows. Is the game fair if

(i)   the expected value $\mu = \mathbb{E}X_1$ satisfies $\mu = 0$ or is this the case if

(ii) the sum $S_n := X_1 + \cdots + X_n$ fulfills

$$\lim_{n \to \infty} \mathbb{P}\{S_n \le 0\} = \lim_{n \to \infty} \mathbb{P}\{S_n \ge 0\} = \frac{1}{2}? \tag{7.26}$$

In the sequel we have to exclude the trivial case $\mathbb{P}\{X_j = 0\} = 1$, that is, in each game one neither wins nor loses some money. Of course, then eq. (7.26) does not hold.

At a first glance one might believe that the two possible definitions of fairness describe the same fact. But this is not so as one may see in an example in [Fel68],

Chapter X, §4. There one finds a sequence of independent random variables $X_1, X_2, \ldots$ with $\mathbb{E}X_1 = 0$, however

$$\lim_{n\to\infty} \mathbb{P}\{S_n \le 0\} = 1.$$

In particular, this tells us that, in general, condition (i) does not imply condition (ii).

The next result clarifies the relation between these two definitions of fairness in the case that the random variables possess a finite second moment.

Proposition 7.2.25. *Let $X_1, X_2, \ldots$ be independent and identically distributed with expected value μ. Assume $\mathbb{P}\{X_j = 0\} < 1$.*
1. *Then eq. (7.26) always implies $\mu = 0$. That is, a fair game in the sense of (ii) also satisfies condition (i).*
2. *Conversely, if $\mathbb{E}|X_1|^2 < \infty$, then (ii) is a consequence of (i). Hence, assuming the existence of a second moment, conditions (i) and (ii) are equivalent.*

Proof: We prove the contraposition of the first statement. Thus, suppose that (i) does not hold, that is, we have $\mu \ne 0$. Without losing generality we may assume $\mu > 0$. Otherwise, investigate $-X_1, -X_2, \ldots$. An application of Proposition 7.1.27 with $\varepsilon = \mu/2$ yields

$$\lim_{n\to\infty} \mathbb{P}\left\{ \left| \frac{S_n}{n} - \mu \right| \le \frac{\mu}{2} \right\} = 1. \tag{7.27}$$

Since $\left| \frac{S_n}{n} - \mu \right| \le \mu/2$ implies $\frac{S_n}{n} \ge \mu/2$, hence $S_n \ge 0$, it follows that

$$\mathbb{P}\left\{ \left| \frac{S_n}{n} - \mu \right| \le \frac{\mu}{2} \right\} \le \mathbb{P}\{S_n \ge 0\}.$$

Consequently, from eq. (7.27) we derive

$$\lim_{n\to\infty} \mathbb{P}\{S_n \ge 0\} = 1,$$

hence eq. (7.26) cannot be valid. This proves the first part of the proposition.

We prove now the second assertion. Thus, suppose $\mu = 0$ as well as the existence of the variance $\sigma^2 = \mathbb{V}X_1$. Note that $\sigma^2 > 0$. Why? If a random variable X satisfies $\mathbb{E}X = 0$ and $\mathbb{V}X = 0$, then necessarily $\mathbb{P}\{X = 0\} = 1$. But, since we assumed $\mathbb{P}\{X_1 = 0\} < 1$, we cannot have $\sigma^2 = \mathbb{V}X_1 = 0$.

Thus, Proposition 7.2.6 applies and leads to

$$\lim_{n\to\infty} \mathbb{P}\{S_n \ge 0\} = \lim_{n\to\infty} \left\{ \frac{S_n}{\sigma\sqrt{n}} \ge 0 \right\} = \frac{1}{\sqrt{2\pi}} \int_0^\infty e^{-x^2/2}\, dx = \frac{1}{2}.$$

The proof for $\mathbb{P}\{S_n \le 0\} \to 1/2$ follows in the same way, thus eq. (7.26) is valid. This completes the proof. ∎

(2) **How fast does** $\frac{S_n - n\mu}{\sigma\sqrt{n}}$ **converge** to a normally distributed random variable? Before we answer this question, we have to determine how this speed is measured. In view of Proposition 7.2.6 we use the following quantity depending on $n \geq 1$:

$$\sup_{t \in \mathbb{R}} \left| \mathbb{P}\left\{ \frac{S_n - n\mu}{\sigma\sqrt{n}} \leq t \right\} - \frac{1}{\sqrt{2\pi}} \int_{-\infty}^{t} e^{-x^2/2}\, dx \right|.$$

Doing so, the following classical result holds (see [Dur10], Section 3.4.4, for a proof).

Proposition 7.2.26 (Berry–Esséen theorem). *Let $X_1, X_2, \ldots$ be independent identically distributed random variables with finite third moment, that is, with $\mathbb{E}|X_1|^3 < \infty$. If $\mu = \mathbb{E}X_1$ and $\sigma^2 = \mathbb{V}X_1 > 0$, then it follows that*

$$\sup_{t \in \mathbb{R}} \left| \mathbb{P}\left\{ \frac{S_n - n\mu}{\sigma\sqrt{n}} \leq t \right\} - \frac{1}{\sqrt{2\pi}} \int_{-\infty}^{t} e^{-x^2/2}\, dx \right| \leq C \frac{\mathbb{E}|X_1|^3}{\sigma^3} n^{-1/2}. \qquad (7.28)$$

Here $C > 0$ denotes a universal constant.

Remark 7.2.27. The order $n^{-1/2}$ in estimate (7.28) is optimal and cannot be improved. This can be seen by the following example. Take independent random variables $X_1, X_2, \ldots$ with $\mathbb{P}\{X_j = -1\} = \mathbb{P}\{X_j = 1\} = 1/2$. Hence, in this case $\mu = 0$ and $\sigma^2 = 1$. Then one has

$$\liminf_{n \to \infty} n^{1/2} \sup_{t \in \mathbb{R}} \left| \mathbb{P}\left\{ \frac{S_n}{\sqrt{n}} \leq t \right\} - \frac{1}{\sqrt{2\pi}} \int_{-\infty}^{t} e^{-x^2/2}\, dx \right| > 0. \qquad (7.29)$$

Assertion (7.29) is a consequence of the fact that, if n is even, then the function $t \mapsto \mathbb{P}\left\{ \frac{S_n}{\sqrt{n}} \leq t \right\}$ has a jump of order $n^{-1/2}$ at zero. This follows by the calculations in Example 4.1.7. On the other hand, $t \mapsto \Phi(t)$ is continuous, hence the maximal difference between these two functions is at least the half of the height of the jump.

Remark 7.2.28. The exact value of the constant $C > 0$ appearing in estimate (7.28) is, in spite of intensive investigations, still unknown. At present, the best-known estimates are $0.40973 < C < 0.478$.

7.3 Problems

Problem 7.1. Let $A_1, A_2, \ldots$ and $B_1, B_2, \ldots$ be two sequences of events in a probability space $(\Omega, \mathcal{A}, \mathbb{P})$. Prove that

$$\limsup_{n \to \infty} (A_n \cup B_n) = \limsup_{n \to \infty} (A_n) \cup \limsup_{n \to \infty} (B_n).$$

Is this also valid for the intersection? That is, does one have

$$\limsup_{n\to\infty}(A_n \cap B_n) = \limsup_{n\to\infty}(A_n) \cap \limsup_{n\to\infty}(B_n)?$$

Problem 7.2. Let $(X_n)_{n\geq 1}$ be a sequence of independent E_λ-distributed random variables. Characterize sequences $(c_n)_{n\geq 1}$ of positive real numbers for which

$$\mathbb{P}\{X_n \geq c_n \text{ infinitely often}\} = 1?$$

Problem 7.3. Let $f : [0, 1] \to \mathbb{R}$ be a continuous function. Its **Bernstein polynomial** B_n^f of degree n is defined by

$$B_n^f(x) := \sum_{k=0}^{n} f\left(\frac{k}{n}\right)\binom{n}{k}x^k(1-x)^{n-k}, \quad 0 \leq x \leq 1.$$

Show that Proposition 7.1.29 implies the following. If $\mathbb{P}$ is the uniform distribution on $[0, 1]$, then

$$\mathbb{P}\left\{x \in [0, 1] : \lim_{n\to\infty} B_n^f(x) = f(x)\right\} = 1.$$

Remark: Using methods from Calculus, one may even show the uniform convergence, that is,

$$\lim_{n\to\infty} \sup_{0\leq x\leq 1} |B_n^f(x) - f(x)| = 0.$$

Problem 7.4. Roll a fair die 180 times. What is the probability that the number "6" occurs less than or equal to 25 times. Determine this probability by the following three methods:

- Directly via the binomial distribution.
- Approximative by virtue of the central limit theorem.
- Approximative by applying the continuity correction.

Problem 7.5. Toss a fair coin 16 times. Compute the probability to observe exactly eight times "head" by the following methods.

- Directly via the binomial distribution.
- Approximative by applying the continuity correction.

Why does one not get a reasonable result using the normal approximation directly, that is, without continuity correction?

Problem 7.6. Let $X_1, X_2, \ldots$ be a sequence of independent G_p-distributed random variables, that is, for some $0 < p < 1$ one has

$$\mathbb{P}\{X_j = k\} = p(1-p)^{k-1}, \quad k = 1, 2, \ldots$$

1. What does the central limit theorem tell us in this case about the behavior of the sums $S_n = X_1 + \cdots X_n$?
2. For two real numbers $a < b$ set

$$I_{n,a,b} := \left\{ k \geq 0 : a \leq \frac{pk - n(1-p)}{\sqrt{n(1-p)}} \leq b \right\}.$$

Show that

$$\lim_{n \to \infty} \sum_{k \in I_{n,a,b}} \binom{-n}{k} p^n (1-p)^k = \frac{1}{\sqrt{2\pi}} \int_a^b e^{-x^2/2} dx.$$

Hint: Use Corollary 4.6.9 and investigate $S_n - n$.

8 Mathematical Statistics

8.1 Statistical Models

8.1.1 Nonparametric Statistical Models

The main objective of Probability Theory is to describe and analyze random experiments by means of a suitable probability space $(\Omega, \mathcal{A}, \mathbb{P})$. Here it is always assumed that the probability space is known, in particular, that the describing probability measure, $\mathbb{P}$, is identified.

Probability Theory:
Description of a random experiment and its properties by a probability space. The distribution of the outcomes is assumed to be *known*.

Mathematical Statistics deals mainly with the reverse question: one executes an experiment, that is, one draws a sample (e.g., one takes a series of measurements of an item or one interrogates several people), and, on the basis of the observed sample, one wants to derive as much information as possible about the (unknown) underlying probability measure $\mathbb{P}$. Sometimes the precise knowledge of $\mathbb{P}$ is not needed; it may suffice to know a certain parameter of $\mathbb{P}$.

Mathematical Statistics:
As a result of a statistical experiment, a (random) sample is observed. On its basis, conclusions are drawn about the unknown underlying probability distribution.

Let us state the mathematical formulation of the task: first we mention that it is standard practice in Mathematical Statistics to denote the describing probability space by $(\mathcal{X}, \mathcal{F}, \mathbb{P})$. As before, $\mathcal{X}$ is the sample space (the set that contains all possible outcomes of the experiment), and $\mathcal{F}$ is a suitable σ-field of events. The probability measure $\mathbb{P}$ describes the experiment, that is, $\mathbb{P}(A)$ is the probability of observing a sample belonging to A, but recall that $\mathbb{P}$ is unknown.

Based on theoretical considerations or on long-time experience, quite often we are able to restrict the entirety of probability measures in question. Mathematically, that means that we choose a set **P** of probability measures on $(\mathcal{X}, \mathcal{F})$, which contains what we believe to be the "correct" $\mathbb{P}$. Thereby, it is not impossible that **P** is the set of **all** probability measures, but for most statistical methods it is very advantageous to take **P** as small as possible. On the other hand, the set **P** cannot be chosen too small, because we have to be sure that the "correct" $\mathbb{P}$ is really contained in **P**. Otherwise, the obtained results are either false or imprecise.

> **Definition 8.1.1.** A subset **P** of probability measures on $(\mathcal{X}, \mathcal{F})$ is called a **distribution assumption**, that is, one assumes that the underlying $\mathbb{P}$ belongs to **P**.

After having fixed the distribution assumption **P**, one now regards only probability measures $\mathbb{P} \in \mathbf{P}$ or, equivalently, measures not in **P** are sorted out.

To get information about the unknown probability measure, one performs a statistical experiment or analyzes some given data. In both cases, the result is a random **sample** $x \in \mathcal{X}$. The task of Mathematical Statistics is to get information about $\mathbb{P} \in \mathbf{P}$, based on the observed sample $x \in \mathcal{X}$. A suitable way to describe the problem is as follows.

> **Definition 8.1.2.** A (nonparametric) **statistical model** is a collection of probability spaces $(\mathcal{X}, \mathcal{F}, \mathbb{P})$ with $\mathbb{P} \in \mathbf{P}$. Here, $\mathcal{X}$ and $\mathcal{F}$ are fixed, and $\mathbb{P}$ varies through the distribution assumption **P**. One writes for the model
>
> $$(\mathcal{X}, \mathcal{F}, \mathbb{P})_{\mathbb{P} \in \mathbf{P}} \quad \text{or} \quad \{(\mathcal{X}, \mathcal{F}, \mathbb{P}) : \mathbb{P} \in \mathbf{P}\}.$$

Let us illustrate the previous definition with two examples.

Example 8.1.3. In an urn are white and black balls of an unknown ratio. Let $\theta \in [0, 1]$ be the (unknown) proportion of white balls, hence $1 - \theta$ is that of the black ones. In order to get some information about θ, one randomly chooses n balls with replacement. The result of this experiment, or the sample, is a number $k \in \{0, \ldots, n\}$, the frequency of observed white balls. Thus, the sample space is $\mathcal{X} = \{0, \ldots, n\}$ and as σ-field we may choose, as always for finite sample spaces, the powerset $\mathcal{P}(\mathcal{X})$. The possible probability measures describing this experiment are binomial distributions $B_{n,\theta}$ with $0 \le \theta \le 1$. Consequently, the distribution assumption is

$$\mathbf{P} = \{B_{n,\theta} : \theta \in [0, 1]\}.$$

Summing up, the statistical model describing the experiment is

$$(\mathcal{X}, \mathcal{P}(\mathcal{X}), \mathbb{P})_{\mathbb{P} \in \mathbf{P}} \quad \text{where} \quad \mathcal{X} = \{0, \ldots, n\} \quad \text{and} \quad \mathbf{P} = \{B_{n,\theta} : 0 \le \theta \le 1\}.$$

Next, we consider an important example from quality control.

Example 8.1.4. A buyer obtains from a trader a delivery of N machines. Among them $M \le N$ are defective. The buyer does not know the value of M. To determine it, he randomly chooses n machines from the delivery and checks them. The result, or the sample, is the number $0 \leq m \leq n$ of defective machines among the n tested.

Thus, the sample space is $\mathcal{X} = \{0, \ldots, n\}$, $\mathcal{F} = \mathcal{P}(\mathcal{X})$, and the probability measures in question are hypergeometric ones. Therefore, the distribution assumption is

$$\mathbf{P} = \{H_{N,M,n} : M = 0, \ldots, N\},$$

where $H_{N,M,n}$ denotes the hypergeometric distribution with parameters N, M, and n, as introduced in Definition 1.4.25.

Before we proceed further, we consider a particularly interesting case of statistical model, which describes the **n-fold independent repetition** of a single experiment. To explain this model, let us investigate the following easy example.

Example 8.1.5. We are given a die that looks biased. To check this, we roll it n times and record the sequence of numbers appearing in each of the trials. Thus, our sample space is $\mathcal{X} = \{1, \ldots, 6\}^n$, and the observed sample is $x = (x_1, \ldots, x_n)$, with $1 \leq x_k \leq 6$. Let $\theta_1, \ldots, \theta_6$ be the probabilities for 1 to 6. Then we want to check whether $\theta_1 = \cdots = \theta_6 = 1/6$, that is, whether $\mathbb{P}_0$ given by $\mathbb{P}_0(\{k\}) = \theta_k$, $1 \leq k \leq 6$, is the uniform distribution. What are the possible probability measures on $(\mathcal{X}, \mathcal{P}(\mathcal{X}))$ describing the statistical experiment? Since the results of different rolls are independent, the describing measure $\mathbb{P}$ is of the form $\mathbb{P} = \mathbb{P}_0^{\otimes n}$ with

$$\mathbb{P}_0^{\otimes n}(\{x\}) = \mathbb{P}_0(\{x_1\}) \cdots \mathbb{P}_0(\{x_n\}) = \theta_1^{m_1} \cdots \theta_6^{m_6}, \quad x = (x_1, \ldots, x_n),$$

and where the m_ks denote the frequency of the number $1 \leq k \leq 6$ in the sequence x. Consequently, the natural distribution assumption is

$$\mathbf{P} = \{\mathbb{P}_0^{\otimes n} : \mathbb{P}_0 \text{ probability measure on } \{1, \ldots, 6\}\}.$$

Suppose we are given a probability space $(\mathcal{X}_0, \mathcal{F}_0, \mathbb{P}_0)$ with unknown $\mathbb{P}_0 \in \mathbf{P}_0$. Here, $\mathbf{P}_0$ denotes a set of probability measures on $(\mathcal{X}_0, \mathcal{F}_0)$, hopefully containing the "correct" $\mathbb{P}_0$. We call $(\mathcal{X}_0, \mathcal{F}_0, \mathbb{P}_0)_{\mathbb{P}_0 \in \mathbf{P}_0}$ the **initial model**. In Example 8.1.5, the initial model is $\mathcal{X}_0 = \{1, \ldots, 6\}$, while $\mathbf{P}_0$ is the set of all probability measures on $(\mathcal{X}_0, \mathcal{P}(\mathcal{F}_0))$.

In order to determine $\mathbb{P}_0$, we execute n independent trials according to $\mathbb{P}_0$. The result, or the observed sample, is a vector $x = (x_1, \ldots, x_n)$ with $x_i \in \mathcal{X}_0$. Consequently, the natural sample space is $\mathcal{X} = \mathcal{X}_0^n$.

Which statistical model does this experiment describe? To answer this question, let us recall the basic results in Section 1.9, where exactly those problems have been investigated. As σ-field $\mathcal{F}$ we choose the n times product σ-field of $\mathcal{F}_0$, that is,

$$\mathcal{F} = \underbrace{\mathcal{F}_0 \otimes \cdots \otimes \mathcal{F}_0}_{n \text{ times}},$$

and the describing probability measure $\mathbb{P}$ is of the form $\mathbb{P}_0^{\otimes n}$, that ist, it is the n-fold product of $\mathbb{P}_0$. Recall that, according to Definition 1.9.5, the product $\mathbb{P}_0^{\otimes n}$ is the unique probability measure on $(\mathcal{X}, \mathcal{F})$ satisfying

$$\mathbb{P}_0^{\otimes n}(A_1 \times \cdots \times A_n) = \mathbb{P}_0(A_1) \cdots \mathbb{P}_0(A_n),$$

whenever $A_j \in \mathcal{F}_0$.

Since we assumed $\mathbb{P}_0 \in \mathbf{P}_0$, the possible probability measures are $\mathbb{P}_0^{\otimes n}$ with $\mathbb{P}_0 \in \mathbf{P}_0$.

Let us summarize what we obtained until now.

> **Definition 8.1.6.** The statistical model for the **n-fold independent repetition** of an experiment, determined by the initial model $(\mathcal{X}_0, \mathcal{F}_0, \mathbb{P}_0)_{\mathbb{P}_0 \in \mathbf{P}_0}$, is given by
>
> $$(\mathcal{X}, \mathcal{F}, \mathbb{P}_0^{\otimes n})_{\mathbb{P}_0 \in \mathbf{P}_0}$$
>
> where $\mathcal{X} = \mathcal{X}_0^n$, $\mathcal{F}$ denotes the product σ-field of the $\mathcal{F}_0$s, and $\mathbb{P}_0^{\otimes n}$ is the n-fold product measure of $\mathbb{P}_0$.

Remark 8.1.7. Of course, the main goal in the model of n-fold repetition is to get some knowledge about $\mathbb{P}_0$. To obtain the desired information, we perform n independent trials, each time observing a value distributed according to $\mathbb{P}_0$. Altogether, the sample is a vector $x = (x_1, \ldots, x_n)$, which is now distributed according to $\mathbb{P}_0^{\otimes n}$.

The two following examples explain Definition 8.1.6.

Example 8.1.8. A coin is labeled on one side with "0" and on the other side with "1." There is some evidence that the coin is biased. To check this, let us execute the following statistical experiment: toss the coin n times and record the sequence of zeroes and ones. Thus, the observed sample is some $x = (x_1, \ldots, x_n)$, with each x_k being either "0" or "1."

Our initial model is given by $\mathcal{X}_0 = \{0, 1\}$ and $\mathbb{P}_0 = B_{1,\theta}$ for a certain (unknown) $\theta \in [0, 1]$. Then the experiment is described by $\mathcal{X} = \{0, 1\}^n$ and $\mathbf{P} = \{B_{1,\theta}^{\otimes n} : 0 \le \theta \le 1\}$. Note that

$$B_{1,\theta}^{\otimes n}(\{x\}) = \theta^k (1 - \theta)^{n-k}, \quad k = x_1 + \cdots + x_n.$$

Example 8.1.9. A company produces a new type of light bulbs with an unknown distribution of the lifetime. To determine it, n light bulbs are switched on at the same time. Let $t = (t_1, \ldots, t_n)$ be the times when the bulbs burn through. Then our sample is the vector $t \in (0, \infty)^n$.

By long-time experience one knows the lifetime of each light bulb is exponentially distributed. Thus, the initial model is $(\mathbb{R}, \mathcal{B}(\mathbb{R}), \mathbf{P}_0)$ with $\mathbf{P}_0 = \{E_\lambda : \lambda > 0\}$. Consequently, the experiment of testing n light bulbs is described by the model

$$(\mathbb{R}^n, \mathcal{B}(\mathbb{R}^n), \mathbb{P}^{\otimes n})_{\mathbb{P} \in \mathbf{P}_0} = (\mathbb{R}^n, \mathcal{B}(\mathbb{R}^n), E_\lambda^{\otimes n})_{\lambda > 0},$$

where $\mathbf{P}_0 = \{E_\lambda : \lambda > 0\}$. Recall that $E_\lambda^{\otimes n}$ is the probability measure on $(\mathbb{R}^n, \mathcal{B}(\mathbb{R}^n))$ with density $p(t_1, \ldots, t_n) = \lambda^n e^{-\lambda(t_1 + \cdots + t_n)}$ for $t_j \geq 0$.

8.1.2 Parametric Statistical Models

In all of our previous examples there was a parameter that parametrized the probability measures in $\mathbf{P}$ in natural way. In Example 8.1.3, this is the parameter $\theta \in [0, 1]$, in Example 8.1.4, the probability measures are parametrized by $M \in \{0, \ldots, N\}$, in Example 8.1.8 the parameter is also $\theta \in [0, 1]$, and, finally, in Example 8.1.9 the natural parameter is $\lambda > 0$. Therefore, from now on, we assume that there is a **parameter set** Θ such that $\mathbf{P}$ may be represented as

$$\mathbf{P} = \{\mathbb{P}_\theta : \theta \in \Theta\}.$$

Definition 8.1.10. A **parametric statistical model** is defined as

$$(\mathcal{X}, \mathcal{F}, \mathbb{P}_\theta)_{\theta \in \Theta}$$

with parameter set Θ. Equivalently, we suppose that the distribution assumption $\mathbf{P}$, appearing in Definition 8.1.2, may be represented as $\mathbf{P} = \{\mathbb{P}_\theta : \theta \in \Theta\}$.

In this notation, the parameter sets in Examples 8.1.3, 8.1.4, 8.1.8, and 8.1.9 are $\Theta = [0, 1]$, $\Theta = \{0, \ldots, N\}$, $\Theta = [0, 1]$, and $\Theta = (0, \infty)$, respectively.

Remark 8.1.11. It is worthwhile mentioning that the parameter can be quite general; for example, it can be a vector $\theta = (\theta_1, \ldots, \theta_k)$, so that in fact there are k unknown parameters θ_j, combined to a single vector θ. For instance, in Example 8.1.5, the unknown parameters are $\theta_1, \ldots, \theta_6$, thus, the parameter set is given by

$$\Theta = \{\theta = (\theta_1, \ldots, \theta_6) : \theta_k \geq 0, \ \theta_1 + \cdots + \theta_6 = 1\}.$$

Let us present two further examples with slightly more complicated parameter sets.

Example 8.1.12. We are given an item of unknown length. It is measured by an instrument of an unidentified precision. We assume that the instrument is unbiased, that is, on average, it shows the correct value. In view of the central limit theorem we may suppose that the measurements are distributed according to a normal distribution $\mathcal{N}(\mu, \sigma^2)$. Here μ is the "correct" length of the item, and $\sigma > 0$ reflects the precision of the measuring instrument. A small $\sigma > 0$ says that the instrument is quite precise, while large σs correspond to inaccurate instruments. Consequently, by the distribution assumption the initial model is given as

$$\left(\mathbb{R}, \mathcal{B}(\mathbb{R}), \mathcal{N}(\mu, \sigma^2)\right)_{\mu \in \mathbb{R}, \sigma^2 > 0}.$$

In order to determine μ (and maybe also σ) we measure the item n times by the same method. As a result we obtain a random sample $x = (x_1, \ldots, x_n) \in \mathbb{R}^n$. Thus, our model describing this experiment is

$$\left(\mathbb{R}^n, \mathcal{B}(\mathbb{R}^n), \mathcal{N}(\mu, \sigma^2)^{\otimes n}\right)_{(\mu, \sigma^2) \in \mathbb{R} \times (0, \infty)}.$$

Because of eq. (6.9), the model may also be written as

$$\left(\mathbb{R}^n, \mathcal{B}(\mathbb{R}^n), \mathcal{N}(\vec{\mu}, \sigma^2 I_n)\right)_{(\mu, \sigma^2) \in \mathbb{R} \times (0, \infty)}$$

with $\vec{\mu} = (\mu, \ldots, \mu) \in \mathbb{R}^n$, and with diagonal matrix $\sigma^2 I_n$. The unknown parameter is (μ, σ^2), taken from the parameter set $\mathbb{R} \times (0, \infty)$.

Example 8.1.13. Suppose now we have two different items of lengths μ_1 and μ_2. We take m measurements of the first item and n of the second one. Thereby, we use different instruments with maybe different degrees of precision. All measurements are taken independently of each other. As a result we get a vector $(x, y) \in \mathbb{R}^{m+n}$, where $x = (x_1, \ldots, x_m)$ are the values of the first m measurements and $y = (y_1, \ldots, y_n)$ those of the second n one. As before we assume that the x_is are distributed according to $\mathcal{N}(\mu_1, \sigma_1^2)$, and the y_js according to $\mathcal{N}(\mu_2, \sigma_2^2)$. We neither know μ_1 and μ_2 nor σ_1^2 and σ_2^2. Thus, the sample space is $\mathbb{R}^{m+n}$ and the vectors (x, y) are distributed according to $\mathcal{N}((\vec{\mu}_1, \vec{\mu}_2), R_{\sigma_1^2, \sigma_2^2})$ with diagonal matrix $R_{\sigma_1^2, \sigma_2^2}$ having σ_1^2 on its first m entries and σ_2^2 on the remaining n ones.

Note that by Definition 1.9.5,

$$\mathcal{N}((\vec{\mu}_1, \vec{\mu}_2), R_{\sigma_1^2, \sigma_2^2}) = \mathcal{N}(\mu_1, \sigma_1^2)^{\otimes m} \otimes \mathcal{N}(\mu_2, \sigma_2^2)^{\otimes n}.$$

This is valid because, if $A \in \mathcal{B}(\mathbb{R}^m)$ and $B \in \mathcal{B}(\mathbb{R}^n)$, then it follows that

$$\mathcal{N}((\vec{\mu}_1, \vec{\mu}_2), R_{\sigma_1^2, \sigma_2^2})(A \times B) = \mathcal{N}(\mu_1, \sigma_1^2)^{\otimes m}(A) \cdot \mathcal{N}(\mu_2, \sigma_2^2)^{\otimes n}(B).$$

The parameter set in this example is given as $\mathbb{R}^2 \times (0, \infty)^2$, hence the statistical model may be written as

$$\left(\mathbb{R}^{m+n}, \mathcal{B}(\mathbb{R}^{m+n}), \mathcal{N}(\mu_1, \sigma_1^2)^{\otimes m} \otimes \mathcal{N}(\mu_2, \sigma_2^2)^{\otimes n}\right)_{(\mu_1, \mu_2, \sigma_1^2, \sigma_2^2) \in \mathbb{R}^2 \times (0, \infty)^2} .$$

8.2 Statistical Hypothesis Testing

8.2.1 Hypotheses and Tests

We start with a parametric statistical model $(\mathcal{X}, \mathcal{F}, \mathbb{P}_\theta)_{\theta \in \Theta}$. Suppose the parameter set Θ is split up into disjoint subsets Θ_0 and Θ_1. The aim of a test is to decide, on the basis of the observed sample, whether or not the "true" parameter θ belongs to Θ_0 or to Θ_1.

Let us explain the problem with two examples.

Example 8.2.1. Consider once more the situation described in Example 8.1.4. Assume there exists a critical value $M_0 \le N$ such that the buyer accepts the delivery if the number M of defective machines satisfies $M \le M_0$. Otherwise, if $M > M_0$, the buyer rejects it and sends the machines back to the trader. In this example the parameter set is $\Theta = \{0, \dots, N\}$. Letting $\Theta_0 = \{0, \dots, M_0\}$ and $\Theta_1 = \{M_0 + 1, \dots, N\}$, the question about acceptance or rejection of the delivery is equivalent to whether $M \in \Theta_0$ or $M \in \Theta_1$. Assume now the buyer checked n of the N machines and found m defective machines. On the basis of this observation, the buyer has to decide about acceptance or rejection, or, equivalently, about $M \in \Theta_0$ or $M \in \Theta_1$.

Example 8.2.2. Let us consider once more Example 8.1.13. There we had two measuring instruments, both being unbiased. Consequently, the expected values μ_1 and μ_2 are the correct lengths of the two items. The parameter set was $\Theta = \mathbb{R}^2 \times (0, \infty)^2$. Suppose we conjecture that both items are of equal length, that is, we conjecture $\mu_1 = \mu_2$. Letting

$$\Theta_0 := \{(\mu, \mu, \sigma_1^2, \sigma_2^2) : \mu \in \mathbb{R}, \ \sigma_1^2, \sigma_2^2 > 0\}$$

and $\Theta_1 = \Theta \backslash \Theta_0$, to prove or disprove the conjecture, we have to check whether $(\mu_1, \mu_2, \sigma_1^2, \sigma_2^2)$ belongs to Θ_0 or to Θ_1.

On the other hand, if we want to know whether or not the first item is smaller than the second one, then we have to choose

$$\Theta_0 := \{(\mu_1, \mu_2, \sigma_1^2, \sigma_2^2) : -\infty < \mu_1 \le \mu_2 < \infty, \ \sigma_1^2, \sigma_2^2 > 0\}$$

and to check whether or not $(\mu_1, \mu_2, \sigma_1^2, \sigma_2^2)$ belongs to Θ_0.

An exact mathematical formulation of the previous problems is as follows.

Definition 8.2.3. Let $(\mathcal{X}, \mathcal{F}, \mathbb{P}_\theta)_{\theta \in \Theta}$ be a parametric statistical model and suppose $\Theta = \Theta_0 \cup \Theta_1$ with $\Theta_0 \cap \Theta_1 = \emptyset$.

Then the **hypothesis** or, more precisely, **null hypothesis** $\mathbb{H}_0$ says that for the "correct" $\theta \in \Theta$ one has $\theta \in \Theta_0$. This is expressed by writing $\mathbb{H}_0 : \theta \in \Theta_0$.

The **alternative hypothesis** $\mathbb{H}_1$ says $\theta \in \Theta_1$. This is formulated as $\mathbb{H}_1 : \theta \in \Theta_1$.

After the hypothesis is set, one executes a statistical experiment. Here the order is important: first one has to set the hypothesis, then test it, not vice versa. If the hypothesis is chosen on the basis of the observed results, then, of course, the sample will confirm it.

Say the result of the experiment is some sample $x \in \mathcal{X}$. One of the fundamental problems in Mathematical Statistics is to decide, on the basis of the observed sample, about acceptance or rejection of $\mathbb{H}_0$. The mathematical formulation of the problem is as follows.

Definition 8.2.4. A (hypothesis) **test T** for checking $\mathbb{H}_0$ (against $\mathbb{H}_1$) is a disjoint partition $\mathbf{T} = (\mathcal{X}_0, \mathcal{X}_1)$ of the sample space $\mathcal{X}$. The set $\mathcal{X}_0$ is called the **region of acceptance** while $\mathcal{X}_1$ is said to be the **critical region**[1] or **region of rejection**. By mathematical reasoning we have to assume $\mathcal{X}_0 \in \mathcal{F}$, which of course implies $\mathcal{X}_1 \in \mathcal{F}$ as well.

Remark 8.2.5. A hypothesis test $\mathbf{T} = (\mathcal{X}_0, \mathcal{X}_1)$ operates as follows: if the statistical experiment leads to a sample $x \in \mathcal{X}_1$, then we reject $\mathbb{H}_0$. But, if we get an $x \in \mathcal{X}_0$, then this does not contradict the hypothesis, and for now we may furthermore work with it.

Important comment: If we observe an $x \in \mathcal{X}_0$, then this does **not** say that $\mathbb{H}_0$ is correct. It only asserts that we failed to reject it or that there is a lack of evidence against it.

Let us illustrate the procedure with Example 8.2.1.

Example 8.2.6. By the choice of Θ_0 and Θ_1, the hypothesis $\mathbb{H}_0$ is given by

$$\mathbb{H}_0 : 0 \leq M \leq M_0, \quad \text{hence} \quad \mathbb{H}_1 : M_0 < M \leq N.$$

To test $\mathbb{H}_0$ against $\mathbb{H}_1$, the sample space $\mathcal{X} = \{0, \ldots, n\}$ is split up into the two regions $\mathcal{X}_0 := \{0, \ldots, m_0\}$ and $\mathcal{X}_1 := \{m_0 + 1, \ldots, n\}$ with some (for now) arbitrary number $m_0 \in \{0, \ldots, n\}$. If among the checked n machines m are defective with some $m > m_0$,

---

1 Sometimes also called "critical section."

then $m \in \mathcal{X}_1$, hence one rejects $\mathbb{H}_0$. In this case the buyer refuses to take the delivery and sends it back to the trader. On the other hand, if $m \leq m_0$, then $m \in \mathcal{X}_0$, which does not contradict $\mathbb{H}_0$, and the buyer will accept the delivery and pay for it. Of course, the key question is how to choose the value m_0 in a proper way.

Remark 8.2.7. Sometimes tests are also defined as mappings $\varphi : \mathcal{X} \to \{0, 1\}$. The link between these two approaches is immediately clear. Starting with φ the hypothesis test $\mathbf{T} = (\mathcal{X}_0, \mathcal{X}_1)$ is constructed by $\mathcal{X}_0 = \{x \in \mathcal{X} : \varphi(x) = 0\}$ and $\mathcal{X}_1 = \{x \in \mathcal{X} : \varphi(x) = 1\}$. Conversely, if $\mathbf{T} = (\mathcal{X}_0, \mathcal{X}_1)$ is a given test, then set $\varphi(x) = 0$ if $x \in \mathcal{X}_0$ and $\varphi(x) = 1$ for $x \in \mathcal{X}_1$. The advantage of this approach is that it allows us to define so-called **randomized tests**. Here $\varphi : \mathcal{X} \to [0, 1]$. Then, as before, $\mathcal{X}_0 = \{x \in \mathcal{X} : \varphi(x) = 0\}$ and $\mathcal{X}_1 = \{x \in \mathcal{X} : \varphi(x) = 1\}$. If $0 < \varphi(x) < 1$, then

$$\varphi(x) = \mathbb{P}\{\text{reject } \mathbb{H}_0 \text{ if } x \text{ is observed}\}.$$

That is, for certain observations $x \in \mathcal{X}$, an additional random experiment (e.g., tossing a coin) decides whether we accept or reject $\mathbb{H}_0$. Randomized tests are useful in the case of finite or countably infinite sample spaces.

When applying a test $\mathbf{T} = (\mathcal{X}_0, \mathcal{X}_1)$ to check the null hypothesis $\mathbb{H}_0 : \theta \in \Theta_0$, two different types of errors may occur.

Definition 8.2.8. An **error of the first kind** or **type I error** occurs if $\mathbb{H}_0$ is true but one observes a sample $x \in \mathcal{X}_1$, hence rejects $\mathbb{H}_0$.

Type I error = incorrect rejection of a true null hypothesis **!**

In other words, a type I error happens if the "true" θ is in Θ_0, but we observe an $x \in \mathcal{X}_1$.

Definition 8.2.9. An **error of the second kind** or **type II error** occurs if $\mathbb{H}_0$ is false, but the observed sample lies in $\mathcal{X}_0$, hence we do not reject the false hypothesis $\mathbb{H}_0$.

Type II error = failure to reject a false null hypothesis **!**

Consequently, a type II error occurs if the "true" θ is in Θ_1, but the observed sample is an element of the region of acceptance $\mathcal{X}_0$.

Example 8.2.10. In the context of Example 8.2.6 a type I error occurs if the delivery was well, but among the checked machines were more than m_0 defective, so that the buyer rejects the delivery. Since the trader was not able to sell a proper delivery, this error is also called the **risk of the trader.**

On the other hand, a type II error occurs if the delivery is not in good order, but among the checked machines were only a few defective ones (less than or equal to m_0). Thus, the buyer accepts the bad delivery and pays for it. Therefore, this type of error is also called the **risk of the buyer.**

8.2.2 Power Function and Significance Tests

The power of a test is described by its power function defined as follows.

Definition 8.2.11. Let $\mathbf{T} = (\mathcal{X}_0, \mathcal{X}_1)$ be a test for $\mathbb{H}_0 : \theta \in \Theta_0$ against $\mathbb{H}_1 : \theta \in \Theta_1$. The function $\beta_{\mathbf{T}}$ from Θ to $[0, 1]$ defined as

$$\beta_{\mathbf{T}}(\theta) := \mathbb{P}_\theta(\mathcal{X}_1)$$

is called the **power function** of the test $\mathbf{T}$.

Remark 8.2.12. If $\theta \in \Theta_0$, that is, if $\mathbb{H}_0$ is true, then $\beta_{\mathbf{T}}(\theta) = \mathbb{P}_\theta(\mathcal{X}_1)$ is the probability that $\mathcal{X}_1$ occurs or, equivalently, that a type I error happens.

On the contrary, if $\theta \in \Theta_1$, that is, $\mathbb{H}_0$ is false, then $1 - \beta_{\mathbf{T}}(\theta) = \mathbb{P}_\theta(\mathcal{X}_0)$ is the probability that $\mathcal{X}_0$ occurs or, equivalently, that a type II error appears.

Thus, a "good" test should satisfy the following conditions: the power function $\beta_{\mathbf{T}}$ attains small values on Θ_0 and/or $1 - \beta_{\mathbf{T}}$ has small values on Θ_1. Then the probabilities for the occurrence of type I and/or type II errors are not too big.[2]

Example 8.2.13. What is the power function of the test presented in Example 8.2.6? Recall that $\Theta = \{0, \ldots, N\}$ and $\mathcal{X}_1 = \{m_0 + 1, \ldots, n\}$. Hence, $\beta_{\mathbf{T}}$ maps $\{0, \ldots, N\}$ to $[0, 1]$ in the following way:

$$\beta_{\mathbf{T}}(M) = H_{N,M,n}(\mathcal{X}_1) = \sum_{m=m_0+1}^{n} \frac{\binom{M}{m}\binom{N-M}{n-m}}{\binom{N}{n}}. \tag{8.1}$$

---

[2] In the literature the power function is sometimes defined in a slightly different way. If $\theta \in \Theta_0$, then it is as in our Definition 8.2.11 while for $\theta \in \Theta_1$ one defines it as $1 - \beta_{\mathbf{T}}(\theta)$. Moreover, for $1 - \beta_{\mathbf{T}}$ one finds the notations **operation characteristics** or **oc-function.**

Thus, the maximal probability for a type I error is given by

$$\max_{0 \le M \le M_0} \beta_{\mathbf{T}}(M) = \max_{0 \le M \le M_0} \sum_{m=m_0+1}^{n} \frac{\binom{M}{m}\binom{N-M}{n-m}}{\binom{N}{n}},$$

while the maximal probability for a type II error equals

$$\max_{M_0 < M \le N} (1 - \beta_{\mathbf{T}}(M)) = \max_{M_0 < M \le N} \sum_{m=0}^{m_0} \frac{\binom{M}{m}\binom{N-M}{n-m}}{\binom{N}{n}}.$$

Remark 8.2.14. The previous example already illustrates the **dilemma of hypothesis testing**. To minimize the type I error one has to choose m_0 as large as possible. But increasing m_0 enlarges the type II error.

This dilemma occurs always in the theory of hypothesis testing. In order to minimize the probability of a type I error, the critical region $\mathcal{X}_1$ has to be chosen as small as possible. But making $\mathcal{X}_1$ smaller enlarges $\mathcal{X}_0$, hence the probability for the occurrence of a type II error increases. In the extreme case, if $\mathcal{X}_1 = \emptyset$, hence $\mathcal{X}_0 = \mathcal{X}$, then a type I error I cannot occur at all. In the context of Example 8.2.6 that means the buyer accepts all deliveries and the trader takes no risk.

On the other hand, to minimize the occurrence of a type II error, the region of acceptance $\mathcal{X}_0$ has to be as small as possible. In the extreme case, if we choose $\mathcal{X}_0 = \emptyset$, then a type II error cannot occur because we always reject the hypothesis. In the context of Example 8.2.6 this says the buyer rejects all deliveries. In this way he avoids buying any delivery of bad quality, but he also never gets a proper one. Thus the buyer takes no risk.

It is pretty clear that both extreme cases presented above are very absurd. Therefore, one has to find a suitable compromise. The approach for such a compromise is as follows: in a first step one chooses tests where the probability of a type I error is bounded from above. And in a second step, among all these tests satisfying this bound, one takes the one that minimizes the probability of a type II error. More precisely, we will investigate tests satisfying the following condition.

Definition 8.2.15. Suppose we are given a number $\alpha \in (0, 1)$, the so-called **significance level**. A test $\mathbf{T} = (\mathcal{X}_0, \mathcal{X}_1)$ for testing the hypothesis $\mathbb{H}_0 : \theta \in \Theta_0$ against $\mathbb{H}_1 : \theta \in \Theta_1$ is said to be an α-**significance test** (or shorter α-**test**), provided the probability for the occurrence of a type I error is bounded by α. That is, the test has to satisfy

$$\sup_{\theta \in \Theta_0} \beta_{\mathbf{T}}(\theta) = \sup_{\theta \in \Theta_0} \mathbb{P}_\theta(\mathcal{X}_1) \le \alpha.$$

Interpretation: The significance level α is assumed to be small. Typical choices are $\alpha = 0.1$ or $\alpha = 0.01$. Let **T** be an α-significance test and assume that $\mathbb{H}_0$ is true. If we observe now a sample in the critical region $\mathcal{X}_1$, then an event occurred with probability less than or equal to α, that is, a very unlikely event has been observed. Therefore, we can be very sure that this could not happen provided that $\mathbb{H}_0$ would be true, and we reject this hypothesis. The probability that we made a mistake is less than or equal to α, hence very small.

Recall that α-significance tests admit no bound for the probability of a type II error. Therefore, we look for those α-significance tests that minimize the probability for a type II error.

Definition 8.2.16. Let $\mathbf{T}_1$ and $\mathbf{T}_2$ be two α-significance tests for checking $\mathbb{H}_0$ against $\mathbb{H}_1$. If their power functions satisfy

$$\beta_{\mathbf{T}_1}(\theta) \geq \beta_{\mathbf{T}_2}(\theta), \quad \theta \in \Theta_1,$$

then we say that $\mathbf{T}_1$ is (uniformly) **more powerful** than $\mathbf{T}_2$.

A (uniformly) **most powerful** α-test **T** is one that is more powerful than all other α-tests.

Remark 8.2.17. Note that $\beta_{\mathbf{T}_1}(\theta) \geq \beta_{\mathbf{T}_2}(\theta)$ implies $1 - \beta_{\mathbf{T}_1}(\theta) \leq 1 - \beta_{\mathbf{T}_2}(\theta)$, hence if $\mathbf{T}_1$ is more powerful than $\mathbf{T}_2$, then, according to Remark 8.2.12, the probability for the occurrence of a type II error is smaller for $\mathbf{T}_1$ than it is for $\mathbf{T}_2$. Therefore, a most powerful α-test is the one that minimizes the probability of occurrence of a type II error.

Remark 8.2.18. The question about existence and uniqueness of most powerful α-tests is treated in the Neyman–Pearson lemma and its consequences. We will not discuss that problem here; instead we will construct most powerful tests in concrete situations. See [CB02], Chapter 8.3.2, for a detailed discussion of the Neyman–Pearson lemma and its consequences.

We start with the construction of such tests in the hypergeometric case. Here we have the following.

Proposition 8.2.19. *If the statistical model is* $(\mathcal{X}, \mathcal{P}(\mathcal{X}), H_{M,N,n})_{M=0,\ldots,N}$ *with* $\mathcal{X} = \{0, \ldots, n\}$*, then a most powerful α-test for testing* $M \leq M_0$ *against* $M > M_0$ *is given by* $\mathbf{T} = (\mathcal{X}_0, \mathcal{X}_1)$*, where* $\mathcal{X}_0 = \{0, \ldots, m_0\}$*, and* m_0 *is defined by*

$$m_0 := \max\left\{ k \leq n : \sum_{m=k}^{n} \frac{\binom{M_0}{m}\binom{N-M_0}{n-m}}{\binom{N}{n}} > \alpha \right\}$$

$$= \min\left\{ k \leq n : \sum_{m=k+1}^{n} \frac{\binom{M_0}{m}\binom{N-M_0}{n-m}}{\binom{N}{n}} \leq \alpha \right\}.$$

Proof: The proof of Proposition 8.2.19 needs the following lemma.

Lemma 8.2.20. *The power function, defined by eq. (8.1), is a nondecreasing function on the set $\{0, \ldots, N\}$.*

Proof: Suppose we get a delivery of N machines containing M defective ones. Now there are not only defective machines within the delivery, but also $\tilde{M} - M$ false ones for some $\tilde{M} \geq M$. We take a sample of size n and test these machines. Let X be the number of defective machines and let $\tilde{X}$ be the number of machines that are either defective or false. Of course, we have $X \leq \tilde{X}$ implying $\mathbb{P}(X > m_0) \leq \mathbb{P}(\tilde{X} > m_0)$. Note that X is $H_{N,M,n}$-distributed while $\tilde{X}$ is distributed according to $H_{N,\tilde{M},n}$. These observations lead to

$$\beta_T(M) = H_{N,M,n}(\{m_0 + 1, \ldots, n\}) = \mathbb{P}\{X > m_0\} \leq \mathbb{P}\{\tilde{X} > m_0\}$$
$$= H_{N,\tilde{M},n}(\{m_0 + 1, \ldots, n\}) = \beta_T(\tilde{M}).$$

This being true for all $M \leq \tilde{M}$ proves that β_T is nondecreasing. ∎

Let us come back to the proof of Proposition 8.2.19. Set $\mathcal{X}_0 := \{0, \ldots, m_0\}$, thus $\mathcal{X}_1 = \{m_0 + 1, \ldots, n\}$ for some (at the moment arbitrary) $m_0 \leq n$. Because of Lemma 8.2.20, the test $\mathbf{T} = (\mathcal{X}_0, \mathcal{X}_1)$ is an α-significance test if and only if it satisfies

$$\sum_{m=m_0+1}^{n} \frac{\binom{M_0}{m}\binom{N-M_0}{n-m}}{\binom{N}{n}} = H_{N,M_0,n}(\mathcal{X}_1) = \sup_{M \leq M_0} H_{N,M,n}(\mathcal{X}_1) \leq \alpha.$$

To minimize the probability for the occurrence of a type II error, we have to choose $\mathcal{X}_1$ as large as possible or, equivalently, m_0 as small as possible, that is, if we replace m_0 by $m_0 - 1$, then the new test is no longer an α-test. Thus, in order that $\mathbf{T}$ is an α-test that minimizes the probability for a type II error, the number m_0 has to be chosen such that

$$\sum_{m=m_0+1}^{n} \frac{\binom{M_0}{m}\binom{N-M_0}{n-m}}{\binom{N}{n}} \leq \alpha \quad \text{and} \quad \sum_{m=m_0}^{n} \frac{\binom{M_0}{m}\binom{N-M_0}{n-m}}{\binom{N}{n}} > \alpha.$$

This completes the proof. ∎

Example 8.2.21. A buyer gets a delivery of 100 machines. In the case that there are strictly more than 10 defective machines in the delivery, he will reject it. Thus, his hypothesis is $\mathbb{H}_0 : M \leq 10$. In order to test $\mathbb{H}_0$, he chooses 15 machines and checks them. Let m be the number of defective machines among the checked ones. For which m does he reject the delivery with significance level $\alpha = 0.01$?

Answer: We have $N = 100$, $M_0 = 10$, and $n = 15$. Since $\alpha = 0.01$, by

$$\sum_{m=5}^{15} \frac{\binom{10}{m}\binom{90}{15-m}}{\binom{100}{15}} = 0.0063 \cdots < \alpha \quad \text{and} \quad \sum_{m=4}^{15} \frac{\binom{10}{m}\binom{90}{15-m}}{\binom{100}{15}} = 0.04 \cdots > \alpha,$$

it follows that the optimal choice is $m_0 = 4$. Consequently, we have $\mathcal{X}_0 = \{0, \ldots, 4\}$, thus, $\mathcal{X}_1 = \{5, \ldots, 15\}$. If there are 5 or even more defective machines among the tested 15 ones, then the buyer should reject the delivery. The probability that his decision is wrong is less than or equal to 0.01.

What can be said about the probability for a type II error? For this test we have

$$\beta_T(M) = \sum_{m=5}^{15} \frac{\binom{M}{m}\binom{100-M}{15-m}}{\binom{100}{15}},$$

hence

$$1 - \beta_T(M) = \sum_{m=0}^{4} \frac{\binom{M}{m}\binom{100-M}{15-m}}{\binom{100}{15}}.$$

Since β_T is nondecreasing, $1 - \beta_T$ is nonincreasing, and the probability for a type II error becomes maximal for $M = 11$. Recall that $\Theta_0 = \{0, \ldots, 10\}$ and, therefore, $\Theta_1 = \{11, \ldots, 100\}$. Thus, an upper bound for the probability of a type II error is given by

$$1 - \beta_T(M) \le 1 - \beta_T(11) = \sum_{m=0}^{4} \frac{\binom{11}{m}\binom{89}{15-m}}{\binom{100}{15}} = 0.989471, \quad M = 11, \ldots, 100.$$

This tells us that even in the case of most powerful tests the likelihood for a type II error may be quite large. Even if the number of defective machines is big , this error may occur with higher probability. For example, we have

$$1 - \beta_T(20) = 0.853089 \quad \text{or} \quad 1 - \beta_T(40) = 0.197057.$$

Important Remark: An α-significance test provides us with quite precise information when rejecting the hypothesis $\mathbb{H}_0$. In contrast, when we observe a sample $x \in \mathcal{X}_0$, then the only information we get is that we failed to reject $\mathbb{H}_0$, thus, we must continue to regard it as true. Consequently, whenever fixing the null hypothesis, we have to fix it in a way that either a type I error has the most serious consequences or that we can achieve the greatest information by rejecting $\mathbb{H}_0$. Let us explain this with two examples.

Example 8.2.22. A certain type of food sometimes contains a special kind of poison. Suppose there are μ milligrams poison in one kilogram of the food. If $\mu > \mu_0$, then eating this becomes dangerous while for $\mu \le \mu_0$ it is unproblematic. How do we successfully choose the hypothesis when testing some sample of the food? We could take either $\mathbb{H}_0 : \mu > \mu_0$ or $\mathbb{H}_0 : \mu \le \mu_0$. Which is the right choice?

Answer: The correct choice is $\mathbb{H}_0 : \mu > \mu_0$. Why? If we reject $\mathbb{H}_0$, then we can be very sure that the food is not poisoned and may be eaten. The probability that someone will be poisoned is less than α. A type II error occurs if the food is harmless, but we discard it because our test tells us that it is poisoned. That results in a loss for the company that produced it, but no one will suffer from poisoning. If we had chosen $\mathbb{H}_0 : \mu \le \mu_0$, then a type II error occurs if $\mathbb{H}_0$ is false, that is, the food is poisoned, but our test says that it is eatable. Of course, this error is much more serious, and we have no control in regards to its probability.

Example 8.2.23. Suppose the height of 18-year-old males in the US is normally distributed with expected value μ and variance $\sigma^2 > 0$. We want to know whether the average height is above or below 6 feet. There is strong evidence that we will have $\mu \le 6$, but we cannot prove this. To do so, we execute a statistical experiment and choose randomly n males of age 18 and measure their height. Which hypothesis should be checked? If we take $\mathbb{H}_0 : \mu \le 6$, then it is very likely that our experiment will lead to a result that does not contradict this hypothesis, resulting in a small amount of information gained. But, if we work with the hypothesis $\mathbb{H}_0 : \mu > 6$, then a rejection of this hypothesis tells us that $\mathbb{H}_0$ is very likely wrong, and we may say the conjecture is true with high probability, namely that we have $\mu \le 6$. Here the probability that our conclusion is wrong is very small.

8.3 Tests for Binomial Distributed Populations

Because of its importance we present tests for binomial distributed populations in a separate section. The starting point is the problem described in Examples 8.1.3 and 8.1.8. In a single experiment we may observe either "0" or "1," but we do not know the probabilities for the occurrence of these events. To obtain some information about the unknown probabilities we execute n independent trials and record how often "1" occurs. This number is $B_{n,\theta}$-distributed for some $0 \le \theta \le 1$. Hence, the describing statistical model is given by

$$(\mathcal{X}, \mathcal{P}(\mathcal{X}), B_{n,\theta})_{\theta \in [0,1]} \quad \text{where} \quad \mathcal{X} = \{0, \ldots, n\}. \tag{8.2}$$

Two-Sided Tests: We want to check whether the unknown parameter θ satisfies $\theta = \theta_0$ or $\theta \ne \theta_0$ for some given $\theta_0 \in [0, 1]$. Thus, $\Theta_0 = \{\theta_0\}$ and $\Theta_1 = [0, 1] \backslash \{\theta_0\}$. In other words, the null and the alternative hypothesis are

$$\mathbb{H}_0 : \theta = \theta_0 \quad \text{and} \quad \mathbb{H}_1 : \theta \ne \theta_0,$$

respectively.

To construct a suitable α-significance test for checking $\mathbb{H}_0$ we introduce two numbers n_0 and n_1 as follows. Note that these numbers are dependent on θ_0 and, of course, also on α.

$$n_0 := \min \left\{ k \le n : \sum_{j=0}^{k} \binom{n}{j} \theta_0^j (1 - \theta_0)^{n-j} > \alpha/2 \right\}$$

$$= \max \left\{ k \le n : \sum_{j=0}^{k-1} \binom{n}{j} \theta_0^j (1 - \theta_0)^{n-j} \le \alpha/2 \right\} \quad \text{and} \quad (8.3)$$

$$n_1 := \max \left\{ k \le n : \sum_{j=k}^{n} \binom{n}{j} \theta_0^j (1 - \theta_0)^{n-j} > \alpha/2 \right\}$$

$$= \min \left\{ k \le n : \sum_{j=k+1}^{n} \binom{n}{j} \theta_0^j (1 - \theta_0)^{n-j} \le \alpha/2 \right\}. \quad (8.4)$$

Proposition 8.3.1. *Regard the statistical model (8.2) and let $0 < \alpha < 1$ be a significance level. The hypothesis test $\mathbf{T} = (\mathcal{X}_0, \mathcal{X}_1)$ with*

$$\mathcal{X}_0 := \{n_0, n_0 + 1, \ldots, n_1 - 1, n_1\} \quad and \quad \mathcal{X}_1 = \{0, \ldots, n_0 - 1\} \cup \{n_1 + 1, \ldots, n\}$$

is an α-significance test to check $\mathbb{H}_0 : \theta = \theta_0$ against $\mathbb{H}_1 : \theta \ne \theta_0$. Here n_0 and n_1 are defined as in eqs. (8.3) and (8.4).

Proof: Since Θ_0 consists only of the point $\{\theta_0\}$, an arbitrary test $\mathbf{T} = (\mathcal{X}_0, \mathcal{X}_1)$ is an α-significance test if and only if $B_{n,\theta_0}(\mathcal{X}_1) \le \alpha$. Now let $\mathbf{T}$ be as in the formulation of the proposition. By the definition of the numbers n_0 and n_1 we obtain

$$B_{n,\theta_0}(\mathcal{X}_1) = \sum_{j=0}^{n_0-1} \binom{n}{j} \theta_0^j (1 - \theta_0)^{n-j} + \sum_{j=n_1+1}^{n} \binom{n}{j} \theta_0^j (1 - \theta_0)^{n-j} \le \frac{\alpha}{2} + \frac{\alpha}{2} = \alpha,$$

that is, as claimed, the test $\mathbf{T} := (\mathcal{X}_0, \mathcal{X}_1)$ is an α-significance test. Note that the region $\mathcal{X}_1$ is chosen maximal. In fact, neither n_0 can be enlarged nor n_1 can be made smaller. ∎

Remark 8.3.2. In this test the critical region $\mathcal{X}_1$ consists of two parts or tails. Therefore, this type of hypothesis test is called **two-sided test**.

Example 8.3.3. In an urn is an unknown number of white and black balls. Let $\theta \in [0, 1]$ be the proportion of white balls. We conjecture that there are as many white as black balls in the urn. That is, the null hypothesis is $\mathbb{H}_0 : \theta = 0.5$. To test this hypothesis we choose one after the other 100 balls with replacement. In order to determine n_0 and n_1 in this situation let φ be defined as

$$\varphi(k) := \sum_{j=0}^{k-1} \binom{100}{j} \cdot \left(\frac{1}{2}\right)^{100}.$$

Numerical calculations give

$\varphi(37) = 0.00331856, \ \varphi(38) = 0.00601649, \ \varphi(39) = 0.0104894$

$\varphi(40) = 0.0176001, \ \varphi(41) = 0.028444, \ \varphi(42) = 0.044313$

$\varphi(43) = 0.0666053, \ \varphi(44) = 0.096674, \ \varphi(45) = 0.135627$

$\varphi(46) = 0.184101, \ \varphi(47) = 0.242059, \ \varphi(48) = 0.30865,$

$\varphi(49) = 0.382177, \ \varphi(50) = 0.460205.$

If the significance level is chosen as $\alpha = 0.1$ we see that $\varphi(42) \leq 0.05$, but $\varphi(43) > 0.05$. Hence, by the definition of n_0 in eq. (8.3), it follows that $n_0 = 42$. By symmetry, for n_1 defined in eq. (8.4), we get $n_1 = 58$. Consequently, the regions of acceptance and rejection are given by

$$\mathcal{X}_0 = \{42, 43, \ldots, 57, 58\} \quad \text{and} \quad \mathcal{X}_1 = \{0, \ldots, 41\} \cup \{59, \ldots, 100\}.$$

For example, if we observe during 100 trials k white balls for some $k < 42$ or some $k > 58$, then we may be quite sure that our null hypothesis is wrong, that is, the number of white and black balls is significantly different. This assertion is 90% sure.

   If we want to be more secure about the conclusion, we have to choose a smaller significance level. For example, if we take $\alpha = 0.01$, the values of φ imply $n_0 = 37$ and $n_1 = 63$, hence

$$\mathcal{X}_0 = \{37, 38, \ldots, 62, 63\} \quad \text{and} \quad \mathcal{X}_1 = \{0, \ldots, 36\} \cup \{64, \ldots, 100\}.$$

Again we see that a smaller bound for the probability of a type I error leads to an enlargement of $\mathcal{X}_0$, thus, to an increase of the chance for a type II error.

One-Sided Tests: Now the null hypothesis is $\mathbb{H}_0 : \theta \leq \theta_0$ for some $\theta_0 \in [0, 1]$. In the context of Example 8.1.3 we claim that the proportion of white balls in the urn does not exceed θ_0. For instance, if $\theta_0 = 1/2$, then we want to test whether or not the number of white balls is less than or equal to that of black ones.

   Before we present a most powerful test for this situation let us define a number n_0 depending on θ_0 and on the significance level $0 < \alpha < 1$.

$$n_0 := \max \left\{ k \le n : \sum_{j=k}^{n} \binom{n}{j} \theta_0^j (1 - \theta_0)^{n-j} > \alpha \right\} \tag{8.5}$$

$$= \min \left\{ k \le n : \sum_{j=k+1}^{n} \binom{n}{j} \theta_0^j (1 - \theta_0)^{n-j} \le \alpha \right\}.$$

Now we are in a position to state the most powerful one-sided α-test for a binomial distributed population.

Proposition 8.3.4. *Suppose $\mathcal{X} = \{0, \ldots, n\}$, and let $(\mathcal{X}, \mathcal{P}(\mathcal{X}), B_{n,\theta})_{\theta \in [0,1]}$ be the statistical model describing a binomial distributed population. Given $0 < \alpha < 1$, define n_0 by eq. (8.5) and set $\mathcal{X}_0 = \{0, \ldots, n_0\}$, hence $\mathcal{X}_1 = \{n_0 + 1, \ldots, n\}$. Then $\mathbf{T} = (\mathcal{X}_0, \mathcal{X}_1)$ is the most powerful α-test to check the null hypothesis $\mathbb{H}_0 : \theta \le \theta_0$ against $\mathbb{H}_1 : \theta > \theta_0$.*

Proof: With an arbitrary $0 \le n' \le n$ define the region of acceptance $\mathcal{X}_0$ of a test $\mathbf{T}$ by $\mathcal{X}_0 = \{0, \ldots, n'\}$. Then its power function is given by

$$\beta_{\mathbf{T}}(\theta) = B_{n,\theta}(\mathcal{X}_1) = \sum_{j=n'+1}^{n} \binom{n}{j} \theta^j (1 - \theta)^{n-j}, \quad 0 \le \theta \le 1. \tag{8.6}$$

To proceed further we need the following lemma.

Lemma 8.3.5. *The power function (8.6) is nondecreasing in $[0, 1]$.*

Proof: Suppose in an urn there are white, red, and black balls. Their proportions are θ_1, $\theta_2 - \theta_1$ and $1 - \theta_2$ for some $0 \le \theta_1 \le \theta_2 \le 1$. Choose n balls with replacement. Let X be the number of chosen white balls, and Y is the number of balls that were either white or red. Then X is B_{n,θ_1}-distributed, while Y is distributed according to B_{n,θ_2}. Moreover, $X \le Y$, hence it follows that $\mathbb{P}(X > n') \le \mathbb{P}(Y > n')$, which leads to

$$\beta_{\mathbf{T}}(\theta_1) = B_{n,\theta_1}(\{n' + 1, \ldots, n\}) = \mathbb{P}(X > n') \le \mathbb{P}(Y > n')$$
$$= B_{n,\theta_2}(\{n' + 1, \ldots, n\}) = \beta_{\mathbf{T}}(\theta_2).$$

This being true for all $\theta_1 \le \theta_2$ completes the proof. ∎

An application of Lemma 8.3.5 implies that the above test $\mathbf{T}$ is an α-significance test if and only if

$$\sum_{j=n'+1}^{n} \binom{n}{j} \theta_0^j (1 - \theta_0)^{n-j} = \beta_{\mathbf{T}}(\theta_0) = \sup_{\theta \le \theta_0} \beta_{\mathbf{T}}(\theta) \le \alpha.$$

In order to minimize the probability of a type II error, we have to choose $\mathcal{X}_0$ as small as possible. That is, if we replace n' by $n' - 1$, the modified test is no longer an α-test.

Thus, the optimal choice is $n' = n_0$ where n_0 is defined by eq. (8.5). This completes the proof of Proposition 8.3.4. ∎

Example 8.3.6. Let us come back to the problem investigated in Example 8.1.3. Our null hypothesis is $\mathbb{H}_0 : \theta \le 1/2$, that is, we claim that at most half of the balls are white. To test $\mathbb{H}_0$, we choose 100 balls and record their color. Let k be the number of observed white balls. For which k must we reject $\mathbb{H}_0$ with a security of 90%?
    Answer: Since

$$\sum_{k=56}^{100} \binom{100}{k} 2^{-100} = 0.135627 \quad \text{and} \quad \sum_{k=57}^{100} \binom{100}{k} 2^{-100} = 0.096674 ,$$

for $\alpha = 0.1$ the number n_0 in eq. (8.5) equals $n_0 = 56$. Consequently, the region of acceptance for the best 0.1-test is given by $\mathcal{X}_0 = \{0, \ldots, 56\}$. Thus, whenever there are 57 or more white balls among the chosen 100 the hypothesis has to be rejected. The probability for a wrong decision is less than or equal to 0.1.
    Making the significance level smaller, for example, taking $\alpha = 0.01$, this implies $n_0 = 63$. Hence, if the number of white balls is 64 or larger, a rejection of $\mathbb{H}_0$ is 99% sure.

Remark 8.3.7. Example 8.3.6 emphasizes once more the **dilemma** of hypothesis testing. The price one pays for higher security, when rejecting $\mathbb{H}_0$, is the increase of the likelihood of a type II error. For instance, replacing $\alpha = 0.1$ by $\alpha = 0.01$ in the previous example leads to an enlargement of $\mathcal{X}_0$ from $\{0, \ldots, 56\}$ to $\{0, \ldots, 63\}$. Thus, if we observe 60 white balls, we reject $\mathbb{H}_0$ in the former case, but we cannot reject it in the latter one. This once more stresses the fact that an observation of an $x \in \mathcal{X}_0$ does not guarantee that $\mathbb{H}_0$ is true. It only means that the observed sample does not allow us to reject the hypothesis.

8.4 Tests for Normally Distributed Populations

During this section we always assume $\mathcal{X} = \mathbb{R}^n$. That is, our samples are vectors $x = (x_1, \ldots, x_n)$ with $x_j \in \mathbb{R}$. Given a sample $x \in \mathbb{R}^n$, we derive from it the following quantities that will soon play a crucial role.

Definition 8.4.1. If $x = (x_1, \ldots, x_n) \in \mathbb{R}^n$, then we set

$$\bar{x} := \frac{1}{n} \sum_{j=1}^{n} x_j, \quad s_x^2 := \frac{1}{n-1} \sum_{j=1}^{n} (x_j - \bar{x})^2 \text{ and } \sigma_x^2 := \frac{1}{n} \sum_{j=1}^{n} (x_j - \bar{x})^2 . \quad (8.7)$$

The number $\bar{x}$ is said to be the **sample mean** of x, while s_x^2 and σ_x^2 are said to be the **unbiased sample variance** and the **(biased) sample variance** of the vector x, respectively.

Analogously, if $X = (X_1, \ldots, X_n)$ is an n-dimensional random vector,[3] then we define the corresponding expressions pointwise. For instance, we have

$$\bar{X}(\omega) := \frac{1}{n} \sum_{j=1}^{n} X_j(\omega) \quad \text{and} \quad s_X^2(\omega) := \frac{1}{n-1} \sum_{j=1}^{n} (X_j(\omega) - \bar{X}(\omega))^2 .$$

8.4.1 Fisher's Theorem

We are going to prove important properties of normally distributed populations. They turn out to be the basis for all hypothesis tests in the normally distributed case. The starting point is a crucial lemma going back to Ronald Aylmer Fisher.

Lemma 8.4.2 (Fisher's lemma). *Let $Y_1, \ldots, Y_n$ be independent $\mathcal{N}(0, 1)$-distributed random variables and let $B = (\beta_{ij})_{i,j=1}^{n}$ be a unitary $n \times n$ matrix. The random variables $Z_1, \ldots, Z_n$ are defined as*

$$Z_i := \sum_{j=1}^{n} \beta_{ij} Y_j, \quad 1 \le i \le n.$$

They possess the following properties.
(i)   The variables $Z_1, \ldots, Z_n$ are also independent and $\mathcal{N}(0, 1)$-distributed.
(ii) For $m < n$ let the quadratic form Q on $\mathbb{R}^n$ be defined by

$$Q := \sum_{j=1}^{n} Y_j^2 - \sum_{i=1}^{m} Z_i^2 .$$

Then Q is independent of all $Z_1, \ldots, Z_m$ and distributed according to χ_{n-m}^2.

Proof: Assertion (i) was already proven in Proposition 6.1.16.

---

[3] To simplify the notation, now and later on, we denote random vectors by X, not by $\vec{X}$ as we did before. This should not lead to confusion.

Let us verify (ii). The matrix B is unitary, thus it preserves the length of vectors in $\mathbb{R}^n$. Applying this to $Y = (Y_1, \ldots, Y_n)$ and $Z = BY$ gives

$$\sum_{i=1}^{n} Z_i^2 = |Z|_2^2 = |BY|_2^2 = |Y|_2^2 = \sum_{j=1}^{n} Y_j^2,$$

which leads to

$$Q = Z_{m+1}^2 + \cdots + Z_n^2. \tag{8.8}$$

By virtue of (i) the random variables $Z_1, \ldots, Z_n$ are independent, hence by eq. (8.8) and Remark 4.1.10 the quadratic form Q is independent of $Z_1, \ldots, Z_m$.

Recall that $Z_{m+1}, \ldots, Z_n$ are independent $\mathcal{N}(0,1)$-distributed. Thus, in view of eq. (8.8), Proposition 4.6.17 implies that Q is χ_{n-m}^2-distributed. Observe that Q is the sum of $n - m$ squares. ∎

Now we are in a position to state and prove one of the most important results in Mathematical Statistics.

Proposition 8.4.3 (Fisher's theorem). *Suppose $X_1, \ldots, X_n$ are independent and distributed according to $\mathcal{N}(\mu, \sigma^2)$ for some $\mu \in \mathbb{R}$ and some $\sigma^2 > 0$. Then the following are valid:*

$$\sqrt{n}\, \frac{\bar{X} - \mu}{\sigma} \quad is \quad \mathcal{N}(0,1)\text{-distributed}. \tag{8.9}$$

$$(n-1)\frac{s_X^2}{\sigma^2} \quad is \quad \chi_{n-1}^2\text{-distributed}. \tag{8.10}$$

$$\sqrt{n}\, \frac{\bar{X} - \mu}{s_X} \quad is \quad t_{n-1}\text{-distributed}, \tag{8.11}$$

where $s_X := +\sqrt{s_X^2}$. Furthermore, $\bar{X}$ and s_X^2 are independent random variables.

Proof: Let us begin with the proof of assertion (8.9). Since the X_js are independent and $\mathcal{N}(\mu, \sigma^2)$-distributed, by Proposition 4.6.18 their sum $X_1 + \cdots + X_n$ possesses an $\mathcal{N}(n\mu, n\sigma^2)$ distribution. Consequently, an application of Proposition 4.2.3 implies that $\bar{X}$ is $\mathcal{N}(\mu, \sigma^2/n)$-distributed, hence, another application of Proposition 4.2.3 tells us that $\frac{\bar{X} - \mu}{\sigma/\sqrt{n}}$ is standard normal. This completes the proof of statement (8.9).

We turn now to the verification of the remaining assertions. Letting

$$Y_j := \frac{X_j - \mu}{\sigma}, \quad 1 \le j \le n, \tag{8.12}$$

the random variables $Y_1, \ldots, Y_n$ are independent $\mathcal{N}(0, 1)$-distributed. Moreover, their (unbiased) sample variance may be calculated by

$$
s_Y^2 = \frac{1}{n-1} \sum_{j=1}^{n} (Y_j - \bar{Y})^2 = \frac{1}{n-1} \left\{ \sum_{j=1}^{n} Y_j^2 - 2\bar{Y} \sum_{j=1}^{n} Y_j + n\bar{Y}^2 \right\}
$$

$$
= \frac{1}{n-1} \left\{ \sum_{j=1}^{n} Y_j^2 - 2n\bar{Y}^2 + n\bar{Y}^2 \right\} = \frac{1}{n-1} \left\{ \sum_{j=1}^{n} Y_j^2 - (\sqrt{n}\,\bar{Y})^2 \right\}. \tag{8.13}
$$

To proceed further, set $b_1 := (n^{-1/2}, \ldots, n^{-1/2})$, and note that b_1 is a normalized n-dimensional vector, that is, we have $|b_1|_2 = 1$. Let $E \subseteq \mathbb{R}^n$ be the $(n-1)$-dimensional subspace consisting of elements that are perpendicular to b_1. Choosing an orthonormal basis $b_2, \ldots, b_n$ in E, then, by the choice of E, the vectors $b_1, \ldots, b_n$ form an orthonormal basis in $\mathbb{R}^n$. If $b_i = (\beta_{i1}, \ldots, \beta_{in})$, $1 \le i \le n$, let B be the $n \times n$-matrix with entries β_{ij}, that is, the vectors $b_1, \ldots, b_n$ are the rows of B. Since $(b_i)_{i=1}^n$ are orthonormal, B is unitary.

As in Lemma 8.4.2, define $Z_1, \ldots, Z_n$ by

$$
Z_i := \sum_{j=1}^{n} \beta_{ij} Y_j, \quad 1 \le i \le n,
$$

and the quadratic form Q (with $m = 1$) as

$$
Q := \sum_{j=1}^{n} Y_j^2 - Z_1^2 .
$$

Because of Lemma 8.4.2, the quadratic form Q is χ_{n-1}^2-distributed and, furthermore, it is independent of Z_1. By the choice of B and of b_1,

$$
\beta_{11} = \cdots = \beta_{1n} = n^{-1/2},
$$

hence $Z_1 = n^{1/2}\,\bar{Y}$, and by eq. (8.13) this leads to

$$
Q = \sum_{j=1}^{n} Y_j^2 - (n^{1/2}\,\bar{Y})^2 = (n-1)\, s_Y^2 .
$$

This observation implies $(n-1)\, s_Y^2$ is χ_{n-1}^2-distributed and, moreover, $(n-1)s_Y^2$ and Z_1 are independent, thus also s_Y^2 and Z_1.

The choice of the Y_js in eq. (8.12) immediately implies $\bar{Y} = \frac{\bar{X}-\mu}{\sigma}$, hence

$$(n-1)s_Y^2 = \sum_{j=1}^{n}(Y_j - \bar{Y})^2 = \sum_{j=1}^{n}\left(\frac{X_j - \mu}{\sigma} - \frac{\bar{X}-\mu}{\sigma}\right)^2 = \frac{s_X^2}{\sigma^2}(n-1),$$

which proves assertion (8.10).

Recall that $Z_1 = n^{1/2}\bar{Y} = n^{1/2}\frac{\bar{X}-\mu}{\sigma}$, which leads to $\bar{X} = n^{-1/2}\sigma Z_1 + \mu$. Thus, because of Proposition 4.1.9, the independence of $s_Y^2 = s_X^2/\sigma^2$ and Z_1 implies that s_X^2 and $\bar{X}$ are independent as well.

It remains to prove statement (8.11). We already know that $V := \sqrt{n}\,\frac{\bar{X}-\mu}{\sigma}$ is standard normal, and $W := (n-1)s_X^2/\sigma^2$ is χ_{n-1}^2-distributed. Since they are independent, by Proposition 4.7.8, applied with $n-1$, we get

$$\sqrt{n}\,\frac{\bar{X}-\mu}{s_X} = \frac{V}{\sqrt{\frac{1}{n-1}W}} \quad \text{is} \quad t_{n-1}\text{-distributed}.$$

This implies assertion (8.11) and completes the proof of the proposition. ∎

Remark 8.4.4. It is important to mention that the random variables $X_1, \ldots, X_n$ satisfy the assumptions of Proposition 8.4.3 if and only if the vector $(X_1, \ldots, X_n)$ is $\mathcal{N}(\mu, \sigma^2)^{\otimes n}$-distributed or, equivalently, if its probability distribution is $\mathcal{N}(\vec{\mu}, \sigma^2 I_n)$.

8.4.2 Quantiles

Quantiles may be defined in a quite general way. However, we will restrict ourselves to those quantiles that will be used later on. The first quantiles we consider are those of the standard normal distribution.

Definition 8.4.5. Let Φ be the distribution function of $\mathcal{N}(0, 1)$, as it was introduced in Definition 1.62. For a given $\beta \in (0, 1)$, the β-quantile z_β of the standard normal distribution is the unique real number satisfying

$$\Phi(z_\beta) = \beta \quad \text{or, equivalently,} \quad z_\beta = \Phi^{-1}(\beta).$$

Another way to define z_β is as follows. Let X be a standard normal random variable. Then z_β is the unique real number such that

$$\mathbb{P}\{X \le z_\beta\} = \beta.$$

The following properties of z_β will be used later on.

Proposition 8.4.6. *Let X be standard normally distributed. Then the following are valid*
1. *We have $z_\beta \leq 0$ for $\beta \leq 1/2$ and $z_\beta > 0$ for $\beta > 1/2$.*
2. $\mathbb{P}\{X \geq z_\beta\} = 1 - \beta.$
3. *If $0 < \beta < 1$, then $z_{1-\beta} = -z_\beta$.*
4. *For $0 < \alpha < 1$ we have $\mathbb{P}\{|X| \geq z_{1-\alpha/2}\} = \alpha$.*

Proof: The first property easily follows by $\Phi(0) = 1/2$, hence $\Phi(t) > 1/2$ if and only if $t > 0$.

Let X be standard normal. Then $\mathbb{P}\{X \geq z_\beta\} = 1 - \mathbb{P}\{X \leq z_\beta\} = 1 - \beta$, which proves the second assertion.

Since $-X$ is standard normal as well, by 2. it follows that

$$\mathbb{P}\{X \leq -z_\beta\} = \mathbb{P}\{X \geq z_\beta\} = 1 - \beta = \mathbb{P}\{X \leq z_{1-\beta}\},$$

hence $z_{1-\beta} = -z_\beta$ as asserted.

To prove the fourth assertion note that properties 2 and 3 imply

$$\mathbb{P}\{|X| \geq z_{1-\alpha/2}\} = \mathbb{P}\{X \leq -z_{1-\alpha/2} \text{ or } X \geq z_{1-\alpha/2}\}$$
$$= \mathbb{P}\{X \leq -z_{1-\alpha/2}\} + \mathbb{P}\{X \geq z_{1-\alpha/2}\}$$
$$= \mathbb{P}\{X \leq z_{\alpha/2}\} + \mathbb{P}\{X \geq z_{1-\alpha/2}\} = \alpha/2 + \alpha/2 = \alpha.$$

Here we used $1 - \alpha/2 > 1/2$ implying $z_{1-\alpha/2} > 0$, hence the events $\{X \leq -z_{1-\alpha/2}\}$ and $\{X \geq z_{1-\alpha/2}\}$ are disjoint. ∎

The next quantile, needed later on, is that of a χ_n^2 distribution.

Definition 8.4.7. Let X be distributed according to χ_n^2 and let $0 < \beta < 1$. The unique (positive) number $\chi_{n;\beta}^2$ satisfying

$$\mathbb{P}\{X \leq \chi_{n;\beta}^2\} = \beta$$

is called β-quantile of the χ_n^2-distribution.

Two other, equivalent, ways to introduce these quantiles are as follows.
1. If $X_1, \ldots, X_n$ are independent standard normal, then

$$\mathbb{P}\{X_1^2 + \cdots + X_n^2 \leq \chi_{n;\beta}^2\} = \beta.$$

2. The quantile $\chi^2_{n;\beta}$ satisfies

$$\frac{1}{2^{n/2}\Gamma(n/2)} \int_0^{\chi^2_{n;\beta}} x^{n/2-1} e^{-x/2} dx = \beta.$$

For later purposes we mention also the following property. If $0 < \alpha < 1$, then for any χ^2_n-distributed random variable X,

$$\mathbb{P}\{X \notin [\chi^2_{n;\alpha/2}, \chi^2_{n;1-\alpha/2}]\} = \alpha.$$

In a similar way we define now the quantiles of Student's t_n and of Fisher's $F_{m,n}$ distributions. For their descriptions we refer to Definitions 4.7.6 and 4.7.13, respectively.

Definition 8.4.8. Let X be t_n-distributed and let Y be distributed according to $F_{m,n}$. For $\beta \in (0,1)$ the β-quantiles $t_{n;\beta}$ and $F_{m,n;\beta}$ of the t_n- and $F_{m,n}$-distributions are the unique numbers satisfying

$$\mathbb{P}\{X \le t_{n;\beta}\} = \beta \quad \text{and} \quad \mathbb{P}\{Y \le F_{m,n;\beta}\} = \beta.$$

Remark 8.4.9. Let X be t_n distributed. Then $-X$ is t_n distributed as well, hence $\mathbb{P}\{X \le s\} = \mathbb{P}\{-X \le s\}$ for $s \in \mathbb{R}$. Therefore, as in the case of the normal distribution, we get $-t_{n;\beta} = t_{n;1-\beta}$, and also

$$\mathbb{P}\{|X| > t_{n;1-\alpha/2}\} = \mathbb{P}\{|X| \ge t_{n;1-\alpha/2}\} = \alpha. \tag{8.14}$$

Remark 8.4.10. Another way to introduce the quantiles of the $F_{m,n}$-distribution is as follows. Let X and Y be independent and distributed according to χ^2_m and χ^2_n, respectively. The quantile $F_{m,n;\beta}$ is the unique number satisfying

$$\mathbb{P}\left\{\frac{X/m}{Y/n} \le F_{m,n;\beta}\right\} = \beta.$$

If $s > 0$, then

$$\mathbb{P}\left\{\frac{X/m}{Y/n} \le s\right\} = \mathbb{P}\left\{\frac{Y/n}{X/m} \ge \frac{1}{s}\right\} = 1 - \mathbb{P}\left\{\frac{Y/n}{X/m} \le \frac{1}{s}\right\},$$

which immediately implies

$$F_{m,n;\beta} = \frac{1}{F_{n,m;1-\beta}}.$$

8.4.3 Z-Tests or Gauss Tests

Suppose we have an item of unknown length. In order to get some information about its length, we measure the item n times with an instrument of known accuracy. As sample we get a vector $x = (x_1, \ldots, x_n)$, where x_j is the value obtained in the jth measurement. These measurements were executed independently, thus, we may assume that the x_js are independent $\mathcal{N}(\mu, \sigma_0^2)$-distributed with known $\sigma_0^2 > 0$ and unknown length $\mu \in \mathbb{R}$. Therefore, the describing statistical model is

$$\left(\mathbb{R}^n, \mathcal{B}(\mathbb{R}^n), \mathcal{N}(\mu, \sigma_0^2)^{\otimes n}\right)_{\mu \in \mathbb{R}} = \left(\mathbb{R}^n, \mathcal{B}(\mathbb{R}^n), \mathcal{N}(\vec{\mu}, \sigma_0^2 I_n)\right)_{\mu \in \mathbb{R}}.$$

From the hypothesis, two types of tests apply in this case. We start with the so-called **one-sided Z-test** (also called one-sided Gauss test). Here the null hypothesis is $\mathbb{H}_0$: $\mu \le \mu_0$, where $\mu_0 \in \mathbb{R}$ is a given real number. Consequently, the alternative hypothesis is $\mathbb{H}_1$: $\mu > \mu_0$, that is, $\Theta_0 = (-\infty, \mu_0]$ while $\Theta_1 = (\mu_0, \infty)$. In the above context this says that we claim that the length of the item is less than or equal to a given μ_0, and to check this we measure the item n times.

Proposition 8.4.11. *Let $\alpha \in (0, 1)$ be a given security level. Then* $\mathbf{T} = (\mathcal{X}_0, \mathcal{X}_1)$ *with*

$$\mathcal{X}_0 := \left\{ x \in \mathbb{R}^n : \bar{x} \le \mu_0 + n^{-1/2} \sigma_0 z_{1-\alpha} \right\}$$

and with

$$\mathcal{X}_1 := \left\{ x \in \mathbb{R}^n : \bar{x} > \mu_0 + n^{-1/2} \sigma_0 z_{1-\alpha} \right\}$$

is an α-significance test to check $\mathbb{H}_0$ against $\mathbb{H}_1$. Here $z_{1-\alpha}$ denotes the $(1 - \alpha)$-quantile introduced in Definition 8.4.5.

Proof: The assertion of Proposition 8.4.11 says that

$$\sup_{\mu \le \mu_0} \mathbb{P}_\mu(\mathcal{X}_1) = \sup_{\mu \le \mu_0} \mathcal{N}(\mu, \sigma_0^2)^{\otimes n}(\mathcal{X}_1) \le \alpha.$$

To verify this, let us choose an arbitrary $\mu \le \mu_0$ and define $S : \mathbb{R}^n \to \mathbb{R}$ by

$$S(x) := \sqrt{n}\,\frac{\bar{x} - \mu}{\sigma_0}, \quad x \in \mathbb{R}^n. \tag{8.15}$$

Regard S as a random variable on the probability space $(\mathbb{R}^n, \mathcal{B}(\mathbb{R}^n), \mathcal{N}(\mu, \sigma_0^2)^{\otimes n})$. Then, by property (8.9), it is standard normally distributed.[4] Consequently,

---

[4] This fact is crucial. For better understanding, here is a more detailed reasoning. Define random variables X_j on the probability space $(\mathbb{R}^n, \mathcal{B}(\mathbb{R}^n), \mathcal{N}(\mu, \sigma_0^2)^{\otimes n})$ by $X_j(x) = x_j$, where $x = (x_1, \ldots, x_n)$. Then the random vector $X = (X_1, \ldots, X_n)$ is the identity on $\mathbb{R}^n$, hence $\mathcal{N}(\mu, \sigma_0^2)^{\otimes n}$-distributed. In view of Remark 8.4.4 and by

$$S(x) = \sqrt{n}\,\frac{\bar{X}(x) - \mu}{\sigma_0},$$

assertion (8.9) applies for S, that is, it is $\mathcal{N}(0, 1)$-distributed.

$$\mathcal{N}(\mu, \sigma_0^2)^{\otimes n}\{x \in \mathbb{R}^n : S(x) > z_{1-\alpha}\} = \alpha. \tag{8.16}$$

Since $\mu \le \mu_0$, we have

$$\mathcal{X}_1 = \left\{ x \in \mathbb{R}^n : \bar{x} > \mu_0 + n^{-1/2} \sigma_0 z_{1-\alpha} \right\}$$

$$\subseteq \{x \in \mathbb{R}^n : \bar{x} > \mu + n^{-1/2} \sigma_0 z_{1-\alpha}\} = \{x \in \mathbb{R}^n : S(x) > z_{1-\alpha}\},$$

hence, by eq. (8.16), it follows that

$$\mathcal{N}(\mu, \sigma_0^2)^{\otimes n}(\mathcal{X}_1) \le \mathcal{N}(\mu, \sigma_0^2)^{\otimes n}\{x \in \mathbb{R}^n : S(x) > z_{1-\alpha}\} = \alpha.$$

This completes the proof. ∎

What does the power function of the Z-test in Proposition 8.4.11 look like? If S is as in eq. (8.15), then, according to Definition 8.2.11,

$$\beta_T(\mu) = \mathcal{N}(\mu, \sigma_0^2)^{\otimes n}(\mathcal{X}_1) = \mathcal{N}(\mu, \sigma_0^2)^{\otimes n} \left\{ x \in \mathbb{R}^n : \sqrt{n}\,\frac{\bar{x} - \mu_0}{\sigma_0} > z_{1-\alpha} \right\}$$

$$= \mathcal{N}(\mu, \sigma_0^2)^{\otimes n} \left\{ x \in \mathbb{R}^n : S(x) > z_{1-\alpha} + (\mu_0 - \mu)\frac{\sqrt{n}}{\sigma_0} \right\}$$

$$= 1 - \Phi\left(z_{1-\alpha} + (\mu_0 - \mu)\frac{\sqrt{n}}{\sigma_0} \right) = \Phi\left(z_\alpha + (\mu - \mu_0)\frac{\sqrt{n}}{\sigma_0} \right).$$

In particular, β_T is increasing on $\mathbb{R}$ with $\beta_T(\mu_0) = \alpha$. Moreover, we see that $\beta_T(\mu) < \alpha$ if $\mu < \mu_0$, and $\beta_T(\mu) > \alpha$ for $\mu > \mu_0$.

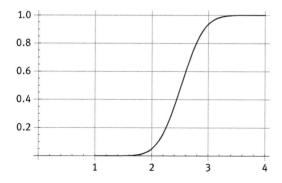

Figure 8.1: Power function of **T** with $\alpha = 0.05$, $\mu_0 = 2$, $\sigma_0 = 1$, and $n = 10$.

While the critical region of a one-sided Z-test is an interval, in the case of the **two-sided Z-test** it is the union of two intervals. Here the null hypothesis is $\mathbb{H}_0 : \mu = \mu_0$, hence the alternative hypothesis is given as $\mathbb{H}_1 : \mu \neq \mu_0$.

Proposition 8.4.12. *The test* $\mathbf{T} = (\mathcal{X}_0, \mathcal{X}_1)$, *where*

$$\mathcal{X}_0 := \left\{ x \in \mathbb{R}^n : \mu_0 - n^{-1/2}\sigma_0 z_{1-\alpha/2} \leq \bar{x} \leq \mu_0 + n^{-1/2}\sigma_0 z_{1-\alpha/2} \right\}$$

and $\mathcal{X}_1 = \mathbb{R}^n \backslash \mathcal{X}_0$, *is an α-significance test to check* $\mathbb{H}_0 : \mu = \mu_0$ *against* $\mathbb{H}_1 : \mu \neq \mu_0$.

Proof: Since here $\Theta_0 = \{\mu_0\}$, the proof becomes easier than in the one-sided case. We only have to verify that

$$\mathcal{N}(\mu_0, \sigma_0^2)^{\otimes n}(\mathcal{X}_1) \leq \alpha. \tag{8.17}$$

Regarding S, defined by

$$S(x) := \sqrt{n}\, \frac{\bar{x} - \mu_0}{\sigma_0},$$

as random variable on $(\mathbb{R}^n, \mathcal{B}(\mathbb{R}^n), \mathcal{N}(\mu_0, \sigma_0^2)^{\otimes n})$, by the same arguments as in the previous proof, it is standard normally distributed. Thus, using assertion 4 of Proposition 8.4.6, we obtain

$$\mathcal{N}(\mu_0, \sigma_0^2)^{\otimes n}(\mathcal{X}_1) = \mathcal{N}(\mu_0, \sigma_0^2)^{\otimes n}\{x \in \mathbb{R}^n : |S(x)| > z_{1-\alpha/2}\} = \alpha.$$

Of course, this completes the proof. ∎

8.4.4 t-Tests

The problem is similar to the one considered in the case of the Z-test. But, there is one important difference. We do no longer assume that the variance is known, which will be so in most cases. Therefore, this test is more realistic than the Z-test.

The starting point is the statistical model

$$(\mathbb{R}^n, \mathcal{B}(\mathbb{R}^n), \mathcal{N}(\mu, \sigma^2)^{\otimes n})_{(\mu,\sigma^2) \in \mathbb{R} \times (0,\infty)}.$$

Observe that the unknown parameter is now a vector $(\mu, \sigma^2) \in \mathbb{R} \times (0, \infty)$. We begin by investigating the **one-sided t-test**. Given some $\mu_0 \in \mathbb{R}$, the null hypothesis is as before, that is, we have $\mathbb{H}_0 : \mu \leq \mu_0$. In the general setting that means $\Theta_0 = (-\infty, \mu_0] \times (0, \infty)$, while $\Theta_1 = (\mu_0, \infty) \times (0, \infty)$

To formulate the next result, let us shortly recall the following notations. If s_x^2 denotes the unbiased sample variance, as defined in eq. (8.7), then we set $s_x := +\sqrt{s_x^2}$. Furthermore, $t_{n-1;1-\alpha}$ denotes the $(1-\alpha)$-quantile of the t_{n-1}-distribution, as introduced in Definition 8.4.8.

Proposition 8.4.13. *Given $\alpha \in (0, 1)$, the regions $\mathcal{X}_0$ and $\mathcal{X}_1$ in $\mathbb{R}^n$ are defined by*

$$\mathcal{X}_0 := \left\{ x \in \mathbb{R}^n : \bar{x} \le \mu_0 + n^{-1/2} s_x\, t_{n-1;1-\alpha} \right\}$$

and $\mathcal{X}_1 = \mathbb{R}^n \backslash \mathcal{X}_0$. With this choice of $\mathcal{X}_0$ and $\mathcal{X}_1$, the test $\mathbf{T} = (\mathcal{X}_0, \mathcal{X}_1)$ is an α-significance test for $\mathbb{H}_0 : \mu \le \mu_0$ against $\mathbb{H}_1 : \mu > \mu_0$.

Proof: Given $\mu \le \mu_0$, define the random variable S on $(\mathbb{R}^n, \mathcal{B}(\mathbb{R}^n), \mathcal{N}(\mu, \sigma^2)^{\otimes n})$ as

$$S(x) := \sqrt{n}\, \frac{\bar{x} - \mu}{s_x}, \quad x \in \mathbb{R}^n.$$

Property (8.11) implies that S is t_{n-1}-distributed, hence by the definition of the quantile $t_{n-1;1-\alpha}$, it follows that

$$\mathcal{N}(\mu, \sigma^2)^{\otimes n}\{x \in \mathbb{R}^n : S(x) > t_{n-1;1-\alpha}\} = \alpha.$$

From $\mu \le \mu_0$ we easily derive

$$\mathcal{X}_1 \subseteq \{x \in \mathbb{R}^n : S(x) > t_{n-1;1-\alpha}\},$$

thus, as asserted,

$$\sup_{\mu \le \mu_0} \mathcal{N}(\mu, \sigma^2)^{\otimes n}(\mathcal{X}_1) \le \mathcal{N}(\mu, \sigma^2)^{\otimes n}\{x \in \mathbb{R}^n : S(x) > t_{n-1;1-\alpha}\} = \alpha.$$

∎

As in the case of the Z-test, the null hypothesis of the **two-sided t-test** is $\mathbb{H}_0 : \mu = \mu_0$ for some $\mu_0 \in \mathbb{R}$. Again, we do not assume that the variance is known.

A two-sided t-test with significance level α may be constructed as follows.

Proposition 8.4.14. *Given $\alpha \in (0, 1)$, define regions $\mathcal{X}_0$ and $\mathcal{X}_1$ in $\mathbb{R}^n$ by*

$$\mathcal{X}_0 := \left\{ x \in \mathbb{R}^n : \sqrt{n}\, \left| \frac{\bar{x} - \mu_0}{s_x} \right| \le t_{n-1;1-\alpha/2} \right\}$$

and $\mathcal{X}_1 = \mathbb{R}^n \backslash \mathcal{X}_0$. Then $\mathbf{T} = (\mathcal{X}_0, \mathcal{X}_1)$ is an α-significance test for $\mathbb{H}_0 : \mu = \mu_0$ against $\mathbb{H}_1 : \mu \ne \mu_0$.

Proposition 8.4.14 is proven by similar methods, as we have used for the proofs of Propositions 8.4.12 and 8.4.13. Therefore, we decline to prove it here.

Example 8.4.15. We claim a certain workpiece has a length of 22 inches. Thus, the null hypothesis is $H_0 : \mu = 22$. To check H_0, we measure the piece 10 times under the same conditions. The 10 values we obtained are (in inches)

$$22.17, \quad 22.11, \quad 22.10, \quad 22.14, \quad 22.02, \quad 21.95, \quad 22.02, \quad 22.08, \quad 21.98, \quad 22.15$$

Do these values allow us to reject the hypothesis or do they confirm it?

We have

$$\bar{x} = 22.072 \quad \text{and} \quad s_x = 0.07554248, \quad \text{hence} \quad \sqrt{10}\,\frac{\bar{x} - 22}{s_x} = 3.013986.$$

If we choose the security level $\alpha = 0.05$, we have to investigate the quantile $t_{9;0.975}$, which equals $t_{9;0.975} = 2.26$. This lets us conclude the observed vector $x = (x_1, \ldots, x_{10})$ belongs to X_1, and we may reject H_0. Consequently, with a security of 95% we may say, $\mu \neq 22$.

Remark 8.4.16. If we plug these 10 values into a mathematical program, the result will be a number $\alpha_0 = 0.00128927$. What does this number tell us? It says the following. If we have chosen a significance level α with $\alpha > \alpha_0$, then we have to reject H_0. But, if the chosen α satisfies $\alpha < \alpha_0$, then we fail to reject H_0. In our case we had $\alpha = 0.05 > 0.00128927 = \alpha_0$, hence we may reject H_0.

8.4.5 χ^2-Tests for the Variance

The aim of this section is to get some information about the (unknown) variance of a normal distribution. Again we have to distinguish between the following two cases. The expected value is known or, otherwise, the expected value is unknown.

Let us start with the former case, that is, we assume that the *expected value is known* to be some $\mu_0 \in \mathbb{R}$. Then the statistical model is $(\mathbb{R}^n, \mathcal{B}(\mathbb{R}^n), \mathcal{N}(\mu_0, \sigma^2)^{\otimes n})_{\sigma^2 > 0}$. In the **one-sided χ^2-test**, the null hypothesis is $H_0 : \sigma^2 \leq \sigma_0^2$, for some given $\sigma_0^2 > 0$, while in the **two-sided χ^2-test** we claim that $H_0 : \sigma^2 = \sigma_0^2$.

Proposition 8.4.17. *In the one-sided setting, an α-significance χ^2-test $\mathbf{T} = (X_0, X_1)$ is given by*

$$X_0 := \left\{ x \in \mathbb{R}^n : \sum_{j=1}^{n} \frac{(x_j - \mu_0)^2}{\sigma_0^2} \leq \chi_{n;1-\alpha}^2 \right\}.$$

For the two-sided case, choose

$$\mathcal{X}_0 := \left\{ x \in \mathbb{R}^n : \chi^2_{n;\alpha/2} \leq \sum_{j=1}^{n} \frac{(x_j - \mu_0)^2}{\sigma_0^2} \leq \chi^2_{n;1-\alpha/2} \right\},$$

to obtain an α-significance test. In both cases the critical region is $\mathcal{X}_1 := \mathbb{R}^n \setminus \mathcal{X}_0$.

Proof: We prove the assertion only in the (slightly more difficult) one-sided case. For an arbitrarily chosen $\sigma^2 \leq \sigma_0^2$, let $\mathcal{N}(\mu_0, \sigma^2)^{\otimes n}$ be the underlying probability measure. We define now the random variables $X_j : \mathbb{R}^n \to \mathbb{R}$ as $X_j(x) = x_j$ for $x = (x_1, \ldots, x_n)$. Then the X_js are independent $\mathcal{N}(\mu_0, \sigma^2)$-distributed. The normalization $Y_j := \frac{X_j - \mu_0}{\sigma}$ leads to independent standard normal Y_js. Thus, if

$$S := \sum_{j=1}^{n} \frac{(X_j - \mu_0)^2}{\sigma^2} = \sum_{j=1}^{n} Y_j^2,$$

then by Proposition 4.6.17, the random variable S is χ_n^2-distributed. By the definition of quantiles, we arrive at

$$\mathcal{N}(\mu_0, \sigma^2)^{\otimes n} \left\{ x \in \mathbb{R}^n : S(x) > \chi_{n;1-\alpha} \right\} = \alpha.$$

Since $\sigma^2 \leq \sigma_0^2$, it follows that

$$\mathcal{X}_1 \subseteq \left\{ x \in \mathbb{R}^n : S(x) > \chi_{n;1-\alpha} \right\},$$

hence $\mathcal{N}(\mu_0, \sigma^2)^{\otimes n}(\mathcal{X}_1) \leq \alpha$. This proves, as asserted, that $\mathbf{T} = (\mathcal{X}_0, \mathcal{X}_1)$ is an α-significance test. ∎

Let us now turn to the case where the *expected value is unknown*. Here the statistical model is given by

$$(\mathbb{R}^n, \mathcal{B}(\mathbb{R}^n), \mathcal{N}(\mu, \sigma^2)^{\otimes n})_{(\mu, \sigma^2) \in \mathbb{R} \times (0, \infty)}.$$

In the **one-sided case**, the parameter set $\Theta = \mathbb{R} \times (0, \infty)$ splits up into $\Theta = \Theta_0 \cup \Theta_1$ with

$$\Theta_0 = \mathbb{R} \times (0, \sigma_0^2] \quad \text{and} \quad \Theta_1 = \mathbb{R} \times (\sigma_0^2, \infty).$$

In the **two-sided case** we have

$$\Theta_0 = \mathbb{R} \times \{\sigma_0^2\} \quad \text{and} \quad \Theta_1 = \mathbb{R} \times \left[(0, \sigma_0^2) \cup (\sigma_0^2, \infty)\right].$$

Proposition 8.4.18. *In the one-sided case, an α-significance test* $\mathbf{T} = (\mathcal{X}_0, \mathcal{X}_1)$ *is given by*

$$\mathcal{X}_0 := \left\{ x \in \mathbb{R}^n : (n-1)\frac{s_x^2}{\sigma_0^2} \leq \chi_{n-1;1-\alpha}^2 \right\}.$$

In the two-sided case, choose the region of acceptance as

$$\mathcal{X}_0 := \left\{ x \in \mathbb{R}^n : \chi_{n;\alpha/2}^2 \leq (n-1)\frac{s_x^2}{\sigma_0^2} \leq \chi_{n-1;1-\alpha/2}^2 \right\}$$

to get an α-significance test. Again, the critical regions are given by $\mathcal{X}_1 := \mathbb{R}^n \backslash \mathcal{X}_0$.

Proof: The proof is very similar to the one of Proposition 8.4.17, but with some important difference. Here we have to set

$$S(x) := (n-1)\frac{s_x^2}{\sigma^2}, \quad x \in \mathbb{R}^n.$$

Then property (8.10) applies, and it lets us conclude that S is χ_{n-1}^2-distributed, provided that $\mathcal{N}(\mu, \sigma^2)^{\otimes n}$ is the true probability measure. After that observation the proof is completed as the one of Proposition 8.4.17. ∎

8.4.6 Two-Sample Z-Tests

The two-sample Z-test compares the parameters of two different populations. Suppose we are given two different series of data, say $x = (x_1, \ldots, x_m)$ and $y = (y_1, \ldots, y_n)$, which were obtained independently by executing m experiments of the first kind and n of the second one. Combine both series to a single vector $(x, y) \in \mathbb{R}^{m+n}$.

A typical example for the described situation is as follows. A farmer grows grain on two different acres. On one acre he added fertilizer, on the other one he did not. Now he wants to figure out whether or not adding fertilizer influenced the amount of grain gathered. Therefore, he measures the amount of grain on the first acre at m different spots and those on the second one at n spots. The aim is to compare the mean values in both series of experiments.

We suppose that the samples $x_1, \ldots, x_m$ of the first population are independent and $N(\mu_1, \sigma_1^2)$-distributed, while the $y_1, \ldots, y_n$ of the second population are independent and $\mathcal{N}(\mu_2, \sigma_2^2)$-distributed. Typical questions are as follows. Do we have $\mu_1 = \mu_2$ or, maybe, only $\mu_1 \leq \mu_2$? One may also ask whether or not $\sigma_1^2 = \sigma_2^2$ or, maybe, only $\sigma_1^2 \leq \sigma_2^2$.

To apply the two-sample Z-test, one has to suppose the variances σ_1^2 and σ_2^2 as known. This reduces the number of parameters from 4 to 2, namely to μ_1 and μ_2 in $\mathbb{R}$. Thus, the describing statistical model is given by

$$\left(\mathbb{R}^{m+n}, \mathcal{B}(\mathbb{R}^{m+n}), \mathcal{N}(\mu_1, \sigma_1^2)^{\otimes m} \otimes \mathcal{N}(\mu_2, \sigma_2^2)^{\otimes n} \right)_{(\mu_1, \mu_2) \in \mathbb{R}^2}. \tag{8.18}$$

Recall that $\mathcal{N}(\mu_1, \sigma_1^2)^{\otimes m} \otimes \mathcal{N}(\mu_2, \sigma_2^2)^{\otimes n}$ denotes the multivariate normal distribution with expected value $(\underbrace{\mu_1, \ldots, \mu_1}_{m}, \underbrace{\mu_2, \ldots, \mu_2}_{n})$ and covariance matrix $R = (r_{ij})_{i,j=1}^{m+n}$, where $r_{ii} = \sigma_1^2$ if $1 \leq i \leq m$, and $r_{ii} = \sigma_2^2$ if $m < i \leq m + n$. Furthermore, $r_{ij} = 0$ if $i \neq j$.

Proposition 8.4.19. *The statistical model is that in (8.18). To test $\mathbb{H}_0 : \mu_1 \leq \mu_2$ against $\mathbb{H}_1 : \mu_1 > \mu_2$, set*

$$\mathcal{X}_0 := \left\{ (x, y) \in \mathbb{R}^{m+n} : \sqrt{\frac{mn}{n\sigma_1^2 + m\sigma_2^2}} \, (\bar{x} - \bar{y}) \leq z_{1-\alpha} \right\}$$

and $\mathcal{X}_1 = \mathbb{R}^{m+n} \setminus \mathcal{X}_0$. Then the test $\mathbf{T} = (\mathcal{X}_0, \mathcal{X}_1)$ is an α-significance test for checking $\mathbb{H}_0$ against $\mathbb{H}_1$.

To test $\mathbb{H}_0 : \mu_1 = \mu_2$ against $\mathbb{H}_1 : \mu_1 \neq \mu_2$, let

$$\mathcal{X}_0 := \left\{ (x, y) \in \mathbb{R}^{m+n} : \sqrt{\frac{mn}{n\sigma_1^2 + m\sigma_2^2}} \, |\bar{x} - \bar{y}| \leq z_{1-\alpha/2} \right\}$$

and $\mathcal{X}_1 = \mathbb{R}^{m+n} \setminus \mathcal{X}_0$. Then the test $\mathbf{T} = (\mathcal{X}_0, \mathcal{X}_1)$ is an α-significance test for checking $\mathbb{H}_0$ against $\mathbb{H}_1$.

Proof: Since the proof of the two-sided case is very similar to that of the one-sided one, we only prove the first assertion. Thus, let us assume that $\mathbb{H}_0$ is valid, that is, we have $\mu_1 \leq \mu_2$. Then we have to verify that

$$\mathcal{N}(\mu_1, \sigma_1^2)^{\otimes m} \otimes \mathcal{N}(\mu_2, \sigma_2^2)^{\otimes n}(\mathcal{X}_1) \leq \alpha. \tag{8.19}$$

To prove this, we investigate the random variables X_i and Y_j defined as $X_i(x, y) = x_i$ and $Y_j(x, y) = y_j$. Since the underlying probability space is

$$(\mathbb{R}^{m+n}, \mathcal{B}(\mathbb{R}^{m+n}), \mathcal{N}(\mu_1, \sigma_1^2)^{\otimes m} \otimes \mathcal{N}(\mu_2, \sigma_2^2)^{\otimes n}),$$

these random variables are independent and distributed according to $\mathcal{N}(\mu_1, \sigma_1^2)$ and $\mathcal{N}(\mu_2, \sigma_2^2)$, respectively. Consequently, $\bar{X}$ is $\mathcal{N}\left(\mu_1, \frac{\sigma_1^2}{m}\right)$-distributed, while $\bar{Y}$ is distributed according to $\mathcal{N}\left(\mu_2, \frac{\sigma_2^2}{n}\right)$. By the construction, $\bar{X}$ and $\bar{Y}$ are independent as well, and moreover since $-\bar{Y}$ is $\mathcal{N}\left(-\mu_2, \frac{\sigma_2^2}{n}\right)$-distributed, we conclude that the distribution of $\bar{X} - \bar{Y}$ equals $\mathcal{N}\left(\mu_1 - \mu_2, \frac{\sigma_1^2}{m} + \frac{\sigma_2^2}{n}\right)$. Therefore, the mapping $S : \mathbb{R}^{m+n} \to \mathbb{R}$ defined by

$$S(x, y) := \left(\frac{\sigma_1^2}{m} + \frac{\sigma_2^2}{n}\right)^{-1/2} \left[(\bar{X}(x, y) - \bar{Y}(x, y)) - (\mu_1 - \mu_2)\right]$$

is standard normal. By the definition of the quantiles, this leads to

$$\mathcal{N}(\mu_1, \sigma_1^2)^{\otimes m} \otimes \mathcal{N}(\mu_2, \sigma_2^2)^{\otimes n}\{(x, y) \in \mathbb{R}^{m+n} : S(x, y) > z_{1-\alpha}\} = \alpha. \tag{8.20}$$

Since we assumed $\mathbb{H}_0$ to be correct, that is, we suppose $\mu_1 \leq \mu_2$, it follows that

$$S(x, y) \geq \left(\frac{\sigma_1^2}{m} + \frac{\sigma_2^2}{n}\right)^{-1/2} [\bar{X}(x, y) - \bar{Y}(x, y)] = \sqrt{\frac{mn}{n\sigma_1^2 + m\sigma_2^2}} [\bar{X}(x, y) - \bar{Y}(x, y)].$$

Hence

$$\mathcal{X}_1 \subseteq \{(x, y) \in \mathbb{R}^{m+n} : S(x, y) > z_{1-\alpha}\},$$

which by eq. (8.20) implies estimate (8.19). This completes the proof of this part of the proposition. ∎

8.4.7 Two-Sample t-Tests

The situation is similar as in the two-sample Z-test, yet with one important difference. The variances σ_1^2 and σ_2^2 of the two populations are no longer known. Instead, we have to assume that they coincide, that is, we suppose

$$\sigma_1^2 = \sigma_2^2 := \sigma^2.$$

Therefore, there are three unknown parameters, the expected values μ_1, μ_2, and the common variance σ^2. Thus, the statistical model describing this situation is given by

$$\left(\mathbb{R}^{m+n}, \mathcal{B}(\mathbb{R}^{m+n}), \mathcal{N}(\mu_1, \sigma^2)^{\otimes m} \otimes \mathcal{N}(\mu_2, \sigma^2)^{\otimes n}\right)_{(\mu_1, \mu_2, \sigma^2) \in \mathbb{R}^2 \times (0, \infty)}. \tag{8.21}$$

To simplify the formulation of the next statement, introduce $T : \mathbb{R}^{m+n} \to \mathbb{R}$ as

$$T(x, y) := \sqrt{\frac{(m+n-2)mn}{m+n}} \frac{\bar{x} - \bar{y}}{\sqrt{(m-1)s_x^2 + (n-1)s_y^2}}, \qquad (x, y) \in \mathbb{R}^{m+n}. \tag{8.22}$$

Proposition 8.4.20. *Let the statistical model be as in (8.21). If*

$$\mathcal{X}_0 := \{(x, y) \in \mathbb{R}^{m+n} : T(x, y) \leq t_{m+n-2;1-\alpha}\}$$

and $\mathcal{X}_1 = \mathbb{R}^{m+n} \backslash \mathcal{X}_0$, then $\mathbf{T} = (\mathcal{X}_0, \mathcal{X}_1)$ is an α-significance test for $\mathbb{H}_0 : \mu_1 \leq \mu_2$ against $\mathbb{H}_1 : \mu_1 > \mu_2$.
On the other hand, the test $\mathbf{T} = (\mathcal{X}_0, \mathcal{X}_1)$ with

$$\mathcal{X}_0 := \{(x, y) \in \mathbb{R}^{m+n} : |T(x, y)| \leq t_{m+n-2;1-\alpha/2}\}$$

and $\mathcal{X}_1 = \mathbb{R}^{m+n} \backslash \mathcal{X}_0$ is an α-significance test for $\mathbb{H}_0 : \mu_1 = \mu_2$ against $\mathbb{H}_1 : \mu_1 \neq \mu_2$.

Proof: This time we prove the two-sided case, that is, the null hypothesis is given by $\mathbb{H}_0 : \mu_1 = \mu_2$.

Let the random vectors $X = (X_1, \ldots, X_m)$ and $Y = (Y_1, \ldots, Y_n)$ on $\mathbb{R}^{m+n}$ be defined with X_is and Y_js as in the proof of Proposition 8.4.19, that is, we have $X(x, y) = x$ and $Y(x, y) = y$. Then by Proposition 4.1.9 and Remark 4.1.10, the unbiased sample variances

$$s_X^2 = \frac{1}{m-1} \sum_{i=1}^{m} (X_i - \bar{X})^2 \quad \text{and} \quad s_Y^2 = \frac{1}{n-1} \sum_{j=1}^{m} (Y_j - \bar{Y})^2$$

are independent as well. Furthermore, by virtue of statement (8.10), the random variables

$$(m-1)\frac{s_X^2}{\sigma^2} \quad \text{and} \quad (n-1)\frac{s_Y^2}{\sigma^2}$$

are distributed according to χ_{m-1}^2 and χ_{n-1}^2, respectively. Proposition 4.6.16 implies that

$$S_{(X,Y)}^2 := \frac{1}{\sigma^2}\left\{ (m-1)s_X^2 + (n-1)s_Y^2 \right\}$$

is χ_{m+n-2}^2-distributed. Since s_X^2 and $\bar{X}$ as well as s_Y^2 and $\bar{Y}$ are independent, by Proposition 8.4.3, this is also so for $S_{(X,Y)}^2$ and $\bar{X} - \bar{Y}$. As in the proof of Proposition 8.4.19, it follows that $\bar{X} - \bar{Y}$ is distributed according to $\mathcal{N}\left(\mu_1 - \mu_2, \frac{\sigma_1^2}{m} + \frac{\sigma_2^2}{n}\right)$. Assume now that $\mathbb{H}_0$ is true, that is, we have $\mu_1 = \mu_2$. Then the last observation implies that $\frac{\sqrt{mn}}{\sigma\sqrt{m+n}}(\bar{X} - \bar{Y})$ is a standard normally distributed random variable and, furthermore, independent of $S_{(X,Y)}^2$. Thus, by Proposition 4.7.8, the distribution of the quotient

$$Z := \sqrt{m+n-2}\ \frac{\frac{\sqrt{mn}}{\sigma\sqrt{m+n}}(\bar{X} - \bar{Y})}{S_{(X,Y)}},$$

where $S_{(X,Y)} := +\sqrt{S_{(X,Y)}^2}$, is t_{m+n-2}-distributed. If T is as in eq. (8.22), then it is not difficult to prove that $Z = T(X, Y)$. Therefore, by the definition of X and Y, the mapping T is a t_{m+n-2}-distributed random variable on $\mathbb{R}^{m+n}$, endowed with the probability measure $\mathbb{P}_{\mu_1,\mu_2,\sigma^2} = \mathcal{N}(\mu_1, \sigma^2)^{\otimes m} \otimes \mathcal{N}(\mu_2, \sigma^2)^{\otimes n}$. By eq. (8.14) this implies

$$\mathbb{P}_{\mu_1,\mu_2,\sigma^2}(\mathcal{X}_1) = \mathbb{P}_{\mu_1,\mu_2,\sigma^2}\left\{ (x, y) \in \mathbb{R}^{m+n} : |T(x, y)| > t_{m+n-2;1-\alpha/2} \right\} = \alpha,$$

as asserted. ∎

8.4.8 F-Tests

In this final section about tests we compare the variances of two normally distributed sample series. Since the proofs of the assertions follow the schemes presented in the previous propositions, we decline to verify them here. We only mention the facts that play a crucial role during the proofs.

1. If $X_1, \ldots, X_m$ and $Y_1, \ldots, Y_n$ are independent and distributed according to $N(\mu_1, \sigma_1^2)$ and $N(\mu_2, \sigma_2^2)$, then

$$V := \frac{1}{\sigma_1^2} \sum_{i=1}^{m} (X_i - \mu_1)^2 \quad \text{and} \quad W := \frac{1}{\sigma_2^2} \sum_{j=1}^{n} (Y_j - \mu_2)^2$$

are χ_m^2 and χ_n^2-distributed and independent. Consequently, the quotient $\frac{V/m}{W/n}$ is $F_{m,n}$-distributed.

2. For $X_1, \ldots, X_m$ and $Y_1, \ldots, Y_n$ independent and standard normal, the random variables

$$(m-1)\frac{s_X^2}{\sigma_1^2} \quad \text{and} \quad (n-1)\frac{s_Y^2}{\sigma_2^2}$$

are independent and distributed according to χ_{m-1}^2 and χ_{n-1}^2, respectively. Thus, assuming $\sigma_1 = \sigma_2$, the quotient s_X^2/s_Y^2 possesses an $F_{m-1,n-1}$-distribution.

When applying an F-test, as before, two different cases have to be considered.

(K) The expected values μ_1 and μ_2 of the two populations are **known**. Then the statistical model is given by

$$\left(\mathbb{R}^{m+n}, \mathcal{B}(\mathbb{R}^{m+n}), N(\mu_1, \sigma_1^2)^{\otimes m} \otimes N(\mu_2, \sigma_2^2)^{\otimes n}\right)_{(\sigma_1^2, \sigma_2^2) \in (0,\infty)^2} .$$

(U) The expected values are **unknown**. This case is described by the statistical model

$$\left(\mathbb{R}^{m+n}, \mathcal{B}(\mathbb{R}^{m+n}), N(\mu_1, \sigma_1^2)^{\otimes m} \otimes N(\mu_2, \sigma_2^2)^{\otimes n}\right)_{(\mu_1, \mu_2, \sigma_1^2, \sigma_2^2) \in \mathbb{R}^2 \times (0,\infty)^2} .$$

In both cases the null hypothesis may either be $\mathbb{H}_0 : \sigma_1^2 \le \sigma_2^2$ in the one-sided case or $\mathbb{H}_0 : \sigma_1^2 = \sigma_2^2$ in the two-sided one. The regions of acceptance in each of the four different cases are given by the following subsets of $\mathbb{R}^{m+n}$, and always $\mathcal{X}_1 = \mathbb{R}^{m+n} \setminus \mathcal{X}_0$.
Case 1: $\mathbb{H}_0 : \sigma_1^2 \le \sigma_2^2$ and μ_1, μ_2 are known.

$$\mathcal{X}_0 := \left\{ (x,y) \in \mathbb{R}^{m+n} : \frac{\frac{1}{m}\sum_{i=1}^{m}(x_i - \mu_1)^2}{\frac{1}{n}\sum_{j=1}^{n}(y_j - \mu_2)^2} \le F_{m,n;1-\alpha} \right\}$$

Case 2: $\mathbb{H}_0 : \sigma_1^2 = \sigma_2^2$ and μ_1, μ_2 are known.

$$\mathcal{X}_0 := \left\{ (x, y) \in \mathbb{R}^{m+n} : F_{m,n;\alpha/2} \le \frac{\frac{1}{m} \sum_{i=1}^{m} (x_i - \mu_1)^2}{\frac{1}{n} \sum_{j=1}^{n} (y_j - \mu_2)^2} \le F_{m,n;1-\alpha/2} \right\}$$

Case 3: $\mathbb{H}_0 : \sigma_1^2 \le \sigma_2^2$ and μ_1, μ_2 are unknown.

$$\mathcal{X}_0 := \left\{ (x, y) \in \mathbb{R}^{m+n} : \frac{s_x^2}{s_y^2} \le F_{m-1,n-1;1-\alpha} \right\}$$

Case 4: $\mathbb{H}_0 : \sigma_1^2 = \sigma_2^2$ and μ_1, μ_2 are unknown.

$$\mathcal{X}_0 := \left\{ (x, y) \in \mathbb{R}^{m+n} : F_{m-1,n-1;\alpha/2} \le \frac{s_x^2}{s_y^2} \le F_{m-1,n-1;1-\alpha/2} \right\}$$

8.5 Point Estimators

Starting point is a parametric statistical model $(\mathcal{X}, \mathcal{F}, \mathbb{P}_\theta)_{\theta \in \Theta}$. Assume we execute a statistical experiment and observe a sample $x \in \mathcal{X}$. The aim of this section is to show how this observation leads to a "good" estimation of the unknown parameter $\theta \in \Theta$.

Example 8.5.1. Suppose the statistical model is $(\mathbb{R}^n, \mathcal{B}(\mathbb{R}^n), \mathcal{N}(\mu, \sigma_0^2)^{\otimes n})_{\mu \in \mathbb{R}}$ for some known $\sigma_0^2 > 0$. Thus, the unknown parameter is the expected value $\mu \in \mathbb{R}$. To estimate it, we execute n independent measurements and get $x = (x_1, \ldots, x_n) \in \mathbb{R}^n$. Knowing this vector x, what is a "good" estimation for μ? An intuitive approach is to define the point estimator $\hat{\mu} : \mathbb{R}^n \to \mathbb{R}$ as

$$\hat{\mu}(x) = \frac{1}{n} \sum_{j=1}^{n} x_j = \bar{x}, \quad x = (x_1, \ldots, x_n) \in \mathbb{R}^n .$$

In other words, if the observed sample is x, then we take its sample mean $\hat{\mu}(x) = \bar{x}$ as estimation for μ. An immediate question is whether $\hat{\mu}$ is a "good" estimator for μ. Or do there exist maybe "better" (more precise) estimators for μ?

Before we investigate such and similar questions, the problem has to be generalized slightly. Sometimes it happens that we are not interested in the concrete value of the parameter $\theta \in \Theta$. We only want to know the value $y(\theta)$ derived from θ. Thus, for some function $y : \Theta \to \mathbb{R}$ we want to find a "good" estimator $\hat{y} : \mathcal{X} \to \mathbb{R}$ for $y(\theta)$. In other words, if we observe a sample $x \in \mathcal{X}$, then we take $\hat{y}(x)$ as estimation for the (unknown) value $y(\theta)$. However, in most cases the function y is not needed. That is, here we have $y(\theta) = \theta$, and we look for a good estimator $\hat{\theta} : \mathcal{X} \to \Theta$ for θ.

Let us state an example where a nontrivial function y plays a role.

Example 8.5.2. Let $(\mathbb{R}^n, \mathcal{B}(\mathbb{R}^n), \mathcal{N}(\mu, \sigma^2)^{\otimes n})_{(\mu,\sigma^2)\in\mathbb{R}\times(0,\infty)}$ be the statistical model. Thus, the unknown parameter is the two-dimensional vector (μ, σ^2). But, in fact we are only interested in μ, not in the pair (μ, σ^2). That is, if

$$y(\mu, \sigma^2) := \mu, \quad (\mu, \sigma^2) \in \mathbb{R} \times (0, \infty),$$

then we want to find an estimation for $y(\mu, \sigma^2)$.

Analogously, if we only want an estimation for σ^2, then we choose y as

$$y(\mu, \sigma^2) := \sigma^2, \quad (\mu, \sigma^2) \in \mathbb{R} \times (0, \infty).$$

After these preliminary considerations, we state now the precise definition of an estimator.

Definition 8.5.3. Let $(\mathcal{X}, \mathcal{F}, \mathbb{P}_\theta)_{\theta\in\Theta}$ be a parametric statistical model and let $y : \Theta \to \mathbb{R}$ be a function of the parameter. A mapping $\hat{y} : \mathcal{X} \to \mathbb{R}$ is said to be a **point estimator** (or simply **estimator**) for $y(\theta)$ if, given $t \in \mathbb{R}$, the set $\{x \in \mathcal{X} : \hat{y}(x) \le t\}$ belongs to the σ-field $\mathcal{F}$. In other words, $\hat{y}$ is a random variable defined on $\mathcal{X}$.

The interpretation of this definition is as follows. If one observes the sample $x \in \mathcal{X}$, then $\hat{y}(x)$ is an estimation for $y(\theta)$. For example, if one measures a workpiece four times and gets 22.03, 21.87, 22.11, and 22, 15 inches as results, then using the estimator $\hat{\mu}$ in Example 8.5.2, the estimation for the mean value equals 22.04 inches.

8.5.1 Maximum Likelihood Estimation

Let $(\mathcal{X}, \mathcal{F}, \mathbb{P}_\theta)_{\theta\in\Theta}$ be a parametric statistical model. There exist several methods to construct "good" point estimators for the unknown parameter θ. In this section we present the probably most important of these methods, the so-called **maximum likelihood principle**.

To understand this principle, the following easy example may be helpful.

Example 8.5.4. Suppose the parameter set consists of two elements, say $\Theta = \{0, 1\}$. Moreover, also the sample space $\mathcal{X}$ has cardinality two, that is, $\mathcal{X} = \{a, b\}$. Then the problem is as follows. Depending on the observation a or b, we have to choose either 0 or 1 as estimation for θ.

For example, let us assume that $\mathbb{P}_0(\{a\}) = 1/4$, hence $\mathbb{P}_0(\{b\}) = 3/4$, and $\mathbb{P}_1(\{a\}) = \mathbb{P}_1(\{b\}) = 1/2$. Say, an experiment has outcome "a." What would be a good estimation for θ in this case? Should we take "0" or "1"? The answer is, we should choose "1." Why? Because the sample "a" fits better to $\mathbb{P}_1$ than to $\mathbb{P}_0$. By the same argument, we should take "0" as an estimation if we observe "b." Thus, the point estimator for θ is given by $\hat{\theta}(a) = 1$ and $\hat{\theta}(b) = 0$.

Which property characterizes the estimator $\hat{\theta}$ in Example 8.5.4? To answer this question, fix $x \in \mathcal{X}$ and look at the function

$$\theta \mapsto \mathbb{P}_\theta(\{x\}), \quad \theta \in \Theta. \tag{8.23}$$

If $x = a$, this function becomes maximal for $\theta = 1$, while for $x = b$ it attains its maximal value at $\theta = 0$. Consequently, the estimator $\hat{\theta}$ could also be defined as follows. For each fixed $x \in \mathcal{X}$, choose as estimation the $\theta \in \Theta$, for which the function (8.23) becomes maximal. But this is exactly the approach of the maximum likelihood principle.

In order to describe this principle in the general setting, we have to introduce the notion of the likelihood function. Let us first assume that the sample space $\mathcal{X}$ consists of **at most countably many elements**.

> **Definition 8.5.5.** The function p from $\Theta \times \mathcal{X}$ to $\mathbb{R}$ defined as
>
> $$p(\theta, x) := \mathbb{P}_\theta(\{x\}), \quad \theta \in \Theta, x \in \mathcal{X},$$
>
> is called **likelihood function** of the statistical model $(\mathcal{X}, \mathcal{P}(\mathcal{X}), \mathbb{P}_\theta)_{\theta\in\Theta}$.

We come now to the case where all probability measures $\mathbb{P}_\theta$ are **continuous**. Thus, we assume that the statistical model is $(\mathbb{R}^n, \mathcal{B}(\mathbb{R}^n), \mathbb{P}_\theta)_{\theta\in\Theta}$ and, moreover, each $\mathbb{P}_\theta$ is continuous, that is, it has a density mapping $\mathbb{R}^n$ to $\mathbb{R}$. This density is not only a function of $x \in \mathbb{R}^n$, it also depends on the probability measure $\mathbb{P}_\theta$, hence on $\theta \in \Theta$. Therefore, we denote the densities by $p(\theta, x)$. In other words, for each $\theta \in \Theta$ and each box $Q \subseteq \mathbb{R}^n$ as in eq. (1.65) we have

$$\mathbb{P}_\theta(Q) = \int_Q p(\theta, x)\,dx = \int_{a_1}^{b_1} \cdots \int_{a_n}^{b_n} p(\theta, x_1, \ldots, x_n)\,dx_n \ldots dx_1 . \tag{8.24}$$

> **Definition 8.5.6.** The function $p : \Theta \times \mathbb{R}^n \to \mathbb{R}$ satisfying eq. (8.24) for all boxes Q and all $\theta \in \Theta$ is said to be the **likelihood function** of the statistical model $(\mathbb{R}^n, \mathcal{B}(\mathbb{R}^n), \mathbb{P}_\theta)_{\theta\in\Theta}$.

For a better understanding of Definitions 8.5.5 and 8.5.6, let us give some examples of likelihood functions.

1. First take $(\mathcal{X}, \mathcal{P}(\mathcal{X}), B_{n,\theta})_{0 \le \theta \le 1}$ with $\mathcal{X} = \{0, \ldots, n\}$ from Section 8.3. Then its likelihood function equals

$$p(\theta, k) = \binom{n}{k} \theta^k (1 - \theta)^{n-k}, \quad \theta \in [0, 1], \; k \in \{0, \ldots, n\}. \tag{8.25}$$

2. Consider the statistical model $(\mathcal{X}, \mathcal{P}(\mathcal{X}), H_{N,M,n})_{M=0,\ldots,N}$ investigated in Example 8.1.4. Then its likelihood function is given by

$$p(M, m) = \frac{\binom{M}{m}\binom{N-M}{n-m}}{\binom{N}{n}}, \quad M = 0, \ldots, N, \; m = 0, \ldots, n. \tag{8.26}$$

3. The likelihood function of the model $(\mathbb{N}_0^n, \mathcal{P}(\mathbb{N}_0^n), \mathrm{Pois}_\lambda^{\otimes n})_{\lambda>0}$ investigated in Example 8.5.21 is

$$p(\lambda, k_1, \ldots, k_n) = \frac{\lambda^{k_1+\cdots+k_n}}{k_1! \cdots k_n!} e^{-\lambda n}, \quad \lambda > 0, \; k_j \in \mathbb{N}_0. \tag{8.27}$$

4. The likelihood function of $(\mathbb{R}^n, \mathcal{B}(\mathbb{R}^n), \mathcal{N}(\mu, \sigma^2)^{\otimes n})_{(\mu,\sigma^2)\in\mathbb{R}\times(0,\infty)}$ from Example 8.1.12 can be calculated by

$$p(\mu, \sigma^2, x) = \frac{1}{(2\pi)^{n/2}\sigma^n} \exp\left(-\frac{|x - \vec{\mu}|^2}{2\sigma^2}\right), \quad \mu \in \mathbb{R}, \sigma^2 > 0. \tag{8.28}$$

Here, as before, let $\vec{\mu} = (\mu, \ldots, \mu)$.

5. The likelihood function of $(\mathbb{R}^n, \mathcal{B}(\mathbb{R}^n), E_\lambda^{\otimes n})_{\lambda>0}$ from Example 8.1.9 may be represented as

$$p(\lambda, t_1, \ldots, t_n) = \begin{cases} \lambda^n e^{-\lambda(t_1+\cdots+t_n)} & : t_j \ge 0, \; \lambda > 0 \\ 0 & : \text{otherwise} \end{cases} \tag{8.29}$$

Definition 8.5.7. Let $(\mathcal{X}, \mathcal{F}, \mathbb{P}_\theta)_{\theta\in\Theta}$ be a parametric statistical model with likelihood function $p : \Theta \times \mathcal{X} \to \mathbb{R}$. An estimator $\hat\theta : \mathcal{X} \to \Theta$ is said to be a **maximum likelihood estimator (MLE)** for $\theta \in \Theta$ provided that, for each $x \in \mathcal{X}$, the following is satisfied:

$$p(\hat\theta(x), x) = \max_{\theta\in\Theta} p(\theta, x)$$

Remark 8.5.8. Another way to define the MLE is as follows[5]:

$$\hat{\theta}(x) = \arg\max_{\theta\in\Theta} p(\theta, x), \quad x \in \mathcal{X}.$$

How does one find the MLE for concrete statistical models? One observation is that the logarithm is an increasing function. Thus, the likelihood function $p(\cdot, x)$ becomes maximal at a certain parameter $\theta \in \Theta$ if $\ln p(\cdot, x)$ does so.

Definition 8.5.9. Let $(\mathcal{X}, \mathcal{F}, \mathbb{P}_\theta)_{\theta\in\Theta}$ be a statistical model and let $p : \Theta \times \mathcal{X} \to \mathbb{R}$ be its likelihood function. Suppose $p(\theta, x) > 0$ for all (θ, x). Then the function

$$L(\theta, x) := \ln p(\theta, x), \quad \theta \in \Theta, x \in \mathcal{X},$$

is called **log-likelihood function** of the model.

Thus, $\hat{\theta}$ is an MLE if and only if

$$\hat{\theta}(x) = \arg\max_{\theta\in\Theta} L(\theta, x), \quad x \in \mathcal{X},$$

or, equivalently, if

$$L(\hat{\theta}(x), x) = \max_{\theta\in\Theta} L(\theta, x).$$

Example 8.5.10. If p is the likelihood function in eq. (8.25), then the log-likelihood function equals

$$L(\theta, k) = c + k \ln\theta + (n - k)\ln(1 - \theta), \quad 0 \le \theta \le 1, \ k = 0, \ldots, n. \tag{8.30}$$

Here $c \in \mathbb{R}$ denotes a certain constant independent of θ.

Example 8.5.11. The log-likelihood function of p in eq. (8.29) is well-defined for $\lambda > 0$ and $t_j \ge 0$. For those λs and t_js it is given by

$$L(\lambda, t_1, \ldots, t_n) = n \ln\lambda - \lambda(t_1 + \cdots + t_n).$$

To proceed further we assume now that the parameter set Θ is a subset of $\mathbb{R}^k$ for some $k \ge 1$. That is, each parameter θ consists of k unknown components, that is, it may be

---

[5] If f is a real-valued function with domain A, then $x = \arg\max_{y\in A} f(y)$ if $x \in A$ and $f(x) \ge f(y)$ for all $y \in A$. In other words, x is one of the points in the domain A where f attains its maximal value.

written as $\theta = (\theta_1, \dots, \theta_k)$ with $\theta_j \in \mathbb{R}$. Furthermore, suppose that for each fixed $x \in \mathcal{X}$ the log-likelihood function $L(\cdot, x)$ is continuously differentiable[6] on Θ. Then points $\theta^* \in \Theta$ where $L(\cdot, x)$ becomes maximal must satisfy

$$\frac{\partial}{\partial \theta_i} L(\theta, x)\Big|_{\theta=\theta^*} = 0, \quad i = 1, \dots, k. \tag{8.31}$$

In particular, this is true for the MLE $\hat{\theta}(x)$. If for each $x \in \mathcal{X}$ the log-likelihood function $L(\cdot, x)$ is continuously differentiable on $\Theta \subseteq \mathbb{R}^k$, then the MLE $\hat{\theta}$ satisfies

!

$$\frac{\partial}{\partial \theta_i} L(\theta, x)\Big|_{\theta=\hat{\theta}(x)} = 0, \quad i = 1, \dots, k.$$

Example 8.5.12. Let us determine the MLE for the log-likelihood function in eq. (8.30). Here we have $\Theta = [0, 1] \subseteq \mathbb{R}$, hence the MLE $\hat{\theta} : \{0, \dots, n\} \to [0, 1]$ has to satisfy

$$\frac{\partial}{\partial \theta} L(\hat{\theta}(k), k) = \frac{k}{\hat{\theta}(k)} - \frac{n-k}{1 - \hat{\theta}(k)} = 0.$$

This easily gives $\hat{\theta}(k) = \frac{k}{n}$, that is, the MLE in this case is defined by

$$\hat{\theta}(k) = \frac{k}{n}, \quad k = 0, \dots, n.$$

Let us interpret this result. In an urn are white and black balls of unknown proportion. Let θ be the proportion of white balls. To estimate θ, draw n balls out of the urn, with replacement. Assume k of the chosen balls are white. Then $\hat{\theta}(k) = \frac{k}{n}$ is the MLE for the unknown proportion θ of white balls.

Example 8.5.13. The logarithm of the likelihood function p in eq. (8.28) equals

$$L(\mu, \sigma^2, x) = L(\mu, \sigma^2, x_1, \dots, x_n) = c - \frac{n}{2} \cdot \ln \sigma^2 - \frac{1}{2\sigma^2} \sum_{j=1}^{n} (x_j - \mu)^2$$

with some constant $c \in \mathbb{R}$, independent of μ and of σ^2. Thus, here $\Theta \subseteq \mathbb{R}^2$, hence, if $\theta^* = (\mu^*, \sigma^{2*})$ denotes the pair satisfying eq. (8.31), then

$$\frac{\partial}{\partial \mu} L(\mu, \sigma^2, x)\Big|_{(\mu,\sigma^2)=(\mu^*,\sigma^{2*})} = 0 \quad \text{and} \quad \frac{\partial}{\partial \sigma^2} L(\mu, \sigma^2, x)\Big|_{(\mu,\sigma^2)=(\mu^*,\sigma^{2*})} = 0.$$

---

6 The partial derivatives exist and are continuous.

Now

$$\frac{\partial}{\partial\mu}L(\mu,\sigma^2,x) = \frac{1}{\sigma^2}\sum_{j=1}^{n}(x_j - \mu) = \frac{1}{\sigma^2}\left[\sum_{j=1}^{n}x_j - n\mu\right],$$

which implies $\mu^* = \frac{1}{n}\sum_{j=1}^{n}x_j = \bar{x}$.

The derivative of L with respect to σ^2, taken at $\mu^* = \bar{x}$, equals

$$\frac{\partial}{\partial\sigma^2}L(\bar{x},\sigma^2,x) = -\frac{n}{2}\cdot\frac{1}{\sigma^2} + \frac{1}{\sigma^4}\sum_{j=1}^{n}(x_j - \bar{x})^2.$$

It becomes zero at σ^{2*} satisfying

$$\sigma^{2*} = \frac{1}{n}\sum_{j=1}^{n}(x_j - \bar{x})^2 = \sigma_x^2,$$

where σ_x^2 was defined in eq. (8.7). Combining these observations, we see that the only pair $\theta^* = (\mu^*,\sigma^{2*})$ satisfying eq. (8.31) is given by $(\bar{x},\sigma_x^2)$. Consequently, as MLE for $\theta = (\mu,\sigma^2)$ we obtain

$$\hat{\mu}(x) = \bar{x} \quad\text{and}\quad \hat{\sigma^2}(x) = \sigma_x^2, \quad x \in \mathbb{R}^n.$$

Remark 8.5.14. Similar calculations as in the previous examples show that the MLE for the likelihood functions in eqs. (8.27) and (8.29) coincide with

$$\hat{\lambda}(k_1,\ldots,k_n) = \frac{1}{n}\sum_{i=1}^{n}k_i \quad\text{and}\quad \hat{\lambda}(t_1,\ldots,t_n) = \frac{1}{\frac{1}{n}\sum_{i=1}^{n}t_i}.$$

Finally, we present two likelihood functions where we have to determine their maximal values directly. Note that the above approach via the log-likelihood function does not apply if the parameter set Θ is either finite or countably infinite. In this case a derivative of $L(\cdot,x)$ does not make sense, hence we cannot determine points where it vanishes.

The first problem is the one we discussed in Remark 1.4.27. A retailer gets a delivery of N machines. Among the N machines are M defective ones. Since M is unknown, the retailer wants a "good" estimate for it. Therefore, he chooses by random n machines and tests them. Suppose he observes m defective machines among the tested. Does this lead to an estimation of the number M of defective machines? The next proposition answers this question.

Proposition 8.5.15. *The statistical model is given by* $(\mathcal{X}, \mathcal{P}(\mathcal{X}), H_{M,N,n})_{M=0,\ldots,N}$. *Then the MLE* $\hat{M}$ *for M is of the form*

$$\hat{M}(m) = \begin{cases} \left[\frac{m(N+1)}{n}\right] & : m < n \\ N & : m = n \end{cases}$$

Here $[\cdot]$ *denotes the integer part of a real number, for example,* $[1.2] = 1$ *or* $[\pi] = 3$.

Proof: The likelihood function p was determined in eq. (8.26) as

$$p(M, m) = \frac{\binom{M}{m}\binom{N-M}{n-m}}{\binom{N}{n}}, \quad M = 0, \ldots, N, \ m = 0, \ldots, n.$$

First note that $p(M, m) \neq 0$ if and only if $M \in \{m, \ldots, N-n+m\}$ and, therefore, it suffices to investigate $p(M, m)$ for Ms in this region. Thus, if $M - 1 \geq m$, then easy calculations lead to

$$\frac{p(M, m)}{p(M - 1, m)} = \frac{M}{M - m} \cdot \frac{N - M + 1 - (n - m)}{N - M + 1}. \tag{8.32}$$

By eq. (8.32) it follows that we have $p(M, m) \geq p(M - 1, m)$ if and only if

$$M(N - M + 1 - (n - m)) \geq (M - m)(N - M + 1).$$

Elementary transformations show the last estimate is equivalent to

$$-nM \geq -mN - m,$$

which happens if and only if $M \leq \frac{m(N+1)}{n}$.

Consequently, $M \mapsto p(M, m)$ is nondecreasing on $\left\{0, \ldots, \left[\frac{m(N+1)}{n}\right]\right\}$, and it is nonincreasing on $\left\{\left[\frac{m(N+1)}{n}\right], \ldots, N\right\}$. Thus, if $m < n$, then the likelihood function $M \mapsto p(M, m)$ becomes maximal for $M^* = \left[\frac{m(N+1)}{n}\right]$, and the MLE is given by

$$\hat{M}(m) = \left[\frac{m(N + 1)}{n}\right], \quad m = 0, \ldots, n - 1.$$

If $m = n$, then $M \mapsto p(M, m)$ is nonincreasing on $\{0, \ldots, N\}$, hence in this case the likelihood function attains its maximal value at $M = N$, that is, $\hat{M}(n) = N$. ∎

Example 8.5.16. A retailer gets a delivery of 100 TV sets for further selling. He chooses by random 15 sets and tests them. If there is exactly one defective TV set among the 15 tested, then the estimation for the number of defective sets in the delivery is 6. If he observes 2 defective sets, the estimation is 13, for 4 it is 26, and if there are even

6 defective TV sets among the 15 chosen, then the estimation is that 40 sets of the delivery are defective.

Finally we come back to the question asked in Remark 1.4.29. In order to estimate the number N of fish in a pond one catches M of them, marks them and puts them back into the pond. After some time one catches fish again, this time n of them. Among them m are marked. Does this number m lead to a "good" estimation of the number of fish in the pond? To describe this problem we choose as statistical model

$$(\mathcal{X}, \mathbb{P}(\mathcal{X}), H_{N,M,n})_{N=0,1,\dots}$$

where $\mathcal{X} = \{0, \dots, n\}$. Here $H_{N,M,n}$ denotes the hypergeometric probability measure introduced in Definition 1.4.25. Thus, in this case the likelihood function is given by

$$p(N, m) = \frac{\binom{M}{m}\binom{N-M}{n-m}}{\binom{N}{n}}, \quad N = 0, 1, \dots, \quad m = 0, \dots, n.$$

In the sequel we have to exclude $m = 0$; in this case there does not exist a reasonable estimation for N.

Proposition 8.5.17. *If $1 \le m \le n$, then the MLE $\hat{N}$ for N is*

$$\hat{N}(m) = \left[\frac{Mn}{m}\right]. \tag{8.33}$$

Proof: The proof is quite similar to that of Proposition 8.5.15. Since

$$\frac{p(N, m)}{p(N-1, m)} = \frac{N-M}{N} \cdot \frac{N-n}{N-M-(n-m)},$$

it easily follows that the inequality $p(N, m) \ge p(N-1, m)$ is valid if and only if $N \le \frac{Mn}{m}$. Therefore, $N \mapsto p(N, m)$ is nondecreasing if $N \le \left[\frac{Mn}{m}\right]$ and nonincreasing for the remaining N. This immediately shows that the MLE is given by eq. (8.33). ∎

Example 8.5.18. An unknown number of balls are in an urn. In order to estimate this number, we choose 50 balls from the urn and mark them. We put back the marked balls and mix the balls in the urn thoroughly. Then we choose another 30 balls from the urn. If there are 7 marked among the 30, then the estimation for the number of balls in the urn is 214. In the case of two marked balls, the estimation equals 750 while in the case of 16 marked balls we estimate that there are 93 balls in the urn.

8.5.2 Unbiased Estimators

Let us come back to the general setting. We are given a function $y : \Theta \to \mathbb{R}$ and look for a "good" estimation for $y(\theta)$. If $\hat{y}(x)$ is the estimation, in most cases it will not be the correct value $y(\theta)$. Sometimes the estimate is larger than $y(\theta)$, sometimes one observes an $x \in \mathcal{X}$ for which $\hat{y}(x)$ is smaller than the true value. For example, if the retailer in Example 8.5.16 gets every week a delivery of 100 TV sets, then sometimes his estimation for the number of defective sets will be bigger than the true value, sometimes smaller. Since he only pays for the nondefective sets, sometimes he pays too much, sometimes too less. Therefore, a crucial condition for a good estimator should be that, on average, it meets the correct value. That is, in the long run the loss and the gain of the retailer should balance. In other words, the estimator should not be biased by a systematic error.

In view of Proposition 7.1.29, this condition for the estimator $\hat{y}$ may be formulated as follows. If $\theta \in \Theta$ is the "true" parameter, then the expected value of $\hat{y}$ should be $y(\theta)$. To make this more precise[7] we need the following notation.

Definition 8.5.19. Let $(\mathcal{X}, \mathcal{F}, \mathbb{P}_\theta)_{\theta \in \Theta}$ be a statistical model and let $X : \mathcal{X} \to \mathbb{R}$ be a random variable. We write $\mathbb{E}_\theta X$ whenever the expected value of X is taken with respect to $\mathbb{P}_\theta$. Similarly, in this case define

$$\mathbb{V}_\theta X = \mathbb{E}_\theta |X - \mathbb{E}_\theta X|^2$$

as variance of X. Of course, we have to assume that the expected value and/or the variance exist.

Remark 8.5.20. If X is discrete with values in $\{t_1, t_2, \ldots\}$, then

$$\mathbb{E}_\theta X = \sum_{j=1}^{\infty} t_j \, \mathbb{P}_\theta\{X = t_j\}.$$

The case of continuous X is slightly more difficult because here we have to describe the density function of X with respect to $\mathbb{P}_\theta$.

To become acquainted with Definition 8.5.19, the two following examples may be helpful. The first one deals with the discrete case, the second with the continuous one.

---

7 How the expected value is defined? Note that we do not have only one probability measure, but many different ones.

Example 8.5.21. Suppose the daily number of customers in a shopping center is Poisson distributed with unknown parameter $\lambda > 0$. To estimate this parameter, we record the number of customers on n different days. Thus, the sample we obtain is a vector $\vec{k} = (k_1, \ldots, k_n)$ with $k_j \in \mathbb{N}_0$, where k_j is the number of customers on day j. The describing statistical model is given by $(\mathbb{N}_0^n, \mathcal{P}(\mathbb{N}_0^n), \text{Pois}_\lambda^{\otimes n})_{\lambda > 0}$ with distribution Pois_λ. Let $X : \mathbb{N}_0^n \to \mathbb{R}$ be defined by

$$X(\vec{k}) = X(k_1, \ldots, k_n) := \frac{1}{n} \sum_{j=1}^n k_j, \quad \vec{k} = (k_1, \ldots, k_n) \in \mathbb{N}_0^n.$$

Which value does $\mathbb{E}_\lambda X$ possess?

Answer: If we choose $\text{Pois}_\lambda^{\otimes n}$ as probability measure, then all X_js defined by $X_j(k_1, \ldots, k_n) := k_j$ are Pois_λ-distributed (and independent, but this is not needed here). Note that X_j is nothing else as the number of customers at day j. Hence, by Proposition 5.1.16, the expected value of X_j is λ, and since $X = \frac{1}{n} \sum_{j=1}^n X_j$, we finally obtain

$$\mathbb{E}_\lambda X = \mathbb{E}_\lambda \left(\frac{1}{n} \sum_{j=1}^n X_j \right) = \frac{1}{n} \sum_{j=1}^n \mathbb{E}_\lambda X_j = \frac{1}{n} n\lambda = \lambda.$$

Example 8.5.22. Take

$$(\mathbb{R}^n, \mathcal{B}(\mathbb{R}^n), \mathcal{N}(\mu, \sigma^2)^{\otimes n})_{(\mu, \sigma^2) \in \mathbb{R} \times (0, \infty)}$$

as the statistical model. Thus, the parameter is of the form (μ, σ^2) for some $\mu \in \mathbb{R}$ and $\sigma^2 > 0$. Define $X : \mathbb{R}^n \to \mathbb{R}$ by $X(x) = \bar{x}$. If the underlying measure is $\mathcal{N}(\mu, \sigma^2)^{\otimes n}$, then[8] X is $\mathcal{N}(\mu, \sigma^2/n)$-distributed. Consequently, in view of Propositions 5.1.34 and 5.2.27 we obtain

$$\mathbb{E}_{\mu, \sigma^2} X = \mu \quad \text{and} \quad V_{\mu, \sigma^2} X = \frac{\sigma^2}{n}.$$

Using the notation introduced in Definition 8.5.19, the above-mentioned requirement for "good" estimators may now be formulated more precisely.

Definition 8.5.23. An estimator $\hat{y} : \mathcal{X} \to \mathbb{R}$ is said to be an **unbiased** estimator for $y : \Theta \to \mathbb{R}$ provided that for each $\theta \in \Theta$

$$\mathbb{E}_\theta |\hat{y}| < \infty \quad \text{and} \quad \mathbb{E}_\theta \hat{y} = y(\theta).$$

---

8 Compare the first part of the proof of Proposition 8.4.3.

Remark 8.5.24. In view of Proposition 7.1.29, an estimator $\hat{y}$ is unbiased if it possesses the following property: observe N independent samples $x^1, \ldots, x^N$ of a statistical experiment. Suppose that $\theta \in \Theta$ is the "true" parameter (according to which the x^js are distributed). Then

$$\mathbb{P}\left\{ \lim_{N \to \infty} \frac{1}{N} \sum_{j=1}^{n} \hat{y}(x^j) = y(\theta) \right\} = 1.$$

Thus, on average, the estimator $\hat{y}$ meets approximately the correct value.

Example 8.5.25. Let us investigate whether the estimator in Example 8.5.12 is unbiased. The statistical model is $(\mathcal{X}, \mathcal{P}(\mathcal{X}), B_{n,\theta})_{0 \le \theta \le 1}$, where $\mathcal{X} = \{0, \ldots, n\}$ and the estimator $\hat{\theta}$ acts as

$$\hat{\theta}(k) = \frac{k}{n}, \quad k = 0, \ldots, n.$$

Setting $Z := n\hat{\theta}$, then Z is the identity on $\mathcal{X}$, hence $B_{n,\theta}$-distributed. Proposition 5.1.13 implies $\mathbb{E}_\theta Z = n\theta$, thus,

$$\mathbb{E}_\theta \hat{\theta} = \mathbb{E}_\theta (Z/n) = \mathbb{E}_\theta Z/n = \theta. \tag{8.34}$$

Equation (8.34) holds for all $\theta \in [0, 1]$, that is, $\hat{\theta}$ is an unbiased estimator for θ.

Example 8.5.26. Next we come back to the problem presented in Example 8.5.21. The number of customers per day is Pois_λ-distributed with an unknown parameter $\lambda > 0$. The data of n days are combined in a vector $\hat{k} = (k_1, \ldots, k_n) \in \mathbb{N}_0^n$. Then the parameter $\lambda > 0$ is estimated by $\hat{\lambda}$ defined as

$$\hat{\lambda}(\vec{k}) = \hat{\lambda}(k_1, \ldots, k_n) := \frac{1}{n} \sum_{j=1}^{n} k_j.$$

Is this estimator for λ unbiased?

Answer: Yes, it is unbiased. Observe that $\hat{\lambda}$ coincides with the random variable X investigated in Example 8.5.21. There we proved $\mathbb{E}_\lambda = \lambda$, hence, if $\lambda > 0$, then we have

$$\mathbb{E}_\lambda \hat{\lambda} = \lambda.$$

Example 8.5.27. We are given certain data $x_1, \ldots, x_n$, which are known to be normally distributed and independent, and where the expected value μ and the variance σ^2 of the underlying probability measure are unknown. Thus, the describing statistical model is

$$(\mathbb{R}^n, \mathcal{B}(\mathbb{R}^n), \mathcal{N}(\mu, \sigma^2)^n)_{(\mu, \sigma^2) \in \Theta} \quad \text{with} \quad \Theta = \mathbb{R} \times (0, \infty).$$

The aim is to find unbiased estimators for μ and for σ^2. Let us begin with estimating μ. That is, if y is defined by $y(\mu, \sigma^2) = \mu$, then we want to construct an unbiased estimator $\hat{y}$ for y. Let us take the MLE $\hat{y}$ defined as

$$\hat{y}(x) := \bar{x} = \frac{1}{n} \sum_{j=1}^{n} x_j, \quad x = (x_1, \ldots, x_n).$$

Due to the calculations in Example 8.5.22 we obtain

$$\mathbb{E}_{\mu,\sigma^2} \hat{y} = \mu = y(\mu, \sigma^2).$$

This holds for all μ and σ^2, hence $\hat{y}$ is an unbiased estimator for $\mu = y(\mu, \sigma^2)$.

How to find a suitable estimator for σ^2? This time the function y has to be chosen as $y(\mu, \sigma^2) := \sigma^2$. With s_x^2 defined in eq. (8.7) set

$$\hat{y}(x) := s_x^2 = \frac{1}{n-1} \sum_{j=1}^{n} (x_j - \bar{x})^2, \quad x \in \mathbb{R}^n.$$

Is this an unbiased estimator for σ^2 ? To answer this we use property (8.10) of Proposition 8.4.3. It asserts that the random variable $x \mapsto (n-1)\frac{s_x^2}{\sigma^2}$ is χ_{n-1}^2-distributed, provided it is defined on $(\mathbb{R}^n, \mathcal{B}(\mathbb{R}^n), \mathcal{N}(\mu, \sigma^2)^{\otimes n})$. Consequently, by Corollary 5.1.30 it follows that

$$\mathbb{E}_{\mu,\sigma^2}\left[(n-1)\frac{s_x^2}{\sigma^2}\right] = n - 1.$$

Using the linearity of the expected value, we finally obtain

$$\mathbb{E}_{\mu,\sigma^2} \hat{y} = \mathbb{E}_{\mu,\sigma^2} s_x^2 = \sigma^2.$$

Therefore, $\hat{y}(x) = s_x^2$ is an unbiased[9] estimator for σ^2.

Remark 8.5.28. Taking the estimator $\hat{y}(x) = \sigma_x^2$ in the previous example, then, in view of $\sigma_x^2 = \frac{n-1}{n} s_x^2$, it follows that

$$\mathbb{E}_{\mu,\sigma^2} \hat{y} = \frac{n-1}{n} \sigma^2.$$

Thus, the estimator $\hat{y}(x) = \sigma_x^2$ is biased. But note that

$$\lim_{n \to \infty} \frac{n-1}{n} \sigma^2 = \sigma^2,$$

---

9 This explains why s_x^2 is called **unbiased** sample variance.

hence, if the sample size n is big, then this estimator is "almost" unbiased. One says in this case the sequence of estimators (in dependence on n) is asymptotically unbiased.

The next example is slightly more involved, but of great interest in application.

Example 8.5.29. The lifetime of light bulbs is supposed to be exponentially distributed with some unknown parameter $\lambda > 0$. To estimate λ we switch on n light bulbs and record the times $t_1, \ldots, t_n$ when they burn out. Thus, the observed sample is a vector $t = (t_1, \ldots, t_n)$ in $(0, \infty)^n$. As estimator for λ we choose

$$\hat{\lambda}(t) := \frac{n}{\sum_{j=1}^{n} t_j} = 1/\bar{t}.$$

Is this an unbiased estimator for λ?

Answer: The statistical model describing this experiment is $(\mathbb{R}^n, \mathcal{B}(\mathbb{R}^n), E_\lambda^{\otimes n})_{\lambda > 0}$. If the random variables X_j are defined by $X_j(t) := t_j$, then they are independent and E_λ-distributed. Because of Proposition 4.6.13, their sum $X := \sum_{j=1}^{n} X_j$ possesses an Erlang distribution with parameters n and λ. An application of eq. (5.22) in Proposition 5.1.36 for $f(x) := \frac{n}{x}$ implies

$$\mathbb{E}_\lambda \hat{\lambda} = \mathbb{E}_\lambda \left(\frac{n}{X} \right) = \int_0^\infty \frac{n}{x} \frac{\lambda^n}{(n-1)!} x^{n-1} e^{-\lambda x} \, dx.$$

A change of variables $s := \lambda x$ transforms the last integral into

$$\frac{\lambda n}{(n-1)!} \int_0^\infty s^{n-2} e^{-s} \, ds = \frac{\lambda n}{(n-1)!} \Gamma(n-1) = \frac{\lambda n}{(n-1)!} \cdot (n-2)! = \lambda \cdot \frac{n}{n-1}.$$

This tells us that, $\hat{\lambda}$ is **not** unbiased estimator for λ. But, as mentioned in Remark 8.5.28 for σ_X^2, the sequence of estimators is asymptotically unbiased as $n \to \infty$.

Remark 8.5.30. If we replace the estimator in Example 8.5.29 by

$$\hat{\lambda}(t) := \frac{n-1}{\sum_{j=1}^{n} t_j} = \frac{1}{\frac{1}{n-1} \sum_{j=1}^{n} t_j}, \qquad t = (t_1, \ldots, t_n),$$

then the previous calculations imply

$$\mathbb{E}_\lambda \hat{\lambda} = \frac{n-1}{n} \cdot \lambda \cdot \frac{n}{n-1} = \lambda.$$

Hence, from this small change we get an unbiased estimator $\hat{\lambda}$ for λ.

Observe that the calculations in Example 8.5.29 were only valid for $n \geq 2$. If $n = 1$, then the expected value of $\hat{\lambda}$ does not exist.

8.5.3 Risk Function

Let $(\mathcal{X}, \mathcal{F}, \mathbb{P}_\theta)_{\theta \in \Theta}$ be a parametric statistical model. Furthermore, $y : \Theta \to \mathbb{R}$ is a function of the parameter and $\hat{y} : \mathcal{X} \to \Theta$ is an estimator for y. Suppose $\theta \in \Theta$ is the true parameter and we observe some $x \in \mathcal{X}$. Then, in general, we will have $y(\theta) \neq \hat{y}(x)$, and the quadratic error $|y(\theta) - \hat{y}(x)|^2$ occurs. Other ways to measure the error are possible and useful, but we restrict ourselves to the quadratic distance. In this way we get the so-called **loss function** $L : \Theta \times \mathcal{X} \to \mathbb{R}$ of $\hat{y}$ defined by

$$L(\theta, x) := |y(\theta) - \hat{y}(x)|^2 .$$

In other words, if θ is the correct parameter and our sample is $x \in \mathcal{X}$, then, using $\hat{y}$ as the estimator, the (quadratic) error or loss will be $L(\theta, x)$. On average, the (quadratic) loss is evaluated by $\mathbb{E}_\theta |y(\theta) - \hat{y}|^2$.

> **Definition 8.5.31.** The function R describing this average loss of $\hat{y}$ is said to be the **risk function** of the estimator $\hat{y}$. It is defined by
>
> $$R(\theta, \hat{y}) := \mathbb{E}_\theta |y(\theta) - \hat{y}|^2 , \quad \theta \in \Theta .$$

Before giving some examples of risk functions, let us rewrite R as follows.

Proposition 8.5.32. *If $\theta \in \Theta$, then it follows that*

$$R(\theta, \hat{y}) = |y(\theta) - \mathbb{E}_\theta \hat{y}|^2 + \mathbb{V}_\theta \hat{y} . \tag{8.35}$$

Proof: The assertion is a consequence of

$$R(\theta, \hat{y}) = \mathbb{E}_\theta \left[y(\theta) - \hat{y} \right]^2 = \mathbb{E}_\theta \left[(y(\theta) - \mathbb{E}_\theta \hat{y}) + (\mathbb{E}_\theta \hat{y} - \hat{y}) \right]^2$$
$$= |y(\theta) - \mathbb{E}_\theta \hat{y}|^2 + 2((y(\theta) - \mathbb{E}_\theta \hat{y}) \mathbb{E}_\theta (\mathbb{E}_\theta \hat{y} - \hat{y}) + \mathbb{V}_\theta \hat{y} .$$

Because of

$$\mathbb{E}_\theta (\mathbb{E}_\theta \hat{y} - \hat{y}) = \mathbb{E}_\theta \hat{y} - \mathbb{E}_\theta \hat{y} = 0 ,$$

this implies eq. (8.35). ∎

> **Definition 8.5.33.** The function $\theta \mapsto |y(\theta) - \mathbb{E}_\theta \hat{y}|^2$, which appears in eq. (8.35), is said to be the **bias** or the **systematic error** of the estimator $\hat{y}$.

Corollary 8.5.34. *A point estimator $\hat{y}$ is unbiased if and only if for all $\theta \in \Theta$ its bias is zero. Moreover, if this is so, then its risk function is given by*

$$R(\theta, \hat{y}) = \mathbb{V}_\theta \hat{y}, \quad \theta \in \Theta.$$

Remark 8.5.35. Another way to formulate eq. (8.35) is as follows. The risk function of an estimator consists of two parts. One part is the systematic error, which does not occur for unbiased estimators. And the second part is given by $\mathbb{V}_\theta \hat{y}$. Thus, the smaller the bias and/or $\mathbb{V}_\theta \hat{y}$ become, the smaller is the risk to get a wrong estimation for $y(\theta)$, and the better is the estimator.

Example 8.5.36. Let us determine the risk functions for the two estimators presented in Example 8.5.27. The estimator $\hat{y}$ for μ was given by $\hat{y}(x) = \bar{x}$. Since this is an unbiased estimator, by Corollary 8.5.34, its risk function is computed as

$$R((\mu, \sigma^2), \hat{y}) = \mathbb{V}_{(\mu,\sigma^2)} \hat{y}.$$

The random variable $x \mapsto \bar{x}$ is $\mathcal{N}(\mu, \sigma^2/n)$-distributed, hence

$$R((\mu, \sigma^2), \hat{y}) = \frac{\sigma^2}{n}.$$

There are two interesting facts about this risk function. First, it does not depend on the parameter μ that we want to estimate. And secondly, if $n \to \infty$, then the risk tends to zero. In other words, the bigger the sample size, the less becomes the risk for a wrong estimation.

Next we evaluate the risk function of the estimator $\hat{y}(x) = s_{\bar{x}}^2$. As we saw in Example 8.5.29, this $\hat{y}$ is also an unbiased estimator for σ^2, hence

$$R((\mu, \sigma^2), \hat{y}) = \mathbb{V}_{(\mu,\sigma^2)} \hat{y}.$$

By eq. (8.10) we know that $\frac{n-1}{\sigma^2} s_{\bar{x}}^2$ is χ_{n-1}^2-distributed, hence Corollary 5.2.26 implies

$$\mathbb{V}_{(\mu,\sigma^2)}\left[\frac{n-1}{\sigma^2} s_{\bar{x}}^2\right] = 2(n-1).$$

From this one easily derives

$$R((\mu, \sigma^2), \hat{y}) = \mathbb{V}_{(\mu,\sigma^2)} s_{\bar{x}}^2 = 2(n-1) \cdot \frac{\sigma^4}{(n-1)^2} = \frac{2\sigma^4}{n-1}.$$

Here, the risk function depends heavily on the parameter σ^2 that we want to estimate. Furthermore, if $n \to \infty$, then also in this case the risk tends to zero.

Example 8.5.37. Finally, regard the statistical model $(\mathcal{X}, \mathcal{P}(\mathcal{X}), B_{n,\theta})_{0 \le \theta \le 1}$, where $\mathcal{X} = \{0, \dots, n\}$. In order to estimate $\theta \in [0, 1]$, we take, as in Example 8.5.25, the estimator $\hat{\theta}(k) = \frac{k}{n}$. There it was shown that the estimator is unbiased, hence, by Corollary 8.5.34, it follows that

$$R(\theta, \hat{\theta}) = \mathbb{V}_\theta \hat{\theta}, \quad 0 \le \theta \le 1.$$

If X is the identity on $\mathcal{X}$, by Proposition 5.2.18, its variance equals $\mathbb{V}_\theta X = n\,\theta(1 - \theta)$. Since $\hat{\theta} = \frac{X}{n}$ this implies

$$R(\theta, \hat{\theta}) = \mathbb{V}_\theta(X/n) = \frac{\mathbb{V}_\theta X}{n^2} = \frac{\theta(1 - \theta)}{n}.$$

Consequently, the risk function becomes maximal for $\theta = 1/2$, while for $\theta = 0$ or $\theta = 1$ it vanishes.

We saw in Corollary 8.5.34 that $R(\theta, \hat{y}) = \mathbb{V}_\theta \hat{y}$ for unbiased $\hat{y}$. Thus, for such estimators inequality (7.2) implies

$$\mathbb{P}_\theta \{x \in \mathcal{X} : |y(\theta) - \hat{y}(x)| > c\} \le \frac{\mathbb{V}_\theta \hat{y}}{c^2},$$

that is, the smaller $\mathbb{V}_\theta \hat{y}$ is, the greater is the chance to estimate a value near the correct one. This observation leads to the following definition.

> **Definition 8.5.38.** Let $\hat{y}_1$ and $\hat{y}_2$ be two unbiased estimators for $y(\theta)$. Then $\hat{y}_1$ is said to be **uniformly better** than $\hat{y}_2$ provided that
>
> $$\mathbb{V}_\theta \hat{y}_1 \le \mathbb{V}_\theta \hat{y}_2 \quad \text{for all } \theta \in \Theta.$$
>
> An unbiased estimator $\hat{y}_*$ is called the **uniformly best estimator** if it is uniformly better than all other unbiased estimators for $y(\theta)$.

Example 8.5.39. We observe values that, for some $b > 0$, are uniformly distributed on $[0, b]$. But the number $b > 0$ is unknown. In order to estimate it, one executes n independent trials and obtains as sample $x = (x_1, \dots, x_n)$. As point estimators for $b > 0$ one may either choose

$$\hat{b}_1(x) := \frac{n+1}{n} \max_{1 \le i \le n} x_i \quad \text{or} \quad \hat{b}_2(x) := \frac{2}{n} \sum_{i=1}^{n} x_i.$$

According to Problem 8.4, the estimators $\hat{b}_1$ and $\hat{b}_2$ are both unbiased. Furthermore, not too difficult calculations show that

$$\mathbb{V}_b \hat{b}_1 = \frac{b^2}{n(n+2)} \quad \text{and} \quad \mathbb{V}_b \hat{b}_2 = \frac{b^2}{3n^2}.$$

Therefore, $\mathbb{V}_b \hat{b}_1 \leq \mathbb{V}_b \hat{b}_2$ for all $b > 0$. This tells us that $\hat{b}_1$ is uniformly better than $\hat{b}_2$.

Remark 8.5.40. A very natural question is whether there exists a lower bound for the precision of an estimator. In other words, are there estimators for which the risk function becomes arbitrarily small? The answer depends heavily on the inherent information in the statistical model. To explain this let us come back once more to Example 8.5.4.

Suppose we had $\mathbb{P}_0(\{a\}) = 1$ and $\mathbb{P}_1(\{b\}) = 1$. Then the occurrence of "a" would tell us with 100% security that $\theta = 0$ is the correct parameter. The risk for the corresponding estimator is then zero. On the contrary, if $\mathbb{P}_0(\{a\}) = \mathbb{P}_0(\{b\}) = 1/2$, then the occurrence of "a" tells us nothing about the correct parameter.

To make the previous observation more precise, we have to introduce some quantity that measures the information contained in a statistical model.

Definition 8.5.41. Let $(\mathcal{X}, \mathcal{F}, \mathbb{P}_\theta)_{\theta \in \Theta}$ be a statistical model with log-likelihood function L introduced in Definition 8.5.9. For simplicity, assume $\Theta \subseteq \mathbb{R}$. Then the function $I : \Theta \to \mathbb{R}$ defined by

$$I(\theta) := \mathbb{E}_\theta \left(\frac{\partial L}{\partial \theta} \right)^2$$

is called the **Fisher information** of the model. Of course, we have to suppose that the derivatives and the expected value exist.

Example 8.5.42. Let us investigate the Fisher information for the model treated in Example 8.5.13. There we had

$$L(\mu, \sigma^2, x) = L(\mu, \sigma^2, x_1, \ldots, x_n) = c - \frac{n}{2} \cdot \ln \sigma^2 - \frac{1}{2\sigma^2} \sum_{j=1}^{n} (x_j - \mu)^2.$$

Fix σ^2 and take the derivative with respect to μ. This leads to

$$\frac{\partial L}{\partial \mu} = \frac{n\bar{x} - n\mu}{\sigma^2},$$

hence

$$\left(\frac{\partial L}{\partial \mu}\right)^2 = \frac{n^2}{\sigma^4}|\bar{x} - \mu|^2.$$

Recall that $\bar{x}$ is $\mathcal{N}(\mu, \sigma^2/n)$-distributed, hence the expected value of $|\bar{x} - \mu|^2$ is nothing else than the variance of $\bar{x}$, that is, it is σ^2/n. Consequently,

$$I(\mu) = \mathbb{E}_{\mu,\sigma^2}\left(\frac{\partial L}{\partial \mu}\right)^2 = \frac{n^2}{\sigma^4}\frac{\sigma^2}{n} = \frac{n}{\sigma^2}.$$

The following result answers the above question: how precise can an estimator become at the most?

Proposition 8.5.43 (Rao–Cramér–Frechet). *Let* $(\mathcal{X}, \mathcal{F}, \mathbb{P}_\theta)_{\theta \in \Theta}$ *be a parametric model for which the Fisher information* $I : \Theta \to \mathbb{R}$ *exists. If* $\hat{\theta}$ *is an unbiased estimator for* θ, *then*

$$\mathbb{V}_\theta \hat{\theta} \geq \frac{1}{I(\theta)}, \quad \theta \in \Theta. \tag{8.36}$$

Remark 8.5.44. Estimators $\hat{\theta}$ that attain the lower bound in estimate (8.36) are said to be **efficient**. That is, for those estimators holds $\mathbb{V}_\theta\hat{\theta} = 1/I(\theta)$ for all $\theta \in \Theta$. In other words, efficient estimators possess the best possible accuracy.

In view of Examples 8.5.36 and 8.5.42, for normally distributed populations the estimator $\hat{\mu}(x) = \bar{x}$ is an efficient estimator for μ. Other efficient estimators are those investigated in Examples 8.5.26 and 8.5.12. On the other hand, the estimator for σ^2 in Example 8.5.27 is not efficient. But it can be shown that s_x^2 is a uniformly best estimator for σ^2, that is, there do not exist efficient estimators in this case.

8.6 Confidence Regions and Intervals

Point estimations provide us with a single value $\theta \in \Theta$. Further work or necessary decisions are then based on this estimated parameter. The disadvantages of this approach are that we have no knowledge about the precision of the obtained value. Is the estimated parameter far away from the true one or maybe very near? To explain the problem, let us come back to the situation described in Example 8.5.16. If the retailer observes 4 defective TV sets among 15 tested, then he estimates that there are 26 defective sets in the delivery of 100. But he does not know how precise his estimation of 26 is. Maybe there are much more defective sets in the delivery, or maybe less than 26. The only information he has is that the estimates are correct on average. But this does not say anything about the accuracy of a single estimate.

This disadvantage of point estimators is avoided when estimating a certain set of parameters, not only a single point. Then the true parameter is contained with great probability in this randomly chosen region. In most cases, these regions will be intervals of real or natural numbers.

Definition 8.6.1. Suppose we are given a parametric statistical model $(\mathcal{X}, \mathcal{F}, \mathbb{P}_\theta)_{\theta \in \Theta}$. A mapping $C : \mathcal{X} \to \mathcal{P}(\Theta)$ is called an **interval estimator**,[10] provided for fixed $\theta \in \Theta$

$$\{x \in \mathcal{X} : \theta \in C(x)\} \in \mathcal{F}. \tag{8.37}$$

Remark 8.6.2. Condition (8.37) is quite technical and will play no role later on. But it is necessary because otherwise the next definition does not make sense.

Definition 8.6.3. Let α be a real number in $(0, 1)$. Suppose an interval estimator $C : \mathcal{X} \to \mathcal{P}(\Theta)$ satisfies for each $\theta \in \Theta$ the condition

$$\mathbb{P}_\theta\{x \in \mathcal{X} : \theta \in C(x)\} \geq 1 - \alpha. \tag{8.38}$$

Then C is said to be a $100(1 - \alpha)\%$ **interval estimator**.[11] The sets $C(x) \subseteq \Theta$ with $x \in \mathcal{X}$ are called $100(1 - \alpha)\%$ **confidence regions** or **confidence intervals**.[12]

How does an interval estimator apply? Suppose $\theta \in \Theta$ is the "true" parameter. In a statistical experiment, one obtains some sample $x \in \mathcal{X}$ distributed according to $\mathbb{P}_\theta$. In dependence of the observed sample x, we choose a set $C(x)$ of parameters. Then with probability greater than or equal to $1 - \alpha$, the observed $x \in \mathcal{X}$ leads to a region $C(x)$ that contains the true parameter θ.

Remark 8.6.4. It is important to say that the region $C(x)$ **is random, not the unknown parameter** $\theta \in \Theta$. Metaphorically speaking, a fish (the true parameter θ) is in a pond at some fixed but unknown spot. We execute a certain statistical experiment to get some information about the place where the fish is situated. In dependence of the result of the experiment, we throw a net into the pond. Doing so, we know that with probability greater than or equal to $1 - \alpha$, the result of the experiment leads to a net that catches the fish. In other words, the position of the fish is not random, it is the observed sample, hence also the thrown net.

---

10 Better notation would be "region estimator" because $C(x) \subseteq \Theta$ may be an arbitrary subset, not necessarily an interval, but "interval estimator" is commonly accepted, therefore, we use it here also.
11 Also $1 - \alpha$ estimator.
12 Also $1 - \alpha$ confidence regions or intervals.

Remark 8.6.5. It is quite self-evident that one should try to choose the confidence regions as small as possible, without violating condition (8.38). If we are not interested in "small" confidence regions, then we could always chose $C(x) = \Theta$. This is not forbidden, but completely useless.

Construction of confidence regions: For a better understanding of the subsequent construction, let us shortly recall the main assertions about hypothesis tests from a slightly different point of view.

Let $(\mathcal{X}, \mathcal{F}, \mathbb{P}_\vartheta)_{\vartheta\in\Theta}$ be a statistical model. We choose a fixed, but arbitrary, $\theta \in \Theta$. With this chosen θ, we formulate the null hypothesis as $\mathbb{H}_0 : \vartheta = \theta$. The alternative hypothesis is then $\mathbb{H}_1 : \vartheta \neq \theta$. Let $\mathbf{T} = (\mathcal{X}_0, \mathcal{X}_1)$ be an α-significance test for $\mathbb{H}_0$ against $\mathbb{H}_1$. Because the hypothesis, hence also the test, depends on the chosen $\theta \in \Theta$, we denote the null hypothesis by $\mathbb{H}_0(\theta)$ and write $\mathbf{T}(\theta) = (\mathcal{X}_0(\theta), \mathcal{X}_1(\theta))$ for the test. That is, $\mathbb{H}_0(\theta) : \vartheta = \theta$ and $\mathbf{T}(\theta)$ is an α-significance test for $\mathbb{H}_0(\theta)$. With this notation set

$$C(x) := \{\theta \in \Theta : x \in \mathcal{X}_0(\theta)\}. \tag{8.39}$$

Example 8.6.6. Choose the hypothesis and the test as in Proposition 8.4.12. The statistical model is then given by $(\mathbb{R}^n, \mathcal{B}(\mathbb{R}^n), \mathcal{N}(v, \sigma_0^2)^{\otimes n})_{v\in\mathbb{R}}$, where this time we denote the unknown expected value by v. For some fixed, but arbitrary, $\mu \in \mathbb{R}$ let

$$\mathbb{H}_0(\mu) : v = \mu \quad \text{and} \quad \mathbb{H}_1(\mu) : v \neq \mu.$$

The α-significance test $\mathbf{T}(\mu)$ constructed in Proposition 8.4.12 possesses the region of acceptance

$$\mathcal{X}_0(\mu) = \left\{ x \in \mathbb{R}^n : \sqrt{n}\left|\frac{\bar{x} - \mu}{\sigma_0}\right| \leq z_{1-\alpha/2} \right\}.$$

Thus, in this case, the set $C(x)$ in eq. (8.39) consists of those $\mu \in \mathbb{R}$ that satisfy the estimate $\sqrt{n}\left|\frac{\bar{x}-\mu}{\sigma_0}\right| \leq z_{1-\alpha/2}$. That is, given $x \in \mathbb{R}^n$, then $C(x)$ is the interval

$$C(x) = \left[\bar{x} - \frac{\sigma_0}{\sqrt{n}} z_{1-\alpha/2}, \bar{x} + \frac{\sigma_0}{\sqrt{n}} z_{1-\alpha/2}\right].$$

Let us come back to the general situation. The statistical model is $(\mathcal{X}, \mathcal{F}, \mathbb{P}_\vartheta)_{\vartheta\in\Theta}$. Given $\theta \in \Theta$ let $\mathbf{T}(\theta)$ be an α-significance test for $\mathbb{H}_0(\theta)$ against $\mathbb{H}_1(\theta)$ where $\mathbb{H}_0(\theta)$ is the hypothesis $\mathbb{H}_0(\theta) : \vartheta = \theta$. Given $x \in \mathcal{X}$ define $C(x) \subseteq \Theta$ by eq. (8.39). Then the following is valid.

Proposition 8.6.7. *Let $\mathbf{T}(\theta)$ be as above an α-significance test for $\mathbb{H}_0(\theta)$ against $\mathbb{H}_1(\theta)$. Define $C(x)$ by eq. (8.39) where $\mathcal{X}_0(\theta)$ denotes the region of acceptance of $\mathbf{T}(\theta)$. Then the mapping $x \mapsto C(x)$ from $\mathcal{X}$ into $\mathcal{P}(\Theta)$ is a $100(1 - \alpha)\%$ interval estimator. Hence, $\{C(x) : x \in \mathcal{X}\}$ is a collection of $100(1 - \alpha)\%$ confidence regions.*

Proof: By assumption, $\mathbf{T}(\theta)$ is an α-significance test for $\mathbb{H}_0(\theta)$. The definition of those tests tells us that

$$\mathbb{P}_\theta(\mathcal{X}_1(\theta)) \le \alpha, \quad \text{hence} \quad \mathbb{P}_\theta(\mathcal{X}_0(\theta)) \ge 1 - \alpha.$$

Given $\theta \in \Theta$ and $x \in \mathcal{X}$, by the construction of $C(x)$, one has $\theta \in C(x)$ if and only if $x \in \mathcal{X}_0(\theta)$. Combining these two observations, given $\theta \in \Theta$, then it follows that

$$\mathbb{P}_\theta\{x \in \mathcal{X} : \theta \in C(x)\} = \mathbb{P}_\theta\{x \in \mathcal{X} : x \in \mathcal{X}_0(\theta)\} = \mathbb{P}_\theta(\mathcal{X}_0(\theta)) \ge 1 - \alpha.$$

This completes the proof. ∎

Example 8.6.8. Proposition 8.6.7 implies that the intervals $C(x)$ in Example 8.6.6 are $1 - \alpha$ confidence intervals. That is, if we execute an experiment or if we analyze some data, then with probability greater than or equal to $1 - \alpha$ we will observe values $x = (x_1, \ldots, x_n)$ such that the "true" parameter μ satisfies

$$\bar{x} - \frac{\sigma_0}{\sqrt{n}} z_{1-\alpha/2} \le \mu \le \bar{x} + \frac{\sigma_0}{\sqrt{n}} z_{1-\alpha/2}.$$

For example, if we choose $\alpha = 0.05$ and observe the nine values

$$10.1, \; 9.2, \; 10.2, \; 10.3, \; 10.1, \; 9.9, \; 10.0, \; 9.7, \; 9.8,$$

then $\bar{x} = 9.9222$. The variance σ_0 is not known, therefore we use its estimation by s_x^2, that is, we take σ_0 as $s_x = 0.330824$. Because of $z_{1-\alpha/2} = z_{0.975} = 1.95996$, with security of 95% we finally get

$$9.7061 \le \mu \le 10.1384.$$

In the next example we describe the confidence intervals generated by the t-test treated in Proposition 8.4.12.

Example 8.6.9. The statistical model is

$$(\mathbb{R}^n, \mathcal{B}(\mathbb{R}^n), \mathcal{N}(\nu, \sigma^2)^{\otimes n})_{(\nu, \sigma^2) \in \mathbb{R} \times (0, \infty)}.$$

By Proposition 8.4.12 an α-significance test $\mathbf{T}(\mu)$ is given by the region of acceptance

$$\mathcal{X}_0(\mu) = \left\{ x \in \mathbb{R}^n : \sqrt{n} \left| \frac{\bar{x} - \mu}{s_x} \right| \le t_{n-1; 1-\alpha/2} \right\}.$$

From this one easily derives

$$C(x) = \left\{ \mu \in \mathbb{R} : \sqrt{n} \left| \frac{\bar{x} - \mu}{s_x} \right| \le t_{n-1;1-\alpha/2} \right\}$$
$$= \left[\bar{x} - \frac{s_x}{\sqrt{n}} t_{n-1;1-\alpha/2} , \bar{x} + \frac{s_x}{\sqrt{n}} t_{n-1;1-\alpha/2} \right].$$

Let us explain the result by the concrete sample investigated in Example 8.4.15. There we had $\bar{x} = 22.072$, $s_x = 0.07554248$, and $n = 10$. For $\alpha = 0.05$, the quantile of t_9 equals $t_{9;0.975} = 2, 26$. From this we derive $[22.016, 22.126]$ as 95% confidence interval.

Verbally this says with a security of 95% we observed those $x_1, \ldots, x_{10}$ for which $\mu \in C(x) = [22.016, 22.126]$.

The next example shows how Proposition 8.6.7 applies in the case of discrete probability measures.

Example 8.6.10. The statistical model is $(\mathcal{X}, \mathcal{P}(\mathcal{X}), B_{n,\theta})_{0 \le \theta \le 1}$ where the sample space is $\mathcal{X} = \{0, \ldots, n\}$. Our aim is to construct confidence regions $C(k) \subseteq [0, 1]$, $k = 0, \ldots, n$, such that

$$B_{n,\theta}\{k \le n : \theta \in C(k)\} \ge 1 - \alpha.$$

In order to get these confidence regions, we use Proposition 8.6.7. As shown in Proposition 8.3.1, the region of acceptance $\mathcal{X}_0(\theta)$ of an α-significance test $\mathbf{T}(\theta)$, where $\mathbb{H}_0 : \vartheta = \theta$, is given by

$$\mathcal{X}_0(\theta) = \{n_0(\theta), \ldots, n_1(\theta)\}.$$

Here, the numbers $n_0(\theta)$ and $n_1(\theta)$ were defined by

$$n_0(\theta) := \min \left\{ k \le n : \sum_{j=0}^{k} \binom{n}{j} \theta^j (1 - \theta)^{n-j} > \alpha/2 \right\}$$

and

$$n_1(\theta) := \max \left\{ k \le n : \sum_{j=k}^{n} \binom{n}{j} \theta^j (1 - \theta)^{n-j} > \alpha/2 \right\}.$$

Applying Proposition 8.6.7, the sets

$$C(k) := \{\theta \in [0, 1] : k \in \mathcal{X}_0(\theta)\} = \{\theta \in [0, 1] : n_0(\theta) \le k \le n_1(\theta)\}, \quad k = 0, \ldots, n,$$

are $1 - \alpha$ confidence regions. By the definition of $n_0(\theta)$ and of $n_1(\theta)$, given $k \leq n$, then a number $\theta \in [0, 1]$ satisfies $n_0(\theta) \leq k \leq n_1(\theta)$ if and only if at the same time

$$B_{n,\theta}(\{0, \ldots, k\}) = \sum_{j=0}^{k} \binom{n}{j} \theta^j (1 - \theta)^{n-j} > \alpha/2 \quad \text{and}$$

$$B_{n,\theta}(\{k, \ldots, n\}) = \sum_{j=k}^{n} \binom{n}{j} \theta^j (1 - \theta)^{n-j} > \alpha/2 .$$

In other words, observing $k \leq n$, then the corresponding $1 - \alpha$ confidence region is given by

$$C(k) = \{\theta : B_{n,\theta}(\{0, \ldots, k\}) > \alpha/2\} \cap \{\theta : B_{n,\theta}(\{k, \ldots, n\}) > \alpha/2\} . \tag{8.40}$$

These sets are called the $100(1 - \alpha)\%$ **Clopper–Pearson** intervals or also **exact confidence intervals** for the binomial distribution.

Let us consider the following concrete example. In an urn are white and black balls with an unknown proportion θ of white balls. In order to get some information about θ, we choose randomly 500 balls with replacement. Say, 220 of the chosen balls are white. What is the 90% confidence interval for θ based on this observation?

Answer: We have $n = 500$ and the observed k equals 220. Consequently, the confidence interval $C(220)$ consists of those $\theta \in [0, 1]$ for which at the same time

$$f(\theta) > \alpha/2 = 0.05 \quad \text{and} \quad g(\theta) > \alpha/2 = 0.05 ,$$

where

$$f(\theta) := \sum_{j=0}^{220} \binom{500}{j} \theta^j (1 - \theta)^{500-j} \quad \text{and} \quad g(\theta) := \sum_{j=220}^{500} \binom{500}{j} \theta^j (1 - \theta)^{500-j} .$$

Numerical calculations tell us that

$$f(0.4777) = 0.0500352 \approx 0.05 \quad \text{as well as} \quad g(0.4028) = 0.0498975 \approx 0.05 .$$

Therefore, a 90% confidence interval $C(220)$ is given by

$$C(220) = (0.4028, 0.4777) .$$

For $n = 1000$ and 440 observed white balls similar calculations lead to the smaller, hence more significant, interval

$$C(440) = (0.4139, 0.4664) .$$

Remark 8.6.11. The previous example already indicates that the determination of the Clopper–Pearson intervals becomes quite complicated for large n. Therefore, one looks for "approximative" intervals. Background for the construction is the central limit theorem in the form presented in Proposition 7.2.15. For S_ns distributed according to $B_{n,\theta}$ it implies

$$\lim_{n\to\infty} \mathbb{P}\left\{ \left| \frac{S_n - n\theta}{\sqrt{n\theta(1-\theta)}} \right| \le z_{1-\alpha/2} \right\} = 1 - \alpha,$$

or, equivalently,

$$\lim_{n\to\infty} B_{n,\theta}\left\{ k \le n : \left| \frac{k - n\theta}{\sqrt{n\theta(1-\theta)}} \right| \le z_{1-\alpha/2} \right\} = 1 - \alpha.$$

Here $z_{1-\alpha/2}$ are the quantiles introduced in Definition 8.4.5. Thus, an "approximative" region of acceptance, testing the hypothesis "the unknown parameter is θ," is given by

$$\mathcal{X}_0(\theta) = \left\{ k \le n : \left| \frac{k}{n} - \theta \right| \le z_{1-\alpha/2}\sqrt{\frac{\theta(1-\theta)}{n}} \right\}. \tag{8.41}$$

An application of Proposition 8.6.7 leads to certain confidence regions, but these are not very useful. Due to the term $\sqrt{\theta(1-\theta)}$ on the right-hand side of eq. (8.41), it is not possible, for a given $k \le n$, to describe explicitly those θs for which $k \in \mathcal{X}_0(\theta)$. To overcome this difficulty, we change $\mathcal{X}_0(\theta)$ another time by replacing θ on the right-hand side by its MLE $\hat{\theta}(k) = \frac{k}{n}$. That is, we replace eq. (8.41) by

$$\tilde{\mathcal{X}}_0(\theta) = \left\{ k \le n : \left| \frac{k}{n} - \theta \right| \le z_{1-\alpha/2}\sqrt{\frac{\frac{k}{n}(1-\frac{k}{n})}{n}} \right\}.$$

Doing so, an application of Proposition 8.6.7 leads to the "approximative" confidence intervals $\tilde{C}(k)$, $k = 0, \ldots, n$, defined as

$$\tilde{C}(k) = \left[\frac{k}{n} - z_{1-\alpha/2}\sqrt{\frac{\frac{k}{n}(1-\frac{k}{n})}{n}}, \frac{k}{n} + z_{1-\alpha/2}\sqrt{\frac{\frac{k}{n}(1-\frac{k}{n})}{n}} \right]. \tag{8.42}$$

Example 8.6.12. We investigate once more Example 8.6.10. Among 500 chosen balls we observed 220 white ones. This observation led to the "exact" 90% confidence interval $C(220) = (0.4028, 0.4777)$.

Let us compare this result with the interval we get by using the approximative approach. Since the quantile $z_{1-\alpha/2}$ for $\alpha = 0.1$ equals $z_{0.95} = 1.64485$, the left and the right endpoints of the interval (8.42) with $k = 220$ are evaluated by

$$\frac{220}{500} - 1.64485 \cdot \sqrt{\frac{220 \cdot 280}{500^3}} = 0.4035 \quad \text{and}$$

$$\frac{220}{500} + 1.64485 \cdot \sqrt{\frac{220 \cdot 280}{500^3}} = 0.4765.$$

Thus, the "approximative" 90% confidence interval is $\tilde{C}(220) = (0.4035, 0.4765)$, which does not defer very much from $C(220) = (0.4028, 0.4777)$.

In the case of 1000 trials and 440 white balls, the endpoints of a confidence interval are evaluated by

$$\frac{440}{1000} - 1.64485 \cdot \sqrt{\frac{440 \cdot 560}{1000^3}} = 0.414181 \quad \text{and}$$

$$\frac{440}{1000} + 1.64485 \cdot \sqrt{\frac{440 \cdot 560}{1000^3}} = 0.4645819.$$

That is, $\tilde{C}(440) = (0.4142, 0.4659)$ compared with $C(440) = (0.4139, 0.4664)$.

Example 8.6.13. A few days before an election 1000 randomly chosen people are questioned for whom they will vote next week, either candidate A or candidate B. 540 of the interviewed people answered that they will vote for candidate A, the remaining 460 favor candidate B. Find a 90% sure confidence interval for the expected result of candidate A in the election.

Solution: We have $n = 1000$, $k = 540$, and $\alpha = 0.1$ The quantile of level 0.95 of the standard normal distribution equals $z_{0.95} = 1.64485$ (compare Example 8.6.12). This leads to $[0.514, 0.566]$ as "approximative" 90% confidence interval for the expected result of candidate A.

If one questions another 1000 randomly chosen people, another confidence interval will occur. But, on average, in 9 of 10 cases a questioning of 1000 people will lead to an interval containing the correct value.

8.7 Problems

Problem 8.1. For some $b > 0$ let $\mathbb{P}_b$ be the uniform distribution on $[0, b]$. The precise value of $b > 0$ is unknown. We claim that $b \leq b_0$ for a certain $b_0 > 0$. Thus, the hypotheses are

$$\mathbb{H}_0 : b \leq b_0 \quad \text{and} \quad \mathbb{H}_1 : b > b_0.$$

To test $\mathbb{H}_0$, we chose randomly n numbers $x_1, \ldots, x_n$ distributed according to $\mathbb{P}_b$. Suppose the region of acceptance $\mathcal{X}_0$ of a hypothesis test $\mathbf{T}_c$ is given by

$$\mathcal{X}_0 := \{(x_1, \ldots, x_n) : \max_{1 \leq i \leq n} x_i \leq c\}$$

for some $c > 0$.
1. Determine those $c > 0$ for which $\mathbf{T}_c$ is an α-significance test of level $\alpha < 1$.
2. Suppose $\mathbf{T}_c$ is an α-significance test. For which of those $c > 0$ does the probability for a type II error become minimal?
3. Determine the power function of the α-test $\mathbf{T}_c$ that minimizes the probability of the occurrence of a type II error.

Problem 8.2. For $\theta > 0$ let $\mathbb{P}_\theta$ be the probability measure with density p_θ defined by

$$p_\theta(s) = \begin{cases} \theta s^{\theta-1} : & s \in (0, 1] \\ 0 : & \text{otherwise} \end{cases}$$

1. Check whether the p_θs are probability density functions.
2. In order to get information about θ we execute n independent trials according to $\mathbb{P}_\theta$. Which statistical model describes this experiment?
3. Find the maximum likelihood estimator for θ.

Problem 8.3. The lifetime of light bulbs is exponentially distributed with unknown parameter $\lambda > 0$. In order to determine λ we switch on n light bulbs and record the number of light bulbs that burn out until a certain time $T > 0$. Determine a statistical model that describes this experiment. Find the MLE for λ.

Problem 8.4. Consider the statistical model in Example 8.5.39, that is, $(\mathbb{R}^n, \mathcal{B}(\mathbb{R}^n), \mathbb{P}_b^{\otimes n})_{b>0}$ with uniform distribution $\mathbb{P}_b$ on $[0, b]$. There are two natural estimators for $b > 0$, namely $\hat{b}_1$ and $\hat{b}_2$ defined by

$$\hat{b}_1(x) := \frac{n+1}{n} \max_{1 \leq i \leq n} x_i \quad \text{and} \quad \hat{b}_2(x) := \frac{2}{n} \sum_{i=1}^{n} x_i, \quad x = (x_1, \ldots, x_n) \in \mathbb{R}^n.$$

Prove that $\hat{b}_1$ and $\hat{b}_2$ possess the following properties.
1. The estimators $\hat{b}_1$ and $\hat{b}_2$ are unbiased.
2. One has

$$\mathbb{V}_b \hat{b}_1 = \frac{b^2}{n(n+2)} \quad \text{and} \quad \mathbb{V}_b \hat{b}_2 = \frac{b^2}{3n^2}.$$

Problem 8.5. In a questioning of 2000 randomly chosen people 1420 answered that they regularly use the Internet. Find an "approximative" 90% confidence interval for the proportion of people using the Internet regularly. Determine the inequalities that describe the exact intervals in eq. (8.40).

Problem 8.6. How do the confidence intervals in eq. (8.40) look like for $k = 0$ or $k = n$?

Problem 8.7. Suppose the statistical model is $(\mathcal{X}, \mathcal{P}(\mathcal{X}), H_{N,M,n})_{M \leq N}$ with $\mathcal{X} = \{0, \ldots, n\}$ and with the hypergeometric distributions $H_{N,M,n}$ introduced in Definition 1.4.25.

1. For some $M_0 \leq M$ the hypotheses are $\mathbb{H}_0 : M = M_0$ against $\mathbb{H}_1 : M \neq M_0$. Find (optimal) numbers $0 \leq m_0 \leq m_1 \leq n$ such that $\mathcal{X}_0 = \{m_0, \ldots, m_1\}$ is the region of acceptance of an α-significance test **T** for $\mathbb{H}_0$ against $\mathbb{H}_1$.
   Hint: Modify the methods developed in Proposition 8.2.19 and compare the construction of two-sided tests for a binomial distributed population.
2. Use Proposition 8.6.7 to derive from $\mathcal{X}_0$ confidence intervals $C(m)$, $0 \leq m \leq n$, of level α for the unknown parameter M.
   Hint: Follow the methods in Example 8.6.10 for the binomial distribution.

Problem 8.8. Use Proposition 8.6.7 to derive from Propositions 8.4.17 and 8.4.18 confidence intervals for the unknown variance of a normal distributed population.

A Appendix

A.1 Notations

Throughout the book we use the following standard notations:

1. The **natural numbers** starting at 1 are always denoted by $\mathbb{N}$. In the case 0 is included we write $\mathbb{N}_0$.
2. As usual the **integers** $\mathbb{Z}$ are given by $\mathbb{Z} = \{\ldots, -2, -1, 0, 1, 2, \ldots\}$.
3. By $\mathbb{R}$ we denote the field of **real numbers** endowed with the usual algebraic operations and its natural order. The subset $\mathbb{Q} \subset \mathbb{R}$ is the union of all **rational numbers**, that is, of numbers m/n where $m, n \in \mathbb{Z}$ and $n \neq 0$.
4. Given $n \geq 1$ let $\mathbb{R}^n$ be the n-**dimensional Euclidean vector space**, that is,

$$\mathbb{R}^n = \{x = (x_1, \ldots, x_n) : x_j \in \mathbb{R}\}.$$

Addition and scalar multiplication in $\mathbb{R}^n$ are carried out coordinate-wise,

$$x + y = (x_1, \ldots, x_n) + (y_1, \ldots, y_n) = (x_1 + y_1, \ldots x_n + y_n)$$

and if $\alpha \in \mathbb{R}$, then

$$\alpha x = (\alpha x_1, \ldots, \alpha x_n).$$

A.2 Elements of Set Theory

Given a set M its **powerset** $\mathcal{P}(M)$ consists of all subsets of M. In the case that M is finite we have $\#(\mathcal{P}(M)) = 2^{\#(M)}$, where $\#(A)$ denotes the **cardinality** (number of elements) of a finite set A.

If A and B are subsets of M, written as $A, B \subseteq M$ or also as $A, B \in \mathcal{P}(M)$, their **union** and their **intersection** are, as usual, defined by

$$A \cup B = \{x \in M : x \in A \text{ or } x \in B\} \text{ and } A \cap B = \{x \in M : x \in A \text{ and } x \in B\}.$$

Of course, it always holds that

$$A \cap B \subseteq A \subseteq A \cup B \quad \text{and} \quad A \cap B \subseteq B \subseteq A \cup B.$$

In the same way, given subsets $A_1, A_2, \ldots$ of M their union $\bigcup_{j=1}^{\infty} A_j$ and their intersection $\bigcap_{j=1}^{\infty} A_j$ is the set of those $x \in M$ that belong to at least one of the A_j or that belong to all A_j, respectively.

Quite often we use the **distributive law** for intersection and union. This asserts

$$A \cap \left(\bigcup_{j=1}^{\infty} B_j \right) = \bigcup_{j=1}^{\infty} (A \cap B_j).$$

Two sets A and B are said to be **disjoint**[1] provided that $A \cap B = \emptyset$. A sequence of sets $A_1, A_2, \ldots$ is called disjoint[2] whenever $A_i \cap A_j = \emptyset$ if $i \neq j$.

An element $x \in M$ belongs to the **set difference** $A \backslash B$ provided that $x \in A$ but $x \notin B$. Using the notion of the **complementary set** $B^c := \{x \in M : x \notin B\}$, the set difference may also be written as

$$A \backslash B = A \cap B^c.$$

Another useful identity is

$$A \backslash B = A \backslash (A \cap B).$$

Conversely, the complementary set may be represented as the set difference $B^c = M \backslash B$. We still mention the obvious $(B^c)^c = B$.

Finally we introduce the **symmetric difference** $A \triangle B$ of two sets A and B as

$$A \triangle B := (A \backslash B) \cup (B \backslash A) = (A \cap B^c) \cup (B \cap A^c) = (A \cup B) \backslash (A \cap B).$$

Note that an element $x \in M$ belongs to $A \triangle B$ if and only if x belongs exactly to one of the sets A or B.

De Morgan's rules are very important and assert the following:

$$\left(\bigcup_{j=1}^{\infty} A_j \right)^c = \bigcap_{j=1}^{\infty} A_j^c \quad \text{and} \quad \left(\bigcap_{j=1}^{\infty} A_j \right)^c = \bigcup_{j=1}^{\infty} A_j^c.$$

Given sets $A_1, \ldots, A_n$ their **Cartesian product** $A_1 \times \cdots \times A_n$ is defined by

$$A_1 \times \cdots \times A_n := \{(a_1, \ldots, a_n) : a_j \in A_j\}.$$

Note that $\#(A_1 \times \cdots \times A_n) = \#(A_1) \cdots \#(A_n)$.

Let S be another set, for example, $S = \mathbb{R}$, and let $f : M \to S$ be some mapping from M to S. Given a subset $B \subseteq S$, we denote the **preimage** of B with respect to f by

$$f^{-1}(B) := \{x \in M : f(x) \in B\}. \tag{A.1}$$

---

1 Sometimes called "mutually exclusive."
2 More precisely, one should say "pairwise disjoint."

In other words, an element $x \in M$ belongs to $f^{-1}(B)$ if and only if its image with respect to f is an element of B.

We summarize some crucial properties of the preimage in a proposition.

Proposition A.2.1. *Let $f : M \to S$ be a mapping from M into another set S.*
(1) $f^{-1}(\emptyset) = \emptyset$ *and* $f^{-1}(S) = M$.
(2) *For any subsets $B_j \subseteq S$ the following equalities are valid:*

$$f^{-1}\left(\bigcup_{j\geq 1} B_j\right) = \bigcup_{j\geq 1} f^{-1}(B_j) \ and \ f^{-1}\left(\bigcap_{j\geq 1} B_j\right) = \bigcap_{j\geq 1} f^{-1}(B_j). \tag{A.2}$$

Proof: We only prove the left-hand equality in eq. (A.2). The right-hand one is proved by the same methods. Furthermore, assertion (1) follows immediately.
Take $x \in f^{-1}\left(\bigcup_{j\geq 1} B_j\right)$. This happens if and only if

$$f(x) \in \bigcup_{j\geq 1} B_j \tag{A.3}$$

is satisfied. But this is equivalent to the existence of a certain $j_0 \geq 1$ with $f(x) \in B_{j_0}$. By definition of the preimage the last statement may be reformulated as follows: there exists a $j_0 \geq 1$ such that $x \in f^{-1}(B_{j_0})$. But this implies

$$x \in \bigcup_{j\geq 1} f^{-1}(B_j). \tag{A.4}$$

Consequently, an element $x \in M$ satisfies condition (A.3) if and only if property (A.4) holds. This proves the left-hand identity in formulas (A.2). ∎

A.3 Combinatorics

A.3.1 Binomial Coefficients

A one-to-one mapping π from $\{1, \ldots, n\}$ to $\{1, \ldots, n\}$ is called a **permutation** (of order n). Any permutation reorders the numbers from 1 to n as $\pi(1), \pi(2), \ldots, \pi(n)$ and, vice versa, each reordering of these numbers generates a permutation. One way to write a permutations is

$$\pi = \begin{pmatrix} 1 & 2 & \cdots & n \\ \pi(1) & \pi(2) & \cdots & \pi(n) \end{pmatrix}$$

For example, if $n = 3$, then $\pi = \begin{pmatrix} 1\,2\,3 \\ 2\,3\,1 \end{pmatrix}$ is equivalent to the order $2, 3, 1$ or to $\pi(1) = 2, \pi(2) = 3$ and $\pi(3) = 1$.

Let S_n be the set of all permutations of order n. Then one may ask for $\#(S_n)$ or, equivalently, for the number of possible orderings of the numbers $\{1, \ldots, n\}$.

To treat this problem we need the following definition.

Definition A.3.1. For $n \in \mathbb{N}$ we define n-**factorial** by setting

$$n! = 1 \cdot 2 \cdots (n-1) \cdot n$$

Furthermore, let $0! = 1$.

Now we may answer the question about the cardinality of S_n.

Proposition A.3.2. *We have*

$$\#(S_n) = n! \tag{A.5}$$

or, equivalently, there are $n!$ different ways to order n distinguishable objects.

Proof: The proof is done by induction over n. If $n = 1$ then $\#(S_1) = 1 = 1!$ and eq. (A.5) is valid.

Now suppose that eq. (A.5) is true for n. In order to prove eq. (A.5) for $n+1$ we split S_{n+1} as follows:

$$S_{n+1} = \bigcup_{k=1}^{n+1} A_k,$$

where

$$A_k = \{\pi \in S_{n+1} : \pi(n+1) = k\}, \quad k = 1, \ldots, n+1.$$

Each $\pi \in A_k$ generates a one-to-one mapping $\tilde{\pi}$ from $\{1, \ldots, n\}$ onto the set $\{1, \ldots, k-1, k+1, \ldots, n\}$ by letting $\tilde{\pi}(j) = \pi(j)$, $1 \leq j \leq n$. Vice versa, each such $\tilde{\pi}$ defines a permutation $\pi \in A_k$ by setting $\pi(j) = \tilde{\pi}(j)$, $j \leq n$, and $\pi(n+1) = k$. Consequently, since eq. (A.5) holds for n we get $\#(A_k) = n!$. Furthermore, the A_ks are disjoint, and

$$\#(S_{n+1}) = \sum_{k=1}^{n+1} \#(A_k) = (n+1) \cdot n! = (n+1)!,$$

hence eq. (A.5) also holds for $n+1$. This completes the proof. ∎

Next we treat a tightly related problem. Say we have n different objects and we want to distribute them into two disjoint groups, one having k elements, the other $n-k$. Hereby it is of no interest in which order the elements are distributed, only the composition of the two sets matters.

Example A.3.3. There are 52 cards in a deck that are distributed to two players, so that each of them gets 26 cards. For this game it is only important which cards each player has, not in which order the cards were received. Here $n = 52$ and $k = n - k = 26$.

The main question is: how many ways can n elements be distributed, say the numbers from 1 to n, into one group of k elements and into another of $n - k$ elements? In the above example, that is how many ways can 52 cards be distributed into two groups of 26.

To answer this question we use the following auxiliary model. Let us take any permutation $\pi \in S_n$. We place the numbers $\pi(1), \ldots, \pi(k)$ into group 1 and the remaining $\pi(k + 1), \ldots, \pi(n)$ into group 2. In this way we obtain all possible distributions but many of them appear several times. Say two permutations π_1 and π_2 are equivalent if (as sets)

$$\{\pi_1(1), \ldots, \pi_1(k)\} = \{\pi_2(1), \ldots, \pi_2(k)\}.$$

Of course, this also implies

$$\{\pi_1(k + 1), \ldots, \pi_1(n)\} = \{\pi_2(k + 1), \ldots, \pi_2(n)\},$$

and two permutations generate the same partition if and only if they are equivalent. Equivalent permutations are achieved by taking one fixed permutation π, then permuting $\{\pi(1), \ldots, \pi(k)\}$ and also $\{\pi(k + 1), \ldots, \pi(n)\}$. Consequently, there are exactly $k!(n - k)!$ permutations that are equivalent to a given one. Summing up, we get that there are $\frac{n!}{k!(n-k)!}$ different classes of equivalent permutations. Setting

$$\binom{n}{k} = \frac{n!}{k! \, (n - k)!}$$

we see the following.

There are $\binom{n}{k}$ different ways to distribute n objects into one group of k and into another one of $n - k$ elements. !

The numbers $\binom{n}{k}$ are called **binomial coefficients,** read "n chosen k." We let $\binom{n}{k} = 0$ in case of $k > n$ or $k < 0$.

Example A.3.4. A digital word of length n consists of n zeroes or ones. Since at every position we may have either 0 or 1, there are 2^n different words of length n. How many of these words possess exactly k ones or, equivalently, $n - k$ zeroes? To answer this put all positions where there is a "1" into a first group and those where there is a "0" into a second one. In this way the numbers from 1 to n are divided into two different groups of size k and $n - k$, respectively. But we already know how many such partitions exist, namely $\binom{n}{k}$. As a consequence we get

---

! There are $\binom{n}{k}$ words of length n possessing exactly k ones and $n - k$ zeroes.

The next proposition summarizes some crucial properties of binomial coefficients.

Proposition A.3.5. *Let n be a natural number, $k = 0, \ldots, n$ and let $r \geq 0$ be an integer. Then the following equations hold:*

$$\binom{n}{k} = \binom{n}{n - k} \tag{A.6}$$

$$\binom{n}{k} = \binom{n-1}{k} + \binom{n-1}{k-1} \quad \text{and} \tag{A.7}$$

$$\binom{n+r}{n} = \sum_{j=0}^{n} \binom{n+r-j-1}{n-j} = \sum_{j=0}^{n} \binom{r+j-1}{j}. \tag{A.8}$$

Proof: Equations (A.6) and (A.7) follow immediately by the definition of the binomial coefficients. Note that eq. (A.7) also holds if $k = n$ because we agreed that $\binom{n-1}{n} = 0$.
An iteration of eq. (A.7) leads to

$$\binom{n}{k} = \sum_{j=0}^{k} \binom{n-j-1}{k-j}.$$

Replacing in the last equation n by $n + r$ as well as k by n we obtain the left-hand identity (A.8). The right-hand equation follows by inverting the summation, that is, one replaces j by $n - j$. ∎

Remark A.3.6. Equation (A.7) allows a graphical interpretation by **Pascal's triangle**. The coefficient $\binom{n}{k}$ in the nth row follows by summing the two values $\binom{n-1}{k-1}$ and $\binom{n-1}{k}$ above $\binom{n}{k}$ in the $(n - 1)$th row.

$$
\begin{array}{ccccccccccc}
 & & & & & 1 & & & & & \\
 & & & & 1 & & 1 & & & & \\
 & & & 1 & & 2 & & 1 & & & \\
 & & 1 & & 3 & & 3 & & 1 & & \\
 & \cdot & & \cdot & & \cdot & & \cdot & & \cdot & \\
 & 1 & \cdot & \cdot & \binom{n-1}{k-1} & & \binom{n-1}{k} & \cdot & \cdot & 1 & \\
1 & \binom{n}{1} & \cdot & \cdot & & \binom{n}{k} & & \cdot & \cdot & \binom{n}{n-1} & 1
\end{array}
$$

Next we state and prove the important binomial theorem.

Proposition A.3.7 (Binomial theorem). *For real numbers a and b and any $n \in \mathbb{N}_0$,*

$$(a + b)^n = \sum_{k=0}^{n} \binom{n}{k} a^k b^{n-k} . \tag{A.9}$$

Proof: The binomial theorem is proved by induction over n. If $n = 0$, then eq. (A.9) holds trivially.

Suppose now that eq. (A.9) has been proven for $n - 1$. Our aim is to verify that it is also true for n. Using that the expansion holds for $n - 1$ we get

$$(a + b)^n = (a + b)^{n-1}(a + b)$$

$$= \sum_{k=0}^{n-1} \binom{n-1}{k} a^{k+1} b^{n-1-k} + \sum_{k=0}^{n-1} \binom{n-1}{k} a^k b^{n-k}$$

$$= a^n + \sum_{k=0}^{n-2} \binom{n-1}{k} a^{k+1} b^{n-1-k} + b^n + \sum_{k=1}^{n-1} \binom{n-1}{k} a^k b^{n-k}$$

$$= a^n + b^n + \sum_{k=1}^{n-1} \left[\binom{n-1}{k-1} + \binom{n-1}{k} \right] a^k b^{n-k}$$

$$= \sum_{k=0}^{n} \binom{n}{k} a^k b^{n-k} ,$$

where we used eq. (A.7) in the last step . ∎

The following property of binomial coefficients plays an important role when introducing the hypergeometric distribution (compare Proposition 1.4.24). It is also used during the investigation of sums of independent binomial distributed random variables (compare Proposition 4.6.1).

Proposition A.3.8 (Vandermonde's identity). *If k, m, and n in $\mathbb{N}_0$, then*

$$\sum_{j=0}^{k} \binom{n}{j} \binom{m}{k-j} = \binom{n+m}{k} . \tag{A.10}$$

Proof: An application of the binomial theorem leads to

$$(1 + x)^{n+m} = \sum_{k=0}^{n+m} \binom{n+m}{k} x^k , \quad x \in \mathbb{R} . \tag{A.11}$$

On the other hand, another use of Proposition A.3.7 implies[3]

$$(1+x)^{n+m} = (1+x)^n (1+x)^m$$

$$= \left[\sum_{j=0}^{n} \binom{n}{j} x^j \right] \left[\sum_{i=0}^{m} \binom{m}{i} x^i \right] = \sum_{j=0}^{n} \sum_{i=0}^{m} \binom{n}{j} \binom{m}{i} x^{i+j}$$

$$= \sum_{k=0}^{n+m} \left[\sum_{i+j=k} \binom{n}{j} \binom{m}{i} \right] x^k = \sum_{k=0}^{n+m} \left[\sum_{j=0}^{k} \binom{n}{j} \binom{m}{k-j} \right] x^k . \qquad (A.12)$$

The coefficients in an expansion of a polynomial are unique. Hence, in view of eqs. (A.11) and (A.12), we get for all $k \le m + n$ the identity

$$\binom{n+m}{k} = \sum_{j=0}^{k} \binom{n}{j} \binom{m}{k-j} .$$

Hereby note that both sides of eq. (A.10) become zero whenever $k > n + m$. This completes the proof. ∎

Our next objective is to generalize the binomial coefficients. In view of

$$\binom{n}{k} = \frac{n(n-1) \cdots (n-k+1)}{k!}$$

for $k \ge 1$ and $n \in \mathbb{N}$ the **generalized binomial coefficient** is introduced as

$$\binom{-n}{k} := \frac{-n(-n-1) \cdots (-n-k+1)}{k!} . \qquad (A.13)$$

The next lemma shows the tight relation between generalized and "ordinary" binomial coefficients.

Lemma A.3.9. *For $k \ge 1$ and $n \in \mathbb{N}$,*

$$\binom{-n}{k} = (-1)^k \binom{n+k-1}{k} .$$

---

3 When passing from line 2 to line 3 the order of summation is changed. One no longer sums over the rectangle $[0, m] \times [0, n]$. Instead one sums along the diagonals, where $i + j = k$.

Proof: By definition of the generalized binomial coefficient we obtain

$$\binom{-n}{k} = \frac{(-n)(-n-1)\cdots(-n-k+1)}{k!}$$

$$= (-1)^k \frac{(n+k-1)(n+k-2)\cdots(n+1)n}{k!} = (-1)^k \binom{n+k-1}{k}.$$

This completes the proof. ∎

For example, Lemma A.3.9 implies $\binom{-1}{k} = (-1)^k$ and $\binom{-n}{1} = -n$.

A.3.2 Drawing Balls out of an Urn

Assume that there are n balls labeled from 1 to n in an urn. We draw k balls out of the urn, thus observing a sequence of length k with entries from $\{1, \ldots, n\}$. How many different results (sequences) may be observed? To answer this question we have to decide the arrangement of drawing. Do we or do we not replace the chosen ball? Is it important in which order the balls were chosen or is it only of importance which balls were chosen at all? Thus, we see that there are four different ways to answer this question (replacement or nonreplacement, recording the order or nonrecording).

Example A.3.10. Let us regard the drawing of two balls out of four, that is, $n = 4$ and $k = 2$. Depending on the different arrangements the following results may be observed. Note, for example, that in the two latter cases $(3, 2)$ does not appear because it is identical to $(2, 3)$.

Replacement and order is important

(1, 1)	(1, 2)	(1, 3)	(1, 4)
(2, 1)	(2, 2)	(2, 3)	(2, 4)
(3, 1)	(3, 2)	(3, 3)	(3, 4)
(4, 1)	(4, 2)	(4, 3)	(4, 4)

16 different results

Nonreplacement and order is important

·	(1, 2)	(1, 3)	(1, 4)
(2, 1)	·	(2, 3)	(2, 4)
(3, 1)	(3, 2)	·	(3, 4)
(4, 1)	(4, 2)	(4, 3)	·

12 different results

Replacement and order is not important

(1, 1)	(1, 2)	(1, 3)	(1, 4)
·	(2, 2)	(2, 3)	(2, 4)
·	·	(3, 3)	(3, 4)
·	·	·	(4, 4)

10 different results

Nonreplacement and order is not important

·	(1, 2)	(1, 3)	(1, 4)
·	·	(2, 3)	(2, 4)
·	·	·	(3, 4)
·	·	·	·

6 different results

Let us come back now to the general situation of n different balls from which we choose k at random.

Case 1 : **Drawing with replacement and taking the order into account.**

We have n different possibilities for the choice of the first ball and since the chosen ball is placed back there are also n possibilities for the second one and so on. Thus, there are n possibilities for each of the k balls, leading to the following result.

!   The number of different results in this case is n^k

Example A.3.11. Letters in Braille, a scripture for blind people, are generated by dots or nondots at six different positions. How many letters may be generated in that way?

Answer: It holds that $n = 2$ (dot or no dot) at $k = 6$ different positions. Hence, the number of possible representable letters is $2^6 = 64$. In fact, there are only 63 possibilities because we have to rule out the case of no dots at all 6 positions.

Case 2 : **Drawing without replacement and taking the order into account.**

This case only makes sense if $k \leq n$. There are n possibilities to choose the first ball. After that there are still $n - 1$ balls in the urn. Hence there are only $n - 1$ possibilities for the second choice, $n - 2$ for the third, and so on. Summing up we get the following.

!   The number of possible results in this case equals

$$n(n - 1) \cdots (n - k + 1) = \frac{n!}{(n - k)!}$$

Example A.3.12. In a lottery 6 numbers are chosen out of 49. Of course, the chosen numbers are not replaced. If we record the numbers as they appear (not putting them in order) how many different sequences of six numbers exist?

Answer: Here we have $n = 49$ and $k = 6$. Hence the wanted number equals

$$\frac{49!}{43!} = 49 \cdots 44 = 10,068,347,520$$

Case 3 : **Drawing with replacement not taking the order into account.**

This case is more complicated and requires a different point of view. We count how often each of the n balls was chosen during the k trials. Let $k_1 \geq 0$ be the frequency of the first ball, $k_2 \geq 0$ that of the second one, and so on. In this way we obtain n non-negative integers $k_1, \ldots, k_n$ satisfying

$$k_1 + \cdots + k_n = k.$$

Indeed, since we choose k balls, the frequencies have to sum to k. Consequently, the number of possible results when drawing k of n balls with replacement and not taking the order into account coincides with

$$\#\{(k_1, \ldots, k_n), k_j \in \mathbb{N}_0, k_1 + \cdots + k_n = k\}. \tag{A.14}$$

In order to determine the cardinality (A.14) we use the following auxiliary model:

Let $B_1, \ldots, B_n$ be n boxes. Given n nonnegative integers $k_1, \ldots, k_n$, summing to k, we place exactly k_1 dots into B_1, k_2 dots into B_2, and so on. At the end we distributed k nondistinguishable dots into n different boxes. Thus, we see that the value of (A.14) coincides with the number of different possibilities to distribute k nondistinguishable dots into n boxes. Now assume that the boxes are glued together; on the very left we put box B_1, on its right we put box B_2 and continue in this way up to box B_n on the very right. In this way we obtain $n + 1$ dividing walls, two outer and $n - 1$ inner ones. Now we get all possible distributions of k dots into n boxes by shuffling the k dots and the $n - 1$ inner dividing walls. For example, if we get the order $w, w, d, d, w \ldots$, then this means that there are no dots in B_1 and B_2, but there are two dots in B_3.

Summing up, we have $N = n + k - 1$ objects, k of them are dots and $n - 1$ are walls. As we know there are $\binom{N}{k}$ different ways to order these N objects. Hence we arrived at the following result.

The number of possibilities to distribute k anonymous dots into n boxes equals

$$\binom{n + k - 1}{k} = \binom{n + k - 1}{n - 1}.$$

It coincides with $\#\{(k_1, \ldots, k_n), k_j \in \mathbb{N}_0, k_1 + \cdots + k_n = k\}$ as well as with the number of different results when choosing k balls out of n **with replacement and not taking order into account**.

Example A.3.13. Dominoes are marked on each half either with no dots, one dot or up to six dots. Hereby the dominoes are symmetric, that is, a tile with three dots on the left-hand side and two ones on the right-hand one is identical with one having two dots on the left-hand side and three dots on the right-hand one. How many different dominoes exist?

Answer: It holds $n = 7$ and $k = 2$, hence the number of different dominoes equals

$$\binom{7 + 2 - 1}{2} = \binom{8}{2} = 28.$$

Case 4 : **Drawing without replacement not taking the order into account.**

Here we also have to assume $k \leq n$. We already investigated this case when we introduced the binomial coefficients. The k chosen numbers are put in group 1, the remaining $n - k$ balls in group 2. As we know there are $\binom{n}{k}$ ways to split the n numbers into such two groups. Hence we obtained the following.

!   **The number of different results in this case is $\binom{n}{k}$**

Example A.3.14. If the order of the six numbers is not taken into account in Example A.3.12, that is, we ignore which number was chosen first, which second, and so on the number of possible results equals

$$\binom{49}{6} = \frac{49 \cdots 43}{6!} = 13{,}983{,}816$$

Let us summarize the four different cases in a table. Here **O** and **NO** stand for recording or nonrecording of the order while **R** and **NR** represent replacement or nonreplacement.

	R	**NR**
O	n^k	$\frac{n!}{(n-k)!}$
NO	$\binom{n+k-1}{k}$	$\binom{n}{k}$

A.3.3 Multinomial Coefficients

The binomial coefficient $\binom{n}{k}$ describes the number of possibilities to distribute n objects into two groups of k and $n - k$ elements. What happens if we have not only two groups but $m \geq 2$? Say the first group has k_1 elements, the second has k_2 elements, and so on, up to the mth group that has k_m elements. Of course, if we distribute n elements the k_j have to satisfy

$$k_1 + \cdots + k_m = n.$$

Using exactly the same arguments as in the case where $m = 2$ we get the following.

!   There exists exactly $\frac{n!}{k_1! \cdots k_m!}$ different ways to distribute n elements into m groups of sizes $k_1, k_2, \ldots, k_m$ where $k_1 + \cdots + k_m = n$.

In accordance with the binomial coefficient we write

$$\binom{n}{k_1, \ldots, k_m} := \frac{n!}{k_1! \cdots k_m!}, \qquad k_1 + \cdots + k_m = n, \tag{A.15}$$

and call $\binom{n}{k_1, \ldots, k_m}$ a **multinomial coefficient**, read "n chosen k_1 up to k_m."

Remark A.3.15. If $m = 2$, then $k_1 + k_2 = n$, and

$$\binom{n}{k_1, k_2} = \binom{n}{k_1, n - k_1} = \binom{n}{k_1} = \binom{n}{k_2}.$$

Example A.3.16. A deck of cards for playing skat consists of 32 cards. Three players each gets 10 cards; the remaining two cards (called "skat") are placed on the table. How many different distributions of the cards exist?

Answer: Let us first define what it means for two distribution of cards to be identical. Say, this happens if each of the three players has exactly the same cards as in the previous game. Therefore, the remaining two cards on the table are also identical. Hence we distribute 32 cards into 4 groups possessing 10, 10, 10, and 2 elements. Consequently, the number of different distributions equals [4]

$$\binom{32}{10, 10, 10, 2} = \frac{32!}{(10!)^3 \, 2!} = 2.753294409 \times 10^{15}.$$

Remark A.3.17. One may also look at multinomial coefficients from a different point of view. Suppose we are given n balls of m different colors. Say there are k_1 balls of color 1, k_2 balls of color 2, up to k_m balls of color m where, of course, $k_1 + \cdots + k_m = n$. Then there exist

$$\binom{n}{k_1, \ldots, k_m}$$

different ways to order these n balls. This is followed by the same arguments as we used in Example A.3.4 for $m = 2$.

For instance, given 3 blue, 4 red and 2 white balls, then there are

$$\binom{9}{3, 4, 2} = \frac{9!}{3! \, 4! \, 2!} = 1260$$

different ways to order them.

Finally, let us still mention that in the literature one sometimes finds another (equivalent) way for the introduction of the multinomial coefficients. Given nonnegative integers $k_1, \ldots, k_m$ with $k_1 + \cdots + k_m = n$, it follows that

$$\binom{n}{k_1, \ldots, k_m} = \binom{n}{k_1}\binom{n - k_1}{k_2}\binom{n - k_1 - k_2}{k_3}\cdots\binom{n - k_1 - \cdots - k_{m-1}}{k_m}. \tag{A.16}$$

A direct proof of this fact is easy and left as an exercise.

---

4 The huge size of this number explains why playing skat never becomes boring.

There is a combinatorial interpretation of the expression on the right-hand side of eq. (A.16). To reorder n balls of m different colors, one chooses first the k_1 positions for balls of color 1. There are $\binom{n}{k_1}$ ways to do this. Thus, there remain $n - k_1$ possible positions for balls of color 2, and there are $\binom{n-k_1}{k_2}$ possible choices for this, and so on. Note that at the end there remain k_m positions for k_m balls; hence, the last term on the right-hand side of eq. (A.16) equals 1.

Let us come now to the announced generalization of Proposition A.3.7.

Proposition A.3.18 (Multinomial theorem). *Let $n \geq 0$. Then for any $m \geq 1$ and real numbers $x_1, \ldots, x_m$,*

$$(x_1 + \cdots + x_m)^n = \sum_{\substack{k_1 + \cdots + k_m = n \\ k_i \geq 0}} \binom{n}{k_1, \ldots, k_m} x_1^{k_1} \cdots x_m^{k_m} . \tag{A.17}$$

Proof: Equality (A.17) is proved by induction. In contrast to the proof of the binomial theorem, now induction is done over m, the number of summands.

If $m = 1$ the assertion is valid by trivial reasons.

Suppose now eq. (A.17) holds for m, all $n \geq 1$ and all real numbers $x_1, \ldots, x_m$. We have to show the validity of eq. (A.17) for $m + 1$ and all $n \geq 1$. Given real numbers $x_1, \ldots, x_{m+1}$ and $n \geq 1$ set $y := x_1 + \cdots + x_m$. Using A.3.7, by the validity of eq. (A.17) for m and all $n - j$, $0 \leq j \leq n$, we obtain

$$(x_1 + \cdots + x_{m+1})^n = (y + x_{m+1})^n = \sum_{j=1}^{n} \frac{n!}{j!\,(n-j)!} x_{m+1}^{j} y^{n-j}$$

$$= \sum_{j=1}^{n} \frac{n!}{j!\,(n-j)!} \sum_{\substack{k_1 + \cdots + k_m = n-j \\ k_i \geq 0}} \frac{(n-j)!}{k_1! \cdots k_m!} x_1^{k_1} \cdots x_m^{k_m} x_{m+1}^{j} .$$

Replacing j by k_{m+1} and combining both sums leads to

$$(x_1 + \cdots + x_{m+1})^n = \sum_{\substack{k_1 + \cdots + k_{m+1} = n \\ k_i \geq 0}} \frac{n!}{k_1! \cdots k_{m+1}!} x_1^{k_1} \cdots x_{m+1}^{k_{m+1}} ,$$

hence eq. (A.17) is also valid for $m + 1$. This completes the proof. ∎

Remark A.3.19. The number of summands in eq. (A.17) equals[5] $\binom{n+m-1}{n}$.

---

[5] Compare case 3 in Section A.3.2.

A.4 Vectors and Matrices

The aim of this section is to summarize results and notations about vectors and matrices used throughout this book. For more detailed reading we refer to any book about Linear Algebra, for example, [Axl15].

Given two vectors x and y in $\mathbb{R}^n$, their[6] **scalar product** is defined as

$$\langle x, y \rangle := \sum_{j=1}^{n} x_j y_j, \quad x = (x_1, \ldots, x_n), \ y = (y_1, \ldots, y_n).$$

If $x \in \mathbb{R}^n$, then

$$|x| := \langle x, x \rangle^{1/2} = \left(\sum_{j=1}^{n} x_j^2 \right)^{1/2}$$

denotes the **Euclidean distance** of x to 0. Thus, $|x|$ may also be regarded as the **length** of the vector x. In particular, we have $|x| > 0$ for all nonzero $x \in \mathbb{R}^n$.

Any matrix $A = (a_{ij})_{i,j=1}^{n}$ of real numbers a_{ij} generates a **linear**[7] **mapping** (also denoted by A) via

$$Ax = \left(\sum_{j=1}^{n} a_{1j} x_j, \ldots, \sum_{j=1}^{n} a_{nj} x_j \right), \quad x = (x_1, \ldots, x_n) \in \mathbb{R}^n. \tag{A.18}$$

Conversely, any linear mapping $A : \mathbb{R}^n \to \mathbb{R}^n$ defines a matrix $(a_{ij})_{i,j=1}^{n}$ by representing $Ae_j \in \mathbb{R}^n$ as

$$Ae_j = (a_{1j}, \ldots, a_{nj}), \quad j = 1, \ldots, n.$$

Here $e_j = (0, \ldots, 0, \underbrace{1}_{j}, 0 \ldots, 0)$ denotes the jth unit vector in $\mathbb{R}^n$. With this generated matrix $(a_{ij})_{i,j=1}^{n}$ the linear mapping A acts as stated in eq. (A.18). Consequently, we may always identify linear mappings in $\mathbb{R}^n$ with $n \times n$-matrices $(a_{ij})_{i,j=1}^{n}$.

A matrix A is said to be **regular**[8] if the generated linear mapping is one-to-one, that is, if $Ax = 0$ implies $x = 0$. This is equivalent to the fact that the determinant $\det(A)$ is nonzero.

---

6 Sometimes also called "dot-product".
7 A mapping $A : \mathbb{R}^n \to \mathbb{R}^n$ is said to be linear if $A(\alpha x + \beta y) = \alpha Ax + \beta Ay$ for all $\alpha, \beta \in \mathbb{R}$ and $x, y \in \mathbb{R}^n$.
8 Sometimes also called **nonsingular** or **invertible**.

Let $A = (\alpha_{ij})_{i,j=1}^{n}$ be an $n \times n$ matrix. Then its **transposed** matrix is defined as $A^T := (\alpha_{ji})_{i,j=1}^{n}$. With this notation it follows for $x, y \in \mathbb{R}^n$ that

$$\langle Ax, y \rangle = \left\langle x, A^T y \right\rangle .$$

A matrix A with $A = A^T$ is said to be **symmetric**. In other words, A satisfies

$$\langle Ax, y \rangle = \langle x, Ay \rangle , \quad x, y \in \mathbb{R}^n .$$

An $n \times n$ matrix $R = (r_{ij})_{i,j=1}^{n}$ is **positive definite** (or shorter, **positive**) provided it is symmetric and that

$$\langle Rx, x \rangle = \sum_{i,j=1}^{n} r_{ij} x_i x_j > 0, \quad x = (x_1, \ldots, x_n) \neq 0 .$$

We will write $R > 0$ in this case. In particular, each positive matrix R is regular and its determinant satisfies $\det(R) > 0$.

Let $A = (\alpha_{ij})_{i,j=1}^{n}$ be an arbitrary regular $n \times n$ matrix. Set

$$R := AA^T , \tag{A.19}$$

that is, the entries r_{ij} of R are computed by

$$r_{ij} = \sum_{k=1}^{n} \alpha_{ik} \alpha_{jk} , \quad 1 \leq i, j \leq n .$$

Proposition A.4.1. *Suppose the matrix R is defined by eq. (A.19) for some regular A. Then it follows that $R > 0$.*

Proof: Because of

$$R^T = \left(AA^T \right)^T = \left(A^T \right)^T A^T = AA^T = R ,$$

the matrix R is symmetric. Furthermore, for $x \in \mathbb{R}^n$ with $x \neq 0$ we obtain

$$\langle Rx, x \rangle = \left\langle AA^T x, x \right\rangle = \left\langle A^T x, A^T x \right\rangle = |A^T x|^2 > 0 .$$

Hereby we used that for a regular A the transposed matrix A^T is regular too. Consequently, if $x \neq 0$, then $A^T x \neq 0$, thus $|A^T x| > 0$. This completes the proof. ∎

The **identity matrix** I_n is defined as $n \times n$ matrix with entries δ_{ij}, $1 \le i, j \le n$, where

$$\delta_{ij} = \begin{cases} 1 & : & i = j \\ 0 & : & i \ne j \end{cases} \qquad (A.20)$$

Of course, $I_n x = x$ for $x \in \mathbb{R}^n$.

Given a regular $n \times n$ matrix A, there is unique matrix B such that $AB = I_n$. B is called the **inverse matrix** of A and denoted by A^{-1}. Recall that also $A^{-1}A = I_n$ and, moreover, $(A^T)^{-1} = (A^{-1})^T$.

An $n \times n$ matrix U is said to be **unitary** or **orthogonal** provided that

$$UU^T = U^TU = I_n$$

with identity matrix I_n. Another way to express this is either that $U^T = U^{-1}$ or, equivalently, that U satisfies

$$\langle Ux, Uy \rangle = \langle x, y \rangle, \quad x, y \in \mathbb{R}^n.$$

In particular, for each $x \in \mathbb{R}^n$ it follows

$$|Ux|^2 = \langle Ux, Ux \rangle = \langle x, x \rangle = |x|^2,$$

that is, U preserves the length of vectors in $\mathbb{R}^n$.

It is easy to see that an $n \times n$ matrix U is unitary if and only if its column vectors $u_1, \ldots, u_n$ form an orthonormal basis in $\mathbb{R}^n$. That is, $\langle u_i, u_j \rangle = \delta_{ij}$ with δ_{ij}s as in (A.20). This characterization of unitary matrices remains valid when we take the column vectors instead of those generated by the rows.

We saw in Proposition A.4.1 that each matrix R of the form (A.19) is positive. Next we prove that conversely, each $R > 0$ may be represented in this way.

Proposition A.4.2. *Let R be an arbitrary positive $n \times n$ matrix. Then there exists a regular matrix A such that $R = AA^T$.*

Proof: Since R is symmetric, we may apply the principal axis transformation for symmetric matrices. It asserts that there exists a diagonal matrix[9] D and a unitary matrix U such that

$$R = UDU^T.$$

Let $\delta_1, \ldots, \delta_n$ be the entries of D at its diagonal. From $R > 0$ we derive $\delta_j > 0$, $1 \le j \le n$. To see this fix $j \le n$ and set $x := Ue_j$ where as above e_j denotes the jth unit vector in $\mathbb{R}^n$.

---

9 The entries d_{ij} of D satisfy $d_{ij} = 0$ if $i \ne j$.

Then $U^T x = e_j$, hence

$$0 < \langle Rx, x \rangle = \langle UDU^T x, x \rangle = \langle DU^T x, U^T x \rangle = \langle De_j, e_j \rangle = \delta_j .$$

Because of $\delta_j > 0$ we may define $D^{1/2}$ as diagonal matrix with entries $\delta_j^{1/2}$ on its diagonal. Setting $A := UD^{1/2}$, because of $(D^{1/2})^T = D^{1/2}$ it follows that

$$R = (UD^{1/2})(UD^{1/2})^T = AA^T .$$

Since $|\det(A)|^2 = \det(A)\det(A^T) = \det(R) > 0$, the matrix A is regular, and this completes the proof. ∎

Remark A.4.3. Note that representation (A.19) is **not** unique. Indeed, whenever $R = A A^T$, then we also have $R = (AV)(AV)^T$ for any unitary matrix V.

A.5 Some Analytic Tools

The aim of this section is to present some special results of Calculus that play an important role in the book. Hereby we restrict ourselves to those topics that are maybe less known and that are not necessarily taught in a basic Calculus course. For a general introduction to Calculus, including those topics as convergence of power series, fundamental theorem of Calculus, mean-value theorem, and so on, we refer to the books [Spi08] and [Ste15].

We start with a result that is used in the proof of Poisson's limit theorem 1.4.22. From Calculus it is well known that for $x \in \mathbb{R}$

$$\lim_{n \to \infty} \left(1 + \frac{x}{n}\right)^n = e^x . \tag{A.21}$$

The probably easiest proof of this fact is via the approach presented in [Spi08]. There the logarithm function is defined by $\ln t = \int_1^t \frac{1}{s} ds$, $t > 0$. Hence, l'Hôpital's rule implies

$$\lim_{t \to \infty} t \ln \left(1 + \frac{x}{t}\right) = x , \quad x \in \mathbb{R} .$$

From this eq. (A.21) easily follows by the continuity of the exponential function.

The next proposition may be viewed as a slight generalization of eq. (A.21).

Proposition A.5.1. Let $(x_n)_{n \geq 1}$ be a sequence of real numbers with $\lim_{n \to \infty} x_n = x$ for some $x \in \mathbb{R}$. Then we get

$$\lim_{n \to \infty} \left(1 + \frac{x_n}{n}\right)^n = e^x .$$

Proof: Because of eq. (A.21) it suffices to verify that

$$\lim_{n \to \infty} \left| \left(1 + \frac{x_n}{n}\right)^n - \left(1 + \frac{x}{n}\right)^n \right| = 0 . \tag{A.22}$$

Since the sequence $(x_n)_{n\geq 1}$ is converging, it is bounded. Consequently, there is a $c > 0$ such that for all $n \geq 1$, we have $|x_n| \leq c$. Of course, we may also assume $|x| \leq c$. Fix for a moment $n \geq 1$ and set

$$a := 1 + \frac{x_n}{n} \quad \text{and} \quad b := 1 + \frac{x}{n}.$$

The choice of $c > 0$ yields $|a| \leq 1 + c/n$ as well as $|b| \leq 1 + c/n$. Hence it follows

$$|a^n - b^n| = |a - b|\, |a^{n-1} + a^{n-2}b + \cdots + ab^{n-2} + b^{n-1}|$$

$$\leq |a - b|\, \left(|a|^{n-1} + |a|^{n-2}|b| + \cdots + |a||b|^{n-2} + |b|^{n-1}\right)$$

$$\leq |a - b|\, n\, \left(1 + \frac{c}{n}\right)^{n-1} \leq C\, n\, |a - b|.$$

Here $C > 0$ is some constant that exists since $(1 + c/n)^{n-1}$ converges to e^c. By the definition of a and b,

$$\left|\left(1 + \frac{x_n}{n}\right)^n - \left(1 + \frac{x}{n}\right)^n\right| \leq C\, n \frac{|x_n - x|}{n} = C\,|x - x_n|.$$

Since $x_n \to x$, this immediately implies eq. (A.22) and proves the proposition. ∎

Our next objective is to present some properties of power series and of the functions generated by them. Hereby we restrict ourselves to such assertions that we will use in this book. For further reading we refer to Part IV in [Spi08].

Let $(a_k)_{k\geq 0}$ be a sequence of real numbers. Then its **radius of convergence** $r \in [0, \infty]$ is defined by

$$r := \frac{1}{\displaystyle\limsup_{k \to \infty} |a_k|^{1/k}} .$$

Hereby we let $1/0 := \infty$ and $1/\infty := 0$. If $0 < r \leq \infty$ and $|x| < r$, then the infinite series

$$f(x) := \sum_{k=0}^{\infty} a_k\, x^k \tag{A.23}$$

converges (even absolutely). Hence the function f generated by eq. (A.23) is well-defined on its **region of convergence** $\{x \in \mathbb{R} : |x| < r\}$. We say that f is represented as a **power series** on $\{x \in \mathbb{R} : |x| < r\}$.

The function f defined by eq. (A.23) is infinitely often differentiable on its region of convergence and

$$f^{(n)}(x) = \sum_{k=n}^{\infty} k(k-1)\cdots(k-n+1)\,a_k\,x^{k-n}$$

$$= \sum_{k=0}^{\infty}(k+n)(k+n-1)\cdots(k+1)\,a_{k+n}\,x^k = n!\sum_{k=0}^{\infty}\binom{n+k}{k}a_{k+n}\,x^k. \qquad (A.24)$$

(compare [Spi08], §27, Thm. 6)

The coefficients $n!\binom{n+k}{k}a_{n+k}$ in the series representation of the nth derivative $f^{(n)}$ possess the same radius of convergence as the original sequence $(a_k)_{k\geq0}$. This is easy to see for $n = 1$. The general case then follows by induction.

Furthermore, eq. (A.24) implies $a_n = f^{(n)}(0)/n!$, which, in particular, tells us that given f, the coefficients $(a_k)_{k\geq0}$ in representation (A.23) are unique.

Proposition A.5.2. *If $n \geq 1$ and $|x| < 1$ then it follows*

$$\frac{1}{(1+x)^n} = \sum_{k=0}^{\infty}\binom{-n}{k}x^k. \qquad (A.25)$$

Proof: Using the formula to add a geometric series and applying $\binom{-1}{k} = (-1)^k$ yields for $|x| < 1$ that

$$\frac{1}{1+x} = \sum_{k=0}^{\infty}(-1)^k\,x^k = \sum_{k=0}^{\infty}\binom{-1}{k}x^k.$$

Consequently Proposition A.5.2 holds for $n = 1$.

Assume now we have proven the proposition for $n - 1$, that is, if $|x| < 1$, then

$$\frac{1}{(1+x)^{n-1}} = \sum_{k=0}^{\infty}\binom{-n+1}{k}x^k.$$

Differentiating this equality on the region $\{x : |x| < 1\}$ implies

$$-\frac{n-1}{(1+x)^n} = \sum_{k=1}^{\infty}\binom{-n+1}{k}k\,x^{k-1} = \sum_{k=0}^{\infty}\binom{-n+1}{k+1}(k+1)\,x^k. \qquad (A.26)$$

Direct calculations give

$$-\frac{k+1}{n-1}\binom{-n+1}{k+1} = -\frac{k+1}{n-1}\cdot\frac{(-n+1)(-n)\cdots(-n+1-(k+1)+1)}{(k+1)!}$$

$$= \frac{(-n)(-n-1)\cdots(-n-k+1)}{k!} = \binom{-n}{k},$$

which together with eq. (A.26) leads to

$$\frac{1}{(1+x)^n} = \sum_{k=0}^{\infty} \binom{-n}{k} x^k .$$

This completes the proof of Proposition A.5.2. ∎

The next proposition may be viewed as a counterpart to eq. (A.10) in the case of generalized binomial coefficients.

Proposition A.5.3. *For $k \geq 0$ and $m, n \in \mathbb{N}$,*

$$\sum_{j=0}^{k} \binom{-n}{j} \binom{-m}{k-j} = \binom{-n-m}{k} .$$

Proof: The proof is similar to that of Proposition A.3.8. Using Proposition A.5.2 we represent the function $(1 + x)^{-n-m}$ as power series in two different ways. On the one hand for $|x| < 1$ we have the representation

$$\frac{1}{(1+x)^{n+m}} = \sum_{k=0}^{\infty} \binom{-n-m}{k} x^k \tag{A.27}$$

and on the other hand

$$\frac{1}{(1+x)^{n+m}} = \left[\sum_{j=0}^{\infty} \binom{-n}{j} x^j \right] \left[\sum_{l=0}^{\infty} \binom{-m}{l} x^l \right]$$

$$= \sum_{k=0}^{\infty} \left[\sum_{j+l=k} \binom{-n}{j} \binom{-m}{l} \right] x^k = \sum_{k=0}^{\infty} \left[\sum_{j=0}^{k} \binom{-n}{j} \binom{-m}{k-j} \right] x^k . \tag{A.28}$$

As observed above the coefficients in a power series are uniquely determined. Thus, the coefficients in eqs. (A.27) and (A.28) have to coincide, which implies

$$\sum_{j=0}^{k} \binom{-n}{j} \binom{-m}{k-j} = \binom{-n-m}{k}$$

as asserted. ∎

Let $f : \mathbb{R}^n \to \mathbb{R}$ be a function. How does one define the integral $\int_{\mathbb{R}^n} f(x) \, dx$? To simplify the notation let us restrict ourselves to the case $n = 2$. The main problems already become clear in this case and the obtained results easily extend to higher dimensions.

The easiest way to introduce the integral of a function of two variables is as follows:

$$\int_{\mathbb{R}^2} f(x)\,dx := \int_{-\infty}^{\infty}\left[\int_{-\infty}^{\infty} f(x_1, x_2)\,dx_2\right] dx_1 .$$

In order for this double integral to be well-defined we have to assume the existence of the inner integral for each fixed $x_1 \in \mathbb{R}$ and then the existence of the integral of the function

$$x_1 \mapsto \int_{-\infty}^{\infty} f(x_1, x_2)\,dx_2 .$$

Doing so the following question arises immediately: why do we not define the integral in reversed order, that is, first integrating via x_1 and then with respect to x_2?

To see the difficulties that may appear let us consider the following example.

Example A.5.4. The function $f : \mathbb{R}^2 \to \mathbb{R}$ is defined as follows (Fig. A.1): If either $x_1 < 0$ or $x_2 < 0$ set $f(x_1, x_2) = 0$. If $x_1, x_2 \geq 0$ define f by

$$f(x_1, x_2) := \begin{cases} +1 & : & x_1 \leq x_2 < x_1 + 1 \\ -1 & : & x_1 + 1 \leq x_2 \leq x_1 + 2 \\ 0 & : & \text{otherwise} \end{cases}$$

We immediately see that

$$\int_0^{\infty} f(x_1, x_2)\,dx_2 = 0 \text{ for all } x_1 \in \mathbb{R}, \quad \text{hence} \quad \int_0^{\infty}\left[\int_0^{\infty} f(x_1, x_2)\,dx_2\right] dx_1 = 0 .$$

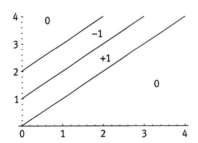

Figure A.1: The function f.

On the other side it follows

$$
\int_0^\infty f(x_1, x_2)\, dx_1 =
\begin{cases}
\int_0^{x_2}(+1)\, dx_1 = x_2 & : \quad 0 \le x_2 < 1 \\
\int_0^{x_2-1}(-1)\, dx_1 + \int_{x_2-1}^{x_2}(+1)\, dx_1 = 2 - x_2 & : \quad 1 \le x_2 \le 2 \\
\int_{x_2-2}^{x_2} f(x_1, x_2)\, dx_1 = 0 & : \quad 2 < x_2 < \infty
\end{cases}
$$

leading to

$$
\int_0^\infty \left[\int_0^\infty f(x_1, x_2)\, dx_1 \right] dx_2 = 1 \ne 0 = \int_0^\infty \left[\int_0^\infty f(x_1, x_2)\, dx_2 \right] dx_1 .
$$

Example A.5.4 shows that neither the definition of the integral of functions of several variables nor the interchange of integrals are unproblematic. Fortunately, we have the following positive result (see [Dur10], Section 1.7, for more information).

Proposition A.5.5 (Fubini's theorem). *If $f(x_1, x_2) \ge 0$ for all $(x_1, x_2) \in \mathbb{R}^2$, then one may interchange the order of integration. In other words,*

$$
\int_{-\infty}^\infty \left[\int_{-\infty}^\infty f(x_1, x_2)\, dx_1 \right] dx_2 = \int_{-\infty}^\infty \left[\int_{-\infty}^\infty f(x_1, x_2)\, dx_2 \right] dx_1 . \qquad (A.29)
$$

Hereby we do not exclude that one of the two, hence also the other, iterated integral is infinite.

Furthermore, in the general case (f may attain also negative values) equality (A.29) holds provided that one of the iterated integrals, for example,

$$
\int_{-\infty}^\infty \left[\int_{-\infty}^\infty |f(x_1, x_2)|\, dx_1 \right] dx_2
$$

is finite. Due to the first part then we also have

$$
\int_{-\infty}^\infty \left[\int_{-\infty}^\infty |f(x_1, x_2)|\, dx_2 \right] dx_1 < \infty .
$$

Whenever a function f on $\mathbb{R}^2$ satisfies one of the two assumptions in Proposition A.5.5, then by

$$
\int_{\mathbb{R}^2} f(x)\, dx := \int_{-\infty}^\infty \left[\int_{-\infty}^\infty f(x_1, x_2)\, dx_1 \right] dx_2 = \int_{-\infty}^\infty \left[\int_{-\infty}^\infty f(x_1, x_2)\, dx_2 \right] dx_1
$$

the integral of f is well-defined. Given a subset $B \subseteq \mathbb{R}^2$ we set

$$\int_B f(x)\, dx := \int_{\mathbb{R}^2} f(x)\, \mathbb{1}_B(x)\, dx ,$$

provided the integral exists. Recall that $\mathbb{1}_B$ denotes the indicator function of B introduced in eq. (3.20).

For example, let K_1 be the unit circle in $\mathbb{R}^2$, that is, $K_1 = \{(x_1, x_2) : x_1^2 + x_2^2 \le 1\}$, then it follows

$$\int_{K_1} f(x)\, dx = \int_{-1}^{1} \int_{-\sqrt{1-x_1^2}}^{\sqrt{1-x_1^2}} f(x_1, x_2)\, dx_2\, dx_1 .$$

Or, if $B = \{(x_1, x_2, x_3) \in \mathbb{R}^3 : x_1 \le x_2 \le x_3\}$, we have

$$\int_B f(x)\, dx = \int_{-\infty}^{\infty} \int_{-\infty}^{x_3} \int_{-\infty}^{x_2} f(x_1, x_2, x_3)\, dx_1\, dx_2\, dx_3 .$$

Remark A.5.6. Proposition A.5.5 is also valid for infinite double series. Let α_{ij} be real numbers either satisfying $\alpha_{ij} \ge 0$ or $\sum_{i=0}^{\infty} \sum_{j=0}^{\infty} |\alpha_{ij}| < \infty$, then this implies

$$\sum_{i=0}^{\infty} \sum_{j=0}^{\infty} \alpha_{ij} = \sum_{j=0}^{\infty} \sum_{i=0}^{\infty} \alpha_{ij} = \sum_{i,j=0}^{\infty} \alpha_{ij} .$$

Even more generally, if the sets $I_k \subseteq \mathbb{N}_0^2$, $k \in \mathbb{N}_0$, form a disjoint partition of $\mathbb{N}_0^2$, then

$$\sum_{i,j=0}^{\infty} \alpha_{ij} = \sum_{k=0}^{\infty} \sum_{(i,j)\in I_k} \alpha_{ij} .$$

For example, if $I_k = \{(i, j) \in \mathbb{N}^2 : i + j = k\}$, then

$$\sum_{i,j=0}^{\infty} \alpha_{ij} = \sum_{k=0}^{\infty} \sum_{(i,j)\in I_k} \alpha_{ij} = \sum_{k=0}^{\infty} \sum_{i=0}^{k} \alpha_{i\,k-i} .$$

Bibliography

[Art64] Emil Artin. *The Gamma Function*. Athena Series: Selected Topics in Mathematics, Holt, Rinehart and Winston, New York-Toronto-London, 1964.

[Axl15] Sheldon Axler. *Linear Algebra Done Right*. Springer International Publishing, Cham Heidelberg, New York, Dordrecht, London, 3rd edition, 2015.

[Bil12] Patrick Billingsley. *Probability and Measure*. John Wiley and Sons, Inc., Hoboken, 4th edition, 2012.

[CB02] George Casella and Roger L. Berger. *Statistical Inference*. Duxburg Press, Pacific Grove, CA, 2nd edition, 2002.

[Mor16] Samuel G. Moreno. *A Short and Elementary Proof of the Basel Problem*. College Math. J. 47 (2016), 134–135.

[Coh13] Donald L. Cohn. *Measure Theory*. Birkhäuser Advanced Texts. Birkhäuser, Springer, New York, 2nd edition, 2013.

[Dud02] Richard M. Dudley. *Real Analysis and Probability*. Cambridge University Press, Cambridge, 2002.

[Dur10] Richard Durrett. *Probability: Theory and Examples*. Cambridge University Press, New York, 4th edition, 2010.

[Fel68] William Feller. *An Introduction to Probability Theory and its Applications*, volume 1. John Wiley and Sons, New York-London-Sydney, 1968.

[Fis11] Hans Fischer. *A History of the Central Limit Theorem*. Ergebnisse der Mathematik und ihrer Grenzgebiete. Springer, New York, 2011.

[Gha05] Saeed Ghahramani. *Fundamentals of Probability*. Pearson Education, Inc., Upper Saddle River, NJ, 3rd edition, 2005.

[GS01a] Geoffrey R. Grimmett and David R. Stirzacker. *One Thousand Exercises in Probability*. Oxford University Press, Oxford, New York, 1st edition, 2001.

[GS01b] Geoffrey R. Grimmett and David R. Stirzacker. *Probability and Random Processes*. Oxford University Press, Oxford, New York, 3rd edition, 2001.

[Kho07] Davar Khoshnevisan. *Probability*. Graduate Studies in Mathematics, 80. American Mathematical Society, New York, 2007.

[Kol33] Andrey Nikolajewitsch Kolmogorov. *Grundbegriffe der Wahrscheinlichkeitsrechnung*. Ergebnisse der Mathematik und ihrer Grenzgebiete. Julius Springer, Berlin, 1933.

[Lag13] Jeffrey C. Lagarias. *Euler's Constant: Euler's Work and Modern Developments*. Bull. Amer. Math. Soc. (N.S.) 50 (2013), 527–628.

[Pao06] Marc S. Paolella. *Fundamental Probability. A Computational Approach*. John Wiley and Sons, Chichester, 2006.

[Ros14] Sheldon Ross. *A First Course in Probability*. Pearson Education Limited, Essex, 9th edition, 2014.

[Spi08] Michael Spivak. *Calculus*. Publish or Perish, Houston, TX, 4th edition, 2008.

[Ste15] James Stewart. *Calculus*. Cengage Learning, Boston, 8th edition, 2015.

Index